Plumbing

Level Two

Trainee Guide
Third Edition

Upper Saddle River, New Jersey
Columbus, Ohio

National Center for Construction Education and Research

President: Don Whyte
Director of Curriculum Revision and Development: Daniele Dixon
Plumbing Project Manager: Daniele Dixon
Production Manager: Jessica Martin
Production Maintenance Supervisor: Debie Ness
Editor: Bethany Harvey
Desktop Publishers: Laura Parker, Debie Ness, Jessica Martin

NCCER would like to acknowledge the contract service provider for this curriculum:
EEI Communications, Alexandria, Virginia.

10 9 8 7 6 5 4 3 2 1
ISBN 0-13-109183-2

Preface

TO THE TRAINEE

Most people are familiar with plumbers who come to their home to unclog a drain or install an appliance. In addition to these activities, however, plumbers install, maintain, and repair many different types of pipe systems. For example, some systems move water to a municipal water treatment plant and then to residential, commercial, and public buildings. Other systems dispose of waste, provide gas to stoves and furnaces, or supply air conditioning. Pipe systems in power plants carry the steam that powers huge turbines. Pipes also are used in manufacturing plants, such as wineries, to move material through production processes.

Plumbers and their associated trades constitute one of the largest construction occupations, holding about 550,000 jobs. Theirs is also among the highest paid construction occupations. Even better, job opportunities are expected to be excellent as demand for skilled craftspeople is expected to outpace the supply of trained plumbers.[1]

We wish you success as you embark on your second year of training in the plumbing craft and hope that you'll continue your training beyond this textbook. As most of the half a million craftspeople employed in this trade can tell you, there are many opportunities awaiting those with the skills and the desire to move forward in the construction industry.

NEW WITH *PLUMBING LEVEL TWO*

This edition of *Plumbing Level Two* presents a new design and features two new modules: *Introduction to Electricity* and *Hangers, Supports, Structural Penetrations, and Fire Stopping*.

Repairing big chillers, installing domestic and chilled water piping, placing fixtures and faucets, fitting and joining carbon-steel piping and even corrugated stainless steel, reading blueprints, and working with a variety of hand tools—your students will learn that there's *much* more to plumbing than unclogging a drain!

We invite you to visit the NCCER website at www.nccer.org for the latest releases, training information, newsletter, and much more. You can also reference the Contren® product catalog online at www.crafttraining.com. Your feedback is welcome. You may email your comments to curriculum@nccer.org or send general comments and inquiries to info@nccer.org.

CONTREN® LEARNING SERIES

The National Center for Construction Education and Research (NCCER) is a not-for-profit 501(c)(3) education foundation established in 1995 by the world's largest and most progressive construction companies and national construction associations. It was founded to address the severe workforce shortage facing the industry and to develop a standardized training process and curricula. Today, NCCER is supported by hundreds of leading construction and maintenance companies, manufacturers, and national associations. The Contren® Learning Series was developed by NCCER in partnership with Prentice Hall, the world's largest educational publisher.

Some features of NCCER's Contren® Learning Series are as follows:

- An industry-proven record of success
- Curricula developed by the industry for the industry
- National standardization, providing portability of learned job skills and educational credits
- Compliance with Apprenticeship, Training, Employer, and Labor Services (ATELS) requirements for related classroom training (CFR 29:29)
- Well-illustrated, up-to-date, and practical information

NCCER also maintains a National Registry that provides transcripts, certificates, and wallet cards to individuals who have successfully completed modules of NCCER's Contren® Learning Series. *Training programs must be delivered by an NCCER Accredited Training Sponsor in order to receive these credentials.*

[1]U.S. Department of Labor, Bureau of Labor Statistics. *Occupational Outlook Handbook, 2004–05 Edition.* www.bls.gov/oco

Special Features of This Book

In an effort to provide a comprehensive user-friendly training resource, we have incorporated many different features for you to use. Whether you are a visual or hands-on learner, this book will provide you with the proper tools to get started in the plumbing industry.

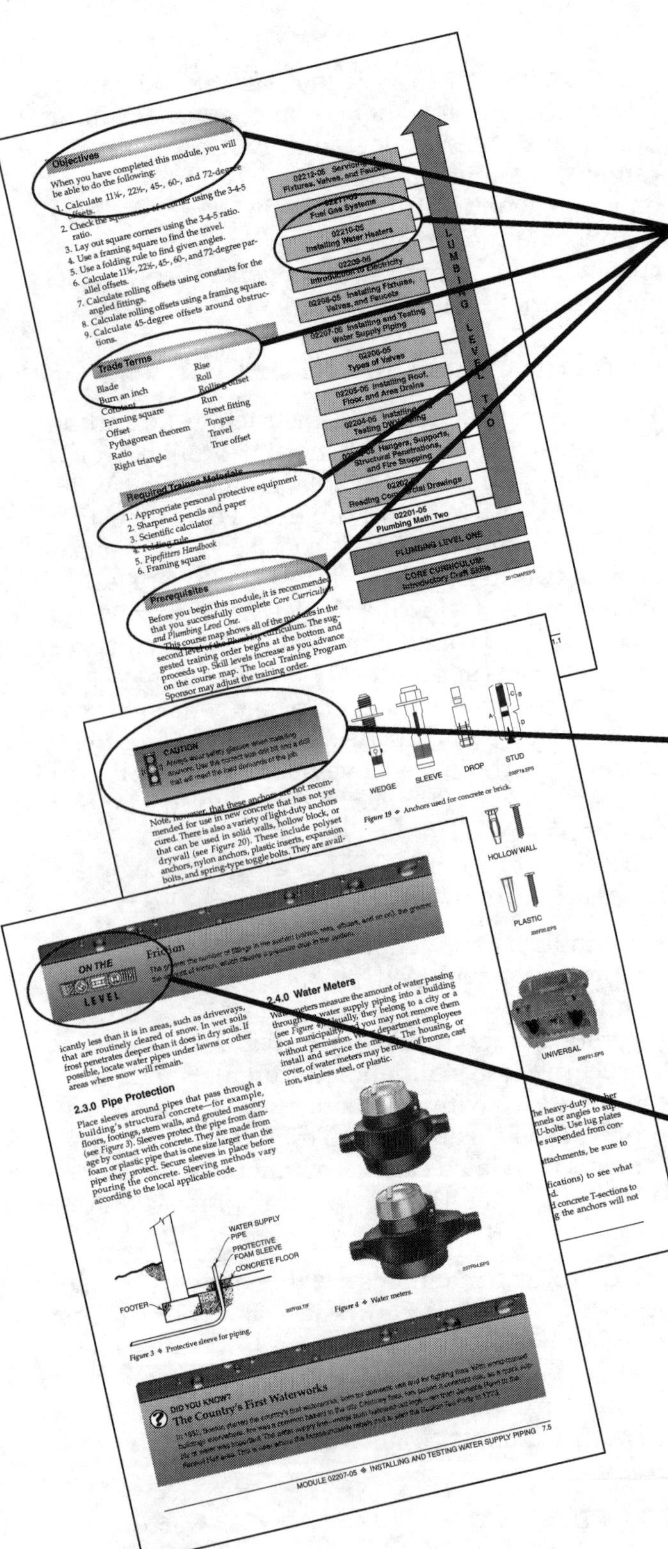

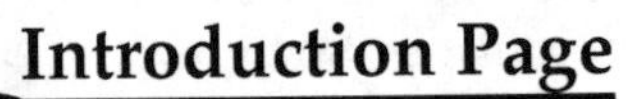

Introduction Page

This page is found at the beginning of each module and lists the Objectives, Trade Terms, Required Trainee Materials, Prerequisites, and Course Map for that module. The Objectives list the skills and knowledge you will need in order to complete the module successfully. The list of Trade Terms identifies important terms you will need to know by the end of the module. Required Trainee Materials list the materials and supplies needed for the module. The Prerequisites for the module are listed and illustrated in the Course Map. The Course Map also gives a visual overview of the entire course and a suggested learning sequence for you to follow.

Notes, Cautions, and Warnings

Safety features are set off from the main text in highlighted boxes and organized into three categories based on the potential danger of the issue being addressed. Notes simply provide additional information on the topic area. Cautions alert you of a danger that does not present potential injury but may cause damage to equipment. Warnings stress a potentially dangerous situation that may cause injury to you or a co-worker.

On the Level

The On the Level features offer technical hints and tips from the plumbing industry. These often include nice-to-know information that you will find helpful. On the Level also presents real-life scenarios similar to those you might encounter on the job site.

Did You Know?

The Did You Know? features introduce historical tidbits or modern information about the plumbing industry. Interesting and sometimes surprising facts about plumbing are also presented.

Illustrations and Photographs

Illustrations and photographs are used throughout each module to provide extra detail. These figures highlight important concepts from the text and provide clarity for complex instructions. Each figure is denoted in the text in *italic type* for easy reference.

Step-by-Step Instructions

Step-by-step instructions are used throughout to guide you through technical procedures and tasks from start to finish. These steps show you not only how to perform a task but also how to do it safely and efficiently.

Trade Terms

Each module presents a list of Trade Terms that are discussed within the text, defined in the Glossary at the end of the module. These terms are denoted in the text with **bold type** upon their first occurrence. To make searches for key information easier, a comprehensive Glossary of Trade Terms from all modules is found at the back of this book.

Review Questions

Review Questions are provided to reinforce the knowledge you have gained. This makes them a useful tool for measuring what you have learned.

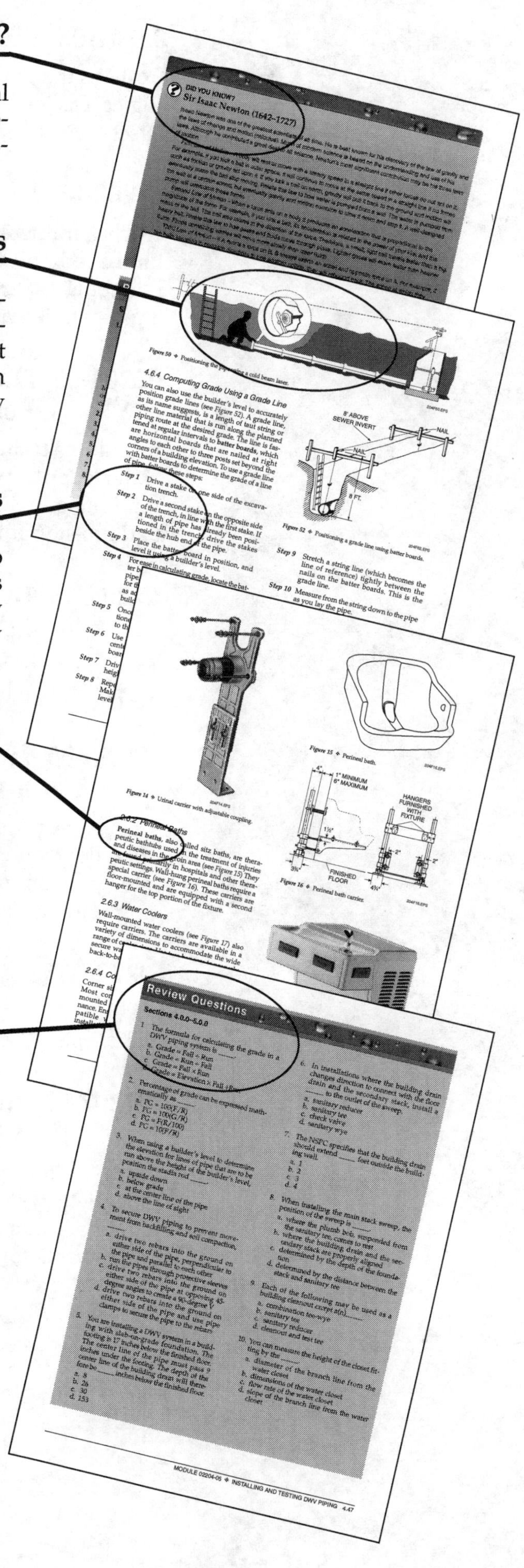

Contents

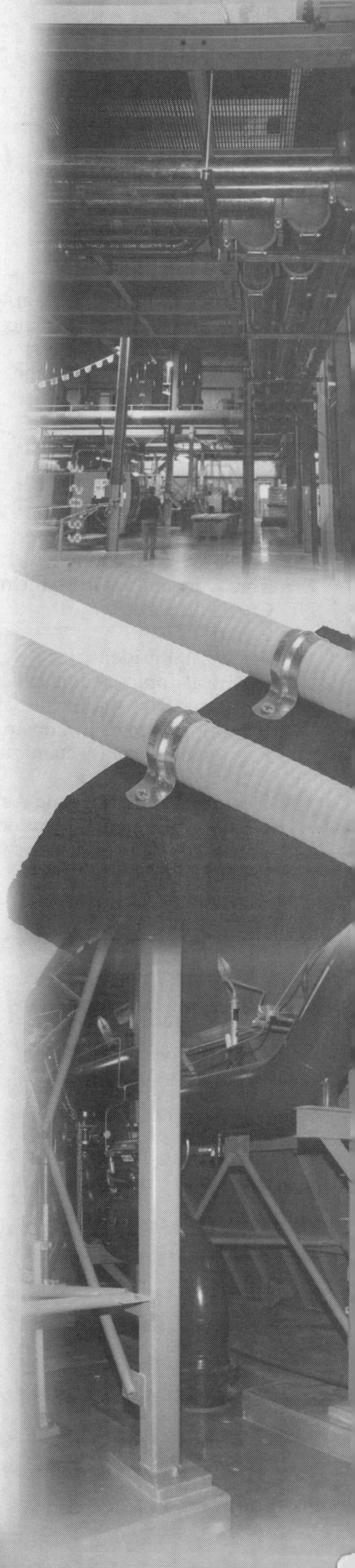

Contren® Curricula

NCCER's training programs comprise more than 40 construction, maintenance, and pipeline areas and include skills assessments, safety training, and management education.

Boilermaking
Carpentry
Carpentry, Residential
Cabinetmaking
Concrete Finishing
Construction Craft Laborer
Construction Technology
Core Curriculum: Introductory Craft Skills
Currículum Básico
Electrical
Electrical, Residential
Electrical Topics, Advanced
Electronic Systems Technician
Exploring Careers in Construction
Fundamentals of Mechanical and Electrical Mathematics
Heating, Ventilating, and Air Conditioning
Heavy Equipment Operations
Highway/Heavy Construction
Instrumentation
Insulating
Ironworking
Maintenance, Industrial
Masonry
Millwright
Mobile Crane Operations
Painting
Painting, Industrial
Pipefitting
Pipelayer
Plumbing
Scaffolding
Sheet Metal
Site Layout
Sprinkler Fitting
Welding

Pipeline

Control Center Operations, Liquid
Corrosion Control
Electrical and Instrumentation
Field Operations, Liquid
Field Operations, Gas
Maintenance
Mechanical

Safety

Field Safety
Orientación de Seguridad
Safety Orientation
Safety Technology

Management

Introductory Skills for the Crew Leader
Project Management
Project Supervision

Acknowledgments

This curriculum was revised as a result of the farsightedness and leadership of the following sponsors:

The Arkord Company
Associated Builders and Contractors of Southern California
EBL Engineers, LLC
G.J. Hopkins, Inc.
Ivey Mechanical Company, LLC
JF Ingram
John J. Muller Plumbing & Heating, Inc.
LeDuc & Dexter, Inc.
M. Davis and Sons
Norfolk Technical Vocational Center
Universal Plumbing & Heating Company
Wat-Kem Mechanical, Inc.

This curriculum would not exist were it not for the dedication and unselfish energy of those volunteers who served on the Authoring Team. A sincere thanks is extended to the following:

Jonathan Byrd
Charlie Chalk
Steve Guy
Richard Kerzetski
Bob Kordulak
Tom LeDuc
Radford Mitchell, Sr.
Bob Muller
Bill Overstreet
Charles Owenby
Charles Robbins
Thomas J. Swafford, CPC
Ray G. Thornton

NCCER PARTNERING ASSOCIATIONS

American Fire Sprinkler Association
American Petroleum Institute
American Society for Training & Development
American Welding Society
Associated Builders & Contractors, Inc.
Association for Career and Technical Education
Associated General Contractors of America
Carolinas AGC, Inc.
Citizens Development Corps
Construction Industry Institute
Construction Users Roundtable
Design-Build Institute of America
Electronic Systems Industry Consortium
Merit Contractors Association of Canada
Metal Building Manufacturers Association
National Association of Minority Contractors
National Association of State Supervisors for Trade and Industrial Education
National Association of Women in Construction
National Insulation Association
National Ready Mixed Concrete Association
National Systems Contractors Association
National Utility Contractors Association
National Technical Honor Society
North American Crane Bureau
North American Technician Excellence
Painting & Decorating Contractors of America
Portland Cement Association
SkillsUSA
Steel Erectors Association of America
Texas Gulf Coast Chapter ABC
U.S. Army Corps of Engineers
University of Florida
Women Construction Owners & Executives, USA
Youth Training and Development Consortium

02201-05

Plumbing Math Two

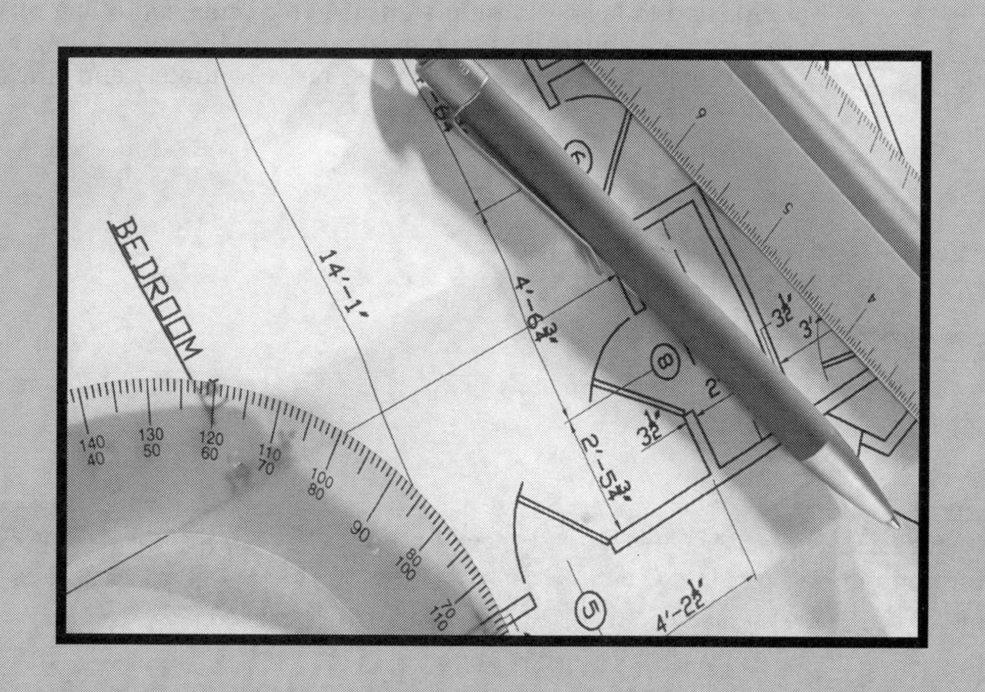

02201-05
Plumbing Math Two

Topics to be presented in this module include:

Overview

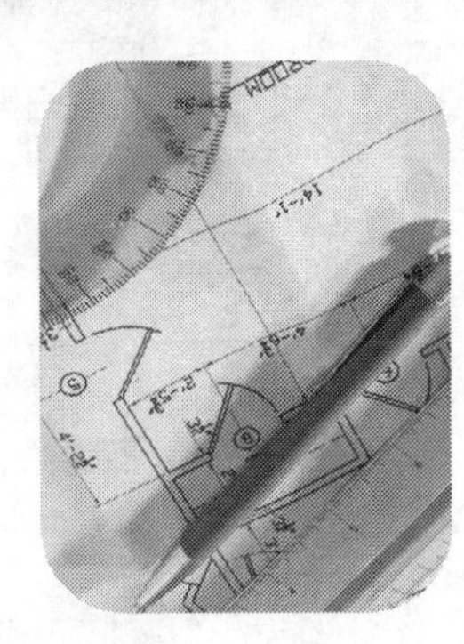

Math is an essential tool in the plumber's toolbox. Plumbers apply math theory to perform numerous tasks in the field, such as calculating cut lengths of pipe when a run changes direction. Plumbers use the 3-4-5 ratio to ensure the squareness of a corner and to lay out a square corner.

Plumbers also use math to calculate simple, parallel, and rolling offsets. Piping offsets form right triangles, and the relationship among the sides of a right triangle is expressed as the Pythagorean theorem. Each leg of the triangle corresponds to a component of the offset: offset, run, and travel. Plumbers calculate travel by multiplying the offset by the constant. Plumbers use a framing square and a wooden rule or tape measure to accurately calculate 45-degree offsets.

Plumbers decide what type of angled fitting is needed to go around an obstruction. To determine how a particular angle turns around the obstruction, plumbers use a folding rule to create angles from 11¼ degrees to 90 degrees. Some installations may require plumbers to calculate two or more parallel runs of pipe and to lay out multiple offsets. To calculate a rolling offset, plumbers use a framing square and tape measure or folding rule. Using math properly ensures that plumbers complete these tasks accurately and efficiently.

Focus Statement

The goal of the plumber is to protect the health, safety, and comfort of the nation job by job.

Code Note

Codes vary among jurisdictions. Because of the variations in code, consult the applicable code whenever regulations are in question. Referring to an incorrect set of codes can cause as much trouble as failing to reference codes altogether. Obtain, review, and familiarize yourself with your local adopted code.

Objectives

When you have completed this module, you will be able to do the following:

1. Calculate 11¼-, 22½-, 45-, 60-, and 72-degree offsets.
2. Check the squareness of a corner using the 3-4-5 ratio.
3. Lay out square corners using the 3-4-5 ratio.
4. Use a framing square to find the travel.
5. Use a folding rule to find given angles.
6. Calculate 11¼-, 22½-, 45-, 60-, and 72-degree parallel offsets.
7. Calculate rolling offsets using constants for the angled fittings.
8. Calculate rolling offsets using a framing square.
9. Calculate 45-degree offsets around obstructions.

Trade Terms

Blade	Rise
Burn an inch	Roll
Constant	Rolling offset
Framing square	Run
Offset	Street fitting
Pythagorean theorem	Tongue
Ratio	Travel
Right triangle	True offset

Required Trainee Materials

1. Appropriate personal protective equipment
2. Sharpened pencils and paper
3. Scientific calculator
4. Folding rule
5. *Pipefitters Handbook*
6. Framing square

Prerequisites

Before you begin this module, it is recommended that you successfully complete *Core Curriculum and Plumbing Level One.*

This course map shows all of the modules in the second level of the *Plumbing* curriculum. The suggested training order begins at the bottom and proceeds up. Skill levels increase as you advance on the course map. The local Training Program Sponsor may adjust the training order.

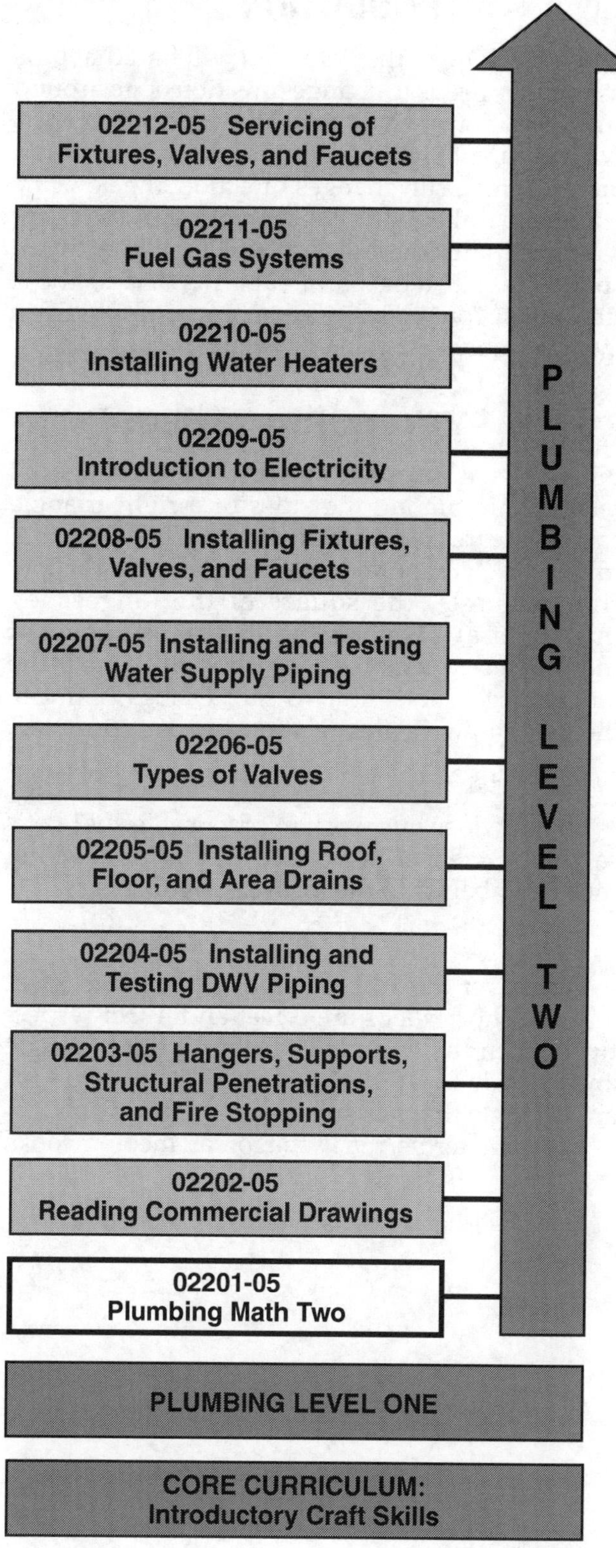

201CMAP.EPS

1.0.0 ◆ INTRODUCTION

When working in the field, you will find that pipe sometimes needs to change direction or go around an obstruction such as a steel beam or other pipes. When a run of pipe changes direction, it creates an **offset.** After a run changes direction, the new line remains parallel with the center line of the original pipe. In this module you will learn how to calculate the cut lengths of pipe needed in these situations. You will also learn how to apply math theory to solve practical problems in the field.

2.0.0 ◆ PYTHAGOREAN THEOREM

Piping offsets form **right triangles,** and the special relationship among the sides of a right triangle can be expressed in a formula. This formula, called the **Pythagorean theorem,** says that in any right triangle, the square of the longest leg (hypotenuse) is equal to the sum of the squares of the other two legs (see *Figure 1*). If we call the longest leg C and the two shorter legs A and B, then we can write this relationship as a formula:

$$C^2 = A^2 + B^2$$

So how does this work in practical terms? Let's say that C = 5 feet, A = 3 feet, and B = 4 feet. Plug these numbers into the formula as follows:

$$5^2 = 3^2 + 4^2$$

Whenever the longest leg of a triangle is 5 feet and the shorter legs are 3 feet and 4 feet respectively, then the triangle is always a right triangle and contains a 90-degree angle. This is true for any triangle that has sides with a 3-4-5 **ratio.**

In piping terms, the Pythagorean theorem looks like this:

$$\text{Travel}^2 = \text{Run}^2 + \text{Offset}^2$$

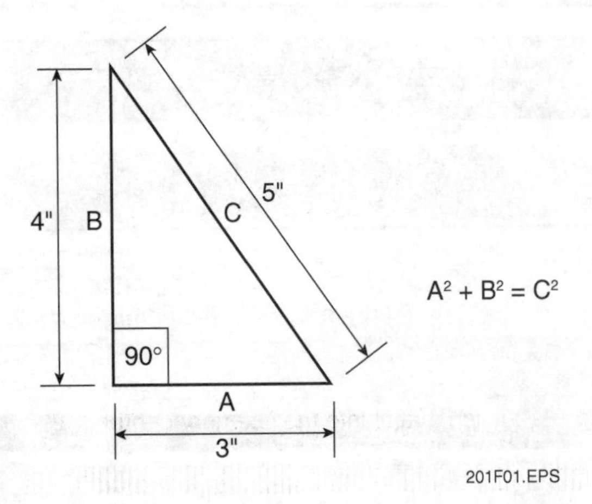

Figure 1 ◆ Right triangle and the Pythagorean theorem.

Because this equation uses one unit, the constant can be multiplied by whatever length you have for the offset or run to determine **travel** in a piping problem.

Example: Say you are using 45-degree elbows and the offset is 36 inches. Calculate the travel.

$$\text{Travel} = 1.414 \times 36" = 50.904" = 50\tfrac{7}{8}"$$

2.1.0 Checking the Squareness of a Corner

In practice, you can use the 3-4-5 ratio to be sure a corner is square. To do this, follow these steps:

Step 1 Measure from the corner 3 feet along one wall, and place a mark at that point. Measure 4 feet along the other wall and mark that point.

Step 2 Use a ruler to measure the straight-line distance between the two points. If that distance is 5 feet, the corner is square. See *Figure 2.*

You can verify this method by using the Pythagorean theorem as shown:

$$3^2 + 4^2 = 5^2$$

$$9 + 16 = 25$$

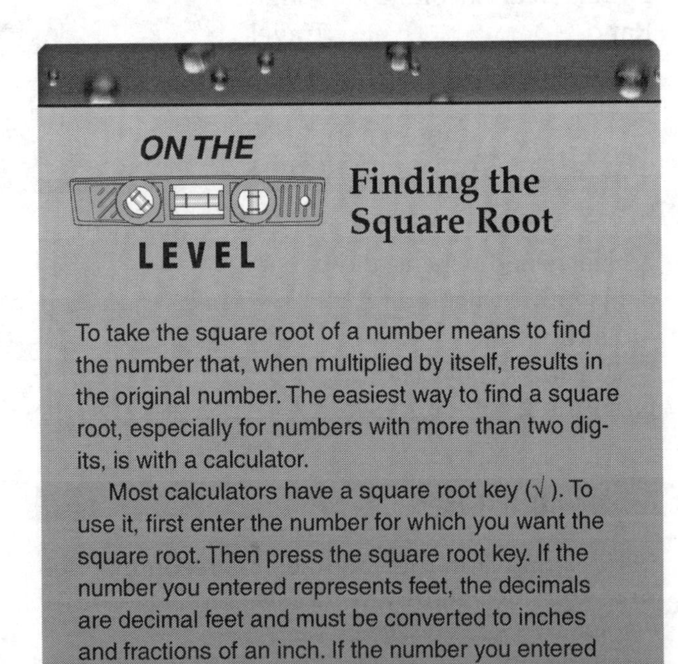

Finding the Square Root

To take the square root of a number means to find the number that, when multiplied by itself, results in the original number. The easiest way to find a square root, especially for numbers with more than two digits, is with a calculator.

Most calculators have a square root key (√). To use it, first enter the number for which you want the square root. Then press the square root key. If the number you entered represents feet, the decimals are decimal feet and must be converted to inches and fractions of an inch. If the number you entered was in inches, the decimals are decimal inches and must be converted to fractions of an inch. Conversion tables are available in math textbooks and in this module. It is important to make all these calculations correctly because an error in math can lead to an error in installation.

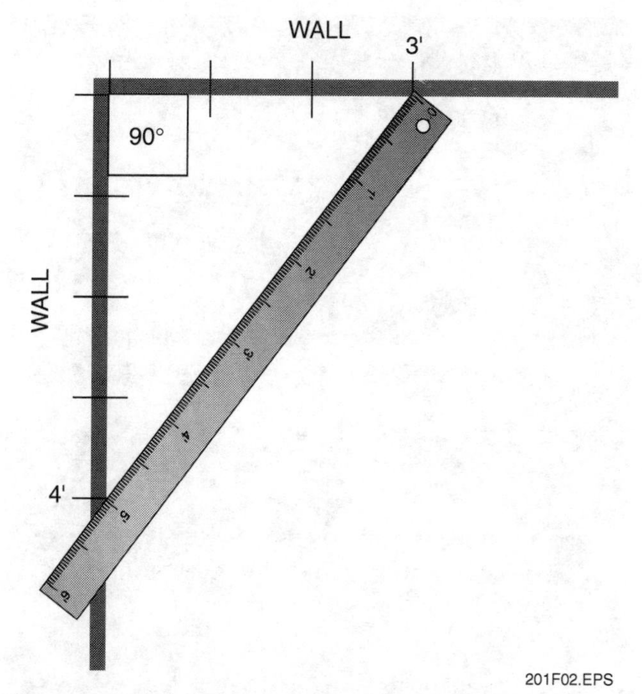

Figure 2 ◆ Checking the squareness of a corner.

2.2.0 Laying Out a Square Corner

You can also use the 3-4-5 ratio to lay out a square corner. Follow these steps:

Step 1 Using two 2 × 4s, mark one 3 feet from the end and the other 4 feet from the end as shown in *Figure 3*.

Step 2 Move the 2 × 4s in relation to each other until you can measure exactly 5 feet between the two marks.

Step 3 When you can measure exactly 5 feet between the marks, the angle must be 90 degrees. You have constructed a square corner.

This procedure works with any multiple of the 3-4-5 ratio. Thus, you could also lay out a square corner using 6 feet, 8 feet, and 10 feet or 30 centimeters, 40 centimeters, and 50 centimeters.

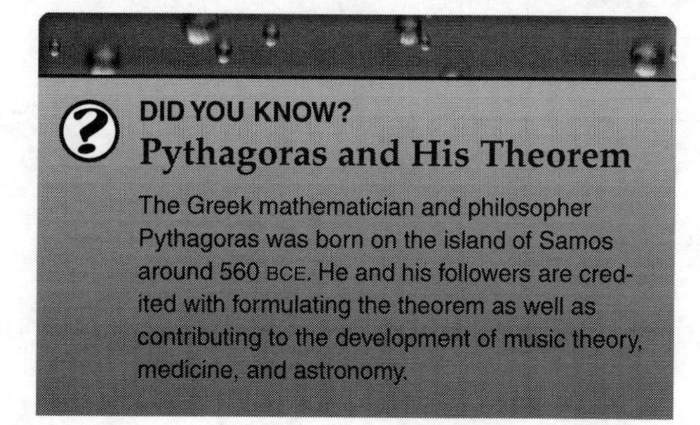

DID YOU KNOW?

Pythagoras and His Theorem

The Greek mathematician and philosopher Pythagoras was born on the island of Samos around 560 BCE. He and his followers are credited with formulating the theorem as well as contributing to the development of music theory, medicine, and astronomy.

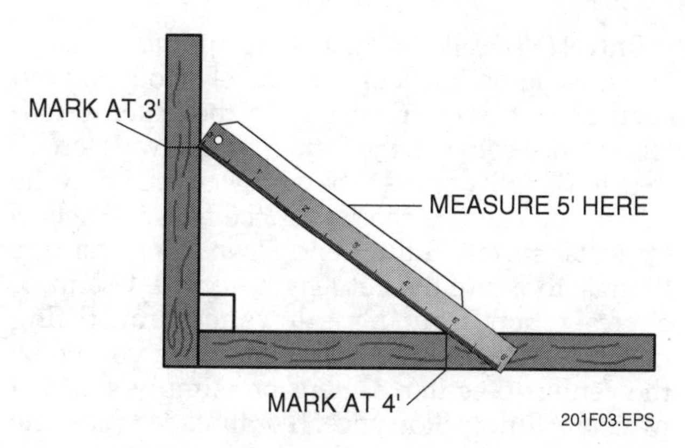

Figure 3 ◆ Laying out a square corner.

3.0.0 ◆ OFFSETS

Piping offsets form right triangles such as those explained by the Pythagorean theorem. Each leg of the triangle corresponds to a component of the offset. *Figure 4* shows the individual components: offset, **run,** and travel. The length of the pipe that runs between the centers of fittings in an offset is called the travel. The distance between the center lines of the two runs of pipe is called the offset. The remaining leg of the right triangle is called the run.

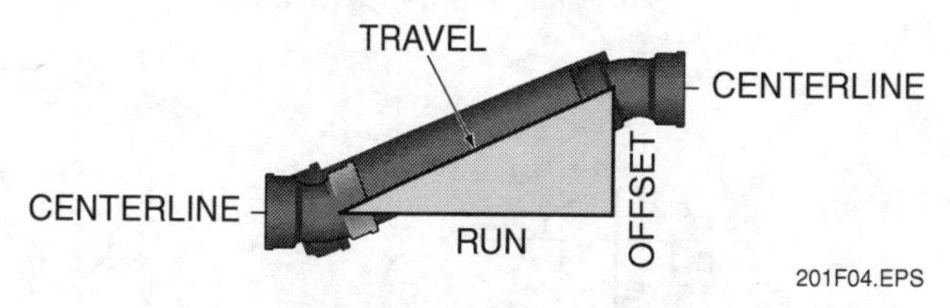

Figure 4 ◆ 22½-degree offset.

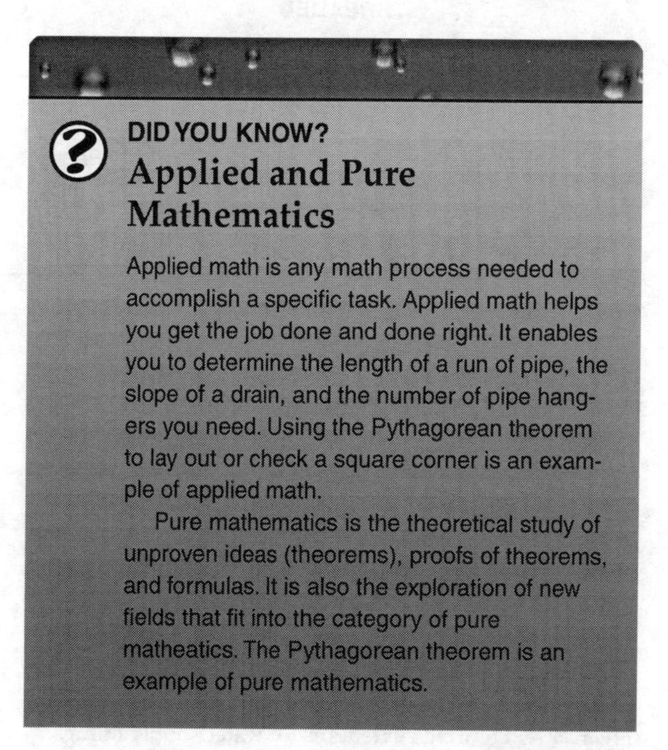

DID YOU KNOW?

Applied and Pure Mathematics

Applied math is any math process needed to accomplish a specific task. Applied math helps you get the job done and done right. It enables you to determine the length of a run of pipe, the slope of a drain, and the number of pipe hangers you need. Using the Pythagorean theorem to lay out or check a square corner is an example of applied math.

Pure mathematics is the theoretical study of unproven ideas (theorems), proofs of theorems, and formulas. It is also the exploration of new fields that fit into the category of pure matheatics. The Pythagorean theorem is an example of pure mathematics.

Travel (also called diagonal) is important because you must know the length of travel before you can cut the pipe to size. Travel is really the center-to-center distance between the fittings. It is always longer than the length of pipe to be cut. Once you know the travel, you can use charts, like the ones in *Figure 5* for water supply fittings and *Figure 6* for drainage fittings, to figure the cut length of pipe. Use these charts when figuring allowances for fitting makeup. To figure the cut length when you know the center-to-center dimension, simply subtract twice the fitting allowance. (For threaded pipe, the fitting allowance equals the center-to-face measure minus the thread-in measure.) Notice the **street fitting** in *Figure 5*. A street fitting is one that has both male and female threads. Street fittings require a separate table because they are measured slightly differently than other fittings.

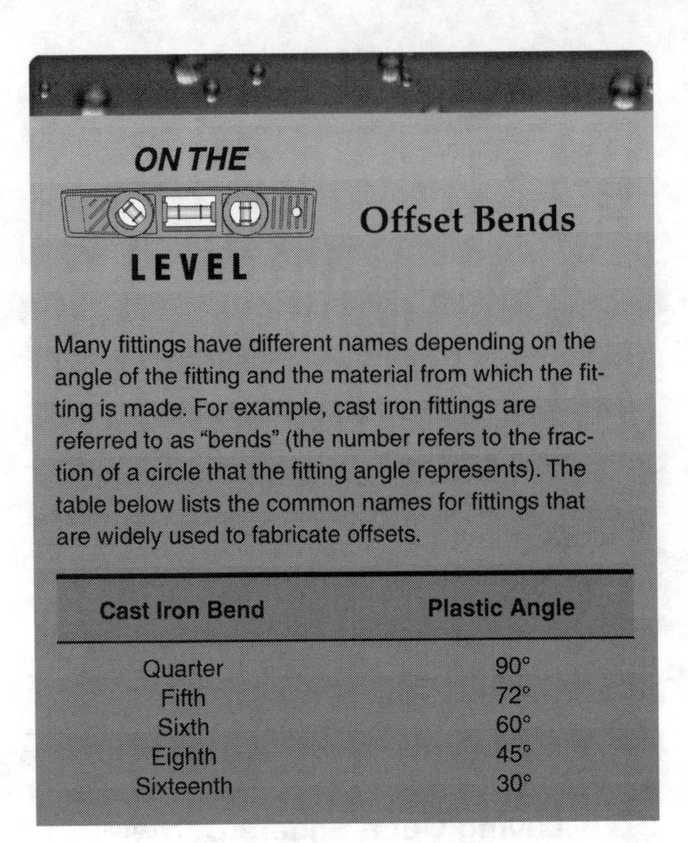

ON THE LEVEL

Offset Bends

Many fittings have different names depending on the angle of the fitting and the material from which the fitting is made. For example, cast iron fittings are referred to as "bends" (the number refers to the fraction of a circle that the fitting angle represents). The table below lists the common names for fittings that are widely used to fabricate offsets.

Cast Iron Bend	Plastic Angle
Quarter	90°
Fifth	72°
Sixth	60°
Eighth	45°
Sixteenth	30°

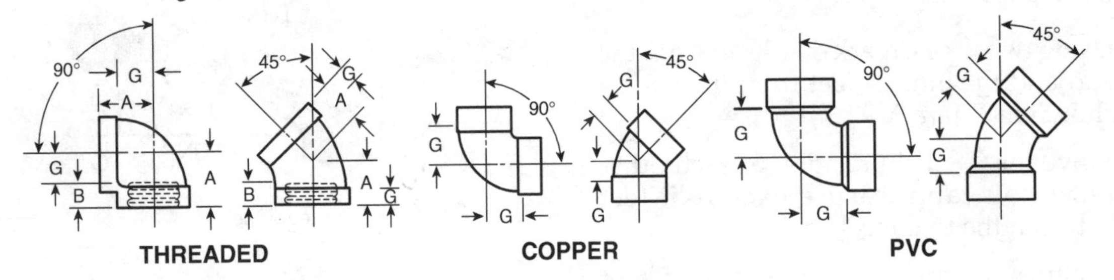

Nominal Pipe Size (inches)	Threaded				Non-Threaded				
	90 A (inches)	45 A (inches)	B (inches)	Thread Engagement (inches, approx.)	Copper 90 G (inches)	Copper 45 G (inches)	PVC 90 G (inches)	PVC 45 G (inches)	Cup/Socket Depth (inches, approx.)
3/8	1	3/4	3/8	3/8	5/16	3/16	3/8	1/4	7/16
1/2	1 1/8	3/4	1/2	1/2	3/8	3/16	1/2	1/4	9/16
3/4	1 3/8	1	1/2	1/2	1/2	1/4	9/16	5/16	9/16
1	1 1/2	1 1/8	1/2	9/16	3/4	5/16	11/16	5/16	11/16
1 1/4	1 3/4	1 5/16	1/2	5/8	1 1/8	7/16	1 9/16	1	11/16
1 1/2	1 15/16	1 7/16	1/2	5/8	1 5/16	9/16	1 3/4	1 1/8	3/4
2	2 1/4	1 11/16	1/2	11/16	1 7/8	3/4	2 5/16	1 1/2	3/4
2 1/2	2 11/16	2 1/16	3/4	15/16	–	–	–	–	1 1/8
3	3 1/16	2 3/16	1	1	2 7/8	1 1/8	3 1/16	1 3/4	1 3/16
4	3 13/16	2 5/8	1	1 1/16	3 3/4	1 1/2	3 7/8	2 3/16	1 5/16

201F05.EPS

Figure 5 ◆ Figuring cut lengths for water supply fittings.

3.1.0 45-Degree Offsets

The 45-degree offset is very common in plumbing. A 45-degree offset creates a 45-degree right triangle. The lengths of the offset and the run on a 45-degree right triangle are equal.

To find the travel, multiply the run or the offset (because they are equal) by a **constant.** Plumbers use constants to simplify the calculations required to find travel. A constant, as its name implies, is a number that always stays the same. Each angled fitting has a constant that, when multiplied by the offset, yields the travel within the piping offset. For a 45-degree right triangle, the constant is 1.414. See the *Appendix* for more information about constants. Consider the following examples:

Example 1: Determine the end-to-end length of 4-inch DWV copper pipe that needs to be cut for the 45-degree offset shown in *Figure 7*.

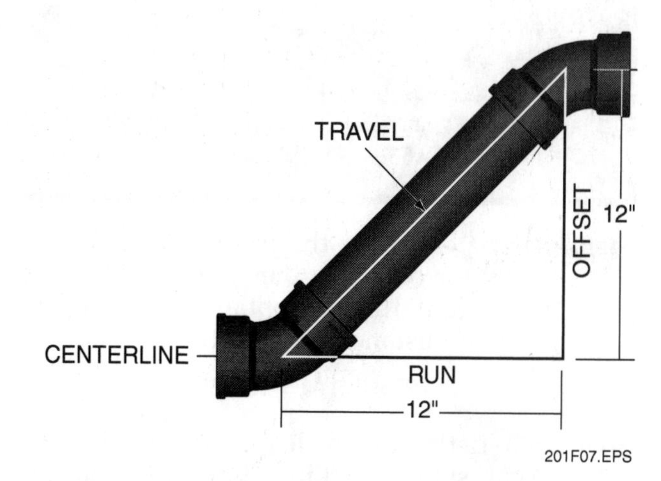

Figure 7 ◆ Figuring the travel.

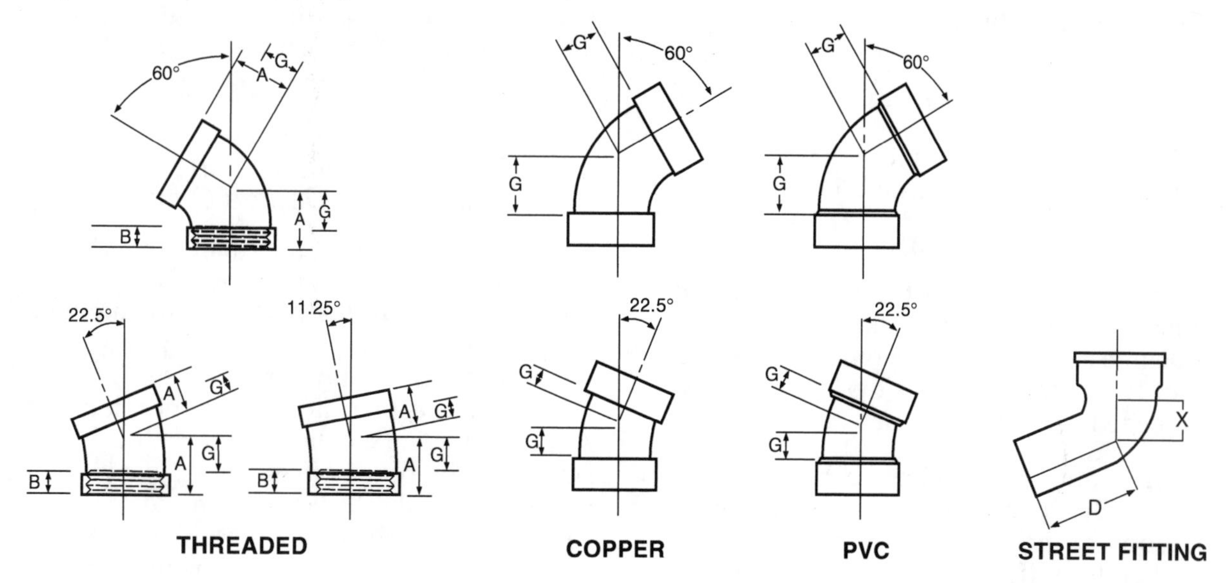

A is Center-to-Face Measure
B is Thread-in Measure
G is Fitting-Allowance Measure

FIFTH BEND (STREET FITTING)

Size	X	D	Weight
2"	2 7/16	5 3/16	4
3"	3 1/16	6 1/16	7
4"	3 7/16	6 15/16	11

Nominal Pipe Size (inches)	Threaded: 60 A (inches)	Threaded: 22½ A (inches)	Threaded: 11¼ A (inches)	Threaded: B (inches)	Threaded: Effective Length (inches, approx.)	Non-Threaded: Copper 90 G (inches)	Non-Threaded: Copper 45 G (inches)	Non-Threaded: PVC 90 G (inches)	Non-Threaded: PVC 45 G (inches)	Cup/Socket Depth (inches, approx.)
1¼	1 9/16	1 1/8	1 1/16	½	5/8	5/8	13/16	–	–	11/16
1½	1¾	1¼	1¼	½	5/8	13/16	¼	1	½	¾
2	2 3/16	1½	1 3/8	½	11/16	1 1/16	5/16	1 5/16	11/16	¾
2½	2½	2	1 5/8	¾	15/16	–	–	–	–	1 1/8
3	2 13/16	1 15/16	1 13/16	1	1	1 5/16	½	1 11/16	13/16	1 13/16
4	3 3/8	2 1/8	2¼	1	1 1/16	2 3/16	11/16	2 1/16	1	1 5/16

201F06.EPS

Figure 6 ◆ Figuring cut lengths for drainage fittings.

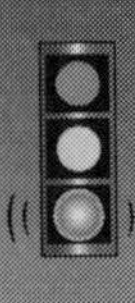

NOTE

When solving the problems in this module, use a calculator and round each answer to the thousandths place, unless otherwise noted. For example, if you multiply 14 by 1.1547, round up the answer, 16.1658, to 16.166.

Step 1 First calculate the travel: Multiply the run (or offset, as they are equal) by the constant 1.414 to obtain the center-to-center distance between the fittings.

$$12" \times 1.414 = 16.968"$$

Step 2 To determine the length of the pipe to be cut, subtract fitting allowances from the travel. To determine the fitting allowance, refer to the sample manufacturer's fitting dimensions table in *Figure 5.* G is the fitting allowance.

$$16.968" - 1.5" - 1.5" = 13.968"$$

Step 3 Convert the answer into the nearest sixteenth of an inch (see *Table 1*).

$$13.968" = 13\tfrac{15}{16}"$$

Example 2: Figure 8 shows a typical installation that requires an offset. Assume that you are using PVC 45-degree elbows, 1¼ inch nominal pipe, and an offset of 18 inches. Determine the length of pipe (pipe B) needed to make the offset.

Step 1 First, determine where pipe A should end. Because you are using 45-degree elbows, the run and offset are equal. Therefore, if the offset is 18 inches, the run must also be 18 inches.

Step 2 To find the travel, multiply the offset by the constant 1.414. Convert to the nearest sixteenth of an inch.

$$18" \times 1.414 = 25.452"$$

$$25.452" = 25\ 7/16"$$

Step 3 To determine the length of the pipe to be cut (pipe B), subtract fitting allowances from the travel. To determine the fitting allowance, subtract the thread-in (B) from the center-to-face measurement (A) (refer to *Figure 5*). (Remember that the 18-inch dimension is taken from the center of the fitting.)

$1\tfrac{5}{16}" - \tfrac{1}{2}" = \tfrac{21}{16} - \tfrac{8}{16} = \tfrac{13}{16}"$ (fitting allowance)

$25\tfrac{7}{16}" - \tfrac{13}{16}" - \tfrac{13}{16}" = \tfrac{407}{16} - \tfrac{13}{16} - \tfrac{13}{16} = \tfrac{381}{16} = 23\tfrac{3}{16}"$ (length of pipe to be cut)

Step 4 Convert the answer to the nearest sixteenth of an inch.

$$23.825" = 23\tfrac{13}{16}"$$

3.1.1 Determining the Starting Point of a 45-Degree Offset

Figure 9 shows a typical situation in which a 45-degree offset must be run around an obstruction. You have already learned how to calculate the simple offset; now you will learn how to determine the starting point for the offset.

In the figure, there are three dimensions labeled A, B, and C. A is the distance from the wall to the starting point of the offset. This is what you need to figure. B is the distance from the corner to the center line of the run of pipe. C is the distance from the corner to the center line of the pipe.

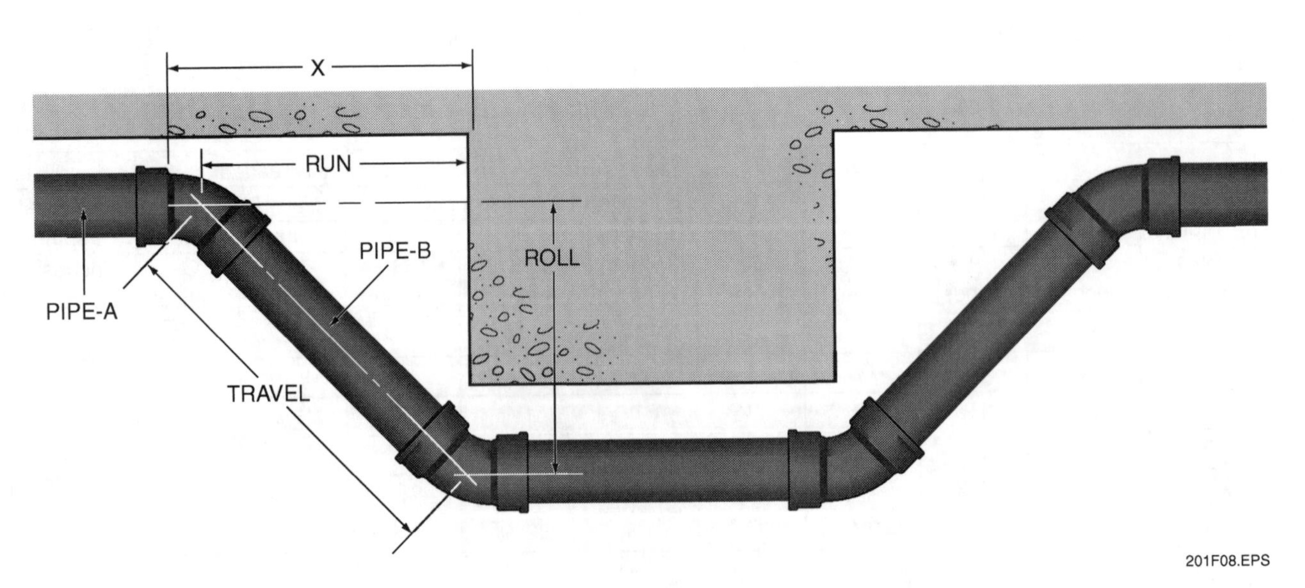

Figure 8 ◆ Calculating pipe offset.

Table 1 Converting to Sixteenths of an Inch

If the decimal falls within this range	Use this fraction (inch)	Which equals this decimal (inch)
0.000–0.031	0	0.000
0.032–0.093	1⁄16	0.063
0.094–0.156	1⁄8	0.125
0.157–0.218	3⁄16	0.188
0.219–0.281	1⁄4	0.250
0.282–0.343	5⁄16	0.313
0.344–0.406	3⁄8	0.375
0.407–0.468	7⁄16	0.438
0.469–0.531	1⁄2	0.500
0.532–0.593	9⁄16	0.563
0.594–0.656	5⁄8	0.625
0.657–0.718	11⁄16	0.688
0.719–0.781	3⁄4	0.750
0.782–0.843	13⁄16	0.813
0.844–0.906	7⁄8	0.875
0.907–0.968	15⁄16	0.938
0.969–1.031	16⁄16	1.000

The formula for finding the starting point of the offset is as follows:

A = B + (C × 1.414)

Example: In the 45-degree offset shown in *Figure 9*, you want the center line of the pipe to clear the corner of the obstruction by 4 inches. You have measured the distance from the corner of the obstruction to the center line of the pipe (B) as 6 inches. Where should you begin the offset?

A = B + (C × 1.414)

A = 6" + (4" × 1.414)

A = 6" + (5.656")

A = 11.656" or 11⅝"

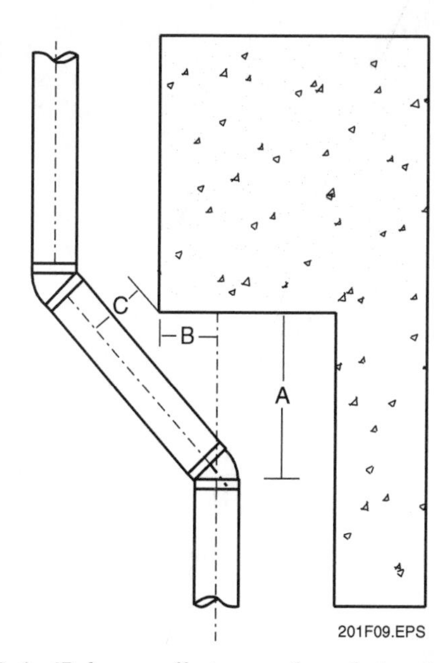

Figure 9 ◆ 45-degree offset around an obstruction.

You should begin the offset 11⅝" from the wall.

Practice Exercises

Section 3.1.1 Study Problems

For each of the following 45-degree simple offsets around an obstruction, find where the offset should begin. Refer to *Figure 9* as a guide. Round final answers to the nearest sixteenth of an inch.

1. A = 12.949
   B = 8"
   C = 3½"
2. A = 13.07"
   B = 6"
   C = 5"
3. A = 19.656"
   B = 14"
   C = 4"
4. A = 18.2165"
   B = 11½"
   C = 4¾"
5. A = 40.486"
   B = 16⅝"
   C = 6⅞"

3.2.0 Figuring Other Offsets

In addition to the 45-degree offset, you need to know how to figure 11¼-, 22½-, 60-, and 72-degree offsets. Each has its own constant. You'll figure travel in exactly the same way for each of these angles: multiply the offset by the constant to find the travel.

Table 2 shows the constants for the common angled fittings used in plumbing offsets when the offset is known. The formula for finding the travel in a piping offset when the offset is known is as follows:

Travel = Offset × Constant

Table 3 shows the constants used to find travel when the run is known. The formula is as follows:

Travel = Run × Constant

Example: Find the travel for an offset of 24.5" with an offset angle of 72 degrees.

Step 1 Refer to *Table 2* to determine the constant. It is 1.05.

Step 2 Multiply the offset by the constant to calculate the travel.

Travel = 24.5" × 1.05 = 25.725"

Step 3 Consult *Table 1* to round the answer to the nearest sixteenth of an inch.

25.725" = 25¾"

Practice Exercises

Section 3.2.0 Study Problems

Find the travel for each of the following piping offsets. Round the answer to the nearest sixteenth of an inch.

1. Offset angle = 60 degrees
   Offset = 18"
   Travel = __________

2. Offset angle = 22½ degrees
   Run = 27¾"
   Travel = __________

3. Offset angle = 11¼ degrees
   Offset = 46⅞"
   Travel = __________

4. Offset angle = 72 degrees
   Run = 37 5/16"
   Travel = __________

5. Offset angle = 45 degrees
   Offset = 14"
   Travel = __________

Table 2 Finding Travel When Offset Is Known

Angled Fitting	Constant
11¼°	5.126
22½°	2.613
45°	1.414
60°	1.1547
72°	1.05

Table 3 Finding Travel When Run Is Known

Angled Fitting	Constant
11¼°	1.019
22½°	1.082
45°	1.414
60°	2.0
72°	3.236

3.3.0 Finding the Travel with a Framing Square

You may find yourself in the field without a calculator or without paper on which to figure the travel for offsets. If that happens, you can use a **framing square** and a wooden rule or tape measure to accurately figure 45-degree offsets.

Because the offset and run are equal, the framing square can show an exact representation of the problem. *Figure 10* illustrates the method.

Step 1 Lay the tape measure or rule at the offset dimension on both the **blade** and **tongue** of the square. Be sure to use either the outer numbers or inner numbers, but do not mix the two.

Step 2 Read the travel from your rule or tape. For greater accuracy, **burn an inch** (grip the folding rule below the 1-inch mark and start measuring at the 1-inch mark) when doing this (see *Figure 11*).

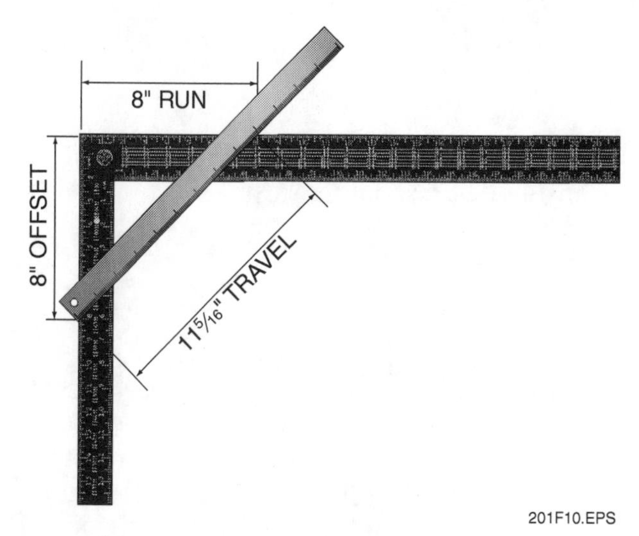

Figure 10 ◆ Finding travel with a framing square.

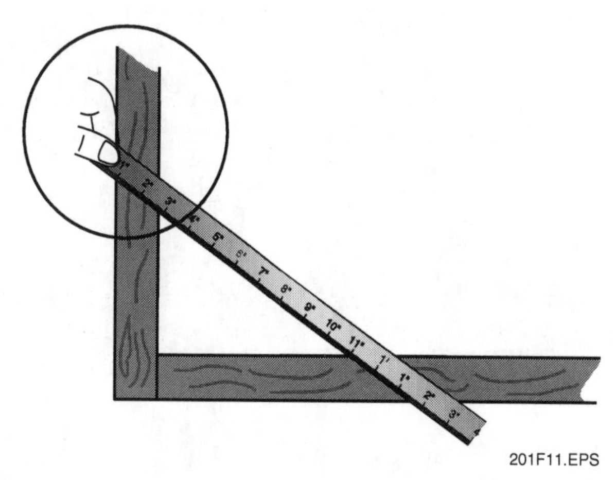

Figure 11 ◆ Burning an inch.

Practice Exercises

Section 3.3.0 Study Problems

Use a framing square and tape measure or wooden rule to figure the travel for the following offsets. Assume the offset angle is 45 degrees.

1. 10 inches = __________
2. 14 inches = __________
3. 8 inches = __________
4. 5½ inches = __________
5. 11¾ inches = __________

Review Questions

Sections 1.0.0–3.0.0

1. In a right triangle, the longest side squared is equal to the sum of the _____.
   a. squares of the two shorter sides
   b. squares of the two longer sides
   c. measurement of the two shorter sides
   d. angles of the two shorter sides

2. Whenever a triangle's sides have a 3-4-5 ratio, the triangle is always a _____.
   a. triangle with sides that add up to 12 units
   b. triangle with two 45-degree angles
   c. right triangle
   d. triangle with two 90-degree angles

3. You are checking a corner of a wall to ensure that it is square. You begin by measuring from the corner 3 feet along one wall and mark that point, then you measure 4 feet along the other wall and mark that point. Then, you _____.
   a. ensure that there are no obstructions between the two points
   b. measure the straight-line distance between the two points
   c. lay a framing square in the corner between the two points
   d. measure and mark a point on the floor that is diagonally opposite the corner

4. You are laying out a square corner. To determine whether the corner is square, you mark a point on one wall 30 inches from the corner and mark a point on the adjacent wall 40 inches from the corner. For the corner to be square, the line drawn between the two points must be _____ inches long.
   a. 20
   b. 30
   c. 40
   d. 50

5. The distance between the center lines of two parallel pipes is called the _____.
   a. travel
   b. angle
   c. offset
   d. engagement

6. The length of pipe that runs between the fittings in an offset is called the _____.
   a. travel
   b. constant
   c. theorem
   d. right triangle

7. A 45-degree offset creates a _____-degree right triangle.
   a. 30
   b. 45
   c. 60
   d. 90

8. The constant to use to calculate travel for any 45-degree angle is _____.
   a. 1.144
   b. 1.141
   c. 1.414
   d. 1.441

9. If the offset and run of a 45-degree offset installation are both 11 inches, the travel is approximately _____ inches. (Convert the answer into the nearest sixteenth of an inch.)
   a. 11½
   b. 15½
   c. 15⁹⁄₁₆
   d. 121 square

Review Questions

10. Refer to *Figure 9*. The formula for finding the starting point of a simple offset that must clear the corner of an obstruction is _____.
    a. A = B + C + 1.414
    b. A = (B + C) × 1.414
    c. A = B + (C × 1.414)
    d. C = A + (B × 1.414)

11. A 45-degree offset must clear the corner of an obstruction by 5⅜ inches. You measure 7⅝ inches from the corner of the obstruction to the center line of the pipe. The offset should begin _____ inches from the wall. (Round your answer to the nearest sixteenth of an inch.)
    a. 9⅞
    b. 13
    c. 15¼
    d. 18⅜

12. A 45-degree offset must clear the corner of an obstruction by 3⅝ inches. You measure 4⅜ inches from the corner of the obstruction to the center line of the pipe. The offset should begin _____ inches from the wall. (Round your answer to the nearest sixteenth of an inch.)
    a. 5⅛
    b. 9½
    c. 9⅛
    d. 11⅜

13. The formula for finding the travel in a piping offset when the offset is known is _____.
    a. Travel = Offset × Constant
    b. Travel = Run × Constant
    c. Travel = Run × Offset
    d. Travel = Offset ÷ Constant

14. Refer to *Table 2* in the text. The travel for a piping length with an offset angle of 72 degrees and an offset of 12 inches is _____ inches. (Round the answer to the nearest sixteenth of an inch.)
    a. 12 5/16
    b. 12⅝
    c. 13 1/16
    d. 13⅞

15. You need to calculate the travel on a 45-degree offset, but all you have is a framing square and a ruler. You measure an 11-inch offset and an 11-inch run. The travel will be approximately _____ inches.
    a. 15½
    b. 22
    c. 22½
    d. 121

4.0.0 ◆ FINDING ANGLES WITH A FOLDING RULE

When you are in the field, you'll sometimes need to visualize how a particular offset looks. For example, when you are deciding what type of angled fitting you'll need to go around an obstruction, it helps to recognize how a particular angle turns around the obstruction. The following method shows you how to use a folding rule to create angles from 11¼ degrees to 90 degrees. To create a 90-degree angle with a folding rule, follow these steps:

Step 1 Open up the rule so that the 1-inch mark is to your left and the 24-inch mark is to your right. You will use only the first four sections of the rule.

Step 2 Moving only the first two sections of the rule, swing the upper corner of the tip of the rule up and over until the tip lines up with the lower edge of the 20¼-inch mark on the fourth section of the rule (see *Figure 12*). You have created a right angle between the first and second sections of the rule.

Step 3 Make sure the third and fourth sections are straight. Note that the 90-degree angle you have created is on the exterior of the triangle, not inside the first and second sections.

Step 4 While holding the corner of the first section in place against the 20¼-inch mark, you can place the triangle near the obstruction to visualize the offset.

Other angles are possible, as shown in *Table 4*. Remember, only the first two sections of the rule may move.

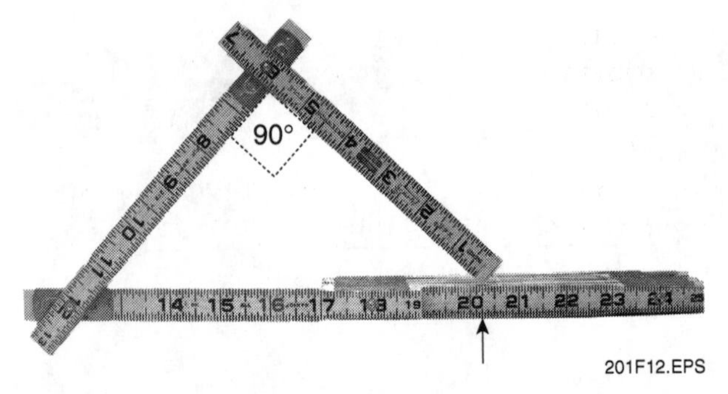

Figure 12 ◆ Finding a 90-degree angle with a folding rule.

Table 4 Forming Angles

To Form This Angle	Touch Tip at This Dimension
11¼°	23$\frac{5}{16}$"
22½°	23¾"
30°	23⅜"
45°	23"
60°	22¼"
72°	21⅝"
90°	20¼"

5.0.0 ◆ OFFSETS ON PARALLEL RUNS OF PIPE

Often, installations require two or more parallel runs of pipe. The offsets can be of different lengths, as illustrated in *Figure 13,* or of equal lengths, as illustrated in *Figure 14.* To lay out multiple identical offsets for parallel runs of pipe, calculate the offset of the first run of pipe. To lay out the second run, you must know the travel between fittings. You have already learned how to use constants to calculate the travel in a simple offset, or an offset in which only one run of pipe is offset. To determine the travel of the horizontal/vertical runs of pipe in the offsets, you need to know how to determine the difference in pipe legs.

The difference between the length of Pipe 1 and the length of Pipe 2 in *Figure 13* is equal to the center-to-center distance between the fittings. To determine this distance, you also need to know the spread, which is the distance between the center lines of the pipes.

To find the difference in pipe legs, multiply the spread by the constant for the angled fitting. The constants for each angled fitting are shown in *Table 5.*

Example: What is the difference in pipe legs in two parallel 22½-degree offsets with a spread of 12 inches?

Difference in pipe legs = Spread × Constant
= 12" × 0.199
= 2.388" or 2⅜"

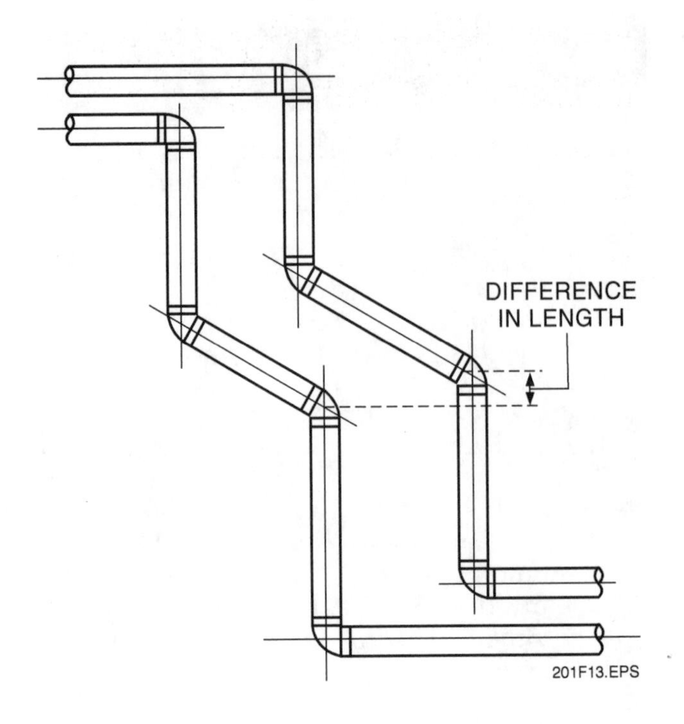

Figure 13 ◆ Parallel runs of pipe with unequal offsets.

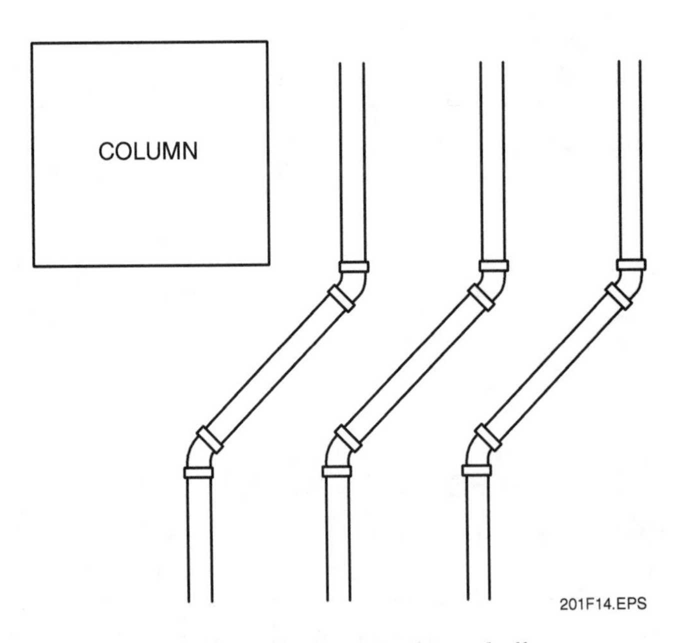

Figure 14 ◆ Parallel runs of pipe with equal offsets.

Table 5 Angled Fittings and Their Constants

Angled Fitting	Constant
11¼°	0.098
22½°	0.199
45°	0.414
60°	0.577
72°	0.727

Practice Exercises

Section 5.0.0 Study Problems

Find the difference in pipe legs in the following offsets for multiple parallel runs of pipe. Round your answer to the nearest sixteenth of an inch.

1. Angled fitting: 11¼ degrees
   Spread: 14"
   Difference: __________

2. Angled fitting: 72 degrees
   Spread: 18"
   Difference: __________

3. Angled fitting: 45 degrees
   Spread: 16¾"
   Difference: __________

4. Angled fitting: 22½ degrees
   Spread: 26⅜"
   Difference: __________

5. Angled fitting: 60 degrees
   Spread: 11⅞"
   Difference: __________

5.1.0 Laying Out Multiple Offsets

Once you know the difference in the lengths of the pipe legs, it is easy to calculate the travel of the parallel pipes in the second offset. To calculate the length of the horizontal/vertical runs of pipe, simply subtract the difference in length of the pipe legs from the center-to-center length of the pipe in the first offset.

Example: Refer to *Figure 13*. In a 45-degree offset, if pipe A measures 15" center-to-center and the spread is 8", find the length of pipe B.

Difference in pipe legs = Spread × Constant
= 8" × 0.414
= 3.312" or 3⁵⁄₁₆"

Length of pipe B = Length of pipe A – Difference in pipe legs
= 15" – 3⁵⁄₁₆"
= 11¹¹⁄₁₆"

6.0.0 ◆ ROLLING OFFSETS

You are already familiar with offsets for single and parallel runs of pipe. Both of these types of offset remain in the same plane after they are offset. If we put an offset into a box, a simple offset would look like *Figure 15*. Notice that the run of pipe enters the box at the bottom of one side and exits at the top of the same side.

In a **rolling offset,** however, something different occurs. A rolling offset results from "rolling," or leaning, a simple offset so that the parallel runs of pipe at either end of the offset are no longer in the same plane. If you imagine a rolling offset placed inside a box, it would look like the offsets illustrated in *Figure 16*. Notice that the first rolling offset enters the box at exactly the same location as the simple offset, but exits the box from the top of a different side, in a different plane.

You are already familiar with the terms offset, travel, and run used for simple offsets. The terms used to describe rolling offsets are illustrated in

A Common Mistake

A contractor was laying out pipe for 40 identical aircraft hangars at an Air Force base in New Mexico. The contractor relied on prefabrication where possible because of the redundancy in the 40 identical plumbing systems. To save time, wherever possible the contractor cut all short pieces of pipe the same length. For two offsets on parallel runs of pipe, the contractor measured a single pipe and multiplied that length by 2 (for the number of offsets) and then by 40 (for the number of hangars).

However, the offsets were of unequal length. The contractor failed to allow for the 12-inch difference in length that resulted (and which was shown on the plans). As a result, the contractor ended up being short 160 pieces of 12-inch-long pipe required to make the spacing work. The contractor admitted his mistake and had 160 pieces of 12-inch pipe headed to the project site the next day. Though in this case the contractor's mistake caused no delays, a similar oversight on another project could possibly cause significant delays. Never rush to save time; you might end up wasting even more time as a result!

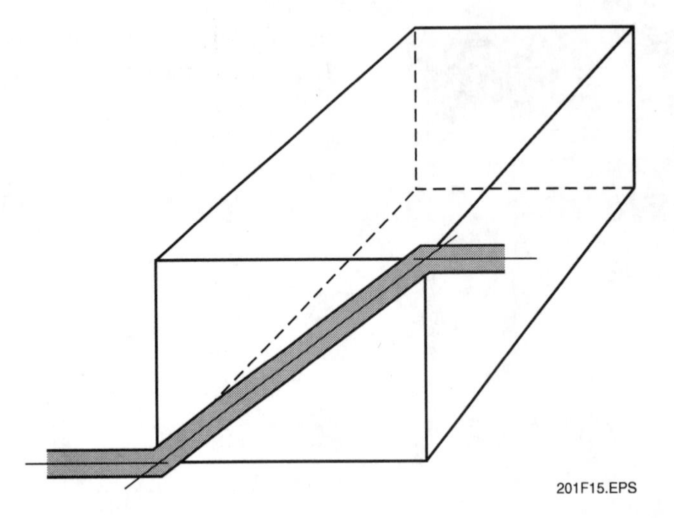

Figure 15 ◆ Simple offset in a box.

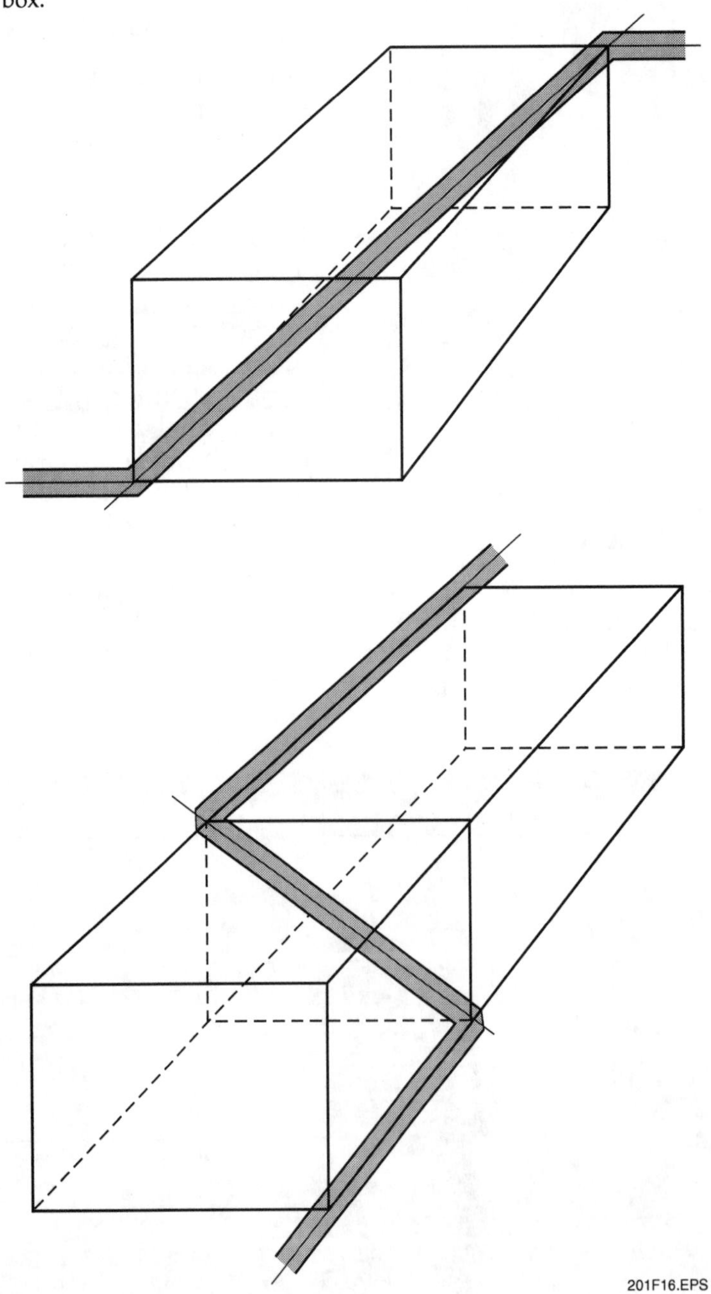

Figure 16 ◆ Two types of rolling offset in boxes.

Figure 17. Whereas simple offsets form a single right triangle, rolling offsets form two right triangles. The **roll** in a rolling offset is that part of the offset that is perpendicular to both the **rise** (the vertical measurement of a run of pipe) and the run. You need to learn how to recognize all the elements of a rolling offset and know how to work with each of them.

6.1.0 Finding True Offset

One of the triangles formed by a rolling offset contains one leg called the offset and another called the rise. The longest side of this right triangle is called the **true offset** (refer to *Figure 17*). Before you can begin to solve rolling offsets, you must know the true offset.

You have already learned that a unique relationship exists among the sides of every right triangle. This relationship can be expressed by the Pythagorean theorem: $C^2 = A^2 + B^2$. Because of this relationship, a missing side can be found when the other two are known. If True offset^2 = Rise^2 + Offset^2, taking the square root of both sides yields this formula:

$$\text{True offset} = \sqrt{\text{Rise}^2 + \text{Offset}^2}$$

The offset and rise form the two short legs of a right triangle. Rise is the vertical measurement; offset is the horizontal measurement. The true offset forms the long leg of the triangle.

Example: If the rise is 3 feet and the offset is 4 feet, figure the true offset.

$$\text{True offset} = \sqrt{3^2 + 4^2}$$

$$\text{True offset} = \sqrt{9 + 16}$$

$$\text{True offset} = \sqrt{25} = 5 \text{ feet}$$

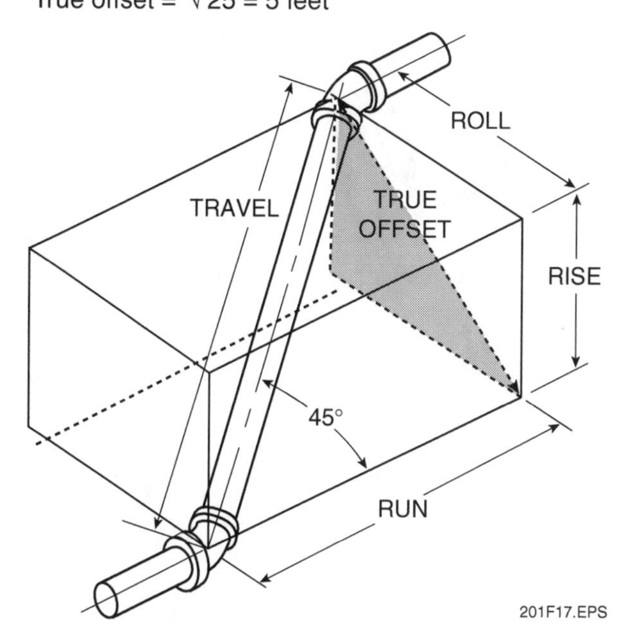

Figure 17 ◆ Rolling offset.

6.2.0 Finding Run and Travel in Rolling Offsets

The second triangle formed by a rolling offset consists of the run, travel, and true offset (see *Figure 18*). You just learned how to calculate the true offset. Now you can easily find the travel and the run by using constants. Remember that you used constants to help you find offsets for both single and parallel runs of pipe.

The constants used to find the travel and the run in rolling offsets are shown in *Table 6*. To find run or travel, multiply the true offset by the appropriate constant for the fitting angle. In each case, you use the same formula but different constants.

Example: Find the travel and run for a rolling offset with a true offset of 5 feet and a fitting angle of 72 degrees.

Travel = 5 × 1.052 = 5.26 feet

Run = 5 × 0.325 = 1.625 feet

Table 6 Constants for Rolling Offsets

Fitting Angle	72°	60°	45°	22½°	11¼°
***Travel* = *True Offset* ×**	1.052	1.155	1.414	2.613	5.126
***Setback* = *True Offset* ×**	0.325	0.577	1.000	2.414	5.027

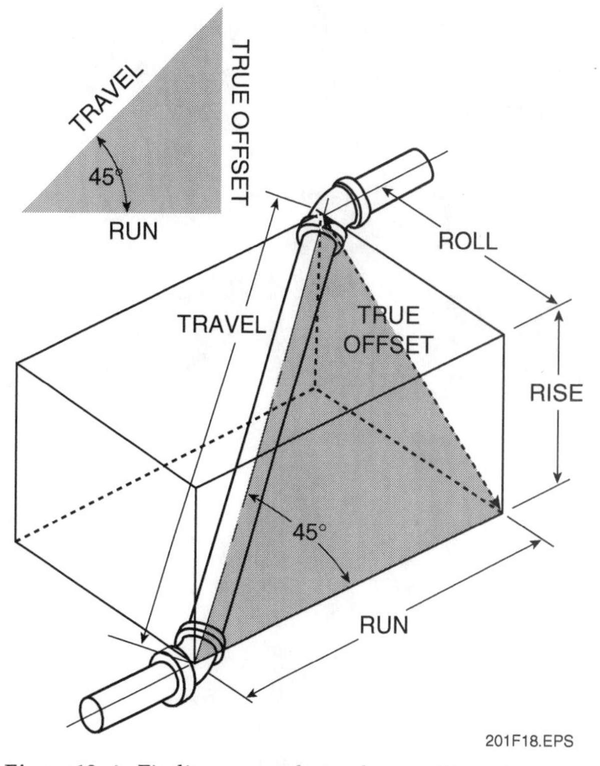

Figure 18 ◆ Finding run and travel in a rolling offset.

Practice Exercises

Sections 6.0.0–6.2.0 Study Problems

Find the true offset, travel, and run for the following rolling offsets. Round decimal answers to the nearest sixteenth of an inch.

1. Fitting angle: 72 degrees
   Rise: 27¾"
   Offset: 36"
   True offset: __________
   Travel: __________
   Run: __________

2. Fitting angle: 11¼ degrees
   Rise: 31¾"
   Offset: 40"
   True offset: __________
   Travel: __________
   Run: __________

3. Fitting angle: 22½ degrees
   Rise: 30¾"
   Offset: 39"
   True offset: __________
   Travel: __________
   Run: __________

4. Fitting angle: 60 degrees
   Rise: 28¾"
   Offset: 37"
   True offset: __________
   Travel: __________
   Run: __________

5. Fitting angle: 45 degrees
   Rise: 29¾"
   Offset: 38"
   True offset: __________
   Travel: __________
   Run: __________

6.3.0 Calculating Rolling Offsets with a Framing Square

You can also calculate rolling offsets using a framing square and tape measure or folding rule. Lay the tape measure or folding rule across the square so that it touches the top of the rise on one side and the far edge of the offset on the other. Multiply the distance between the two points by the constant for the angled fitting.

Suppose a rolling offset has a 15-inch rise and an 8-inch offset, and the fitting angle is 45 degrees. Follow these steps to find the center-to-center length of the pipe:

Step 1 Place the rule or tape as shown in *Figure 19*, and measure the distance between the rise and the offset.

Step 2 Multiply the distance measured on the rule (17 inches) by the constant for the 45-degree fitting (1.414).

$$17 \text{ inches} \times 1.414 = 24 \text{ inches}$$

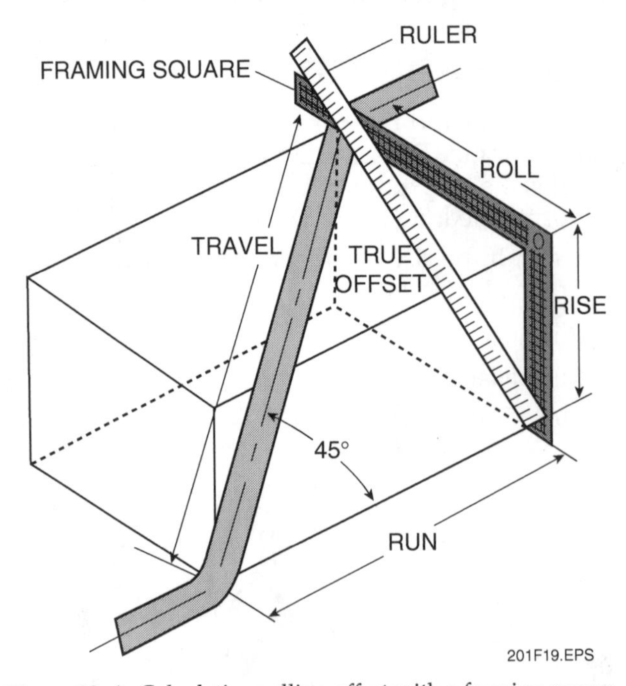

Figure 19 ◆ Calculating rolling offset with a framing square.

DID YOU KNOW?

Sir Isaac Newton (1642–1727)

Isaac Newton was one of the greatest scientists of all time. He is best known for his discovery of the law of gravity and the laws of change and motion (calculus). Much of modern science is based on the understanding and use of his laws. Although he contributed a great deal to all science, Newton's most significant contribution may be his three laws of motion.

First Law of Motion – A body will rest or move with a steady speed in a straight line if other forces do not act on it. For example, if you kick a ball in outer space, it will continue to move at the same speed in a straight line if no forces such as friction or gravity act upon it. If you kick a ball on earth, gravity will pull it back to the ground and friction will eventually make the ball stop moving. Relate this law to how water is pumped from a well. The water is pumped from the well at a certain speed, but eventually gravity and friction combine to slow it down and stop it. A well-designed pump will overcome these forces.

Second Law of Motion – When a force acts on a body it produces an acceleration that is proportional to the magnitude of the force. For example, if you kick a ball, its acceleration is related to the power of your kick and the mass of the ball. The ball also moves in the direction of the force. Therefore, a small, light ball travels faster than a big, heavy ball. Relate this law to how gases and fluids move through pipes. Lighter gases will move faster than heavier fluids. Fluids containing wastes will move more slowly than clear fluids.

Third Law of Motion – If A exerts a force on B, B always exerts an equal and opposite force on A. For example, if two balls traveling in opposite directions to one another collide, they will rebound back. The speed at which they collide is the sum of the two balls' speeds. Relate this law to water hammer. If air in a pipe collides with fluid in a pipe, the two materials will rebound from one another with equal force, creating a shock wave that results in water hammer.

How else does Isaac Newton's work relate to your work as a plumber? He figured out how to find the roots of numbers that you now use when calculating rolling offsets.

Review Questions

Sections 4.0.0–6.0.0

1. Refer to *Table 4* in the text. When finding angles with a folding rule, if you place the tip of the rule at 22¼ inches, it creates a _____.
   a. 22½-degree angle
   b. 45-degree angle
   c. 60-degree angle
   d. 90-degree angle

Match the offset angle with the measurement on the fourth section of a folding rule that indicates that angle.

2. _____ 11¼°
3. _____ 22½°
4. _____ 30°
5. _____ 45°
6. _____ 60°
7. _____ 72°
8. _____ 90°

a. 23¾"
b. 23"
c. 21¾"
d. 21⅝"
e. 23⁵⁄₁₆"
f. 22¼"
g. 20¼"
h. 23⅜"

9. In parallel simple offsets, _____ is the distance between the center lines of the pipes.
   a. fitting
   b. spread
   c. constant
   d. travel

10. The formula for finding the difference in pipe legs in two parallel offsets is _____.
    a. angled fitting × constant
    b. travel × spread
    c. spread × constant
    d. run × constant

11. Refer to *Table 5*. The difference in pipe legs in two parallel 60-degree offsets with a spread of 18 inches is _____ inches. (Round the answer to the nearest sixteenth of an inch.)
    a. 10⅜
    b. 10⅕
    c. 13¹⁄₁₆
    d. 25⁷⁄₁₆

Review Questions

12. Refer to *Table 5*. The difference in pipe legs for two parallel 22½-degree offsets with a spread of 28⅞16 inches is _____ inches. (Round the answer to the nearest sixteenth of an inch.)
    a. 2¹³⁄₁₆
    b. 5⅝
    c. 5¹¹⁄₁₆
    d. 16⁷⁄₁₆

13. In a rolling offset, the run of pipe begins in one plane and ends up in the same plane.
    a. True
    b. False

14. Rolling offsets form _____ right triangle(s).
    a. 0
    b. 1
    c. 2
    d. 3

15. The correct equation for figuring the true offset is _____.
    a. $\sqrt{\text{Rise}^2 + \text{Offset}^2}$
    b. $\sqrt{\text{Rise}^2} + \sqrt{\text{Offset}^2}$
    c. $\sqrt{\text{Rise}^2} \times \sqrt{\text{Offset}^2}$
    d. $\sqrt{\text{Rise} + \text{Offset}}$

16. The true offset for an installation with a rise of 5⅝ inches and an offset of 4⅞ inches is _____ inches.
    a. 3¼
    b. 4⁹⁄₁₆
    c. 7⁷⁄₁₆
    d. 27⁷⁄₁₆

17. Before you can use constants to find the travel or run in a rolling offset, you must know the _____.
    a. true offset
    b. simple offset
    c. parallel offset
    d. rise

18. To find the travel for a rolling offset with a 22½-degree fitting angle, multiply the true offset by _____.
    a. 1.052
    b. 1.155
    c. 2.414
    d. 2.613

19. Refer to *Table 5*. A 45-degree rolling offset with an 8-inch rise and a 15-inch offset will require a pipe that measures _____ inches center-to-center. (Round answer to nearest sixteenth of an inch.)
    a. 18⅞
    b. 23¹⁄₁₆
    c. 24¹⁄₁₆
    d. 27⅜

20. A 45-degree rolling offset with a 9-inch rise and a 12-inch offset will require a pipe that measures _____ inches center to center. (Refer to *Table 5*.)
    a. 15
    b. 16⅞
    c. 18⅜
    d. 21³⁄₁₆

21. To find the run in a rolling offset, multiply the _____ by the constant for the fitting angle.
    a. offset
    b. rise
    c. true offset
    d. offset angle

Math is one of the most important tools in your plumber's toolbox. You learned about the Pythagorean theorem and how to apply it in the field. Tricks of the trade, such as using triangles to square corners or using a folding rule to find angles, help make you a more efficient plumber. You also learned how to calculate simple, parallel, and rolling offsets, as well as how to calculate 45-degree offsets around obstructions.

You'll run into math at every stage of plumbing, whether you use it to figure lengths of pipe, to calculate offsets, or to square a corner. Studying math and applying it every day is a career-long process.

Notes

Trade Terms Introduced in This Module

Blade: The longer of two sides of a framing square.

Burn an inch: Practice of deducting an inch from the reading (measurement) to account for the room required to grip the folding rule with your fingers.

Constant: A number that never changes when used in an equation. For example, when finding the travel of a 45-degree right triangle, always use the constant 1.414.

Framing square: Specialized square with tables and formulas printed on the blade for making quick calculations, such as finding travel and determining area and volume.

Offset: A combination of elbows or bends that brings one section of pipe out of line but into a line parallel with the other section.

Pythagorean theorem: In any right triangle, the longest leg squared is equal to the sum of the other two legs squared. As applied to plumbing, the theorem is $\text{Travel}^2 = \text{Run}^2 + \text{Offset}^2$.

Ratio: A relationship between numbers.

Right triangle: Any triangle with a 90-degree angle.

Rise: The vertical measurement as opposed to the horizontal measurement of a run of pipe.

Roll: In a rolling offset, the displacement of the offset that is perpendicular to both the rise and the run.

Rolling offset: An offset in which the two parallel sections of pipe on either end of the offset are not in the same vertical or horizontal plane.

Run: The distance between where a run of pipe starts to change direction and where it returns to its original direction.

Street fitting: A fitting with male threads on one end and female threads on the other.

Tongue: The shorter of two sides of a framing square.

Travel: The longest leg of a right triangle. It is also the center-to-center distance between fittings.

True offset: The longest of the legs of a triangle formed by a rolling offset.

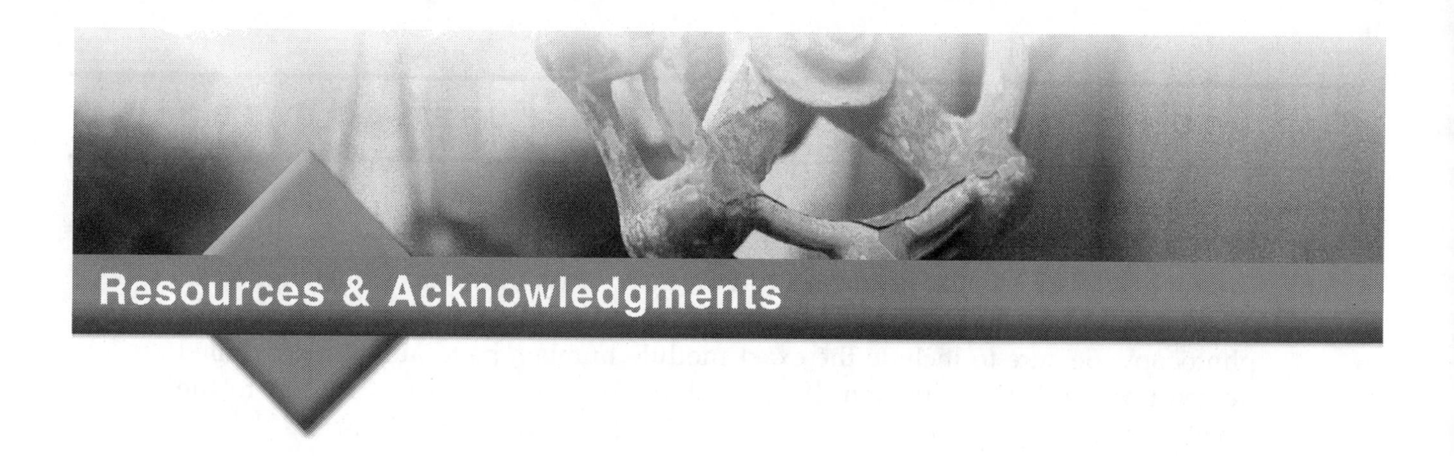

Additional Resources

This module is intended to be a thorough resource for task training. The following reference works are suggested for further study. These are optional materials for continued education rather than for task training.

Pipefitters Handbook. Forrest R. Lindsay. New York: Industrial Press Inc.

References

Pipefitters Handbook. Forrest R. Lindsay. New York: Industrial Press Inc.

Pipe Fitter's Math Guide. Johnny Hamilton. Clinton, NC: Construction Trades Press.

Plumber's and Pipefitter's Calculations Manual. R. Dodge Woodson. McGraw Hill Professional.

Plumber's Quick-Reference Manual: Tables, Charts, and Calculations. R. Dodge Woodson. McGraw Hill Professional.

Plumbing Instant Answers. R. Dodge Woodson. McGraw Hill Professional.

Figure Credits

Cooper Hand Tools 201F12

Jonathan Byrd 201F14

CONTREN® LEARNING SERIES — USER UPDATE

The NCCER makes every effort to keep these textbooks up-to-date and free of technical errors. We appreciate your help in this process. If you have an idea for improving this textbook, or if you find an error, a typographical mistake, or an inaccuracy in NCCER's Contren® textbooks, please write us, using this form or a photocopy. Be sure to include the exact module number, page number, a detailed description, and the correction, if applicable. Your input will be brought to the attention of the Technical Review Committee. Thank you for your assistance.

Instructors – If you found that additional materials were necessary in order to teach this module effectively, please let us know so that we may include them in the Equipment/Materials list in the Annotated Instructor's Guide.

Write: Product Development and Revision
National Center for Construction Education and Research
P.O. Box 141104, Gainesville, FL 32614-1104

Fax: 352-334-0932

E-mail: curriculum@nccer.org

Craft ______ Module Name ______

Copyright Date ______ Module Number ______ Page Number(s) ______

Description

______

______

______

(Optional) Correction

______

______

(Optional) Your Name and Address

______

______

Plumbing Level Two

02202-05

Reading Commercial Drawings

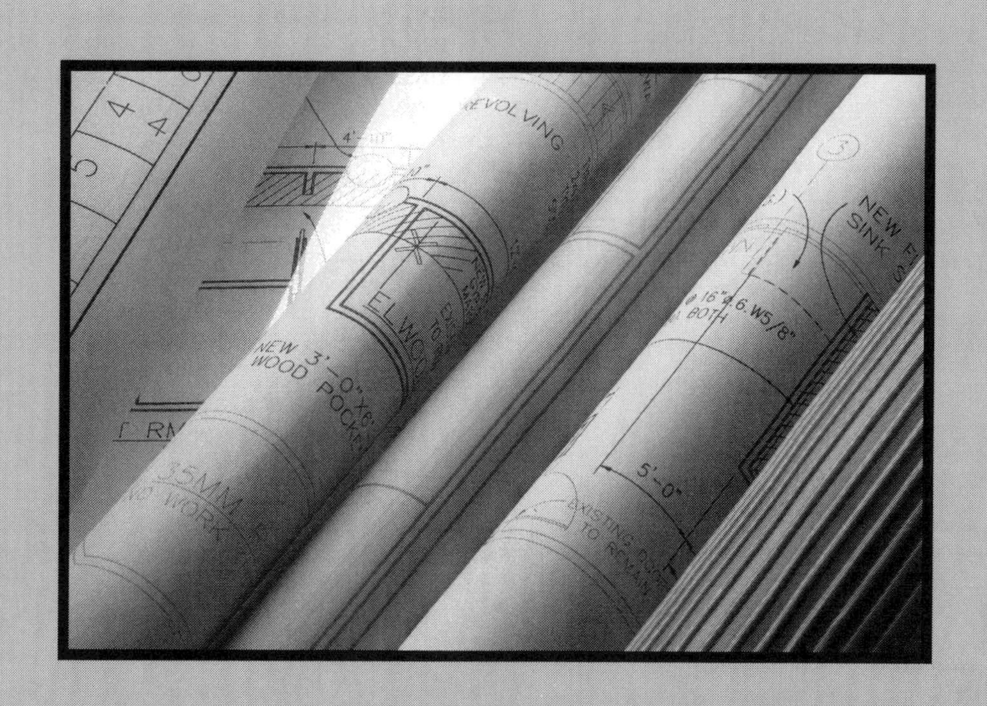

02202-05
Reading Commercial Drawings

Topics to be presented in this module include:

Overview

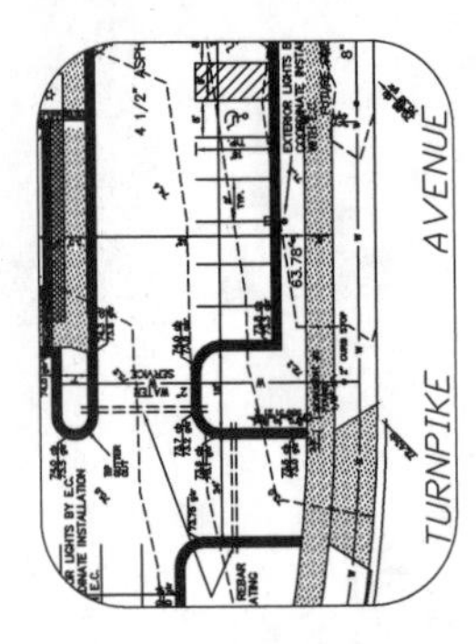

Plumbers interpret and apply information contained in construction drawings. These drawings identify a building's design, location, and dimensions. Typical drawing sets include a variety of drawings. The civil drawings (the site plans) describe the overall shape of the building site. Architectural drawings show the appearance of the structure. Structural drawings show how the framework is to be constructed. Mechanical drawings show the location and size of the components of the heating, ventilating, and air conditioning (HVAC) system. Plumbing drawings show the location and size of the drain, waste, and vent (DWV) and water supply piping. Electrical drawings show the electrical system details.

Changes are often necessary when working with construction drawings. To make such changes, plumbers submit addenda, change orders, requests for information, and clarifications. Drawings are also used to determine the number, type, and locations of fixtures; identify changes and alterations; use the correct plumbing codes, materials, and equipment; and avoid conflicts with other trades. Plumbers use approved submittal data for each fixture, including a cut sheet or drawing of a specific fixture, and material takeoffs, which are like shopping lists for a project. Plumbers must understand how various drawings work together and be able to identify, suggest, and document changes to plans made before and during the construction process. Doing so will help to identify and correct errors before they cause delays.

Focus Statement

The goal of the plumber is to protect the health, safety, and comfort of the nation job by job.

Code Note

Codes vary among jurisdictions. Because of the variations in code, consult the applicable code whenever regulations are in question. Referring to an incorrect set of codes can cause as much trouble as failing to reference codes altogether. Obtain, review, and familiarize yourself with your local adopted code.

Objectives

When you have completed this module, you will be able to do the following:

1. Interpret information from given site plans.
2. Verify dimensions shown on drawings and generate a request for information (RFI) when you find discrepancies.
3. Locate plumbing entry points, walls, and chases.
4. Create an isometric drawing.
5. Do a material takeoff for drainage, waste, and vent (DWV) and water supply systems from information shown on drawings.
6. Use approved submittal data, floor plans, and architectural details to lay out fixture rough-ins, to develop estimates, and to establish general fixture locations.
7. Recognize the need for coordination and shop drawings.

Trade Terms

- Addendum
- Alternate
- Approved submittal data
- Architectural drawing
- As-built drawing
- Change order
- Chase
- Civil drawing
- Clarification
- Contour line
- Convention
- Coordination drawing
- Cut sheet
- Elevation
- Landscape drawing
- Mechanical drawing
- North arrow
- Plan view
- Request for information (RFI)
- Schedule
- Section
- Site plan
- Structural drawing
- Title block

Required Trainee Materials

1. Appropriate personal protective equipment
2. Pencil and paper
3. Copy of local applicable code

Prerequisites

Before you begin this module, it is recommended that you successfully complete *Core Curriculum: Introductory Craft Skills; Plumbing Level One; Plumbing Level Two,* Module 02202-05.

This course map shows all of the modules in the second level of the Plumbing curriculum. The suggested training order begins at the bottom and proceeds up. Skill levels increase as you advance on the course map. The local Training Program Sponsor may adjust the training order.

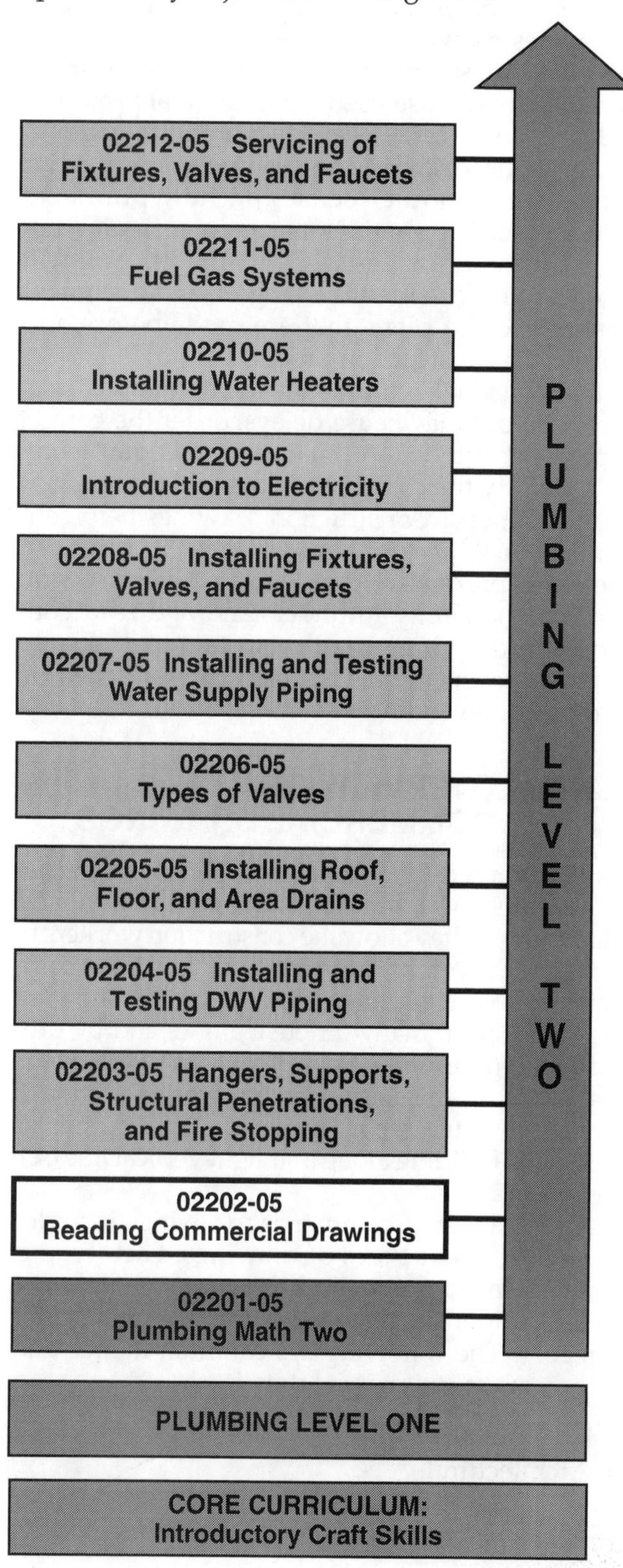

202CMAP.EPS

1.0.0 ◆ INTRODUCTION

Commercial projects are more complicated than residential projects. The buildings may have multiple stories. Different construction trades are on the site at different times. Often the work of one trade can't begin until another trade successfully completes its work. For these reasons, construction drawings for commercial projects are more complex than those used for residential projects.

In *Plumbing Level One*, you were introduced to construction drawings and learned how to read them. This module focuses on how plumbers interpret and apply the information contained in construction drawings while on the job. You will also learn how to identify, suggest, and document changes to the plans that are made before and during the construction process.

Drawings help contractors identify and correct errors before they cause delays, order the correct types and numbers of materials and equipment, and identify the correct locations and dimensions of installations. Construction drawings help contractors to plan work in advance, resulting in increased productivity and reduced costs. As you can see, the ability to understand and read construction drawings is an important skill. It will help you to ensure that a project runs on schedule and on—or even under—budget.

2.0.0 ◆ OVERVIEW OF COMMERCIAL DRAWINGS

As you learned in *Plumbing Level One*, commercial drawings show a building's design, location, and dimensions. They allow the construction workers to visualize the complete project and its various components before starting. They do this by showing the layout from many different angles and by providing **sections** and details of the various elements that make up the completed building. A section is what you would see if you could slice a particular part of the building along a straight vertical line (see *Figure 1*).

A typical set of commercial drawings includes several different types of drawings. Each type is referred to by name, and the pages usually are coded to identify the drawing with a particular category. The following are common names and categories of commercial drawings:

- Civil–C
- Architectural–A
- Structural–S
- Mechanical–M
- Plumbing–P
- Electrical–E

Depending on the size of the project, other types of drawings may be included in the set. For example, commercial drawing sets for office buildings often include **landscape drawings.** Landscape drawings show the arrangement of decorative elements such as trees and grass, water features, and paths surrounding the building. These drawings are designated with the letter L.

Most drawings are prepared according to a standard format so that they can be read and used by any construction professional. However, some companies may have developed their own unique ways to denote the elements of a drawing. Don't expect all drawings to follow the same procedure outlined in this module. Learn your company's procedures as well.

2.1.0 Civil Drawings—The Site Plans

The **civil drawings,** also called the **site plans,** describe the overall shape of the building site. In addition, these drawings show a building's location and other structures planned for the site (see *Figure 2*). Pages on civil drawings are numbered C-1, C-2, and so on.

Notice the numbered dashed lines in *Figure 2*. These are **contour lines.** They show lines of equal elevation on the land. The numbers refer to the elevation in feet above sea level. Contour lines are used to indicate the grade. Dashed contour lines show the existing grade; solid contour lines show the desired grade when construction is finished.

Plumbers use civil drawings to locate existing utilities. The drawings indicate the location and depth of existing utility lines and valves. Use this information when installing building drains and lines to ensure proper connections and to avoid damaging existing lines.

Most site plans are drawn to a relatively small scale such as 1" = 20'-0". A small scale generally is used so that the entire site can be shown on one page. However, the small scale limits the amount of information that can be shown. Therefore, other drawings that provide sufficient detail are included in the set.

2.2.0 Architectural Drawings

Architectural drawings show the shape, size, and appearance of the structure using **plan views, elevations,** sections, and details. Plan views show the horizontal projections of a building; elevations show vertical projections (see *Figure 3*). Pages on these drawings are numbered A-1, A-2, and so on.

Obtain all dimensions from the architectural drawings. Do not obtain measurements by measuring an architectural drawing. Use the scale

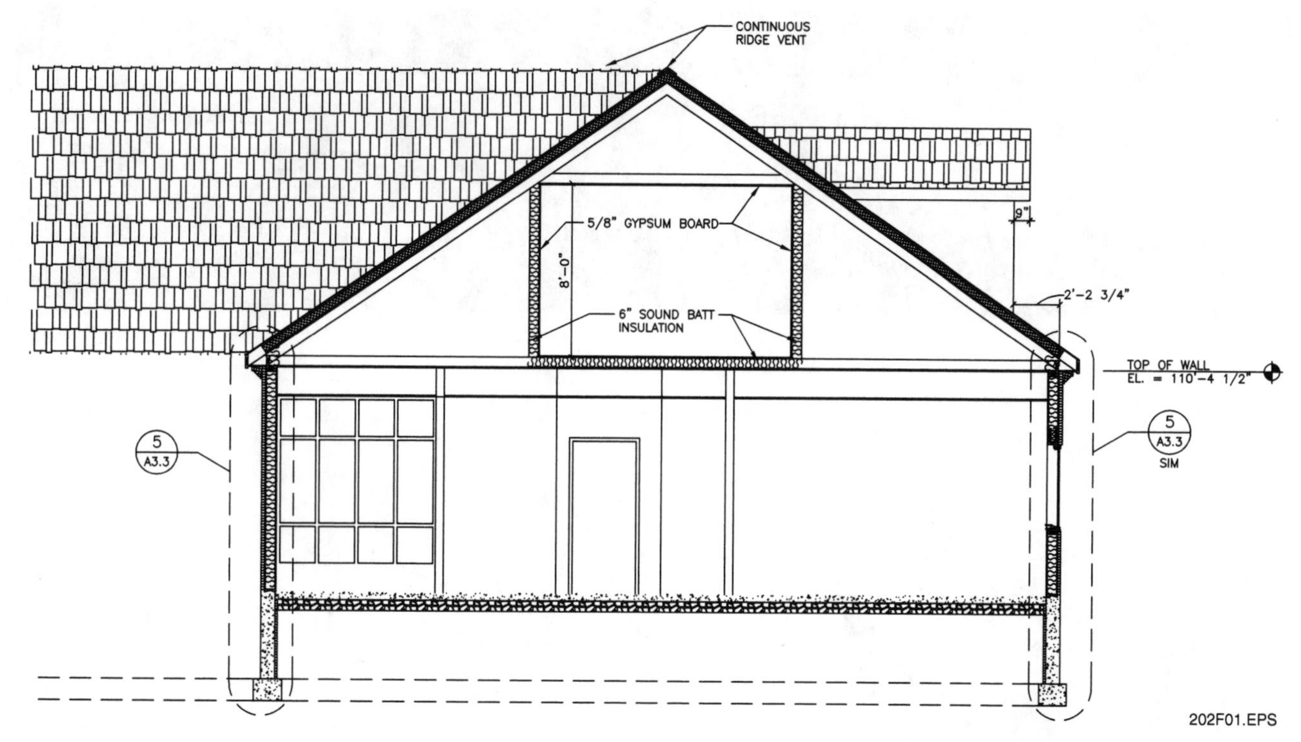

Figure 1 ◆ Section of a building.

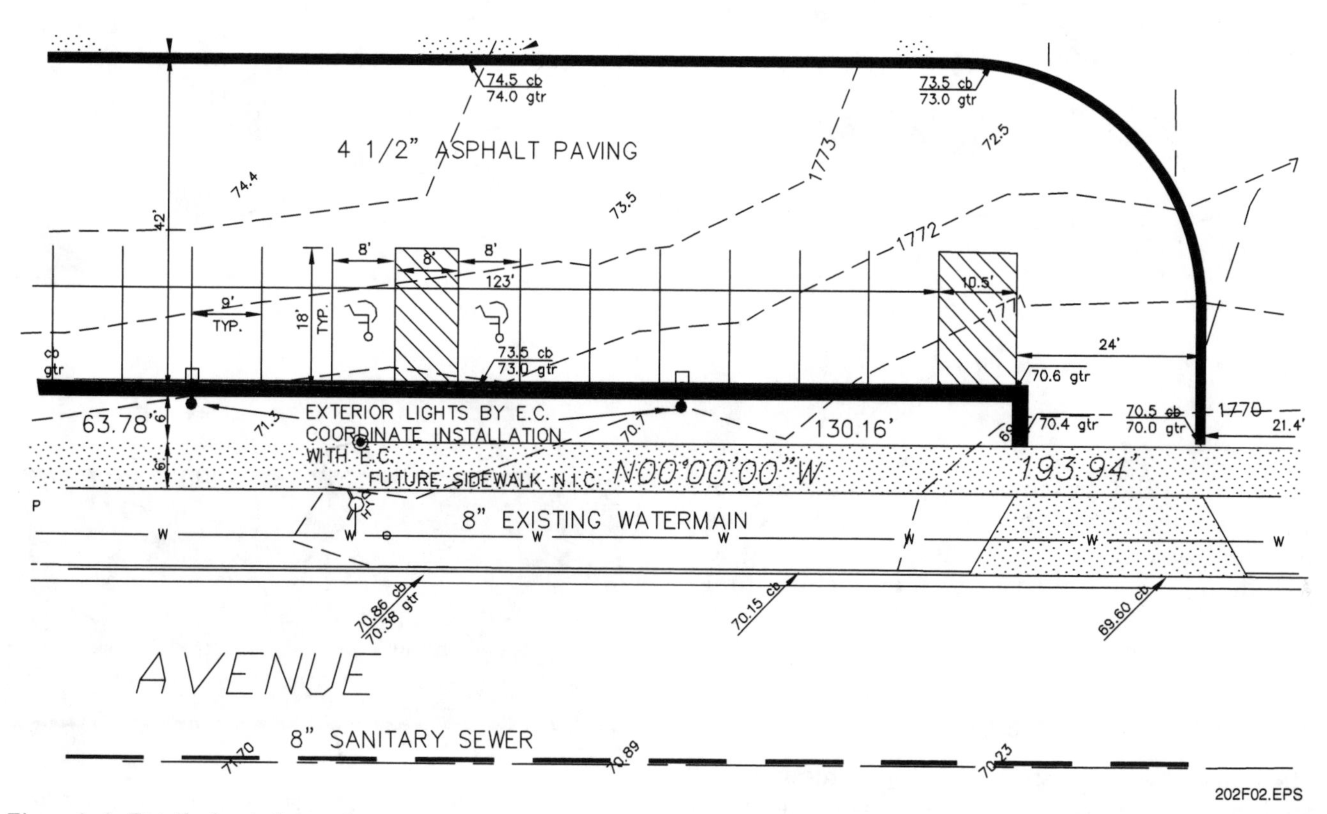

Figure 2 ◆ Detail of a civil drawing.

Studying Plans

When you look at a floor plan, first locate any rooms that require plumbing: kitchen, bathrooms, and utility rooms. Review these areas to determine the piping and fixtures needed for each. Then study the rest of the floor plan carefully for other areas that might require water supply or DWV piping.

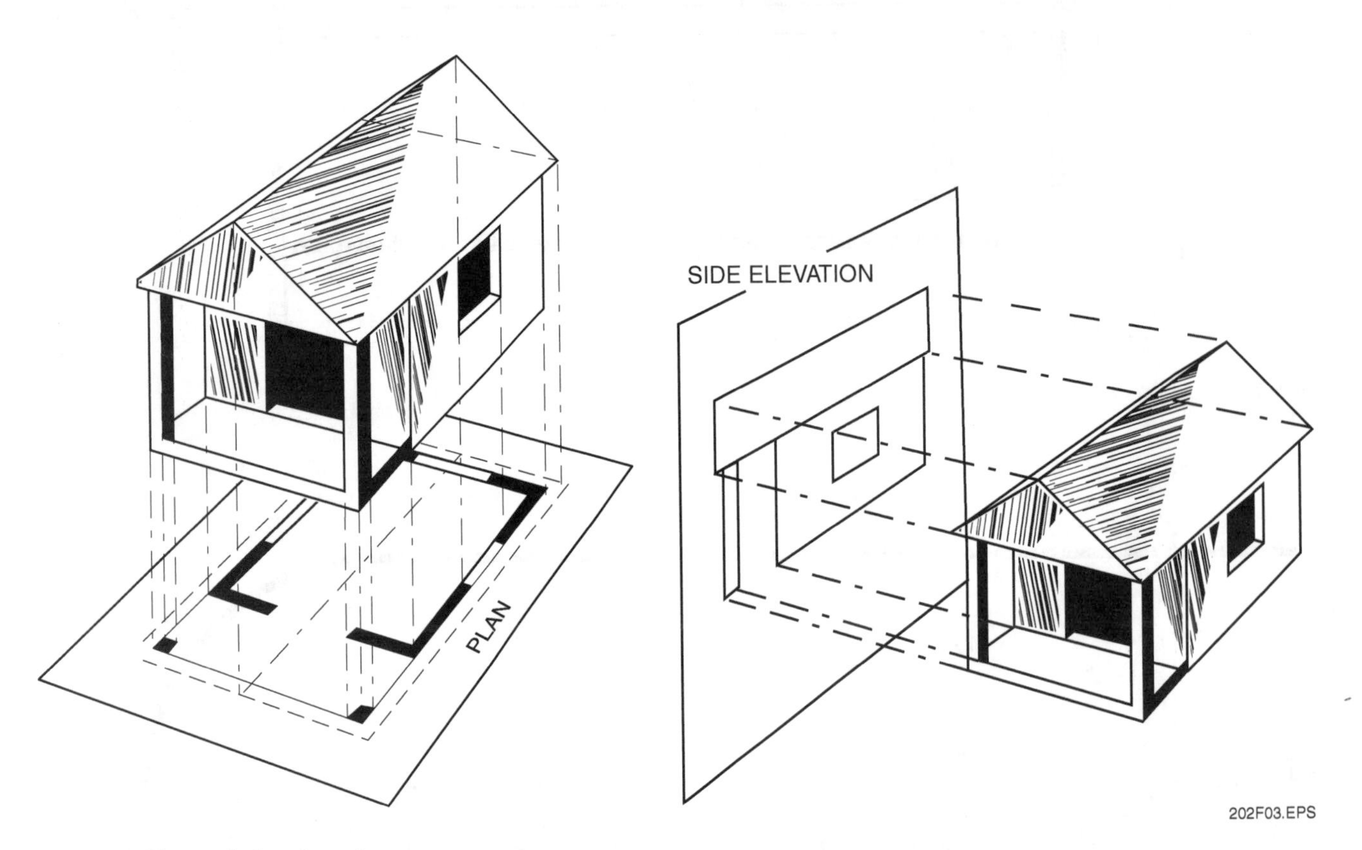

Figure 3 ◆ Plan and elevation views.

provided on the drawing. If a drawing does not include a scale, the contractor should ask the architect or engineer to revise the drawing to include one. The dimensions on an architectural drawing help the contractor purchase the correct fixtures and equipment to install.

2.3.0 Structural Drawings

Structural drawings show how the framework of the building is to be constructed. The framework may include structural steel, post-tension cables, cast-in-place reinforced steel, cast-in-place concrete, precast concrete, or other building materials (see *Figure 4*). Pages on these drawings are numbered S-1, S-2, and so on.

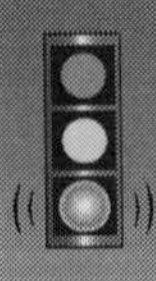

NOTE

Holes for cast-in-place reinforced steel are not shown in structural drawings. In projects where cast-in-place reinforced steel is used, the coordination drawings will include a sleeve layout drawing that shows the locations for the holes.

The location of structural piping supports affects the location of water supply and drainage systems. You will need to refer to these drawings before installing pipe. Structural drawings will help you determine the correct pipe supports, hangers, and connectors to use with the building's

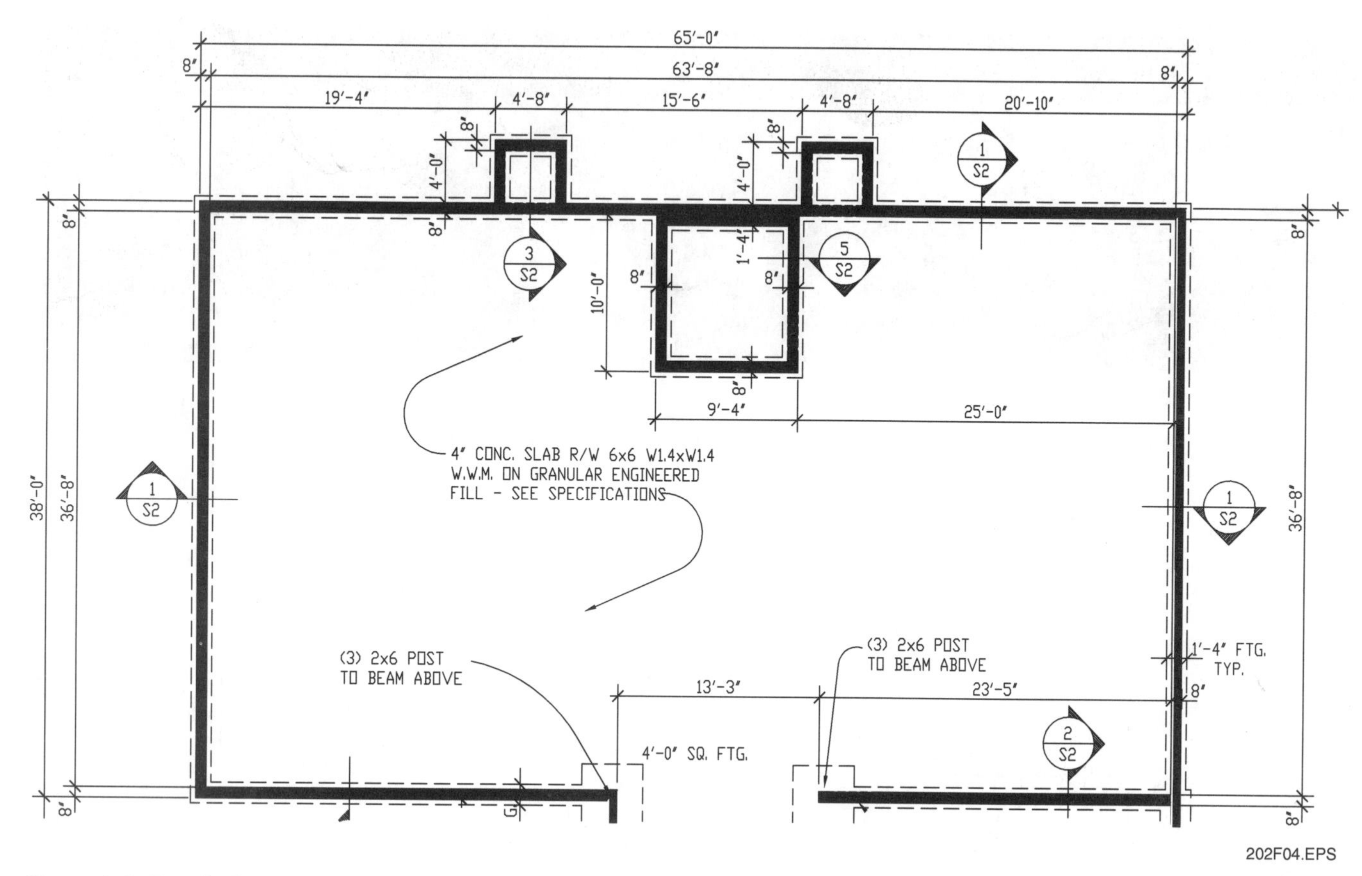

Figure 4 ◆ Detail of a structural drawing.

structural materials. Study structural drawings carefully to obtain as much information about the layout as possible.

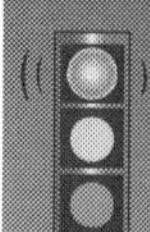

WARNING!

Most codes do not allow structural steel to be cut. In those jurisdictions where cutting is permitted, you must obtain prior permission from the structural engineer and possibly the code administrator. Improper cutting of structural steel could seriously weaken the structure, causing partial or total collapse. This could cause injuries and even death, as well as destruction of property.

2.4.0 Mechanical Drawings

Mechanical drawings show the location and size of the components of the heating, ventilating, and air-conditioning (HVAC) system. They also may show elevators, conveyors, and escalators (see *Figure 5*). Pages on these drawings are numbered M-1, M-2, and so on. Be aware that mechanical drawings do not show conflicts (for example, electrical conduit hung in the intended location of water supply piping). Be sure to coordinate your work with other trades. The construction manager establishes the elevations, and each trade must follow those elevations precisely. To do that requires each trade to identify and work out potential conflicts with equipment, fittings, and structural members.

DID YOU KNOW?

A Museum for Architecture

The Athenaeum of Philadelphia is a museum of American architecture. The museum, which has collected the work of about 1,000 American architects, has 150,000 drawings, 50,000 photographs, and many architectural documents.

ON THE LEVEL

Plan Size

The drawings for a small residence may be combined. The electrical plan, for example, may be included on the floor plan. So there may be no more than three or four plan sheets. In contrast, the plans for a large commercial building may contain more than 100 pages. Each page shows a different aspect of the plan—the floor plan, the electrical plan, the plumbing plan, the HVAC plan, and so on.

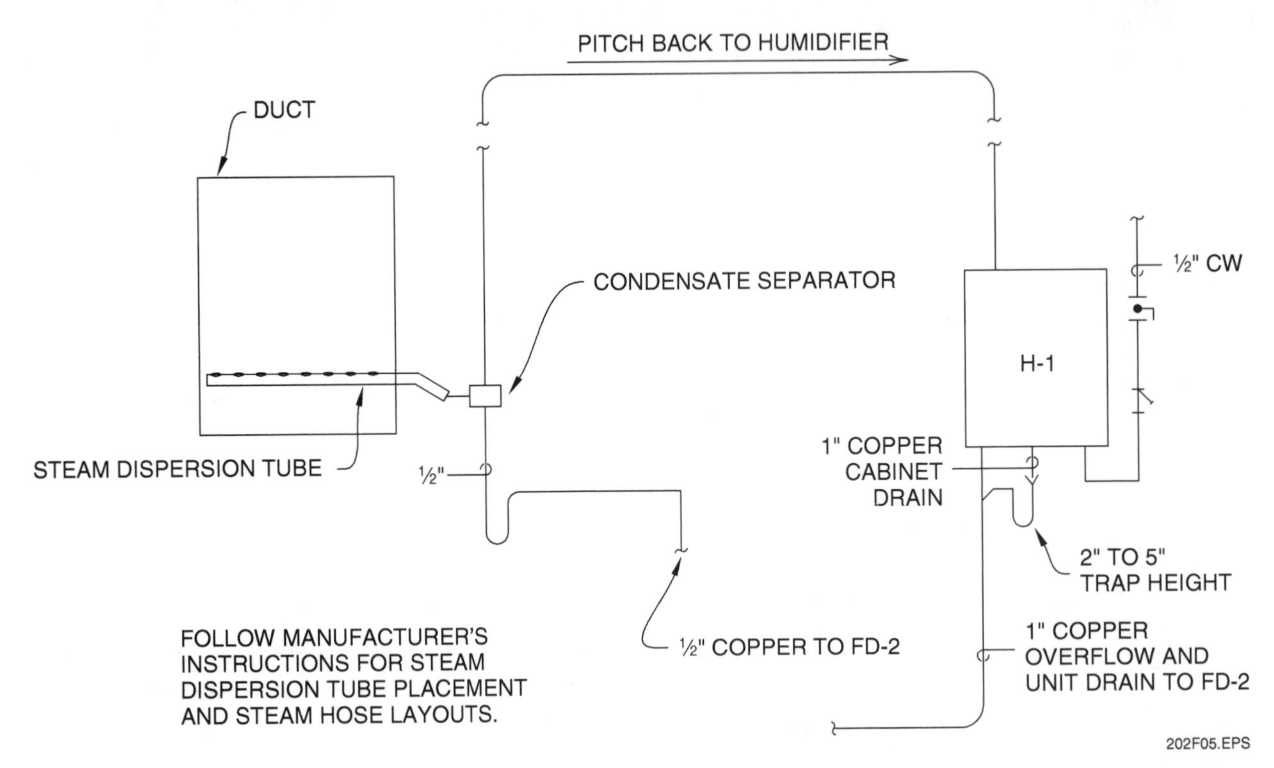

Figure 5 ◆ Detail of a mechanical drawing.

Check the mechanical drawings to determine what limitations mechanical equipment will impose on the plumbing system. You may be required to install the piping that connects to some of the mechanical equipment.

2.5.0 Plumbing Drawings

Plumbing drawings show the location and size of the drain, waste, and vent (DWV) and water supply piping. These drawings include a special floor plan, which omits the detailed information used by other trades and provides additional details that plumbers need. Frequently, riser diagrams are included on plumbing drawings to show the installation of the vertical piping system (see *Figure 6*). Plumbing drawings do not show dimensions. Refer to the architectural drawings for height and center-to-center dimensions. Pages on these drawings are numbered P-1, P-2, and so on. You will learn more about how to use plumbing drawings on the job elsewhere in this module.

2.6.0 Electrical Drawings

Electrical drawings show the details of the electrical system. They typically show the location of fixtures, controls, receptacles, circuit breakers, and wiring for larger electrical devices. They are sometimes superimposed on the architectural drawing that shows the building's floor plan (see *Figure 7*). Pages on these drawings are numbered E-1, E-2, and so on.

Plumbers seldom refer to the electrical drawings for details related to their work. However, because piping and electrical conduit often run through buildings in areas that are close to one another, plumbers and electricians usually coordinate their work to avoid conflicts.

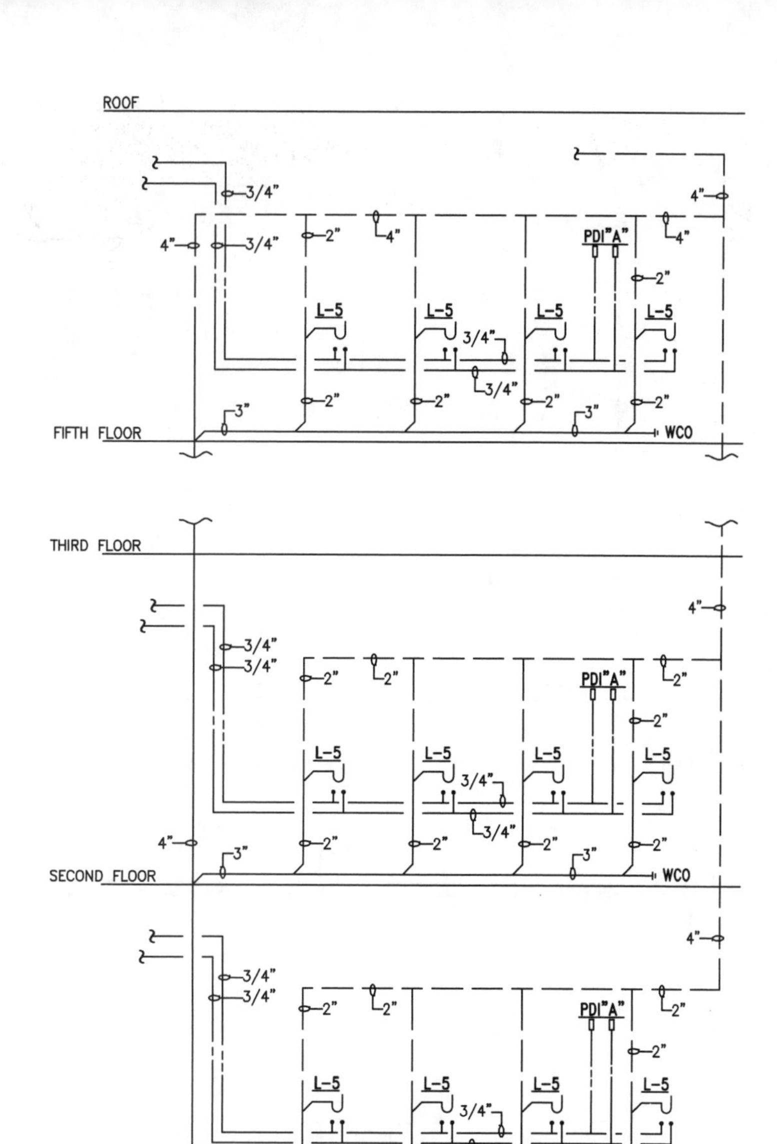

Figure 6 ◆ Detail of a plumbing riser diagram.

3.0.0 ◆ WORKING WITH CONSTRUCTION DRAWINGS

When working with any type of drawing found in a set of commercial drawings, keep the following points in mind:

- Read the **title block** (see *Figure 8*). The title block gives you critical information about the drawing, such as the scale, the last revision date, the drawing number, and the name of the architect or engineer. If you have to remove a sheet from a set of drawings, be sure to return the sheet folded so that the title block faces up. Match the construction drawings with the information on the contract, and then match up the specifications and any addenda as well.

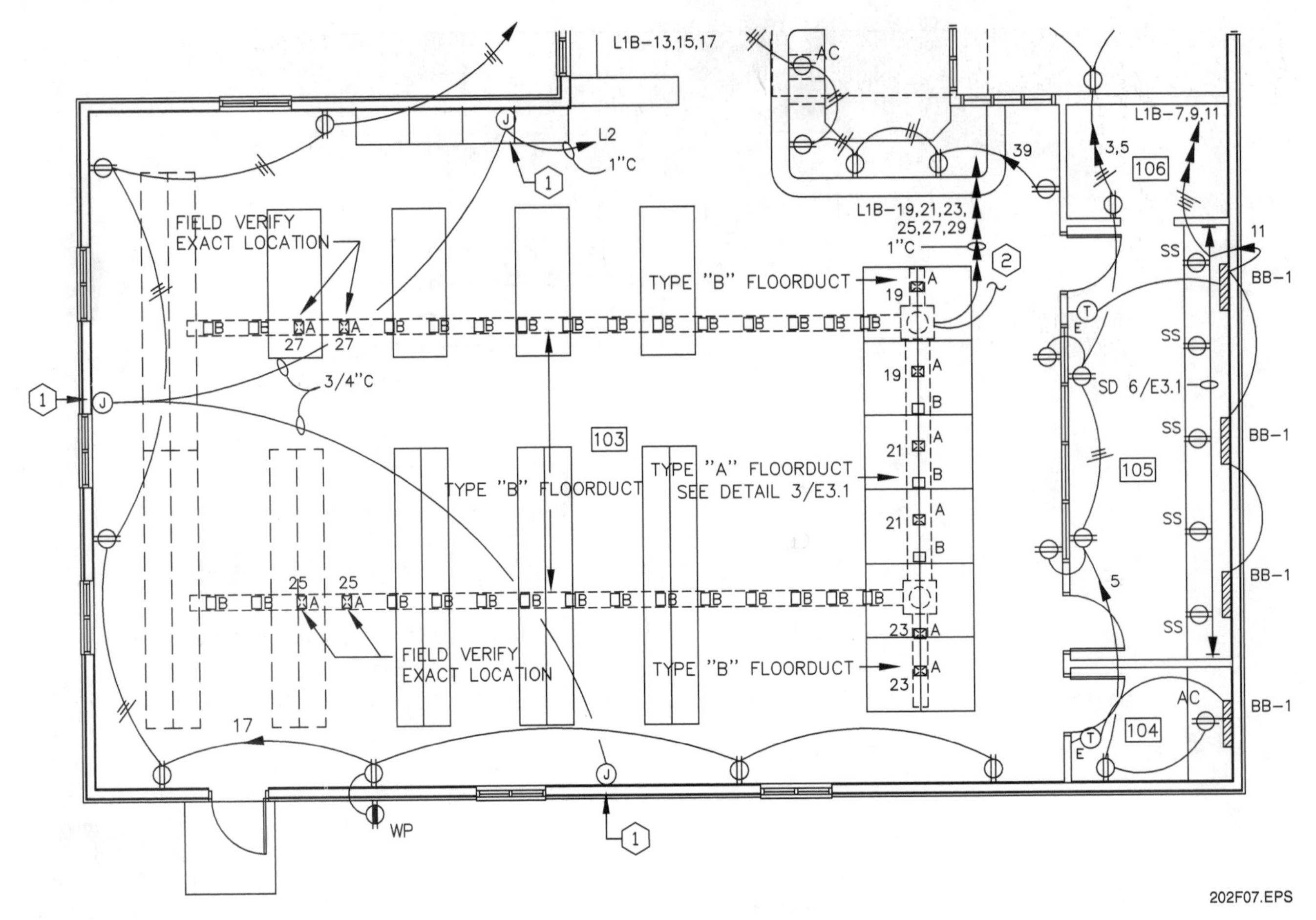

Figure 7 ◆ Detail of an electrical drawing.

CODE INFORMATION

			REVISIONS
CODE USED:	1997 UBC		
OCCUPANCY GROUP:	B		
TYPE OF CONSTRUCTION:	TYPE V N		DRAWN BY JLF/LSG
BUILDING AREA:			
MAIN FLOOR: MECHANICAL ROOM:	5,890 324	SQ. FT. SQ. FT.	DATE 5-30-00
TOTAL:	6,214	SQ. FT.	PROJECT NO. 9918

I HEREBY CERTIFY THAT THIS PLAN, SPECIFICATION, OR REPORT WAS PREPARED BY ME OR UNDER MY DIRECT SUPERVISION AND THAT I AM A DULY REGISTERED ARCHITECT UNDER THE LAWS OF THE STATE OF NORTH DAKOTA.

DATE: 5-30-2000 REG. NO. 578

DRAWING DESCRIPTION: TITLE SHEET

A 0.1

PLOT SCALE: 1=1 FILE NAME: 9918_A_0-1.DWG

© 2000 Ritterbush – Ellig – Hulsing P.C. Architects – Planners

202F08.EPS

Figure 8 ◆ Title block.

- Find the **north arrow** (see *Figure 9*). Always orient yourself to the structure. If you know where north is, you'll be able to describe the locations of walls and other parts of the building accurately. A north arrow should appear on each sketch that is used to describe the building, especially on drawings that specify changes.

Figure 9 ◆ North arrow.

- Recognize that drawings work together as a set. Architects and engineers draw plans, elevations, and sections because it takes more than one type of view to communicate the whole project. Learn how to use more than one drawing, when necessary, to find the information you need.

Review Questions

Sections 1.0.0–3.0.0

1. Drawings numbered L-1, L-2, and so on are _____ drawings.
   a. legal
   b. lateral projection
   c. landscape
   d. listed

2. A civil drawing shows the _____.
   a. shape of the building site, the building location, and planned structures
   b. shape of the building site, the floor plan, and a schedule of elevations
   c. topography of the site, the HVAC plan, and the plumbing and piping plan
   d. property lines, a list of the finish schedules, and the governing codes

3. Site plans are drawn using a small scale so that _____.
   a. more than one drawing can be included on a page
   b. additional detail can be included
   c. the entire site can be shown on one page
   d. plan and elevation views can be included

4. Refer to the illustration in *Figure 10.* The drawing element indicated is a(n) _____.
   a. easement
   b. proposed grade
   c. exterior wall
   d. existing contour line

5. Plumbers must be familiar with structural drawings because they show _____.
   a. where the structural members might affect the piping
   b. where they are allowed to modify the structural members
   c. what types of hangers and piping support to use
   d. what types of structural members must not be modified

6. You will find information about the HVAC system on the _____.
   a. civil drawings
   b. structural drawings
   c. plumbing drawings
   d. mechanical drawings

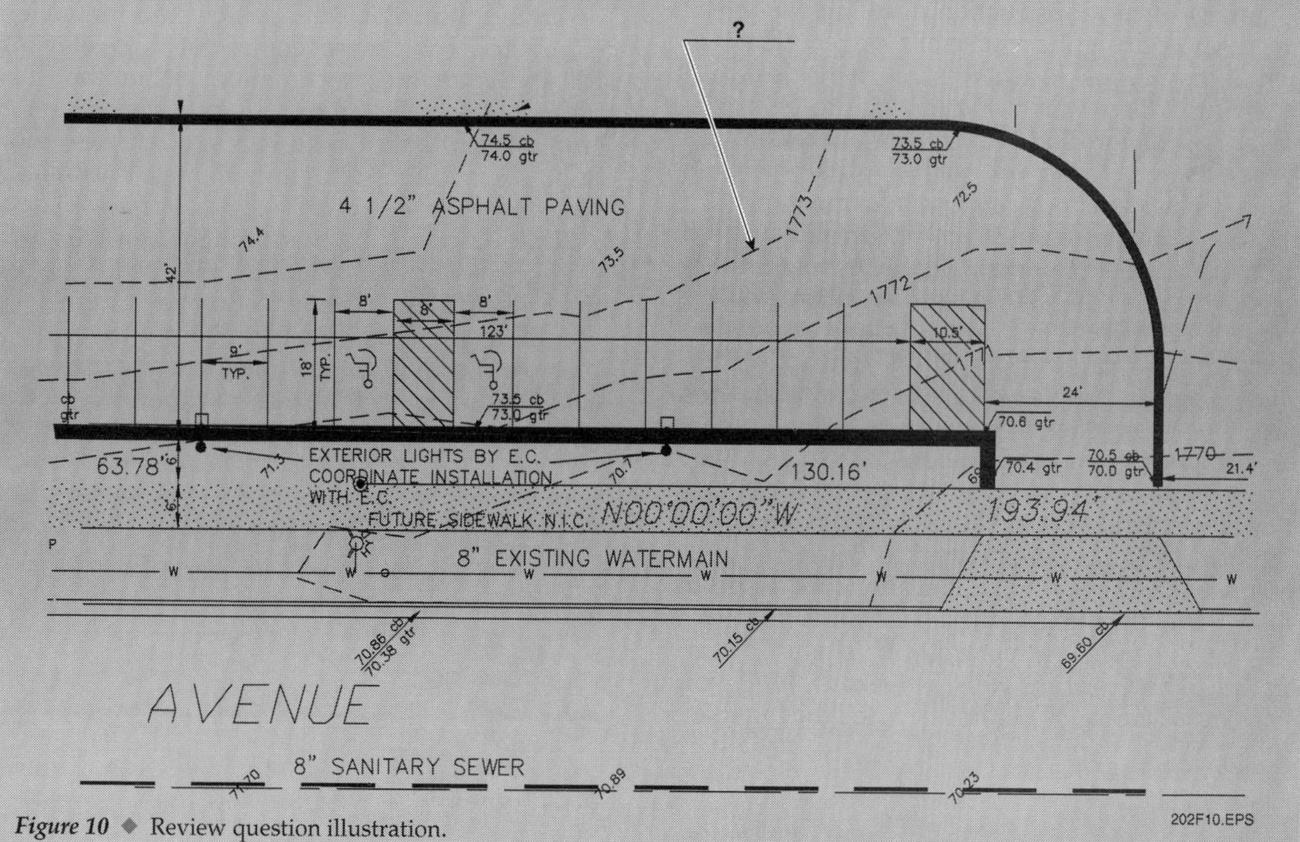

Figure 10 ◆ Review question illustration.

Review Questions

7. You should know how to read mechanical drawings because _____.
   a. you will have to locate and size the HVAC units
   b. they are included in your set of plumbing plan drawings
   c. you may have to install the piping that connects to the HVAC units
   d. you may have to review them for errors

8. Riser diagrams are frequently included on architectural drawings.
   a. True
   b. False

9. The title block on drawings shows the _____.
   a. north arrow
   b. last revision date
   c. number of elevations
   d. current date

Match the definition to the correct view.

10. __B__ A vertical slice through a building
11. __A__ The vertical projection of a building
12. __D__ The horizontal projection of a building

   a. elevation
   b. section
   c. isometric
   d. plan

4.0.0 ◆ DOCUMENTING CHANGES TO CONSTRUCTION DRAWINGS

Changes to the construction drawings are often necessary during the course of a project. The customer may decide to add features to the design. A drawing will contain what appears to be a mistake. A problem that requires an alteration to a drawing may crop up on the job site. When changes happen, they must be documented.

Several different contractual documents are used to make changes to construction drawings. These include **addenda**, **change orders**, **requests for information (RFI)**, and **clarifications.** An addendum (the singular form of addenda) is a notification of a change in the plan before the project has been bid. Once construction has begun, change orders are used to indicate modifications to the plan. When a plan appears to contain an error, or when a question arises about a certain detail, an RFI pointing out the error or question is issued to the architect or engineer (see *Figure 11*). Clarifications are issued in response to the error or question in the RFI.

Workers follow a hierarchy, or chain of command, when issuing RFIs. For example, if you notice a discrepancy on the plans, tell the foreman. The foreman writes the RFI, using specific details and making sure to put the time and date on it. The foreman passes the RFI to the superintendent or project manager, who passes it to the general contractor. The general contractor then relays the RFI to the architect or engineer. While the RFI is being addressed, work in the area must be stopped until the issue is resolved. Any work done prior to the solution is the responsibility of the contractor.

RFIs may require a change in the work. If so, the change may or may not affect the cost of the contract. Regardless of the effect on costs, remember that all changes are departures from the contract plans and should be documented. That way, you can demonstrate that you were authorized to make the change specified in the clarification.

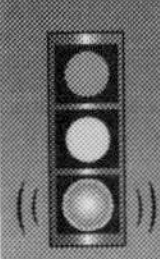

NOTE

Ensure that all drawings submitted with an RFI include a north arrow. This will help avoid confusion as to the correct placement of building and system components.

DATE ________________ RFI NO. ________________

PROJECT NAME ________________________________ PROJECT NO. ____________

REQUEST: REF D.W.G.NO. ________ REV. ________ OTHER ________

BY: ______________________________ REPLY BY (DATE): ________________

REPLY:

DATE: ________________________

202F11.EPS

Figure 11 ◆ Request for information form.

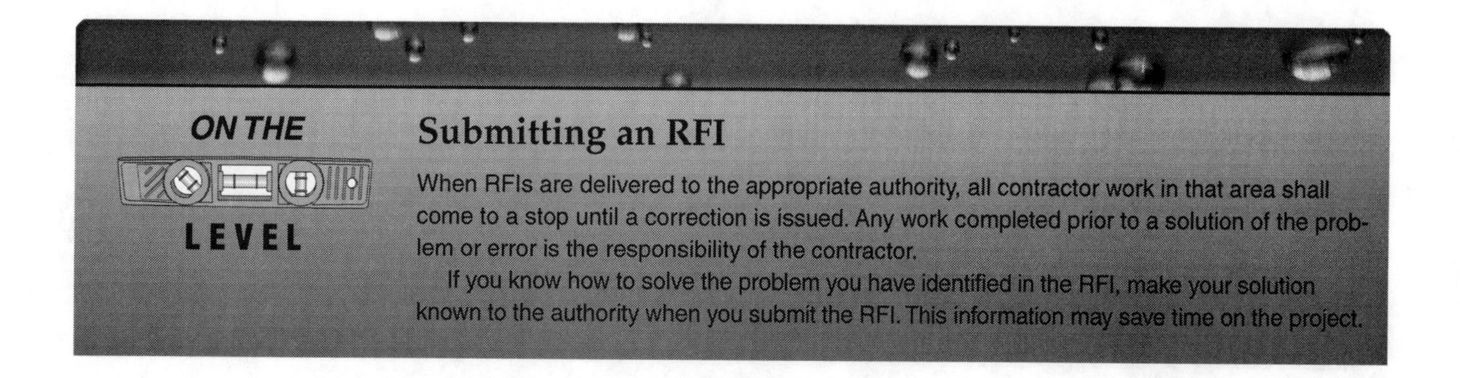

Submitting an RFI

When RFIs are delivered to the appropriate authority, all contractor work in that area shall come to a stop until a correction is issued. Any work completed prior to a solution of the problem or error is the responsibility of the contractor.

If you know how to solve the problem you have identified in the RFI, make your solution known to the authority when you submit the RFI. This information may save time on the project.

5.0.0 ◆ THE WORKSHEET DRAWINGS

On the job, you'll work with a wide variety of drawings as well as documents and information that support those drawings. Drawings that you will commonly use include the site plan, floor plans, supply and waste plumbing drawings, and isometric drawings. You will use the information on the drawings to do the following activities, among others:

- Determine the correct numbers and types of fixtures and fittings
- Ensure that fixtures and fittings are installed in the correct location
- Identify changes and alterations to the design
- Use the correct plumbing codes, plumbing materials, and equipment
- Avoid conflicts with other trades and with the building structure

This section introduces you to the various types of drawings you will use on the job and shows you how to use them effectively on the job.

5.1.0 Floor Plan and Schedules

Figure 12 is a floor plan for a light commercial construction building, the new home for the Germans from Russia Heritage Society, located in Bismarck, North Dakota. The building provides workrooms and offices for staff and visitors, as well as storage space for the society's records, historical artifacts, and library. The floor plan describes the size and shape of the floor and includes necessary dimensions. Floor plans are generally drawn to a scale of 1/8" = 1'-0", although sometimes 1/4" = 1'-0" is used. Plumbers refer to floor plans for the number and location of plumbing fixtures.

Notice that all the rooms and doors on the plan in *Figure 12* are numbered. These numbers are assigned by the designer. They are keyed to a corresponding **schedule,** which is a detailed list of components to be installed in the building. Contractors use the schedules to obtain the information they need to put the project out for bids. Plans may include schedules for doors, windows, flooring, finishing, and wall treatments. Depending on the complexity of the project, each of these schedules could require a separate page or pages. Designers use a standard labeling and marking system to show information on floor plans and in the schedules. Such standardized systems are called **conventions.**

The drawing in *Figure 12* includes a room finish schedule and a door schedule. The room finish schedule lists the treatments required for the floors, baseboards, walls, and ceilings. It also includes height information and a remarks section. The door schedule (see *Figure 13*) provides the height and width, materials, and frame of each type of door used in the building. For example, door 1121 at the entrance to vault room 112 is a hollow metal (HM) door with a 3-hour fire rating. The size of the door is always shown as width by height by thickness. Door thickness may sometimes be listed only at the top of the size column if it is standard for all doors.

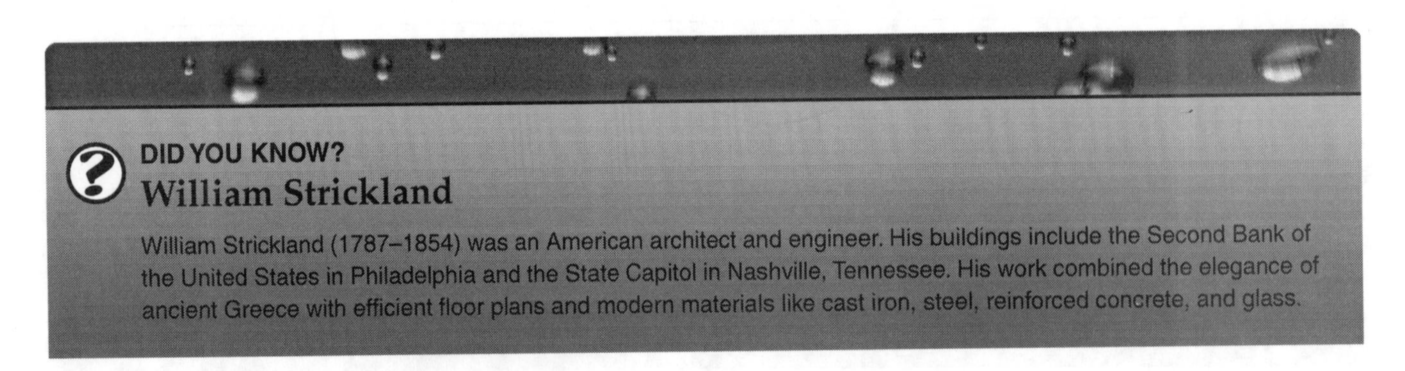

DID YOU KNOW?

William Strickland

William Strickland (1787–1854) was an American architect and engineer. His buildings include the Second Bank of the United States in Philadelphia and the State Capitol in Nashville, Tennessee. His work combined the elegance of ancient Greece with efficient floor plans and modern materials like cast iron, steel, reinforced concrete, and glass.

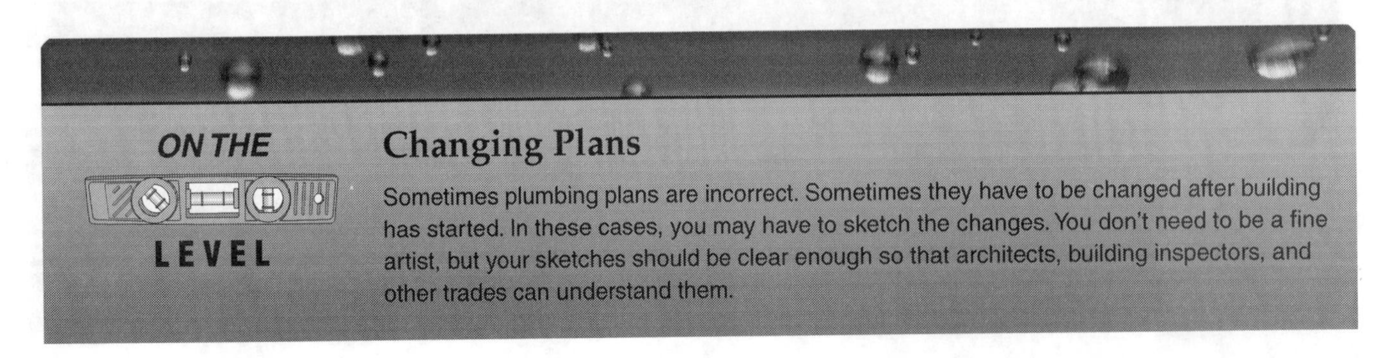

ON THE LEVEL

Changing Plans

Sometimes plumbing plans are incorrect. Sometimes they have to be changed after building has started. In these cases, you may have to sketch the changes. You don't need to be a fine artist, but your sketches should be clear enough so that architects, building inspectors, and other trades can understand them.

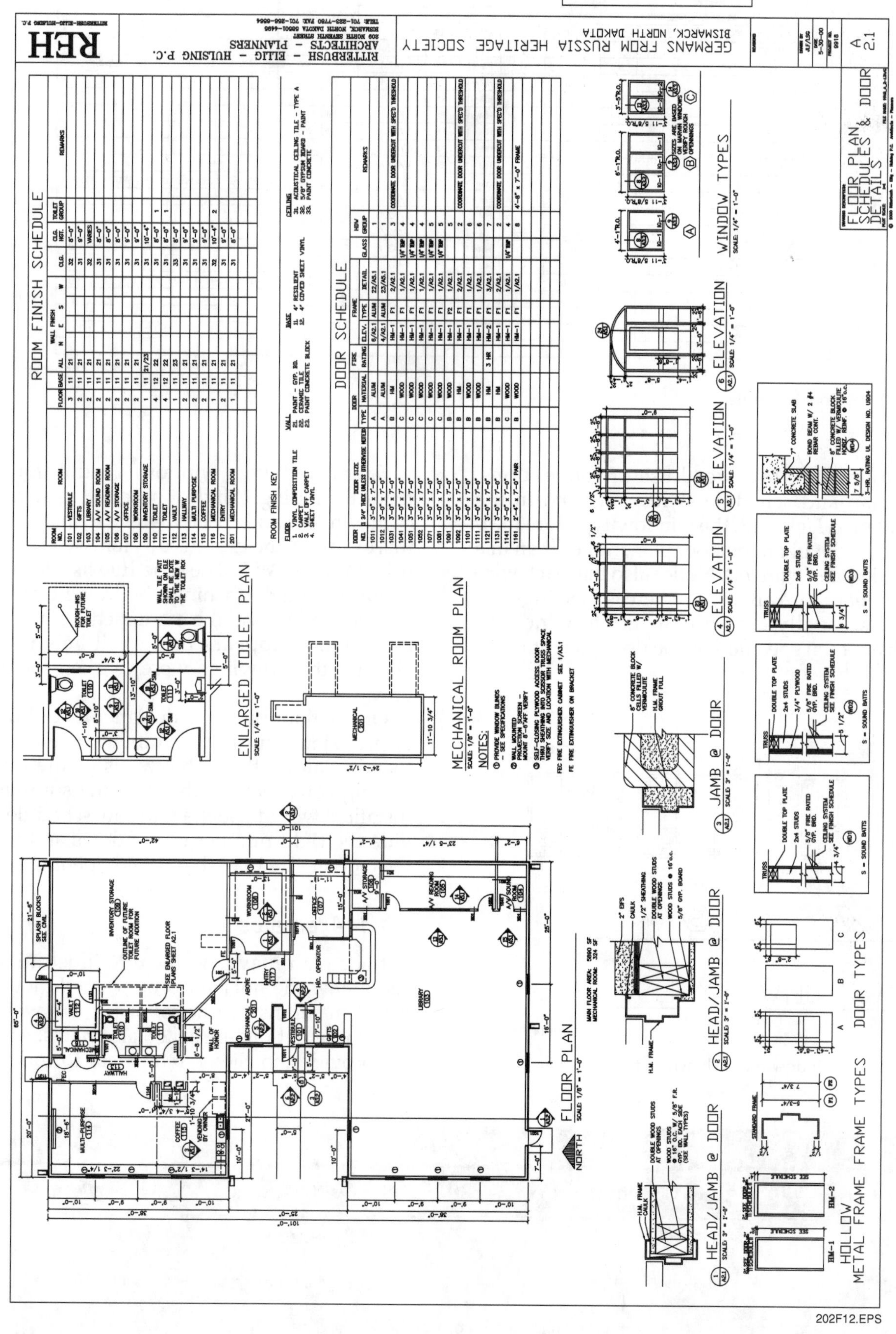

202F12.EPS

Figure 12 ◆ Floor plan and schedules.

DOOR SCHEDULE

DOOR NO.	DOOR SIZE (1 3/4" THICK UNLESS OTHERWISE NOTED)	DOOR		FIRE RATING	FRAME			GLASS	HDW GROUP	REMARKS
		TYPE	MATERIAL		ELEV.	TYPE	DETAIL			
1011	3'-0" x 7'-0"	A	ALUM		6/A2.1	ALUM	22/A5.1		1	
1012	3'-0" x 7'-0"	A	ALUM		4/A2.1	ALUM	23/A5.1		1	
1031	3'-0" x 7'-0"	B	HM		HM-1	F1	2/A2.1		3	COORDINATE DOOR UNDERCUT WITH SPEC'D THRESHOLD
1041	3'-0" x 7'-0"	C	WOOD		HM-1	F1	1/A2.1	1/4" TEMP	4	
1051	3'-0" x 7'-0"	C	WOOD		HM-1	F1	1/A2.1	1/4" TEMP	4	
1052	3'-0" x 7'-0"	C	WOOD		HM-1	F1	1/A2.1	1/4" TEMP	4	
1071	3'-0" x 7'-0"	C	WOOD		HM-1	F1	1/A2.1	1/4" TEMP	5	
1081	3'-0" x 7'-0"	C	WOOD		HM-1	F1	1/A2.1	1/4" TEMP	5	
1091	3'-0" x 7'-0"	B	WOOD		HM-1	F2	1/A2.1		5	
1092	3'-0" x 7'-0"	B	HM		HM-1	F1	2/A2.1		2	COORDINATE DOOR UNDERCUT WITH SPEC'D THRESHOLD
1101	3'-0" x 7'-0"	B	WOOD		HM-1	F1	1/A2.1		6	
1111	3'-0" x 7'-0"	B	WOOD		HM-1	F1	1/A2.1		6	
1121	3'-0" x 7'-0"	B	HM	3 HR	HM-2	F1	3/A2.1		7	
1131	3'-0" x 7'-0"	B	HM		HM-1	F1	2/A2.1		2	COORDINATE DOOR UNDERCUT WITH SPEC'D THRESHO
1141	3'-0" x 7'-0"	C	WOOD		HM-1	F1	1/A2.1	1/4" TEMP	4	
1161	2'-4" x 7'-0" PAIR	B	WOOD		HM-1	F1	1/A2.1		8	4'-8" x 7'-0" FRAME

202F13.EPS

Figure 13 ◆ Door schedule.

Despite the widespread use of conventions, floor plan and schedule styles can vary. In the floor plan in *Figure 12,* the window types are indicated by a detail on the floor plan, rather than in a schedule (see *Figure 14*). Notice also that each window is marked A, B, or C, whereas labeling conventions usually assign numbers to windows, which are then specified on the floor plan or in the appropriate schedule.

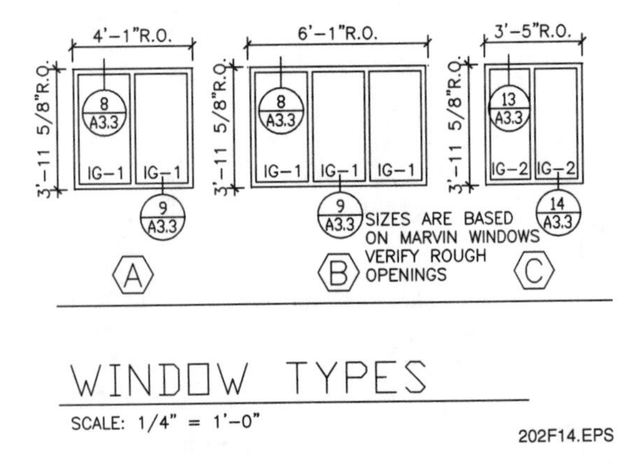

Figure 14 ◆ Window types shown on the floor plan.

Table 1 is a typical window schedule for a hypothetical building. Window schedules list the size of each window and specify where to install them. In this example, the design calls for two sizes of window, W-1 and W-2. The schedule lists information about the rough opening, the frame material, the model number, and the manufacturer's name. (No information appears in the sill height column because the drawings were marked for bidding only.)

On the floor plan, the designer would write the appropriate window number at each window location along the outer walls of the building. During construction, the window suppliers and installers would use the window schedule to provide the right number of windows and to match the correct size to each marked opening.

5.1.1 Alternates

A floor plan may also include illustrations that show how the finished project would look if different materials, methods, or layouts were used. These illustrations are called **alternates.** Alternates are provided to clients who want to compare the

Table 1 Window Schedule

Window No.	Rough Openings	Sill Ht.	Frame Material	Remarks
W-1	3'-5½" × 4'-5¾"		Vinyl-clad wood	Pella fixed casement Model 3648CC
W-2	3'-5½" × 6'-1¾"		Vinyl-clad wood	Pella fixed casement Model 3668CC

Windows shall be dark bronze/brown vinyl clad finish.

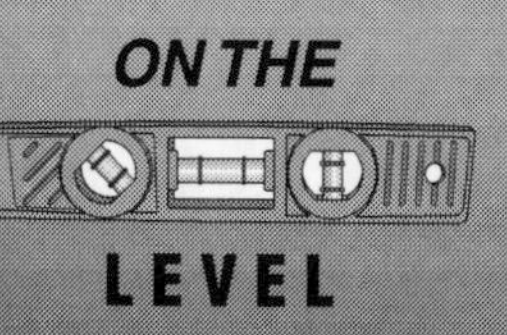

Windows and Plumbing

Generally, restrooms in commercial buildings do not include windows. However, other rooms requiring plumbing, such as kitchens, may have windows. Windows in rooms that require plumbing can affect the position of the vent stack.

The Detail Plan

Whether you are plumbing a commercial building or a residence, one detail drawing you'll find useful is the kitchen cabinet plan. You can determine the location and type of sink from this drawing. In addition, if a window is planned over the kitchen sink, for example, you'll need to adjust the location of the vent stack or the local vent. While kitchens and baths in most modern commercial buildings have no exterior walls or windows, you may still encounter this situation.

costs among different designs. By requesting alternate bids, clients can plan more effectively and get the most for the money budgeted for the project. Alternates can affect the plumbing for a project. You must be aware of them from the outset and plan accordingly.

5.1.2 Entry Points, Walls, and Chases

You will need to locate the plumbing entry points. This information can be obtained from the floor plan. Trace the route from the building sewer to the main stack, and from there to each fixture.

Interior and exterior walls are usually drawn differently on the floor plan. Refer to the key on the drawing to identify the walls correctly. Exterior walls run around the perimeter of the building. Interior walls are drawn as a single, solid line. Rooms on a floor plan are easy to identify. They usually include a doorway and in many cases are labeled with the name and number of the room.

A **chase** is a hollow, enclosed area built between two back-to-back restrooms. They are commonly used in commercial restrooms. All the plumbing connections for the restrooms fit into the area enclosed by the chase. The chase can also be used to route HVAC and exhaust ducts that exit the restrooms.

In residential installations where a chase is not shown in the plans, the contractor or the owner should determine the location of the chase. For example, if a 10-inch chase is required, the contractor or owner can decide to relocate one wall the complete distance or move two adjacent walls 5 inches each.

5.2.0 Using the Plumbing Plans

A set of construction drawings will contain plumbing plans that show the water supply and DWV systems. The bigger the project, the more plumbing drawings will be available. Large and complex projects can have separate drawings for each piping system, such as vent, waste, supply, etc. Residential projects, because of their relatively small scale, usually have one plumbing drawing for the entire building.

The plumbing plans share some information in common with the floor plan. The overall dimensions are the same as those on the floor plan, but the plumbing plans do not include detailed interior dimensions.

Many plumbing plans include isometric drawings of the supply and DWV systems. The isometric drawing shows the general route of the water supply piping, the piping sizes, and the general locations of the urinals, water closets, and lavatories.

5.2.1 Schedules

Plumbing plans typically include schedules for the individual plumbing systems. Plumbing schedules establish the governing plumbing codes, specify piping materials, and call out insulation requirements. The first column contains the letter or number key that corresponds to the fixture's location in the isometric drawing. The schedule also specifies the type of fixture, the manufacturer's name, the model number, and an alternate manufacturer and provides a space for notes.

Generally, the owner, the owner's agent, the architect, or the engineer will specify the name of the manufacturer and the preferred model. The plumbing fixtures schedule may include a list of suitable alternate models. This gives the contractor or plumber more flexibility in getting fixtures to complete the job. *Table 2* is a detail of a hypothetical plumbing fixtures schedule.

5.2.2 Approved Submittal Data

Before roughing in the fixtures shown on the plumbing plan, you will need to obtain **approved submittal data** for each fixture from the purchasing agent or the project manager. Approved submittal data, which includes a **cut sheet** or a drawing of a specific fixture, is installation information about the fixture that has been specified for use by the architect or engineer. It includes rough-in dimensions, manufacturer's specifications, and other information (see *Figure 15*). The fixture should be labeled according to the materials and equipment schedules in the plans.

Submittal data for a fixture is submitted to the contractor and then to the owner for approval. When approved, this submittal data becomes the approved submittal data for that fixture.

5.2.3 Isometric Drawings

The word *isometric* comes from Latin words that mean "equal measurement." Isometric drawings are three-dimensional renderings of fixtures, appliances, or installations in which all vertical lines are depicted vertically and all horizontal lines are projected at a 30-degree angle and appear to go back into the horizon (see *Figure 16*). Dimensions in isometric drawings are usually not to scale. Dashed lines indicate elements that are hidden from view.

As a plumber, you must be able to read and create isometric sketches of piping assemblies. The piping system typically needs to fit in or near the walls, though you don't have to include walls in

Table 2 Plumbing Fixtures Schedule

No.	Item	Manufacturer	Model	Alt. MFG.	Notes
WC-1	Water closet	American Standard	2109.405	Eljer, Kohler	W/Church 5330.063 Supply & Stop
WC-2	Water closet Handicap	American Standard	2108.408	Eljer, Kohler	W/Church 5330.063 Supply & Stop
UR-1	Urinal	American Standard	6560.015	Eljer, Kohler	Sloan Royal 186 FV

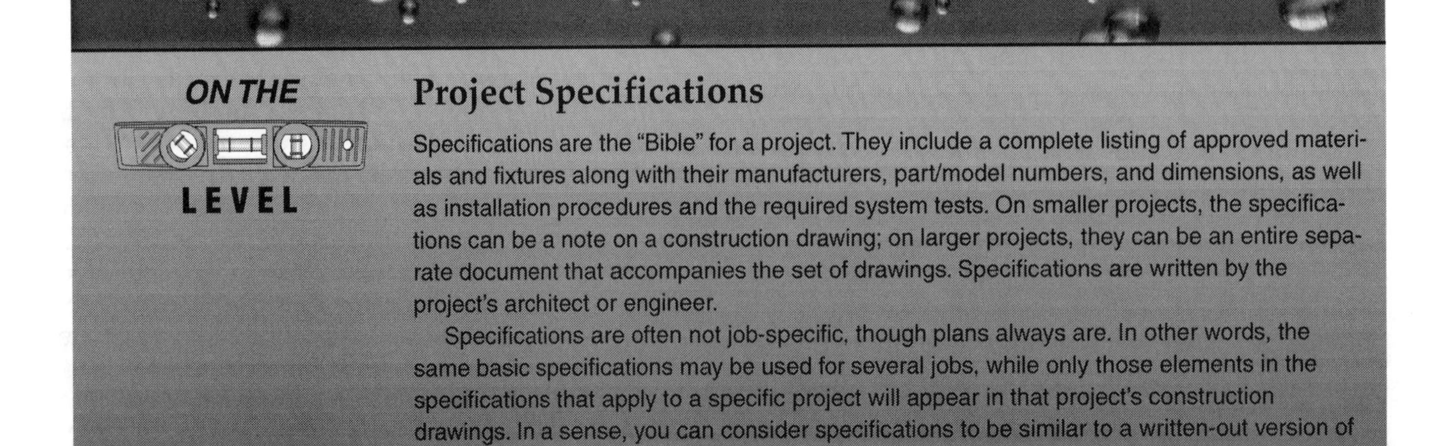

Project Specifications

Specifications are the "Bible" for a project. They include a complete listing of approved materials and fixtures along with their manufacturers, part/model numbers, and dimensions, as well as installation procedures and the required system tests. On smaller projects, the specifications can be a note on a construction drawing; on larger projects, they can be an entire separate document that accompanies the set of drawings. Specifications are written by the project's architect or engineer.

Specifications are often not job-specific, though plans always are. In other words, the same basic specifications may be used for several jobs, while only those elements in the specifications that apply to a specific project will appear in that project's construction drawings. In a sense, you can consider specifications to be similar to a written-out version of the schedules in the construction drawings.

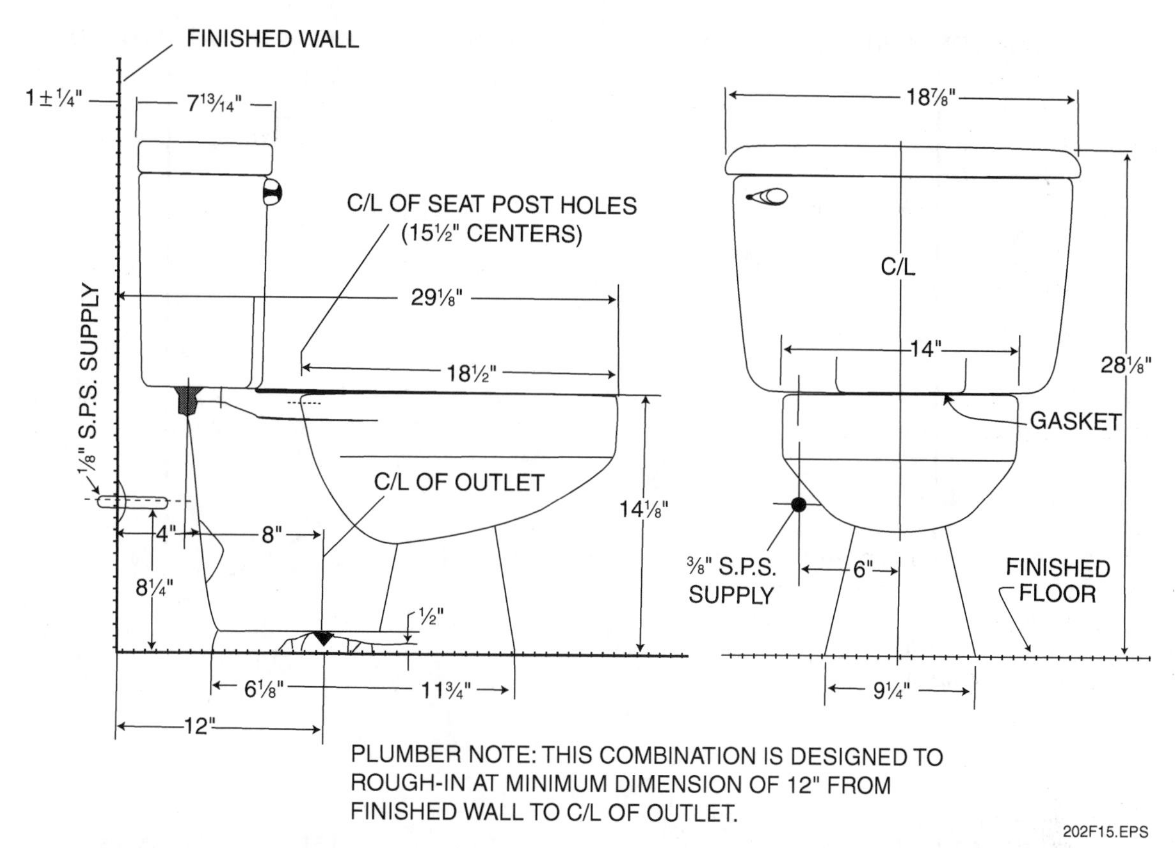

Figure 15 ◆ Approved submittal data.

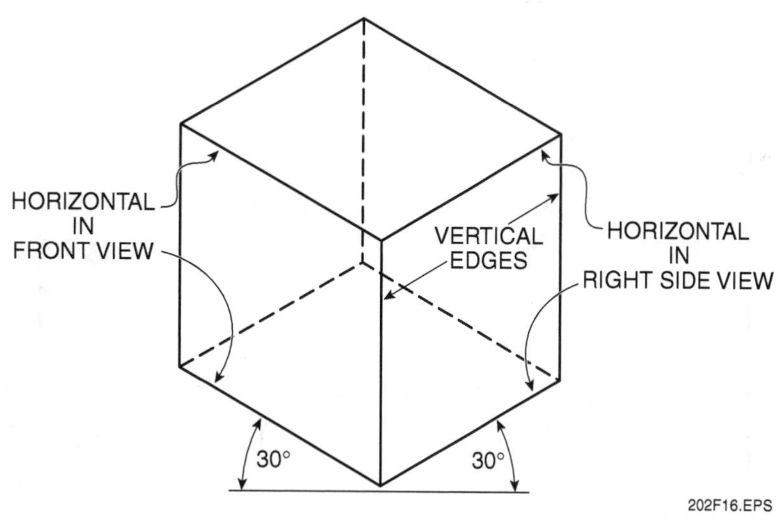

Figure 16 ◆ Simple isometric sketch.

an isometric sketch. Ensure that fittings in the sketch include all required bends. Follow these steps for creating an isometric drawing of a plumbing installation (see *Figure 17*):

Step 1 Begin by drawing the stack as a vertical line.

Step 2 Add the building drain at a 30-degree angle to the bottom of the stack.

Step 3 Add branch piping. Remember that branches are drawn on the vertical stack line in the relative positions that they will actually be installed.

Step 4 Finally, add fittings along the branches and stack. All wyes, tees, and other connections to the stack and drain should indicate the direction of flow. Indicate the vent above the highest fixture branch.

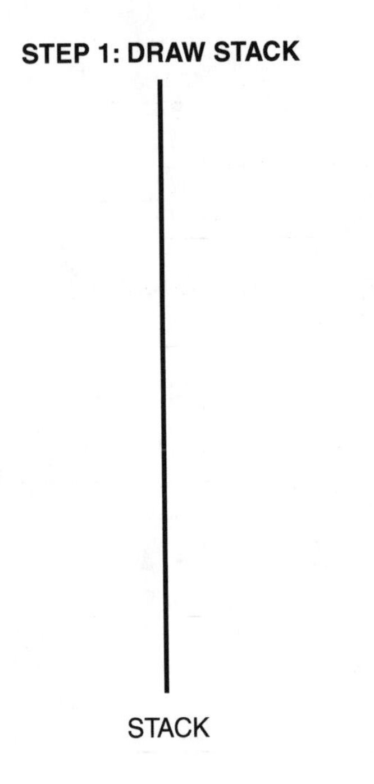

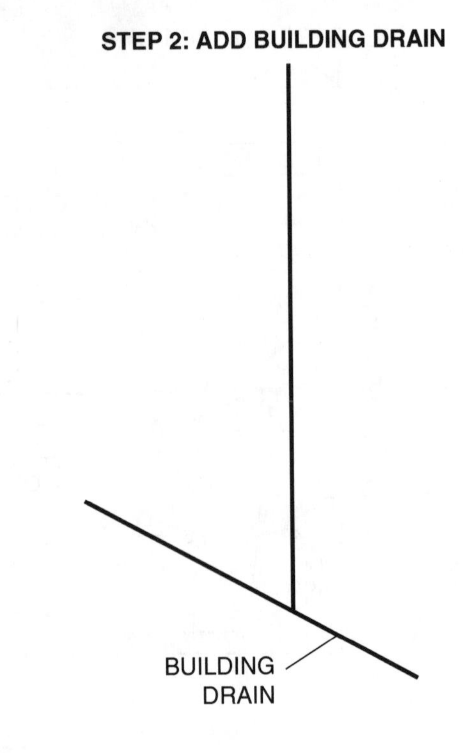

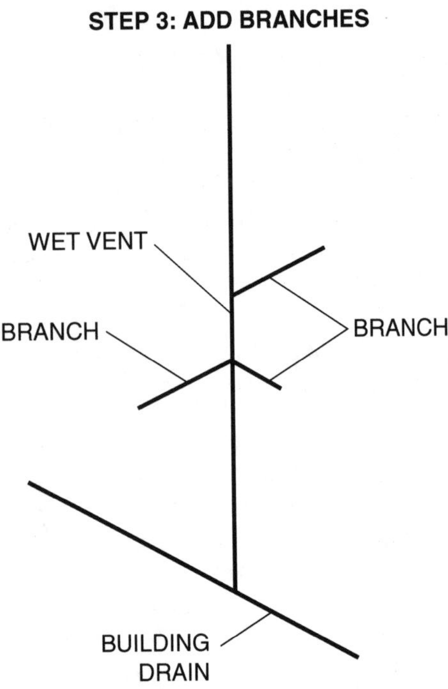

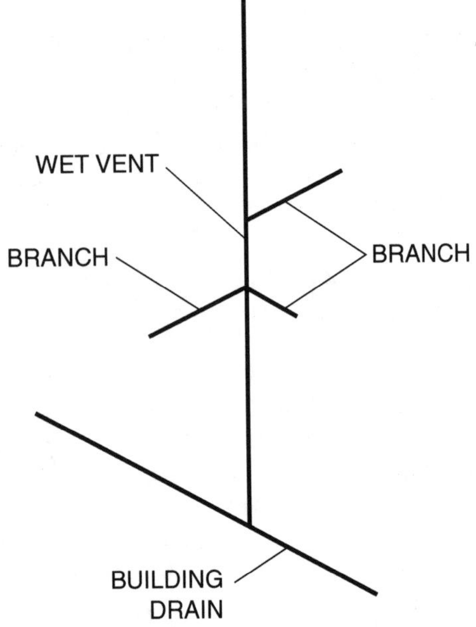

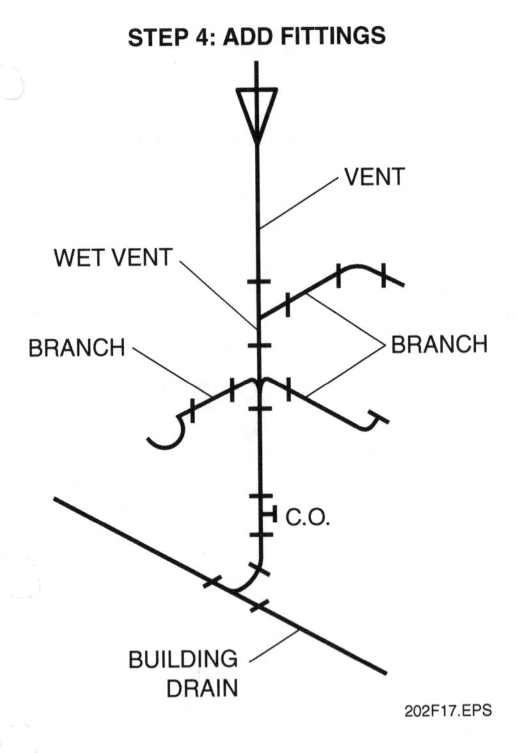

Figure 17 ◆ Steps for creating an isometric drawing.

5.3.0 The Material Takeoff

A material takeoff (see *Figure 18*) is like a shopping list for a project. Refer to the isometric drawings and the plumbing plans to determine how many of each item (fixtures, piping, fittings, and other materials) you will need to complete the project. Include a description of each item and an estimate of material and labor costs. In most cases, you will be able to scale the pipe quantities from the plumbing plans to determine the total bill of materials. However, you should use actual dimensions to rough-in the fixtures or equipment.

Project:		Estimate No.
Location:		Sheet No.
Architect/Engineer:		Date:
Summary by:	Prices by:	Checked by:

Quantity	Description	Unit Price Material Cost	Total Estimated	Unit Price Labor Cost	Total Estimated	Total

202F18.EPS

Figure 18 ◆ Material takeoff.

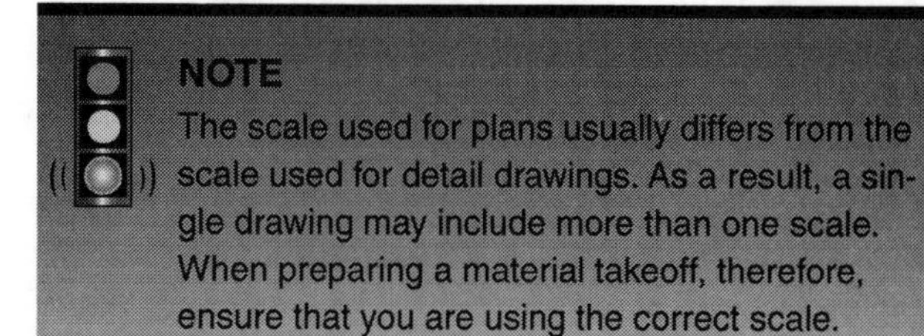

NOTE
The scale used for plans usually differs from the scale used for detail drawings. As a result, a single drawing may include more than one scale. When preparing a material takeoff, therefore, ensure that you are using the correct scale.

5.4.0 Coordination Drawings

Because the information in commercial drawings is more specific than that in residential drawings, contractors on a project often cooperate to produce an additional set of drawings called **coordination drawings.** Many contracts specifically require the preparation of coordination drawings.

A coordination drawing is a dimensioned drawing that indicates the proposed routing of the components of a plumbing system. Coordination drawings show how the various building components and all the materials and equipment in the building will fit together.

Coordination drawings may include elevations and sections. They are based as nearly as possible on the construction drawings. They usually include information about the equipment being provided for the installation as well as the clearances required.

Use coordination drawings to find and resolve conflicts between the plumbing system and building structures or the installations of other trades. Once coordination drawings are approved, contractors who fail to comply with the drawings may be forced to move their work at their expense. Many contracts identify the contractor who makes the coordination drawings as having the right of way. Refer to the project's specifications to identify any right-of-way issues that apply on the project.

5.5.0 As-Built Drawings

When a project is completed, the client will request a set of **as-built drawings** from the contractor. As-built drawings are the drawings of record for that project. They show the project's final configuration, including all of the approved changes from addenda, change orders, and RFIs. Typically, the client will review the as-built drawings against the plans and approved change notices, as well as against the actual completed project, to ensure their accuracy.

As-built drawings serve as the permanent record of a project. They are consulted when maintenance or servicing is required. They are also used when an expansion, addition, or modification is planned for the structure. In addition to providing copies of the as-built drawings to the client and, if required, to the proper municipal agencies, the contractor may retain a set for reference and for potential future liability issues.

Review Questions

Sections 4.0.0–5.0.0

Match the type of document with its correct definition.

1. _____ Clarification
2. _____ Change order
3. _____ Request for information (RFI)
4. _____ Addendum

   a. points out an error on, or question about, the plans
   b. indicates a change in the plan before the project has been bid
   c. responds to a notice of error or a question about the plans
   d. illustrates changes resulting from the use of different materials or fittings
   e. indicates modifications to the plan once construction has begun

5. Floor plans are generally drawn to a scale of _____.
   a. ½" = 1'-0"
   b. ⅓" = 1'-0"
   c. ⅕" = 1'-0"
   d. ⅛" = 1'-0"

For Review Questions 6–9, refer to the door schedule in *Figure 13.*

6. All doors in the schedule have the same size *except* for door number _____.
   a. 1051
   b. 1052
   c. 1121
   d. 1161

7. There are _____ types of doors listed that include glass.
   a. four
   b. six
   c. eight
   d. ten

8. You can obtain all of the following information from this schedule *except* for the _____.
   a. number of doors used in the building
   b. location of detail drawings for each type
   c. instructions for dealing with special requirements
   d. materials used for doors and frames

9. The annotation 5/A2.1 in the Frame/Elev. column stands for _____.
   a. five doors total, Architectural drawing 2.1
   b. detail drawing 2.1, Architectural drawing 5
   c. detail drawing 5, Architectural drawing 2.1
   d. page 5, Architectural drawing 2.1

10. Plumbing entry points are shown on the _____ plan.
    a. civil
    b. floor
    c. plumbing
    d. landscape

11. You are working on a residential installation. The plans call for two back-to-back bathroom installations. There is insufficient wall space between the rooms for the plumbing. You should _____.
    a. install a chase
    b. determine the total amount of space needed for the installation and move each wall apart a total of half the distance
    c. relocate the bathroom installation in one of the rooms to provide proper clearance
    d. let the contractor or the owner decide on the wall relocation

12. Plumbing plans do not include detailed interior dimensions.
    a. True
    b. False

Review Questions

13. Approved submittal data is often referred to as a _____.
    a. specification
    b. cut sheet
    c. schedule
    d. detail drawing

14. In an isometric drawing, all vertical lines are depicted vertically, and all horizontal lines are projected at a _____ angle.
    a. 15-degree
    b. 30-degree
    c. 45-degree
    d. 60-degree

15. When preparing a material takeoff, you will be able to determine the correct number of items to order for the plumbing installations by referring to the _____ and _____.
    a. isometric drawings; plumbing plans
    b. isometric drawings; schedules
    c. detail drawings; plumbing plans
    d. detail drawings; approved submittal data

16. Estimates of material and labor costs are generally included on approved submittal data.
    a. True
    b. False

17. Each of the following is a correct statement about coordination drawings *except* _____.
    a. many contracts for commercial projects require them
    b. they show how the various building components and all the materials and equipment in the building will fit together
    c. you can use them to find conflicts between the plumbing system and building structures or other installations
    d. they can be used to compare the costs among different designs

18. _____ drawings serve as the permanent record of a project.
    a. As-built
    b. Coordination
    c. Civil
    d. Site

To understand the scope of a plumbing project, you must know how to read and interpret commercial drawings. These include civil, architectural, structural, mechanical, plumbing, and electrical drawings, as well as coordination drawings and approved submittal data of materials, fixtures, and equipment. Each of these drawings gives you different, yet related, information. These drawings, especially the architectural, structural, and plumbing drawings, affect how you will do your work. To become an accomplished drawing reader, practice reading these plans. Drawing reading is a tool, like mathematics, that you need in your plumber's toolbox.

Notes

Trade Terms Introduced in This Module

Addendum: A notification issued prior to a bid for changes in a construction drawing.

Alternate: An illustration on a plan showing changes resulting from the use of different materials, methods, or layouts.

Approved submittal data: A drawing that includes rough-in dimensions, manufacturer's specifications, and other information for a fixture that has been approved for use in an installation.

Architectural drawing: A construction drawing that uses plan views, elevations, sections, and details to show the overall shape, size, and appearance of a building.

As-built drawing: A drawing of record for a project, showing the final configuration that incorporates all of the approved changes from addenda, change orders, and RFIs.

Change order: A document issued to indicate modifications to a plan once construction has begun.

Chase: A hollow, enclosed area between two back-to-back commercial restrooms used to house plumbing connections.

Civil drawing: A construction drawing that provides the overall shape of the building site, the location of the building, and other planned structures.

Clarification: Documents issued in response to an RFI.

Contour line: An element of a civil drawing that indicates areas of identical elevation, used to determine grade.

Convention: A standard labeling and marking system used to provide information in plans and schedules.

Coordination drawing: A drawing that is used to identify the proposed locations of all trade work, in order to identify potential conflicts.

Cut sheet: Another term for approved submittal data.

Elevation: A drawing that shows a vertical projection of a building.

Landscape drawing: A construction drawing that indicates how decorative elements will be arranged around the building.

Mechanical drawing: A construction drawing that indicates the location and size of the HVAC system, as well as other features such as elevators, conveyors, and escalators.

North arrow: An element on a construction drawing that indicates compass north.

Plan view: A drawing that shows the horizontal projection of a building.

Request for information (RFI): A document issued to the project architect or engineer to identify an error on, or ask a question about, a construction drawing.

Schedule: A detailed list of components included on a construction drawing.

Section: A hypothetical view that appears to be a vertical slice through a building.

Site plan: Another term for civil drawing.

Structural drawing: A construction drawing that indicates the structure of a building's framework.

Title block: An element of a construction drawing that includes information such as scale, revision date, drawing number, and the architect's or engineer's name.

PRACTICE PROBLEM I: THE SITE PLAN

The illustration in *Figure A-1* shows a site plan for the office building of the Germans from Russia Heritage Society in Bismarck, North Dakota. Use this plan to answer the following questions.

1. What is the scale?
   NOT TO SCALE

2. Where would you find the client's name?
   TITLE BLOCK

3. How can you tell which part of the drawing shows the building and which part shows the parking lot?
   HEAVY BLK. BORDER LINES

4. Locate the north arrow. Which wall of the building—north, south, east, or west—is the gas meter attached to?
   EAST

5. What type of line designates the property line? Where did you find this information?
   DASHED LINES

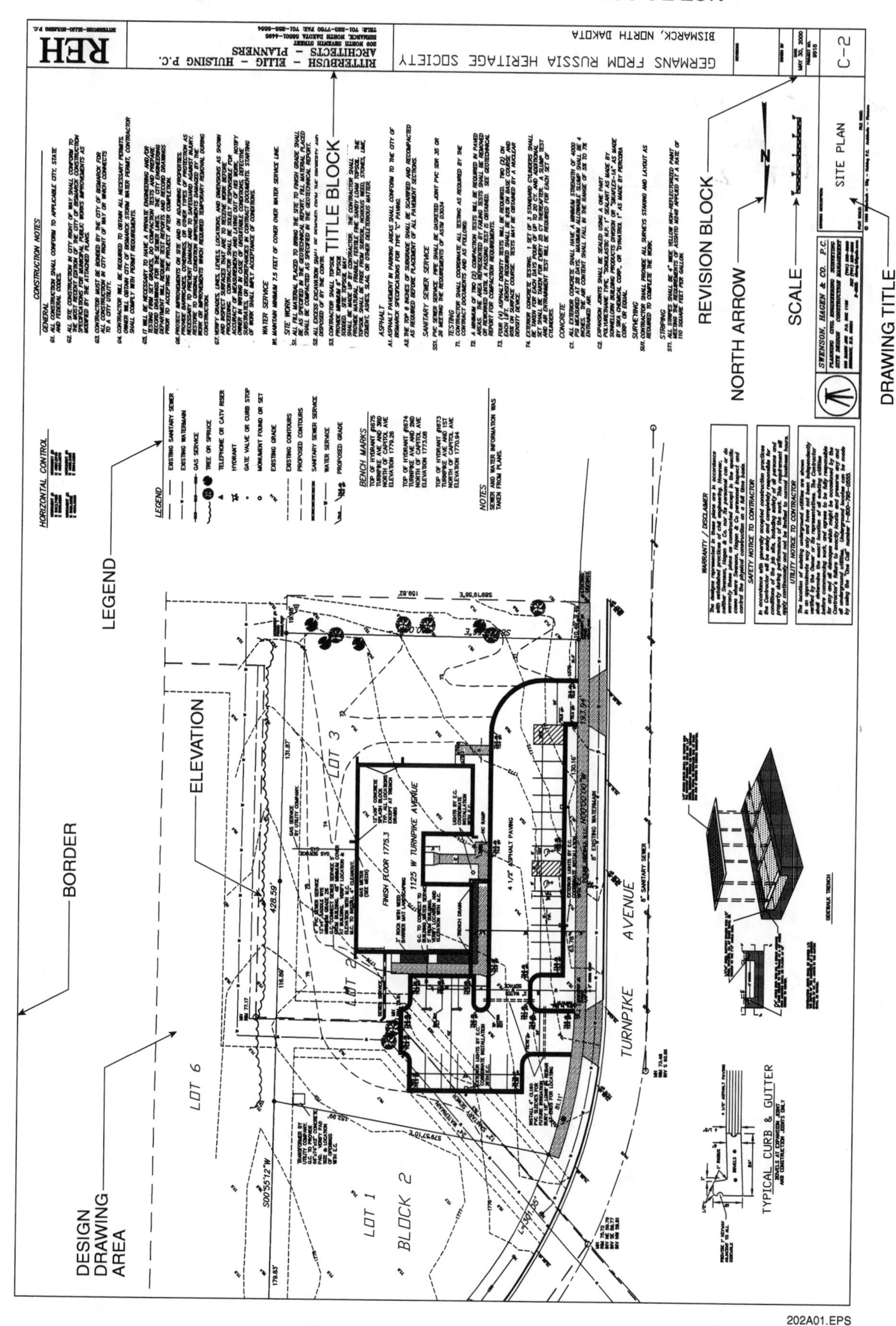

202A01.EPS

Figure A-1 ◆ Practice problem site plan.

PRACTICE PROBLEM II: THE PLUMBING PLAN

Refer to *Figure A-2* to answer the following questions.

1. What is the diameter of the fixture vent piping on each floor?

   2 INCHES
   ____

2. Where are the waste cleanouts located?

   LOT 2
   ____

3. What is the dimension of the pipe that extends directly to the sewer?

   ____

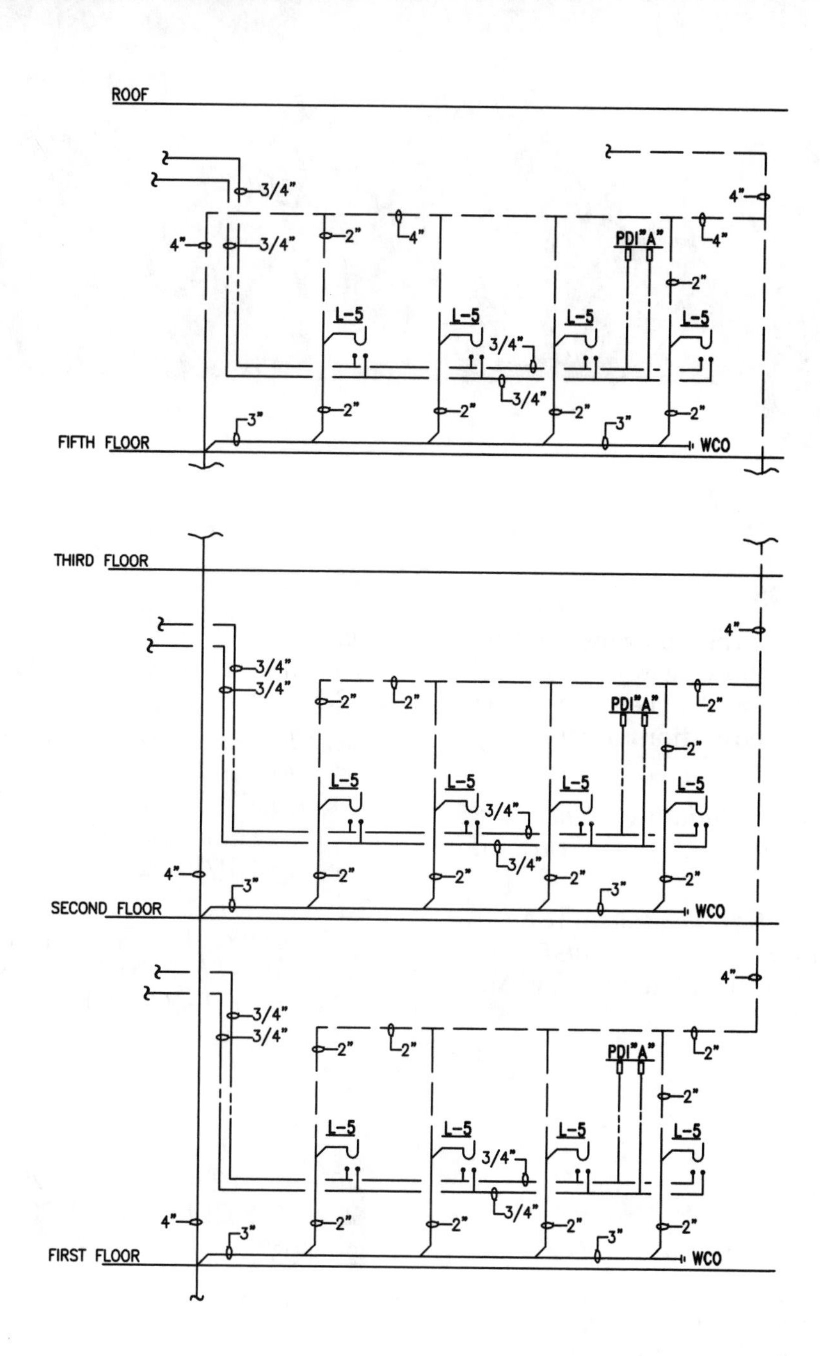

202A02.EPS

Figure A-2 ◆ Practice problem detail of a plumbing riser diagram.

Additional Resources

This module is intended to be a thorough resource for task training. The following reference works are suggested for further study. These are optional materials for continued education rather than for task training.

Blueprint Reading for the Building Trades. 1985. John E. Traister. Carlsbad, CA: Craftsman Book Company.

A Manual of Construction Documentation: An Illustrated Guide to Preparing Construction Drawings. 1989. Glenn E. Wiggins. New York: Whitney Library of Design.

References

Dictionary of Architecture and Construction, Third Edition. 2000. Cyril M. Harris, ed. New York: McGraw-Hill.

Facilities Maintenance and Engineering Procedure, Subject: As-Built Drawings, Publication FMEP-P-0300B, Revision No. 2. SAIC-Frederick, Inc. http://home.ncifcrf.gov/fme/pdfs/procedures/P300B/Procedures.pdf

The NAPHCC Plumbing Apprentice Student Workbook, Year One, Third Edition. 1995. Ruth H. Boutelle, Patrick J. Higgins, and Richard E. White. Falls Church, VA: Plumbing-Heating-Cooling Contractors—National Association.

Figure Credits

Ritterbush-Ellig-Hulsing P.C. — 202F01, 202F02, 202F04, 202F05, 202F07, 202F08, 202F12, 202F13, 202F14, 202A01, 202A02

Reproduced with permission of The McGraw-Hill Companies *Dictionary of Architecture and Construction, Third Edition, ed. Cyril M. Harris, 2000.* — 202F03

Ivey Mechanical Company — 202F06

CONTREN® LEARNING SERIES — USER UPDATE

The NCCER makes every effort to keep these textbooks up-to-date and free of technical errors. We appreciate your help in this process. If you have an idea for improving this textbook, or if you find an error, a typographical mistake, or an inaccuracy in NCCER's Contren® textbooks, please write us, using this form or a photocopy. Be sure to include the exact module number, page number, a detailed description, and the correction, if applicable. Your input will be brought to the attention of the Technical Review Committee. Thank you for your assistance.

Instructors – If you found that additional materials were necessary in order to teach this module effectively, please let us know so that we may include them in the Equipment/Materials list in the Annotated Instructor's Guide.

Write: Product Development and Revision
National Center for Construction Education and Research
P.O. Box 141104, Gainesville, FL 32614-1104

Fax: 352-334-0932

E-mail: curriculum@nccer.org

Craft Module Name

Copyright Date Module Number Page Number(s)

Description

(Optional) Correction

(Optional) Your Name and Address

Plumbing Level Two

02203-05

Hangers, Supports, Structural Penetrations, and Fire Stopping

02203-05

Hangers, Supports, Structural Penetrations, and Fire Stopping

Topics to be presented in this module include:

Overview

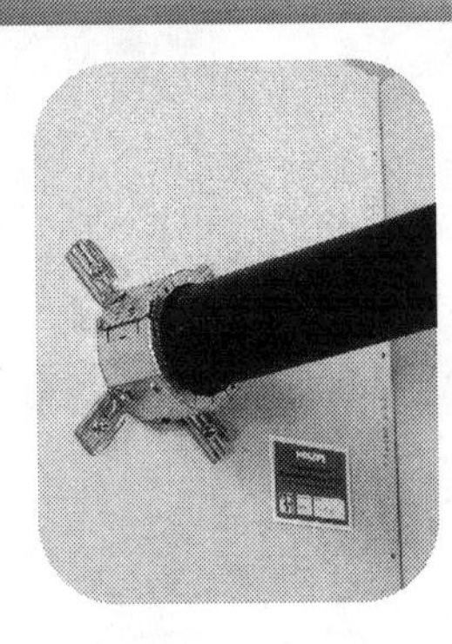

When installing plumbing systems for water supply and drain, waste, and vent (DWV) service, plumbers design each system to accommodate the existing building structure. DWV and water supply pipes require adequate support. Plumbers install pipe hangers and supports and modify structural members according to code requirements and manufacturer's specifications. Hangers and supports, which are made from a variety of materials and in a variety of finishes, hold pipe in horizontal or vertical positions.

Plumbers use powder-actuated anchor systems to anchor static loads to steel and concrete beams, walls, and other structural members, while anchors are used to mount into concrete, brick, or hollow walls. Plumbers use pipe rollers to support piping subject to temperature changes, which causes pipe to expand, contract, or even move sideways. Plumbers also support closet bends, stack bases, and multiple runs of pipe. When structural members lie in the path of a run of pipe, or the space available between them is insufficient, plumbers modify structural members by drilling, notching, boxing, furring, or building a chase.

Fire codes require the installation of fire-stopping material in penetrations through fire-rated structural members and through fire-rated walls, floors, and ceilings. When performing this work, plumbers always follow local applicable code and the requirements in the project plans and specifications. When attaching things to, modifying, or protecting structural members, plumbers must consider the health, safety, and comfort of the building's occupants.

Focus Statement

The goal of the plumber is to protect the health, safety, and comfort of the nation job by job.

Code Note

Codes vary among jurisdictions. Because of the variations in code, consult the applicable code whenever regulations are in question. Referring to an incorrect set of codes can cause as much trouble as failing to reference codes altogether. Obtain, review, and familiarize yourself with your local adopted code.

Objectives

When you have completed this module, you will be able to do the following:

1. Identify the hangers and supports used to install DWV and water supply systems and explain their applications.
2. Install pipe hangers and supports correctly according to local applicable codes and manufacturer's specifications.
3. Modify structural members using the appropriate tools and without weakening the structure.
4. Identify and install common types of fire-stopping materials used in penetrations through fire-rated structural members, walls, floors, and ceilings.

Trade Terms

Fire stopping
I-beam
Intumescent
Structural penetration

Required Trainee Materials

1. Appropriate personal protective equipment
2. Pencil and paper
3. Copy of local applicable code

Prerequisites

Before you begin this module, it is recommended that you successfully complete *Core Curriculum; Plumbing Level One; Plumbing Level Two,* Modules 02201-05 and 02202-05.

This course map shows all of the modules in the second level of the *Plumbing* curriculum. The suggested training order begins at the bottom and proceeds up. Skill levels increase as you advance on the course map. The local Training Program Sponsor may adjust the training order.

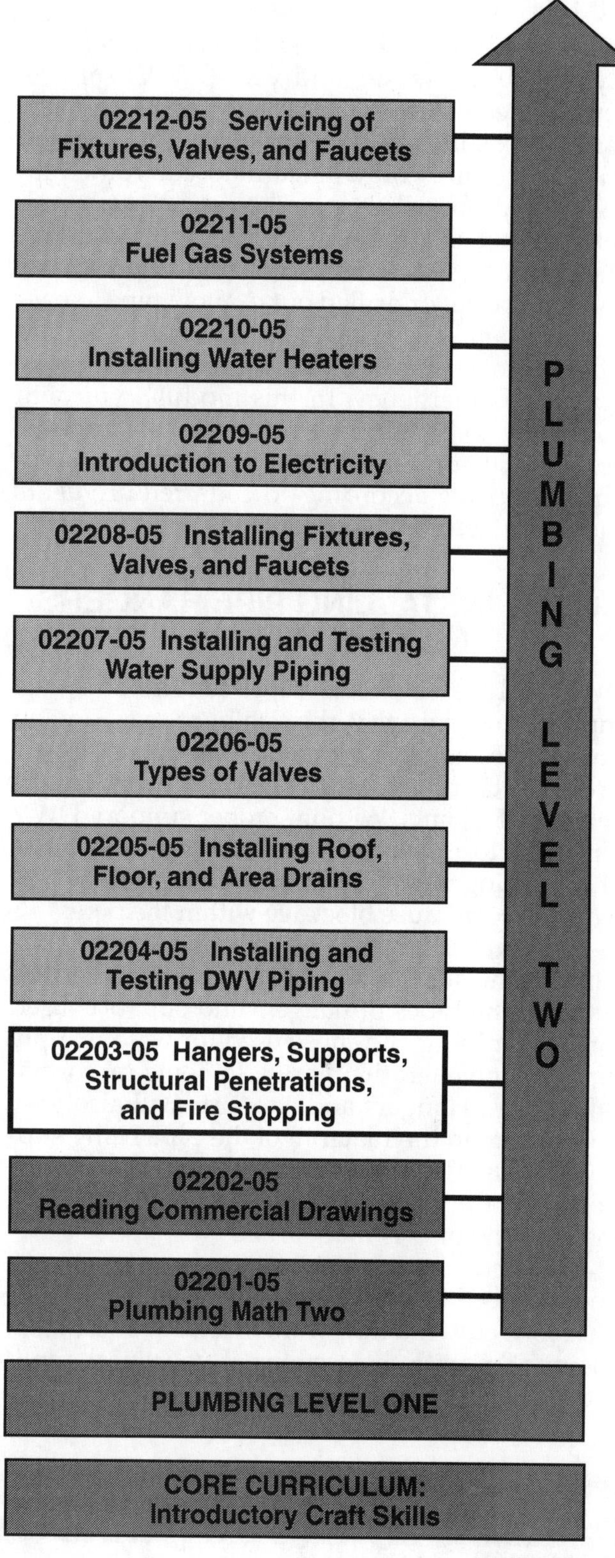

203CMAP.EPS

1.0.0 ◆ INTRODUCTION

When installing plumbing systems for water supply and drain, waste, and vent (DWV) service, you will need to design each system to accommodate the existing building structure. Elsewhere in this curriculum, you will learn how to route piping around obstacles created by structural members, but not all structural members are obstacles. They can be used to hang and support pipe runs. They can be cut or drilled out to allow pipe to pass through them. They can be fitted with devices that will prevent or limit the passage of fire and smoke in case of emergency. In this module, you will learn how to install pipe hangers and supports, modify structural members, and install **fire-stopping** materials according to code requirements and manufacturer's specifications.

2.0.0 ◆ INSTALLING PIPE HANGERS AND SUPPORTS

DWV and water supply pipes require adequate support to ensure that they will not sag. Sagging causes stress on pipe joints, which increases the probability of leaks, breaks, or even cracks between the joints. Without proper support, DWV pipes can lose their proper pitch and form traps. The resulting traps fill with liquid and solid waste, which in turn cause blockage within the pipe.

Pipes are supported and held in place by pipe hangers and supports. There are many similarities between the types of hangers and supports used for DWV piping and those used for water supply piping. However, there are also some important differences. Hangers and supports will also vary depending on the material of the pipe being supported. Always use the proper hangers and supports for the type of pipe being used and the material the pipe is made from.

This section reviews the methods and materials you will need to learn in order to hang and support pipe correctly. You may want to read this section in conjunction with the sections of your local applicable code that deal with hangers and supports. Always follow your local code requirements when selecting and installing hangers and supports. Be sure to follow the manufacturer's specifications closely when installing hangers and supports.

The terms "pipe hangers" and "pipe supports" are interchangeable. They can be used either separately or together to mean the same thing. This module uses "pipe hangers and supports" to refer to the entire range of devices used to hold pipe in place while providing adequate support for the pipe's weight.

2.1.0 Types of Pipe Hangers and Supports

Hangers and supports are designed to hold pipe in either a horizontal or vertical position. They are made from various materials, including carbon steel, malleable iron, steel, cast iron, and plastic. Pipe hangers and supports come in a variety of finishes, including copper plate, black iron, and electrogalvanized steel.

The style and finish of the hanger you use will depend upon the type of pipe and its application. Ensure that you use hangers and supports that are specifically designed for use with the piping material. Mixing material types can damage the pipe. For example, plastic pipe expands and contracts at a different rate than metal pipe. Metal straps used to support plastic pipe will therefore restrict normal expansion and contraction. Furthermore, they will abrade the pipe surface. When installing plastic pipe, use only suitable hangers and supports (see *Figure 1*).

Because hangers and supports are frequently used in combination with each other, clearly defined categories are hard to establish. In this curriculum, the basic components used to hang and support pipe are divided into the following three broad categories:

- Pipe attachments
- Connectors
- Structural attachments

ON THE LEVEL

Keep Pipes Clean

Make every effort to keep water supply piping as clean as possible before installation. Sand, gravel, insects, debris, or mud trapped in the pipes will flow through the piping system, lodge in the valves and faucets, and damage the washers and O-rings. The faucets will then leak. Store pipe in a clean, dry area and cap all ends of pipe at the end of each workday. Be sure to flush the line as thoroughly as possible to remove any sand, mud, or gravel from the system.

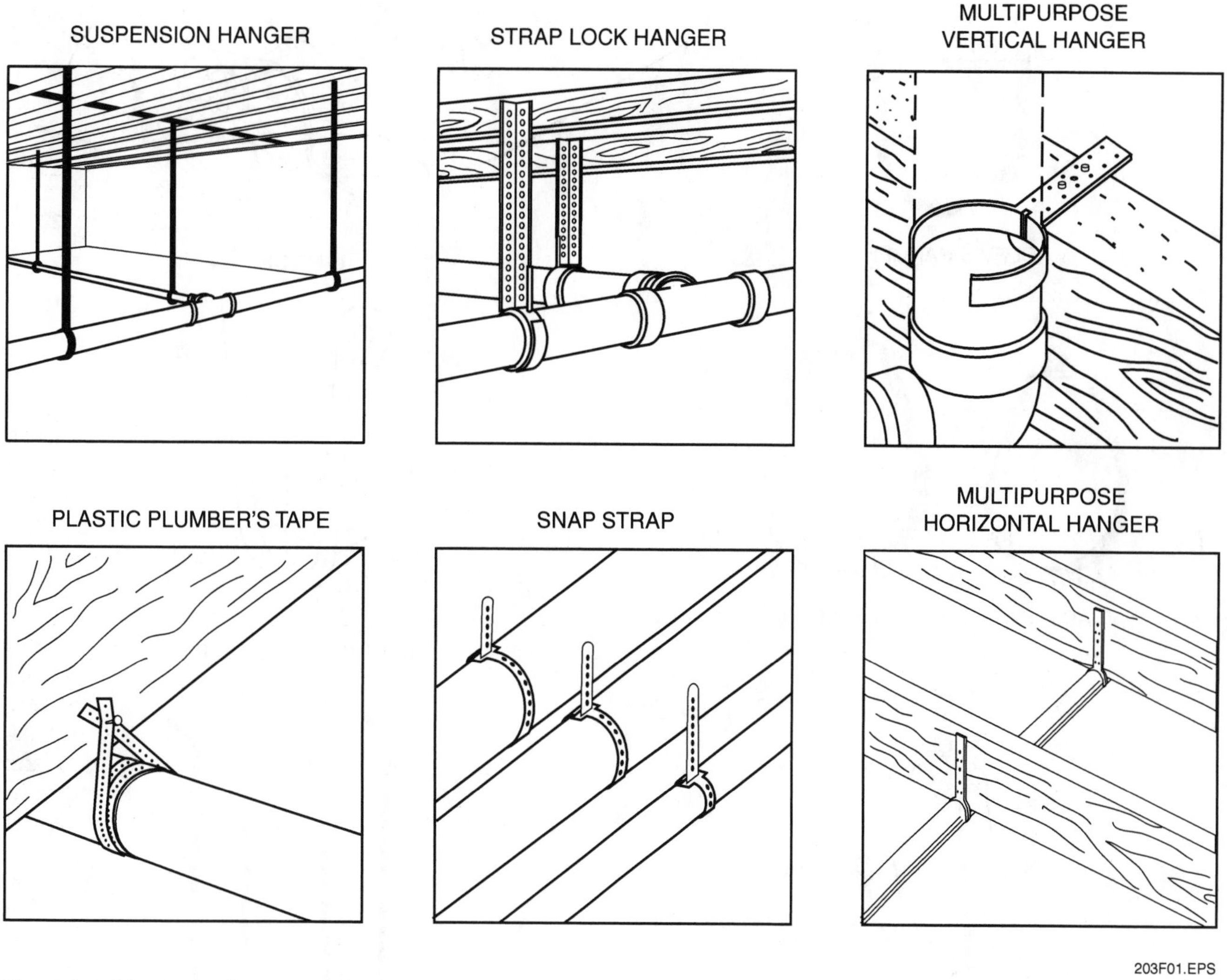

SUSPENSION HANGER

STRAP LOCK HANGER

MULTIPURPOSE VERTICAL HANGER

PLASTIC PLUMBER'S TAPE

SNAP STRAP

MULTIPURPOSE HORIZONTAL HANGER

203F01.EPS

Figure 1 ◆ Plastic pipe hangers and supports.

2.1.1 Pipe Attachments

A pipe attachment is the part of the hanger that touches or connects directly to the pipe. They may be designed for either heavy duty or light duty, for use on either covered or plain pipe. The styles illustrated in *Figure 2* are recommended for use with hot and cold stationary piping without insulation. They are designed to be used when suspending pipe from steel **I-beams**.

Piping can be supported on wood frame construction with several styles of pipe attachments. Common supports and hangers for metal pipes include tin straps, U-hooks, and hold-down clips (see *Figure 3*). Plastic pipe may be supported on wood frame construction using plastic clamps (see *Figure 4*).

Several styles of pipe attachments are available for supporting piping from the walls or the sides of beams and columns (see *Figures 5* and *6*). They include one-hole clamps, steel brackets, offset clamps, and various styles of clips and straps, as well as combinations using extension split-clamp hangers and wall plates.

Riser clamps support vertical pipe at each floor level (see *Figure 7*). Riser clamps are available in different finishes that will accommodate steel, copper, and DWV pipe. Mount the riser clamp so the bracket supports the pipe weight directly on the floor.

CAUTION

Refer to your local code for proper spacing of hangers and supports. Hanger spacing must never exceed the maximum distance called for in the manufacturer's specifications or your local applicable code.

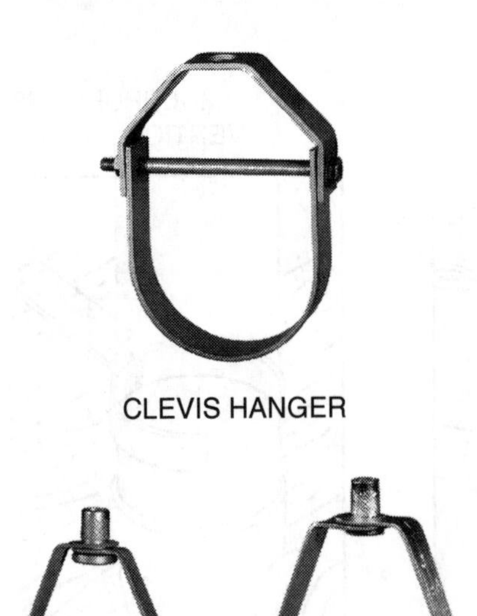

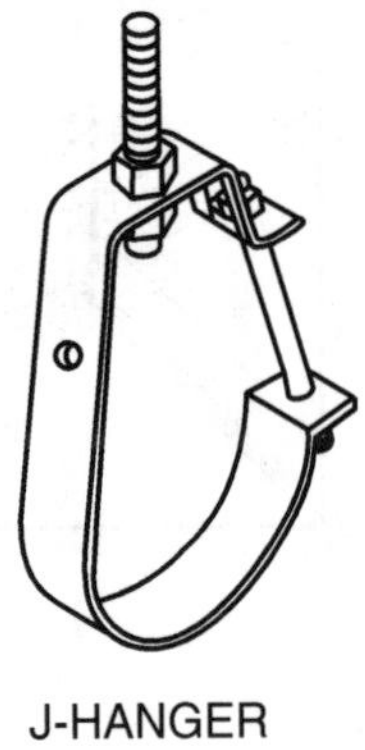

203F02.EPS

Figure 2 ◆ Hangers used to support pipe horizontally from the ceiling.

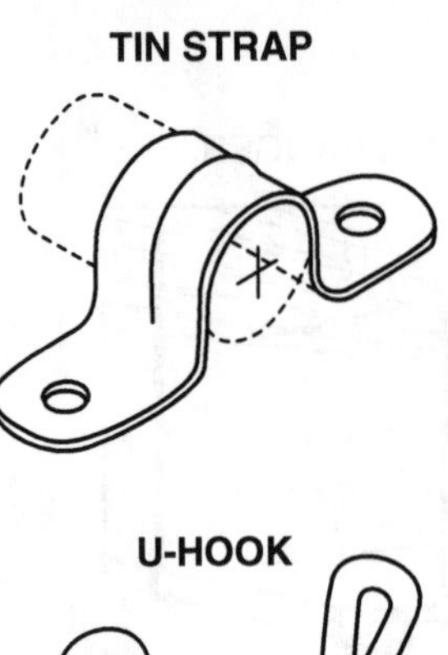

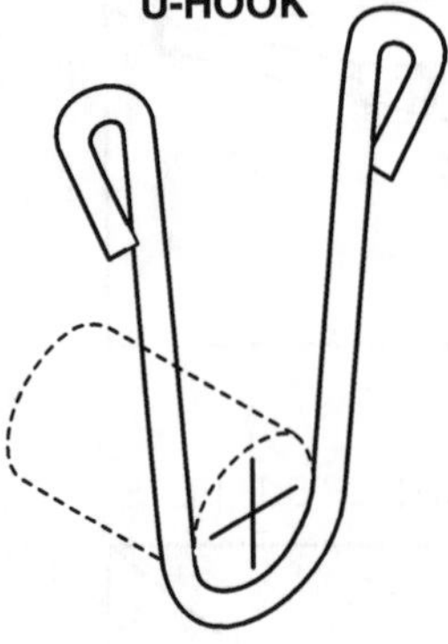

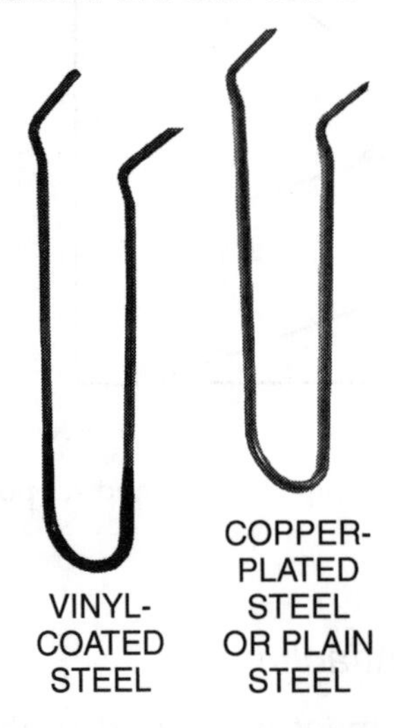

203F03.EPS

Figure 3 ◆ Common supports for metal pipe on wood frame construction.

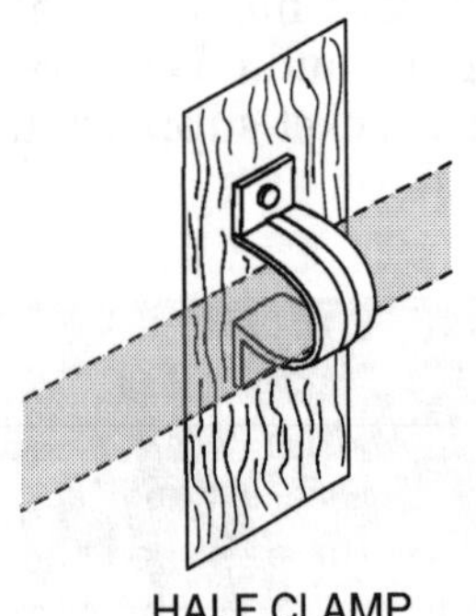

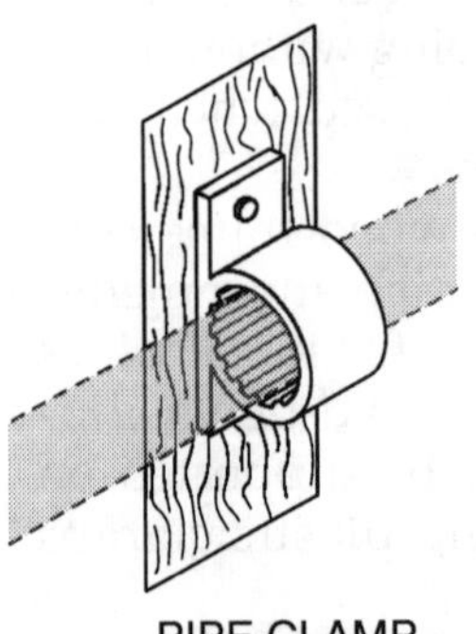

203F04.EPS

Figure 4 ◆ ABS plastic pipe clamps for wood.

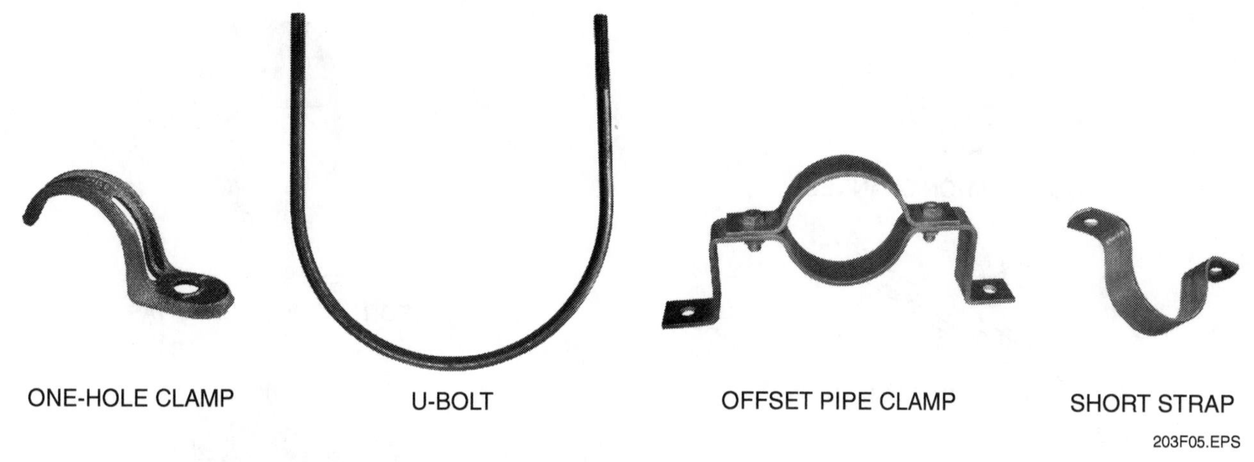

Figure 5 ◆ Attachments for use on walls or beams and columns.

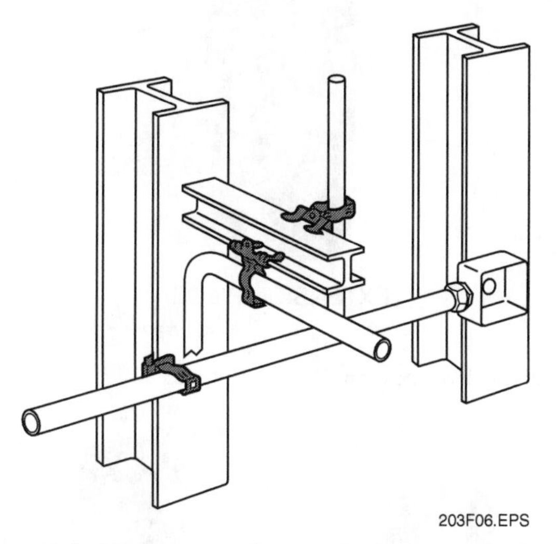

Figure 6 ◆ Clip-type attachments for beams and columns.

Figure 7 ◆ Riser clamps.

Other pipe attachments include universal pipe clamps (see *Figure 8*) and standard 1⅝-inch or 1½-inch channels (see *Figure 9*). The clamps are available in standard finishes of mild and electrogalvanized steel. Aluminum, copper-plated, and stainless steel finishes are also available from some manufacturers, but these finishes usually have to be special ordered. Insert the notched steel clamps by twisting them into position along the slotted side of the channel. Align the pipes as close to one another as the couplings allow.

2.1.2 Connectors

The connector portion of the hanger is the intermediate attachment that links the pipe attachment to the structural attachment. These intermediate attachments can be divided into two groups: rods and bolts, and other rod attachments.

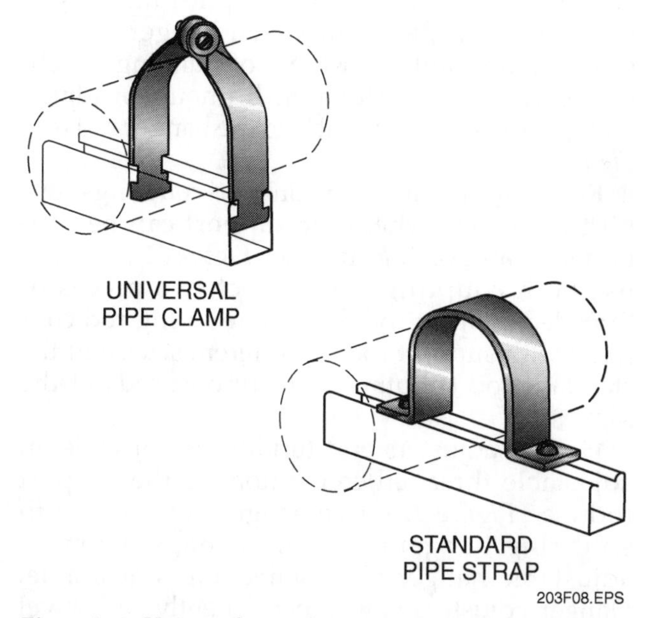

Figure 8 ◆ Universal pipe clamps.

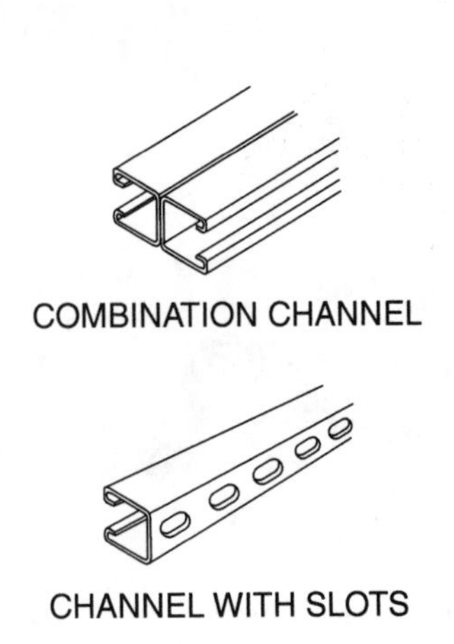

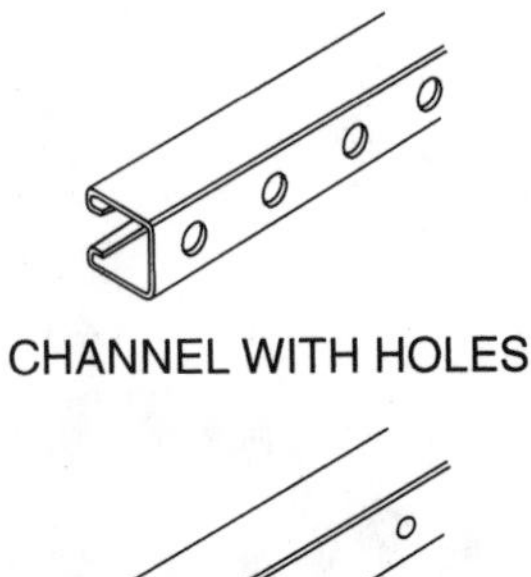

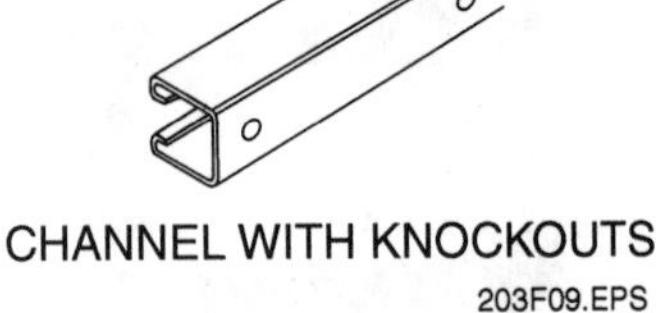

Figure 9 ◆ Channels for use with universal pipe clamps.

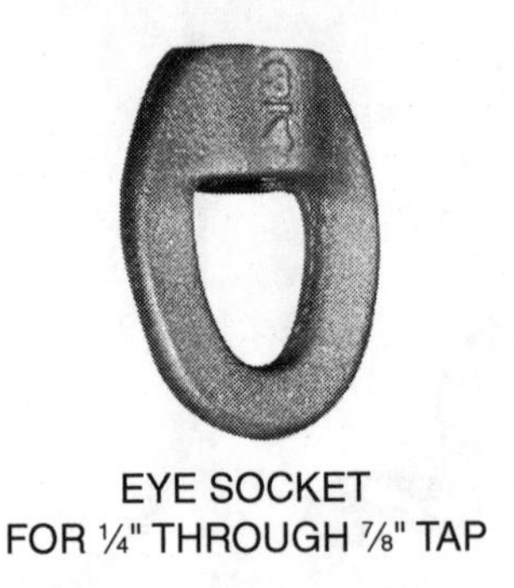

Figure 10 ◆ Eye socket and extension piece.

Rod attachments include eye sockets, extension pieces, rod couplings, reducing rod couplings, hanger adjusters, turnbuckles, clevises, and eye rods. All are available in several sizes that will meet most installation requirements.

The eye socket provides for a nonadjustable threaded connection. Use it in conjunction with a hanger rod when installing a split-ring hanger. Use extension pieces to attach hanger rods to beam clamps and other types of building attachments. They provide for a small amount of adjustment, approximately 1 inch, to the hanger rod (see *Figure 10*).

Rod couplings and reducing rod couplings support pipe runs where the support can be connected to an existing stud (see *Figure 11*). You can use both couplings to connect two pieces of threaded support rod. Use the reducing rod coupling with support rods of different sizes and the standard rod coupling with support rods of the same size.

Hanger adjusters and turnbuckles provide an adjustable threaded connection for the support rod (see *Figure 12*). Use hanger adjusters with split-ring hangers or beam clamps when an adjustable hanger rod connection is desirable. Hanger adjusters have a permanently set swivel in the body that allows smooth adjustment during installation. Use turnbuckles to connect two

Figure 11 ◆ Rod coupling and reducing rod coupling.

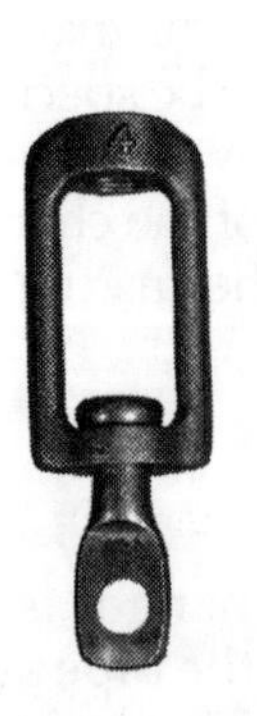

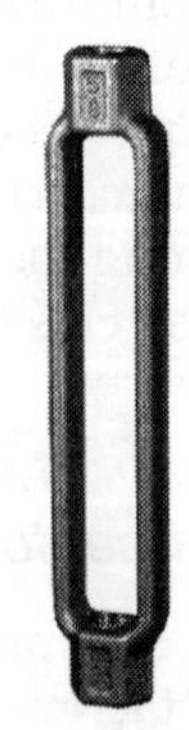

Figure 12 ◆ Hanger adjuster and turnbuckle.

hanger support rods together. They provide 6 inches of adjustment.

Use weldless eye nuts and forged steel clevises on high-temperature piping installations (see *Figure 13*). Install the eye nut where a flexible connection is required. Use the clevis to connect the support rod to the welded lug or structural steel on heavy-duty piping installations.

The hanger rod portion of the connector includes the eye rods and the machine-threaded rods (see *Figure 14*). Use them to connect the rod attachments to the pipe attachments to form a hanger assembly. Eye rods are available with and without welded eyes. Use eye rods with welded eyes for installations requiring additional strength. Eye rods without welded eyes can be used in lighter installations. Machine-threaded rods are available with continuous threads that run the complete length. They can be cut to the required hanger length on the job. This eliminates the need for field threading. You can purchase hanger rods either with right-hand threads on both ends or with one right-hand thread and one left-hand thread.

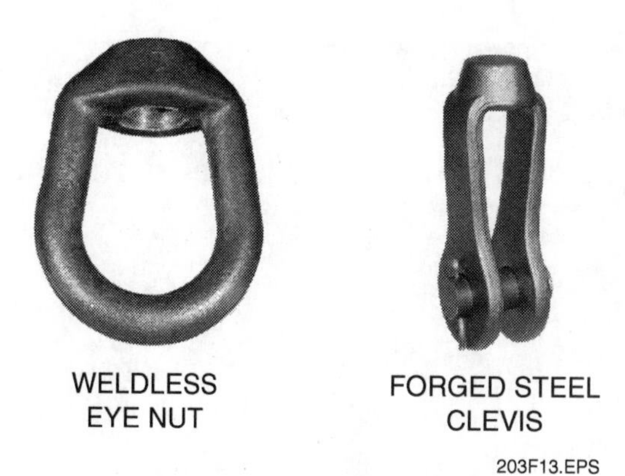

Figure 13 ◆ Weldless eye nut and forged steel clevis.

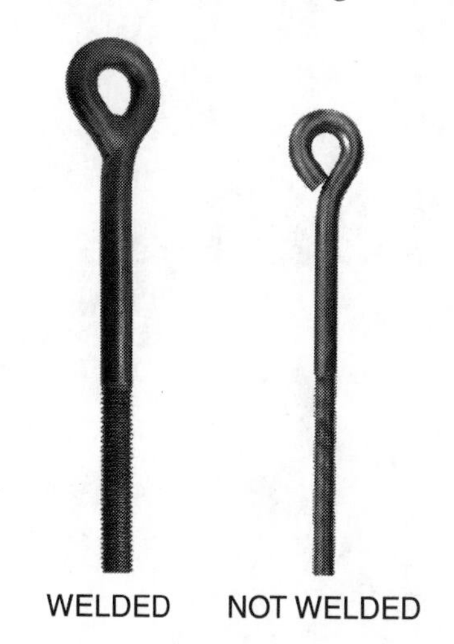

Figure 14 ◆ Hanger rods.

2.1.3 Structural Attachments

Structural attachments, also called anchors or anchoring devices, actually hold the pipe hanger assembly securely to the structure of the building. Common structural attachments include the following:

- Powder-actuated anchors
- Concrete inserts
- Beam clamps
- C-clamps
- Beam attachments
- Brackets
- Ceiling flanges
- Plates
- Plate washers
- Lug plates

2.2.0 Powder-Actuated Fastening Systems

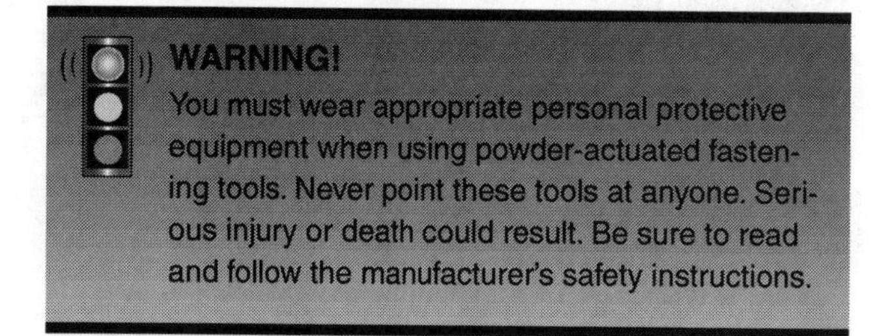

WARNING!
You must wear appropriate personal protective equipment when using powder-actuated fastening tools. Never point these tools at anyone. Serious injury or death could result. Be sure to read and follow the manufacturer's safety instructions.

Powder-actuated anchor systems are widely used for anchoring static (stationary and vibration-free) loads to steel and concrete beams, walls, and other structural members (see *Figure 15*). They work by firing a booster charge at a specially designed piston. The expanding gas from the fired booster charge drives the piston down the barrel of the tool, where it hammers a fastener into the structural member. Fasteners are either steel pins (see *Figure 16*) or threaded steel studs (see *Figure 17*).

The piston controls the speed and direction of the fastener, ensuring that it penetrates the building material safely. The piston is stopped inside the tool when the fastener reaches full penetration (see *Figure 18*). Pipe hangers, brackets, and other supports can then be attached to the fastener.

Refer to the published guidelines from organizations such as Underwriter's Laboratories (UL) and your local building codes to identify the specific types of loads and load limits that are allowable in your area. OSHA requires manufacturers to certify plumbers who are allowed to use their powder-actuated systems. Certification can be arranged through factory sales representatives.

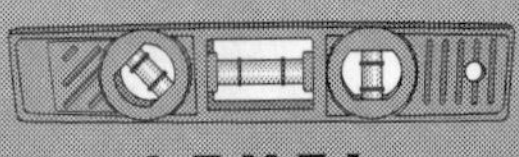

Determining the Length of the Hanger Rod

The figure shows a 6" pipe in a hanger hanging 16'-3" above the floor. When hanging pipe on a job, you must be able to determine how long the hanger and rod need to be to hang the pipe at the proper height.

1. What is the distance from the bottom of the top deck to the bottom of the pipe hanger? (Answer: 4'-3")

   20'-6"
   –16'-3"
   4'-3"

2. In the figure, the pipe hanger take-out measurement is 10". What is the distance between the bottom of the top deck and the top of the hanger? (Answer: 3'-5")

   4'-3"
   – 10"
   3'-5"

3. Use the information and figure above. Knowing that the hanger rod will go into the anchor 1", how long does the hanger rod need to be to extend into the hanger 2½"? (Answer: 3'-8½")

   (1" + 2½" = 3½")
   3'-5" + 3½" = 3'-8½"

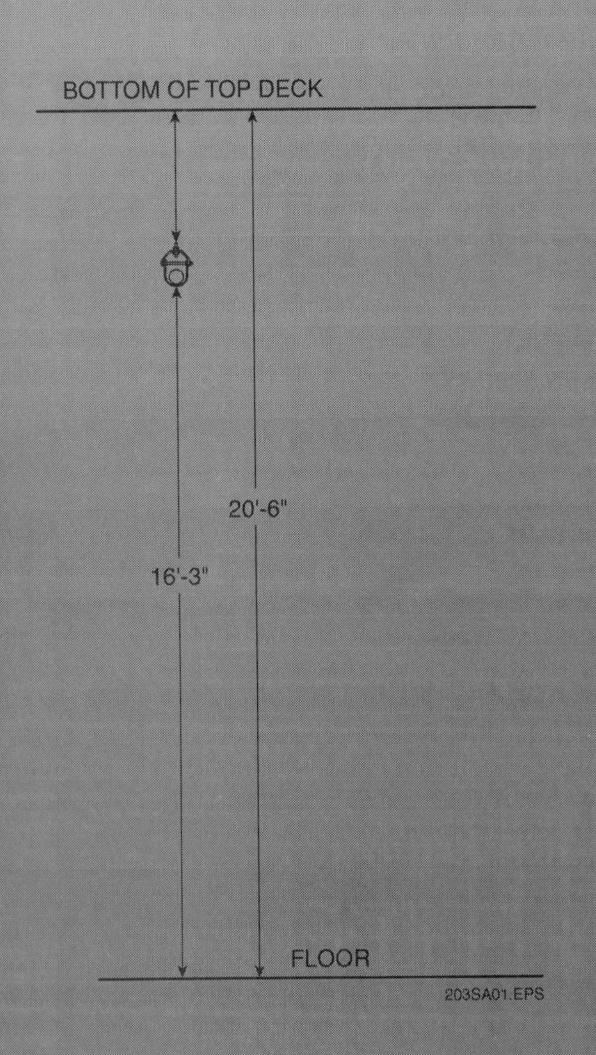

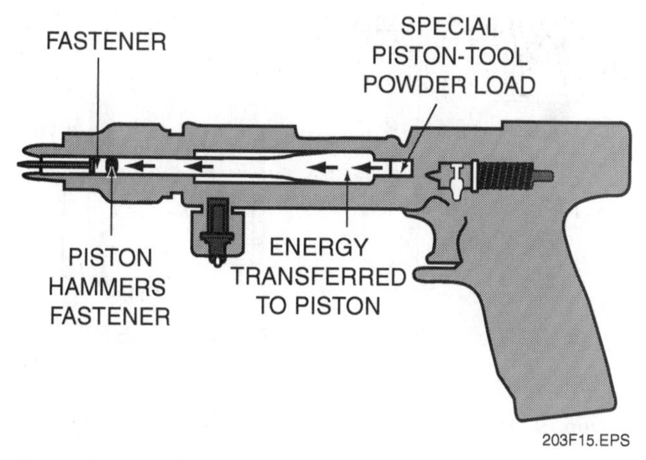

Figure 15 ◆ Powder-actuated tool.

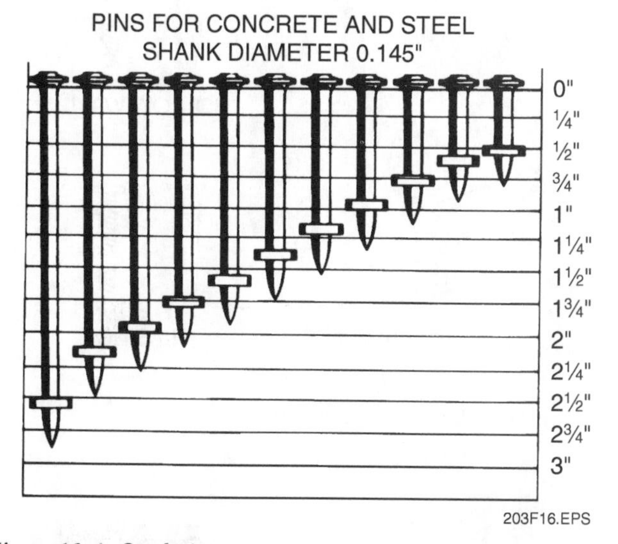

Figure 16 ◆ Steel pins.

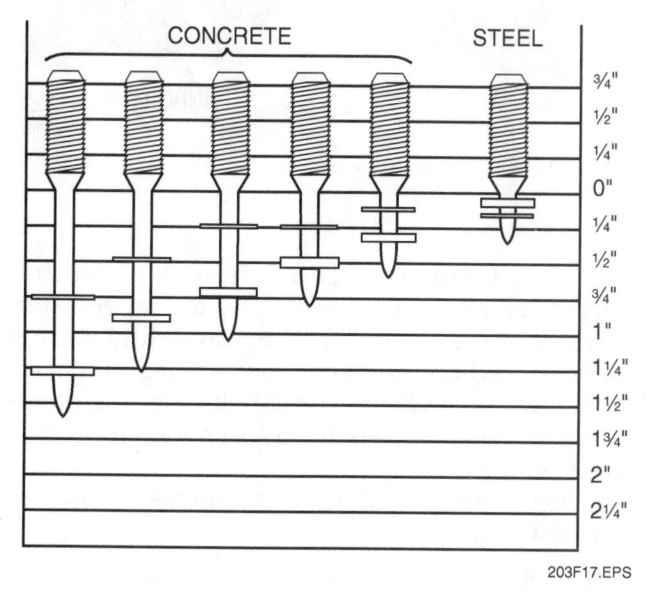

Figure 17 ◆ Threaded steel studs.

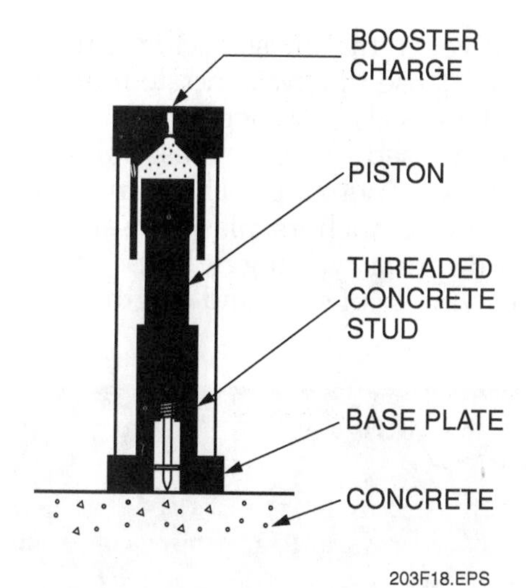

Figure 18 ◆ How the powder-actuated tool works.

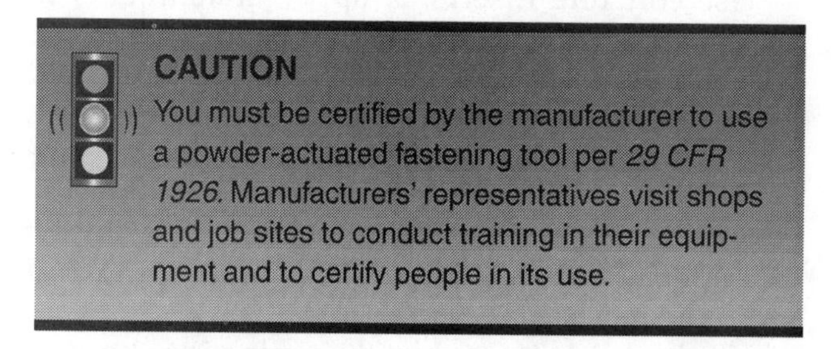

To operate a powder-actuated fastening tool, follow these steps:

Step 1 Feed the pin or stud into the piston.

Step 2 Feed the powder booster into position.

Step 3 Position the item to be fastened in front of the tool, and press the tool and item against the mounting surface. This pressure releases the tool's safety lock.

Step 4 Pull the trigger handle to fire the booster charge.

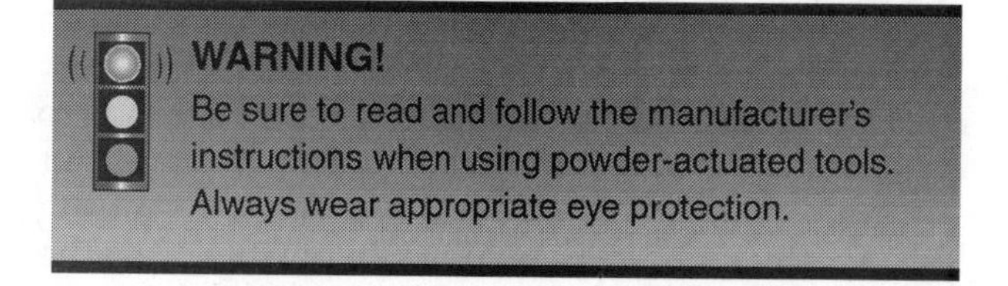

2.3.0 Structural Attachments for Concrete, Brick, and Hollow Walls

Anchors for mounting into concrete, brick, or hollow walls are available in several styles (see *Figure 19*). Use wedge anchors, sleeve anchors, stud anchors, nondrilling anchors, or self-drilling anchors to fasten the piping system to concrete.

Note, however, that these anchors are not recommended for use in new concrete that has not yet cured. There is also a variety of light-duty anchors that can be used in solid walls, hollow block, or drywall (see *Figure 20*). These include polyset anchors, nylon anchors, plastic inserts, expansion bolts, and spring-type toggle bolts. They are available in a variety of sizes and lengths.

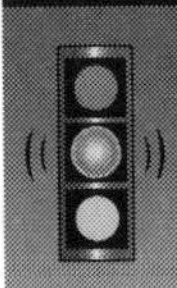

CAUTION

Always wear safety glasses when installing anchors. Use the correct size drill bit and a drill that will meet the load demands of the job.

Use concrete inserts as upper attachments to suspend pipe from a concrete structure (see *Figure 21*). Standard universal steel inserts are fabricated from heavy gauge steel. They are designed with one case size that can be used with all sizes of support rods up through ¾ inch. After installing the inserts and removing the knockout plate, insert the special nuts. Then suspend a hanging rod from the insert.

Use beam clamps when the piping is to be supported from the building steel (see *Figure 22*). Attach them to the bottom flange of I-beams. All beam clamps are fitted with jaws that lock in position on the beam when fully and properly adjusted. Be sure to select a beam clamp that fits the thickness of the I-beam.

Use C-clamps in installations where the pipe support is to be attached to I-beams, channels, or wide flange beams and it is desirable to have the support rod offset from the beam. Secure the clamp to the flange with a cup-pointed, hardened setscrew. After the clamp is in place, thread a support rod into the tapped hole at the base. Install retaining straps to prevent the C-clamp from moving.

Use beam attachments to attach support rods to a structural member. Weld or bolt them in either an upright or inverted position. When installing beam attachments in an inverted position, use the hanger rod to make small vertical adjustments.

Brackets and clips used to attach threaded connections to wood beams are available in several styles. Secure them in place using bolts or screws. You may also install them on concrete structures by either bolting or welding them to the steel beam.

Ceiling flanges and plates are recommended for suspending pipelines from wood beams or ceilings (see *Figure 23*). The ceiling plate gives a finished appearance where the rod or pipe enters the ceiling.

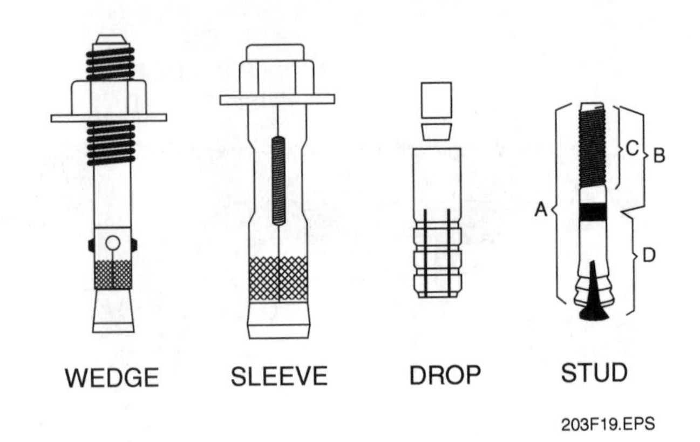

Figure 19 ◆ Anchors used for concrete or brick.

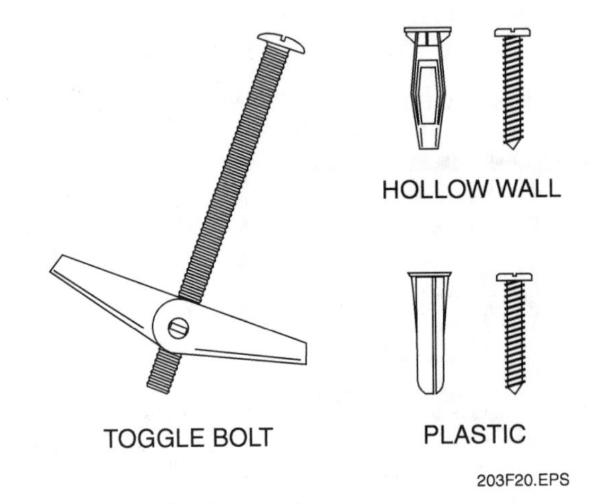

Figure 20 ◆ Light-duty anchors.

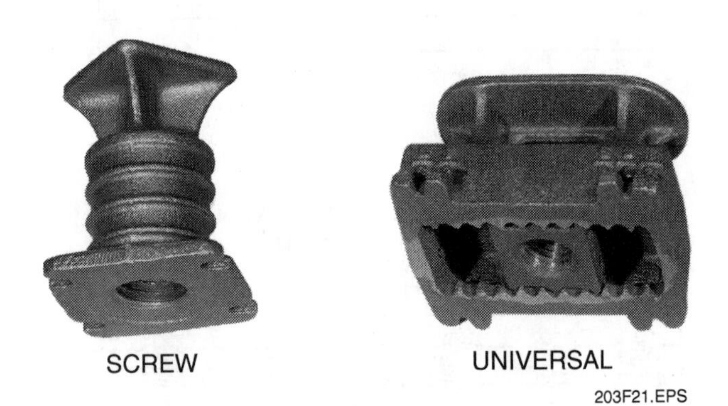

Figure 21 ◆ Concrete inserts.

Use structural attachments such as washer plates, lug plates, or clevis plates to suspend support rods (see *Figure 24*). The heavy-duty washer plate is used on top of channels or angles to support the pipe with rods or U-bolts. Use lug plates when support rods are to be suspended from concrete ceilings.

When using structural attachments, be sure to do the following:

- Check the specs (specifications) to see what type of hanger is required.

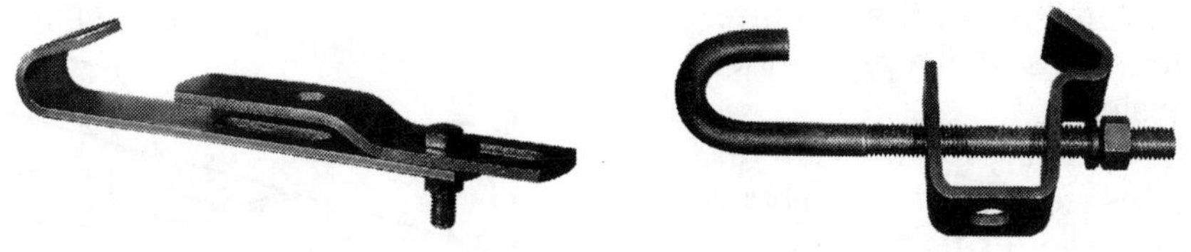

Figure 22 ◆ Beam clamps.

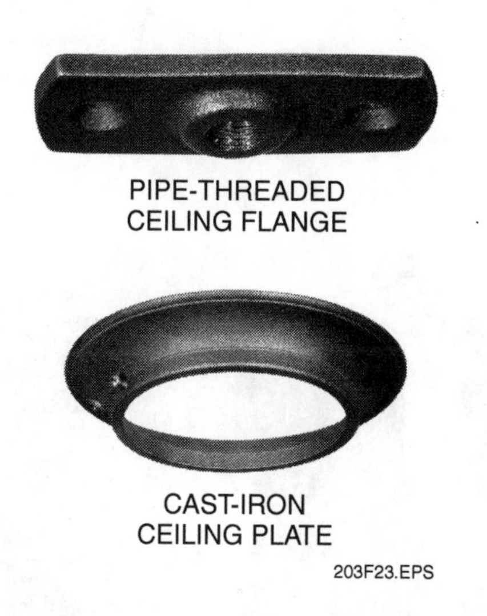

Figure 23 ◆ Ceiling flange and plate.

- Check pre-/post-stressed concrete T-sections to make sure that installing the anchors will not weaken the structure.
- Clear all locations for concrete inserts with the architect or structural engineer in charge of the building.
- Check the specs before welding attachments onto the structure.
- Avoid using powder-actuated anchors.
- Make sure that the anchor is properly inserted.

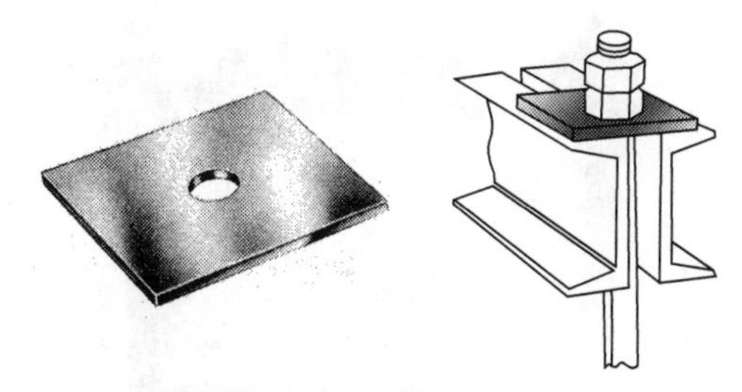

Figure 24 ◆ Structural attachments for suspending support rods.

2.4.0 Special Hangers and Accessories

Temperature changes can cause pipes to expand and contract as well as to move sideways. Use pipe rollers to support piping that is subject to these types of movement (see *Figure 25*). Pipe rollers allow pipes to expand and to move laterally. You can also use rollers as a rolling guide to feed the piping into place during pipe system assembly.

Field-made pipe alignment guides, also called shoe guides, also can be used to allow pipe to expand and contract. Shoe guides can be made from materials readily available on the job site (see *Figure 26*). Shoe guides consist of a pipe shoe, a shoe guide, and a pipe sleeve.

Spring hangers allow vertical pipe movement caused by thermal changes (see *Figure 27*). These hangers are available in several styles and weights. Light-duty spring hangers provide a flexible spring support for light-duty loads where the vertical movement does not exceed 1¼ inches. In installations where you want to prevent vibration noise and other sounds from being transmitted into the building, use vibration control hangers to suspend the piping.

Spring cushion hangers are designed for use with a single pipe run in installations where the vertical movement does not exceed 1¼ inches. Use constant support hangers where constant, accurate support is needed on piping systems that move vertically because of temperature changes. It is good practice to first install the pipe with rigid hangers. After the pipe is installed, replace the rigid hangers with the proper spring hangers.

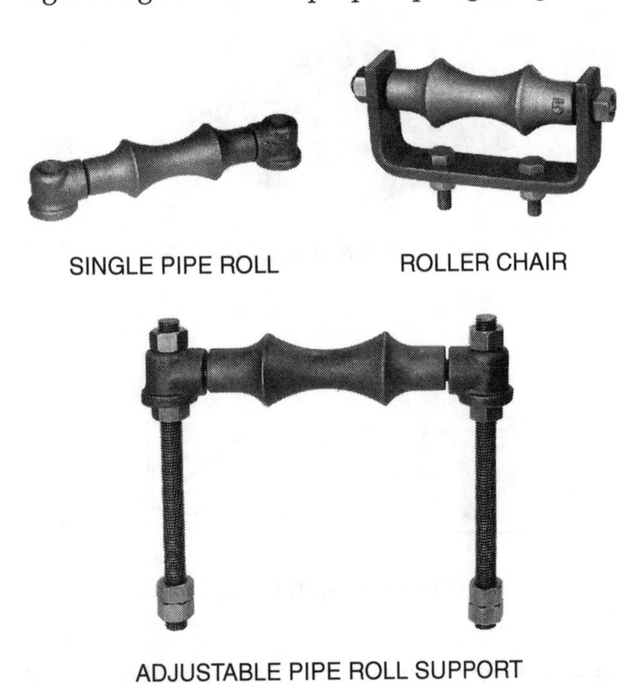

Figure 25 ◆ Pipe rollers.

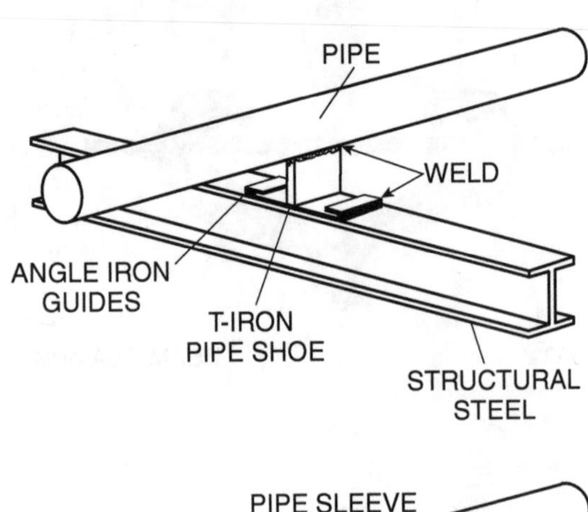

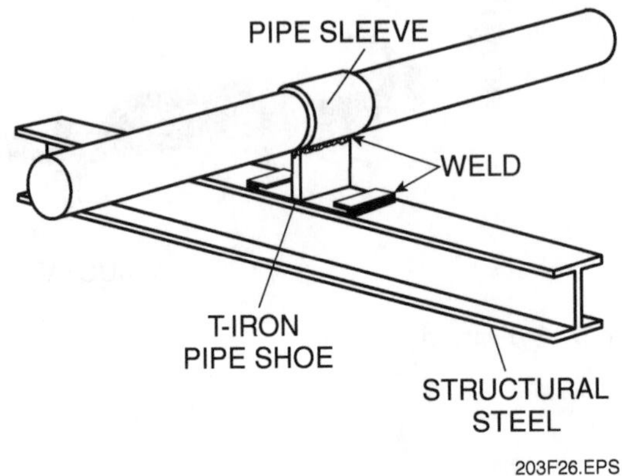

Figure 26 ◆ Field-made pipe alignment guides.

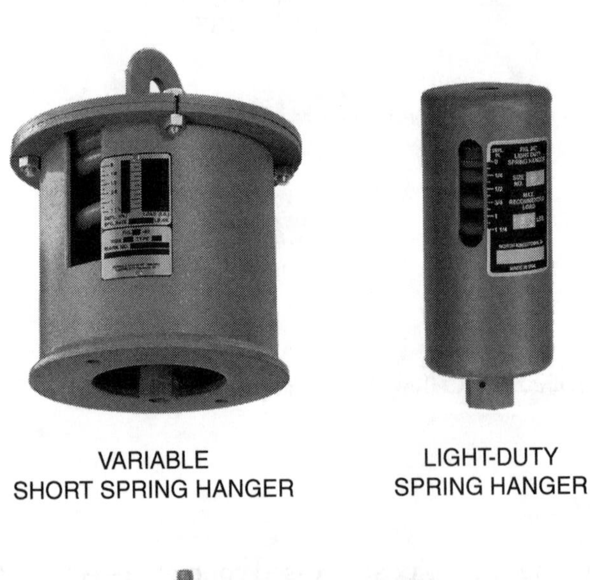

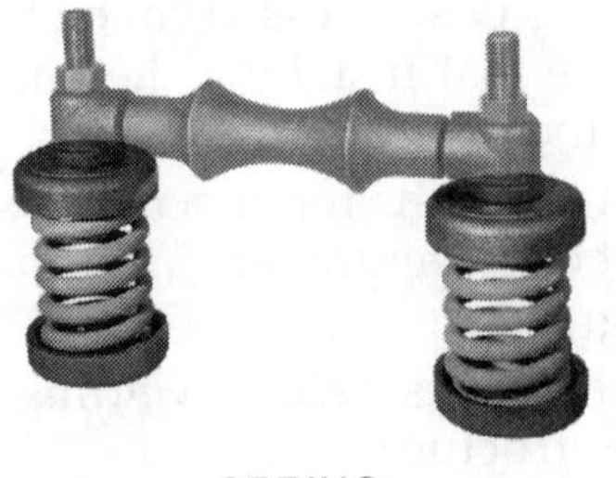

Figure 27 ◆ Spring hangers.

Protection saddles and insulation protection shields are used to protect pipe in hangers (see *Figure 28*). Protection saddles are designed for use on high-temperature lines or where heat losses are to be kept to a minimum. They are also designed to transmit the pipe load to the supporting unit without damaging the covering. Insulation protection shields are recommended when low compression strength and vapor-barrier-type insulation such as foam or fiberglass is installed. The protection shield prevents the hanger unit from cutting, crushing, or otherwise damaging the insulation or the vapor barrier.

2.5.0 Pipe Hanger Locations

Your local applicable code governs the location of pipe hangers. Factors governing pipe hanger locations are pipe size, piping layout, concentrated loads of heavy valves and fittings, and structural building steel available for piping support.

Support concentrated piping loads as close as possible to the load. Support terminal points close to the equipment. Locate hangers right next to any change in piping direction. When installing pipe supports and hangers, be familiar with the engineer's specifications as well as your local applicable code. For sprinkler system installation, follow the appropriate National Fire Protection Association (NFPA) specifications.

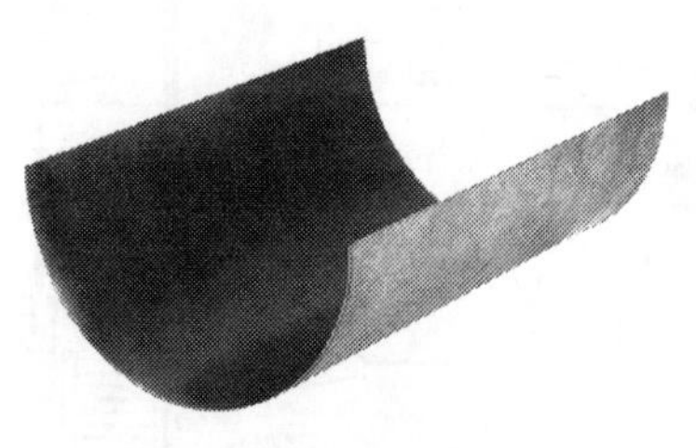

Figure 28 ◆ Protection saddles and insulation shields.

2.6.0 Supporting Vertical Piping

Support vertical piping at sufficient intervals to keep the pipe in alignment. Methods for supporting vertical piping depend upon many factors, including the following:

- Whether the pipe is to be supported
- Pipe materials
- Seismic code requirements

Local code requirements specify the support systems that can be used with vertical pipe runs. Refer to your local code or the project specifications to determine the proper spacing between supports.

You can support vertical pipe at each floor line with riser clamps (refer to *Figure 7*). You may need to install hangers that are attached to the walls or vertical structural members between floors. These hangers will maintain alignment and support part of the vertical load of the pipe. These vertical runs between floor levels can be supported using either an extension ring hanger and wall plate (see *Figure 29*) or a one-hole strap (see *Figure 30*).

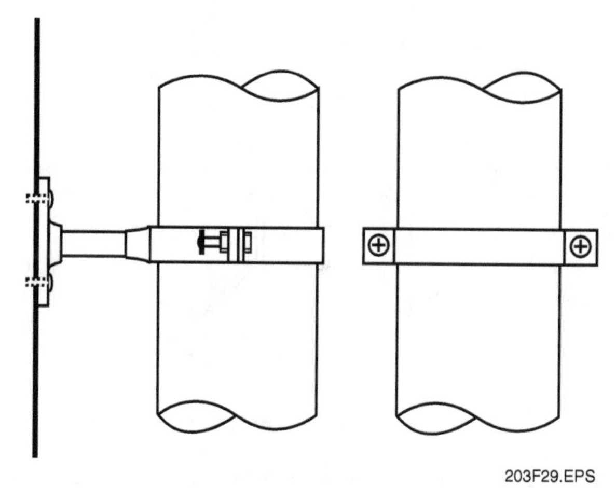

Figure 29 ◆ Extension ring hanger.

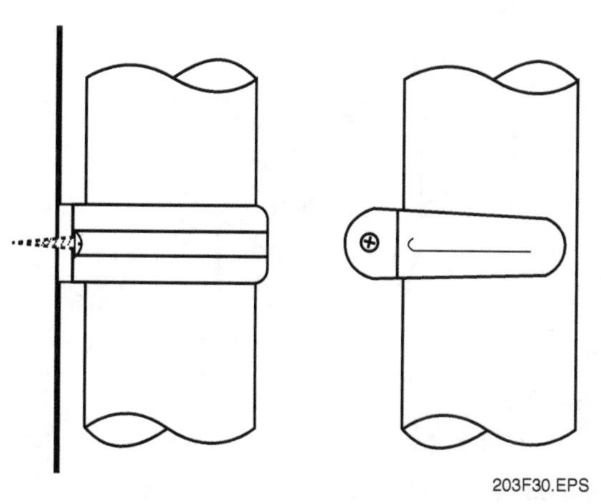

Figure 30 ◆ One-hole strap.

2.7.0 Supporting Horizontal Piping

Support horizontal or sloping pipe at intervals that are close enough to prevent sagging and to keep the pipe in alignment. Sagging pipes produce traps that allow deposits to accumulate in low spots. Traps also can allow air or vapor to accumulate in high spots. As a general rule, each length of pipe should be independently supported so that it does not depend on the neighboring pipe for support. Calculate the slope on which supports are placed and the distance between supports so that each point of support is lower than the nearest upstream point (see *Figure 31*).

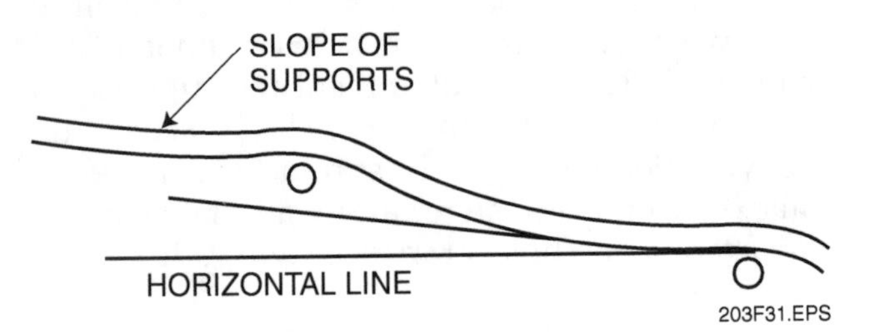

Figure 31 ◆ Slope of supports.

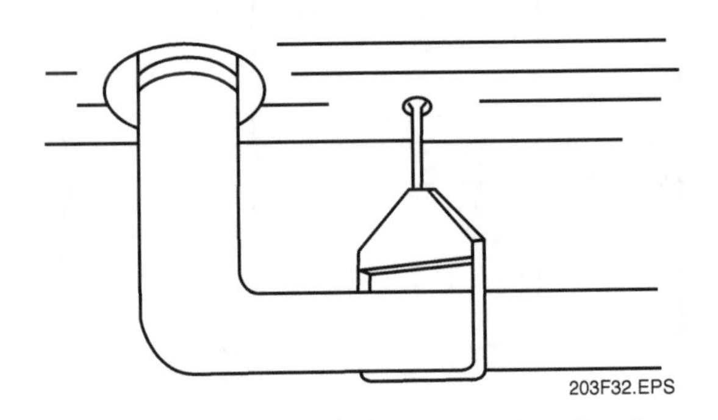

Figure 32 ◆ Clevis hanger supporting a closet bend.

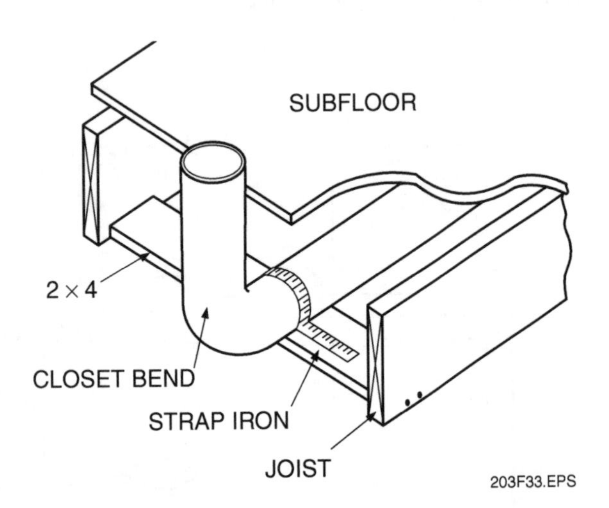

Figure 33 ◆ Closet bend braced in wood frame construction.

2.8.0 Supporting Closet Bends in DWV Systems

Support closet bends horizontally and vertically to prevent movement in either direction. Use a clevis hanger (see *Figure 32*) or a wood frame brace (see *Figure 33*). The type of clevis or strapping material used will depend upon the type of pipe being supported. Consult the manufacturer's specifications and your local applicable code.

2.9.0 Supporting Stack Bases in DWV Systems

Provide adequate support at the stack base. When the stack base is underground, you can support it by placing a brick or concrete support under the fitting at the base of the stack (see *Figure 34*). Support aboveground stack bases with a hanger placed on the base fitting or as close to it as possible (see *Figure 35*).

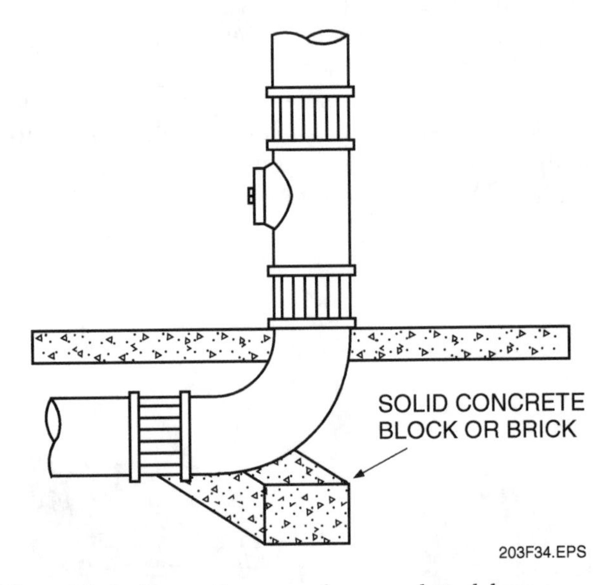

Figure 34 ◆ Supporting an underground stack base.

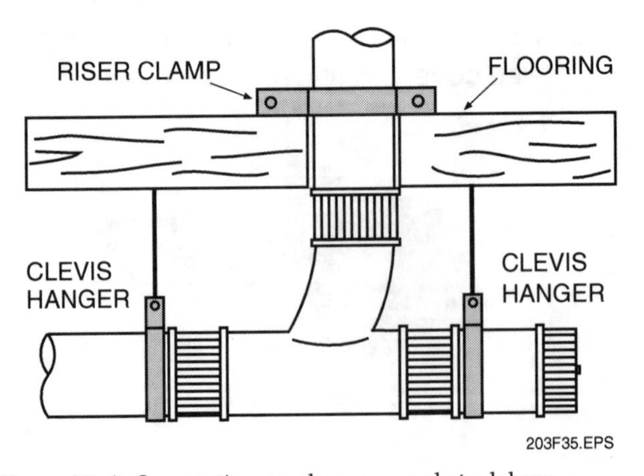

Figure 35 ◆ Supporting an aboveground stack base.

2.10.0 Supporting Multiple Side-by-Side Runs of Pipe

In installations where multiple pipelines are run side by side, various methods are available to support the pipe in one unit, such as a frame or trapeze assembly (see *Figures 36, 37, 38,* and *39*). These assemblies make it possible to hang, attach, mount, frame, and support the piping system as one unit.

By using strut clips, pipe can be added to or removed from trapeze and frame systems made from channels without disturbing pipe that has already been installed (see *Figure 40*). For trapeze assemblies that are not made from channels, side-by-side runs of pipe can be supported directly using multipurpose clips (see *Figure 41*).

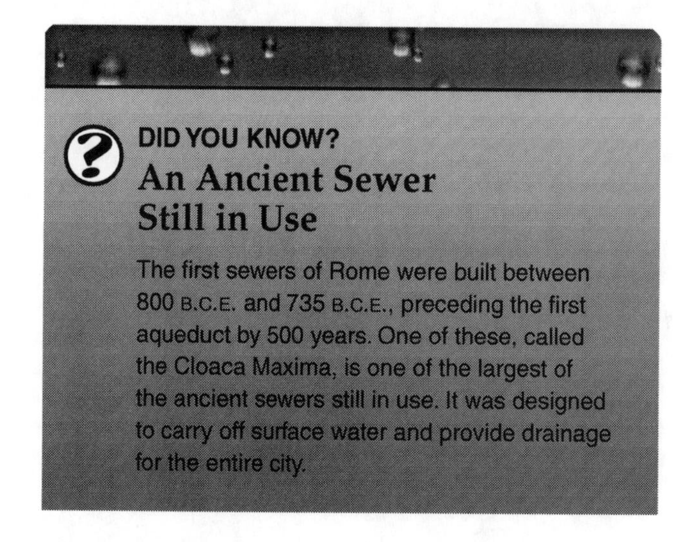

DID YOU KNOW?
An Ancient Sewer Still in Use
The first sewers of Rome were built between 800 B.C.E. and 735 B.C.E., preceding the first aqueduct by 500 years. One of these, called the Cloaca Maxima, is one of the largest of the ancient sewers still in use. It was designed to carry off surface water and provide drainage for the entire city.

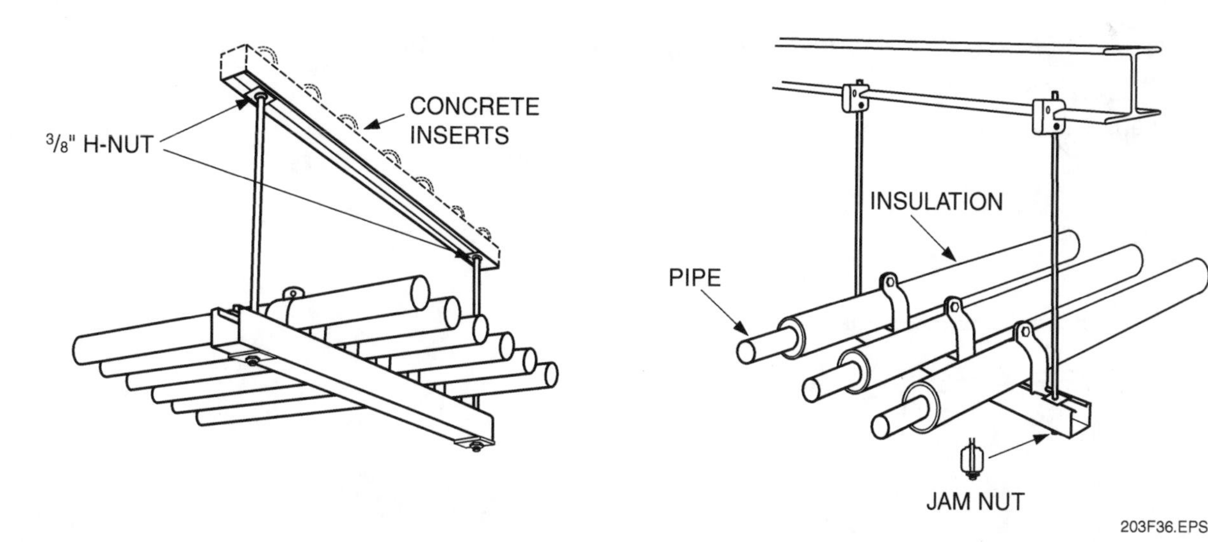

Figure 36 ◆ Applications using trapeze hangers made from channel assemblies.

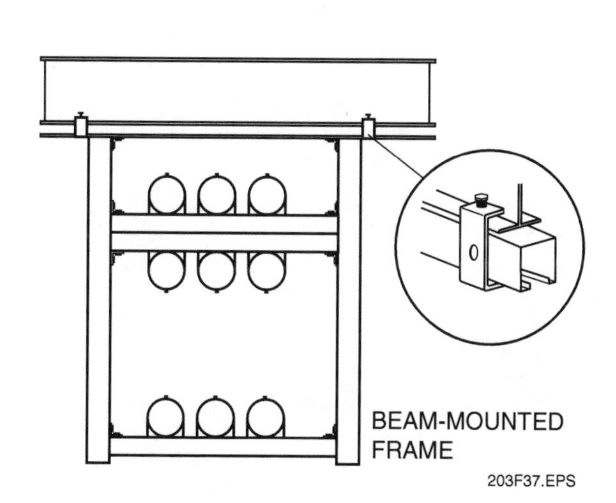

Figure 37 ◆ Beam-mounted channel frame using beam clamp.

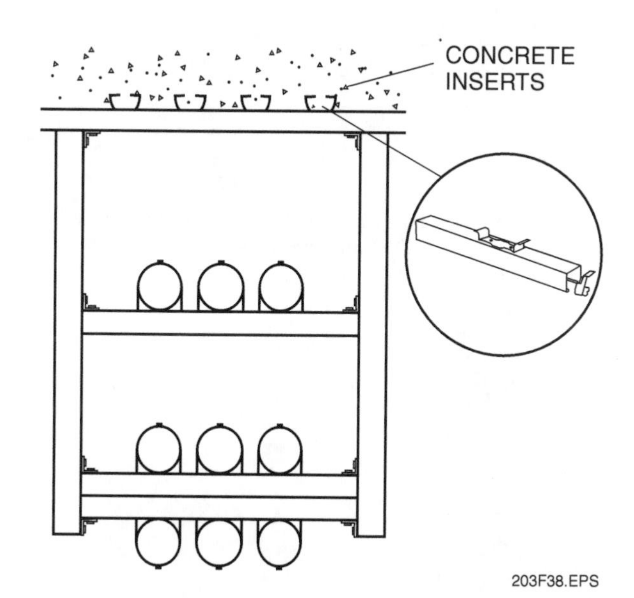

Figure 38 ◆ Ceiling-mounted channel frame using concrete inserts.

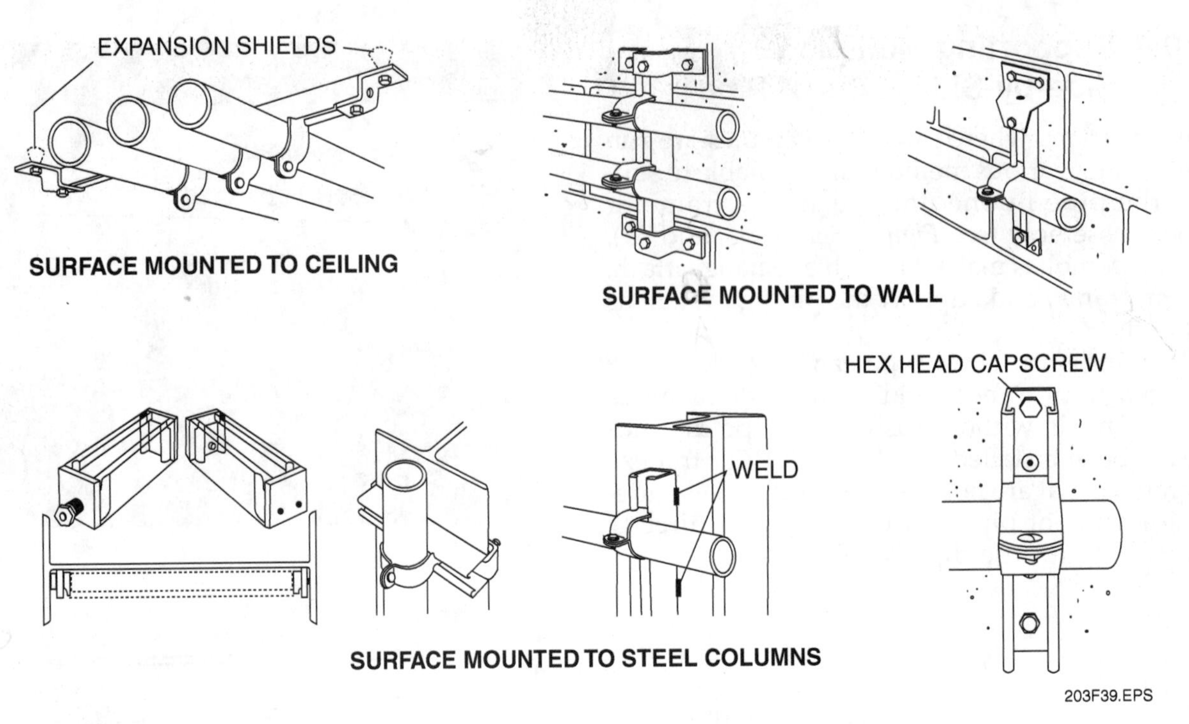

Figure 39 ◆ Surface-mounted channel frames.

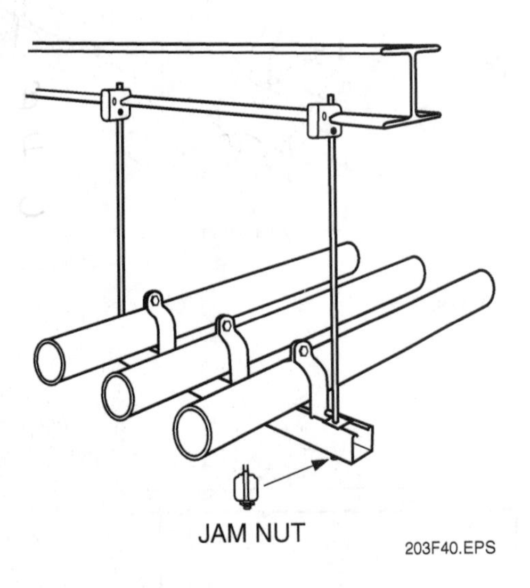

Figure 40 ◆ Side-by-side runs of piping supported with strut clips and channels.

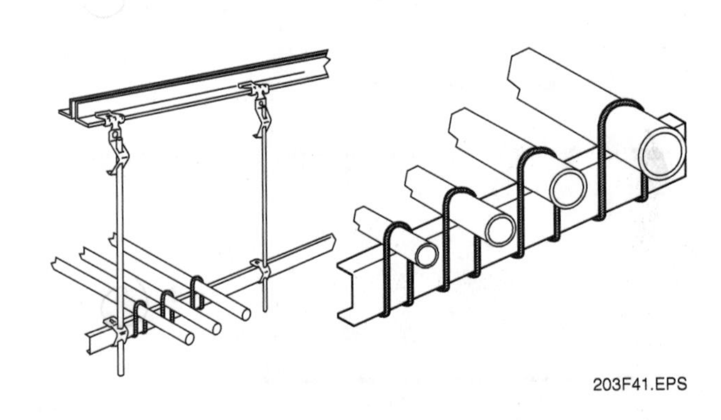

Figure 41 ◆ Supporting pipe directly on steel trapeze using multipurpose clips.

Plumbing Code Requirements for Hangers and Supports

Plumbing codes regulate the maximum interval allowed between supports. Plumbing codes vary from area to area, but each may be based on one of several model codes. Check your local code. Following is a sample of model code requirements for hanger and support locations as they apply to each pipe listed.

2003 National Standard Plumbing Code (NSPC)

Vertical pipe of the following materials shall be supported according to manufacturer's recommendations, but no less than the distances listed below:

- Cast-iron soil pipe: at base and at each story height.
- Steel threaded pipe: at every other story height.
- Copper tube: at each story height but not more than 10-foot intervals.
- Lead pipe: 4-foot intervals.
- Plastic pipe: per *NSPC Section 8.7.*
- Flexible plastic tubing: each story height and at mid-story.
- Stainless steel drainage pipe: at each story height.

Horizontal pipe of the following materials shall be supported according to manufacturer's recommendations, but not less than the distances listed below:

- Copper tube (1¼ inch size and smaller): 6-foot intervals.
- Copper tube (1½ inch size and larger): 10-foot intervals.
- Lead pipe: on continuous metal or wood strips for its entire length.
- Plastic pipe: per *NSPC Section 8.7.*
- Flexible plastic tubing: 32 inches.
- Stainless steel drainage pipe: 10-foot intervals and at changes of direction and branch connections.

International Plumbing Code (IPC)

Piping Material	Maximum Horizontal Spacing (feet)	Maximum Vertical Spacing (feet)
ABS pipe	4	10[b]
Aluminum tubing	10	15
Brass pipe	10	10
Cast-iron pipe[a]	5	15
Copper or copper-alloy pipe	12	10
Copper or copper-alloy tubing, 1¼ inch diameter and smaller	6	10
Copper or copper-alloy tubing, 1½ inch diameter and larger	10	10
Cross-linked polyethylene (PEX) pipe	2.67 (32 inches)	10[b]
Cross-linked polyethylene/aluminum/crosslinked polyethylene (PEX-AL-PEX) pipe	2⅔ (32 inches)	4
CPVC pipe or tubing, 1 inch or smaller	3	10[b]
CPVC pipe or tubing, 1¼ inches or smaller	4	10[b]
Steel pipe	12	15
Lead pipe	Continuous	4
PB pipe or tubing	2.67 (32 inches)	4
Polyethylene/aluminum/polyethylene (PE-AL-PE) pipe	2.67 (32 inches)	4
PVC pipe	4	10[b]
Stainless steel drainage systems	10	10[b]

Notes: a.) The maximum horizontal spacing of cast-iron pipe hangers shall be increased to 10 feet where 10-foot lengths of pipe are installed. b.) Mid-story guide for sizes 2 inches and smaller.

Other Requirements

The Manufacturers Standardization Society of the Valve and Fitting Industry, Inc. publishes a standard that is often referred to by the model building codes when it comes to hangers and supports. This standard, which is the industry source, is titled *Pipe Hangers and Supports: Selection and Application* and is available from the Society as *SP-69-83.*

In addition, some of the model plumbing codes also contain information about specialty piping. You must know and follow your local applicable plumbing code.

Review Questions

Sections 1.0.0–2.0.0

1. Metal straps on plastic DWV pipe _____.
   a. are recommended for all residential DWV systems
   b. cannot be attached with powder-actuated fastening systems
   c. are restricted to single-family homes by most building codes
   d. do not allow for thermal expansion and contraction of the pipe

2. Vertical pipes should be supported at each floor using _____.
   a. riser clamps
   b. connectors
   c. universal pipe clamps
   d. perforated band irons

Match the following pipe attachments to their applications.

3. _____ Beam clamp
4. _____ C-clamp
5. _____ Beam attachment
6. _____ Bracket
7. _____ Washer plate

   a. used to support pipe from I-beams or channels
   b. used to attach a support rod to a structural member
   c. used to suspend a support rod from a channel or angle
   d. used to suspend piping from the building steel
   e. used as an upper attachment to suspend pipe from concrete ceilings
   f. used to attach threaded connections to wood beams

8. Use a(n) _____ to provide a nonadjustable threaded connection to a hanger rod when installing a split-ring hanger.
   a. extension piece
   b. rod coupling
   c. turnbuckle
   d. eye socket

Match each basic component used to hang and support pipe with its correct description.

9. _____ Pipe attachment
10. _____ Connector
11. _____ Structural attachment

   a. The part that anchors the assembly securely to the structure.
   b. The part that allows pipes to expand and move laterally.
   c. The part that touches or connects directly to the pipe.
   d. The intermediate part that links the other two components.

12. When suspending pipelines from wood beams or ceilings using a powder-actuated fastening tool, use _____ and plates.
    a. concrete inserts
    b. beam clamps
    c. wedge anchors
    d. ceiling flanges

13. You should install a hanger right next to any change of piping direction.
    a. True
    b. False

14. Horizontal pipe supports are necessary to prevent sagging and keep pipe _____.
    a. parallel
    b. properly vented
    c. well insulated
    d. in alignment

15. Closet bends can be supported using either a(n) _____ hanger or a wood frame brace.
    a. spring
    b. extension ring
    c. clevis
    d. strap lock

16. To support side-by-side runs of pipe on a trapeze assembly that is not made from a channel, use _____.
    a. multipurpose clips
    b. tin straps
    c. U-hooks
    d. short clips

3.0.0 ◆ STRUCTURAL PENETRATIONS

Occasionally, structural members lie in the path of a run of pipe, or the space available between structural members is not sufficient to allow pipe to be run between them. In such cases, plumbers must modify structural members. This section deals with the five most common methods of modifying structural members:

- Drilling
- Notching
- Boxing
- Furring
- Building a chase

Collectively, these methods can be called **structural penetrations** or structural modifications. When modifying structural members, ensure that the structure is not weakened. If questions concerning the safety and strength of the structure arise, consult your local applicable code or the local plumbing inspector.

> **WARNING!**
> Do not modify steel structural members without written permission from the authority having jurisdiction. You should also obtain permission from the architect or engineer for the project. Unauthorized modifications to steel structural members could weaken the building's structural integrity.

3.1.0 Drilling

Floor joists may be drilled to allow passage for a run of pipe (see *Figure 42*). Holes cannot be larger than one-third the depth of the floor joist. For example, the largest hole that could be drilled in the center of a 2 × 10 floor joist would be $3\frac{5}{16}$ inches in diameter. Holes of this size must be drilled in the middle of the joist, rather than through the upper or lower third of the joist. Only one such hole is permissible per joist. Off-center holes that are smaller than the maximum depth should be drilled no less than 2 inches from the top or bottom of the joist. Do not drill holes within 6 inches of either end of a joist.

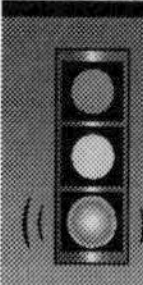

> **NOTE**
> Drilling holes in several adjacent floor joists will significantly weaken the structure. Route pipes to require a minimum of drilling through joists.

Holes drilled in the floor joists must be larger than the pipes that pass through them. This provides necessary clearance for the pipe. As a rule of thumb, you should drill a 3-inch hole for a 2-inch pipe, and a 2-inch hole for a $1\frac{1}{2}$-inch pipe.

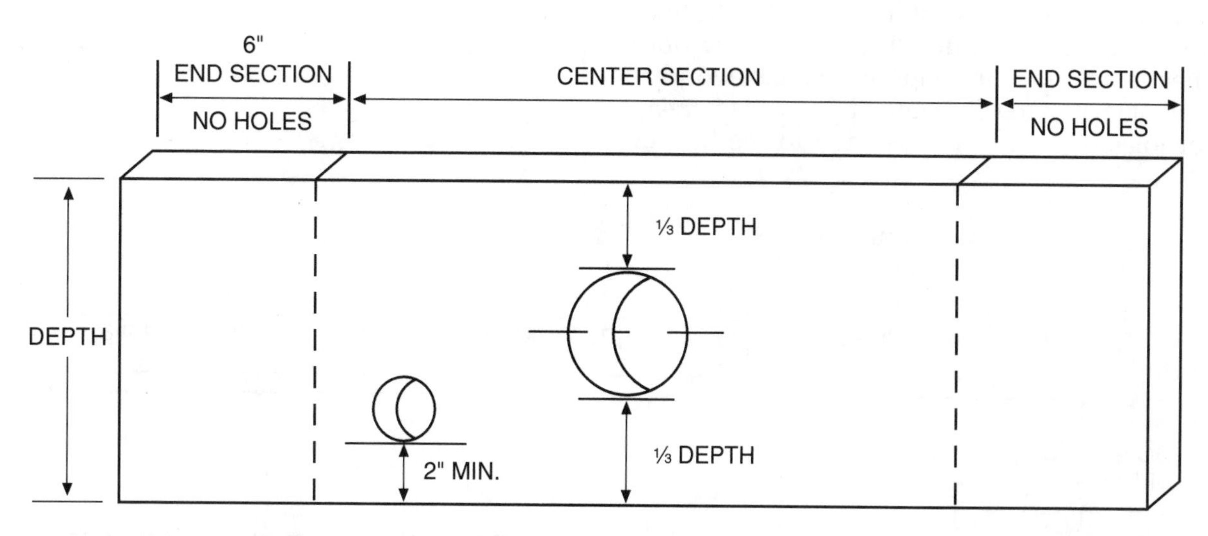

Figure 42 ◆ Drilling floor joists.

3.2.0 Notching

Structural members may also be notched to provide clearance for plumbing pipes and fittings. As with drilling, refer to your local applicable code to determine the amount and type of notching that is permitted. Because notching presents a greater threat to a building's structural strength than drilling, notching should be done only when absolutely necessary.

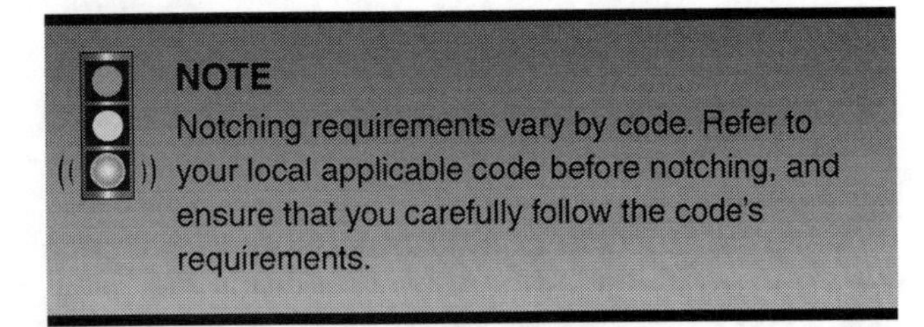

NOTE
Notching requirements vary by code. Refer to your local applicable code before notching, and ensure that you carefully follow the code's requirements.

Notching is not permitted in the center third of structural members (see *Figure 43*). Notching can safely be done only near the ends of the floor joists. Also, the depth of the notch cannot exceed one-fourth of the depth of the board (that is, the distance from the top to the bottom of the board). In cases where you are forced to notch a structural member more than one fourth of its depth, you should use bracing. In most cases, the braces consist of short lengths of 2 × 4s nailed to both sides of the notched member (see *Figure 44*). Bracing significantly strengthens the member.

3.3.0 Boxing Floor Joists

In some cases, neither drilling nor notching will provide adequate plumbing clearance. For example, a floor joist interferes with the installation of the closet flange. In such cases, you can box the floor joist. This involves cutting the end of the floor joist back to clear the fitting. Then, construct a double header to bridge the gap between the two adjacent floor joists and nail it into place (see *Figure 45*). Floor joists may also be fastened to the header using strap hangers (see *Figure 46*). To further strengthen the floor, double the adjacent joists.

3.4.0 Furring Strips

The space inside the framed wall must be wide enough to allow passage of the DWV stack. In most cases, 2 × 4 stud framing does not provide sufficient space. Therefore, 2 × 6 or 2 × 8 studs are usually placed in walls where the DWV stack is located. In cases where the wall has previously been framed with 2 × 4 studs, add furring strips (see *Figure 47*). This procedure provides the space required for the DWV stack. When adding furring strips, remember to locate the fixtures from the furred wall and not from the original wall.

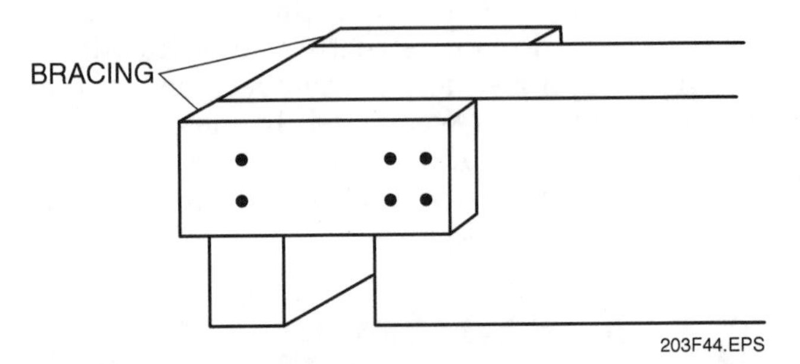

Figure 44 ◆ Bracing structural members.

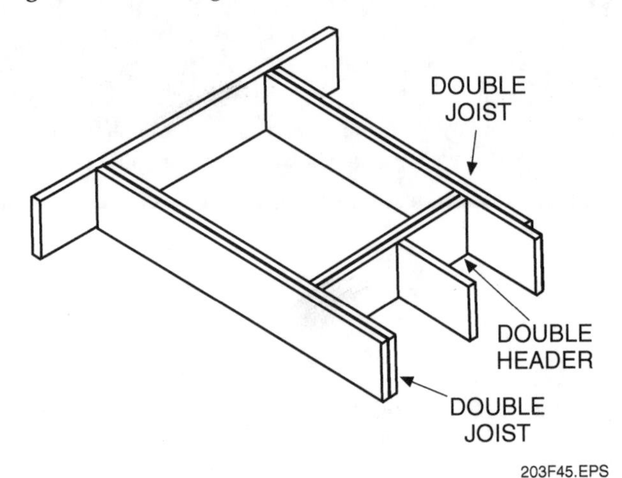

Figure 45 ◆ Boxing floor joists.

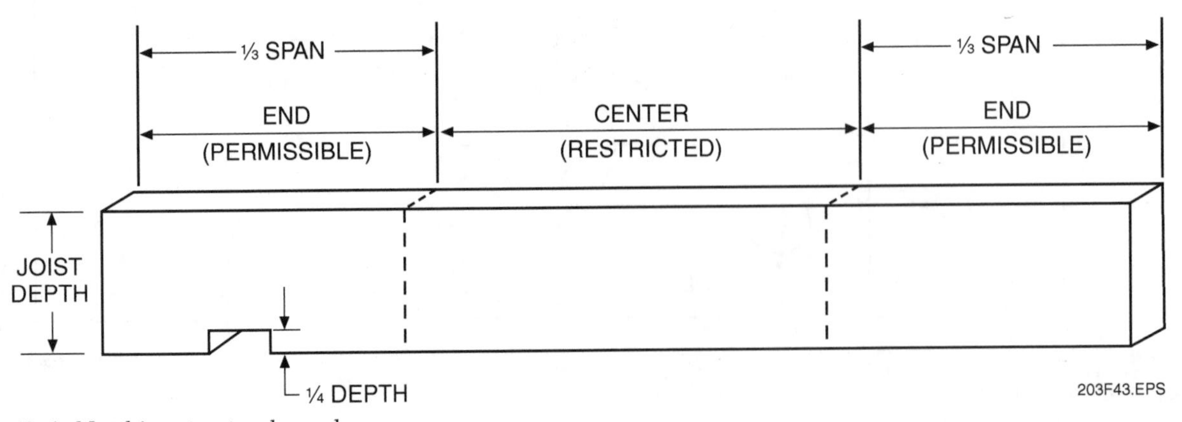

Figure 43 ◆ Notching structural members.

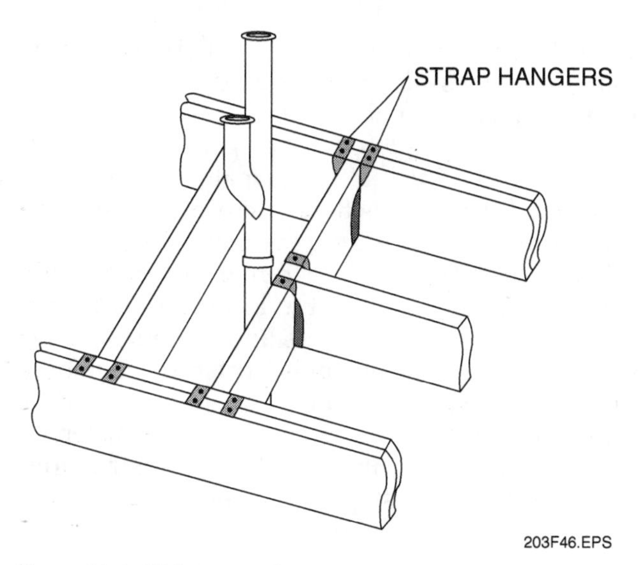

Figure 46 ◆ Using strap hangers.

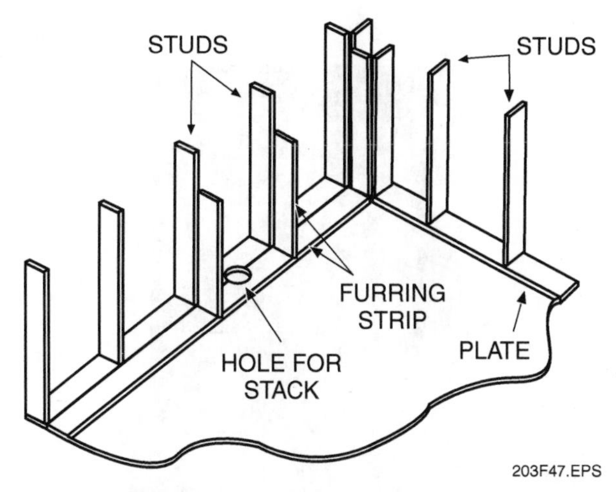

Figure 47 ◆ Using furring strips.

3.5.0 Building a Chase

In some instances, even furring strips will not provide the necessary clearance for plumbing pipes and fixtures. This happens when a line of fixtures, such as water closets, are placed along one wall. In these situations, build a chase (see *Figure 48*). The chase is a small sealed space that provides the necessary clearance for pipes, fittings, carriers, and other pieces of plumbing. The chase may or may not extend to the ceiling and adjacent walls. Usually, it is permanently sealed, so you must properly install the plumbing before the wall finishing material is applied. Make cleanouts accessible with removable cover plates flush-mounted on the chase wall.

4.0.0 ◆ FIRE STOPPING

Fire codes require the installation of fire-stopping material in penetrations through fire-rated structural members, as well as through fire-rated walls, floors, and ceilings. Fire-stopping systems are designed to completely seal structural penetrations in the event of a fire.

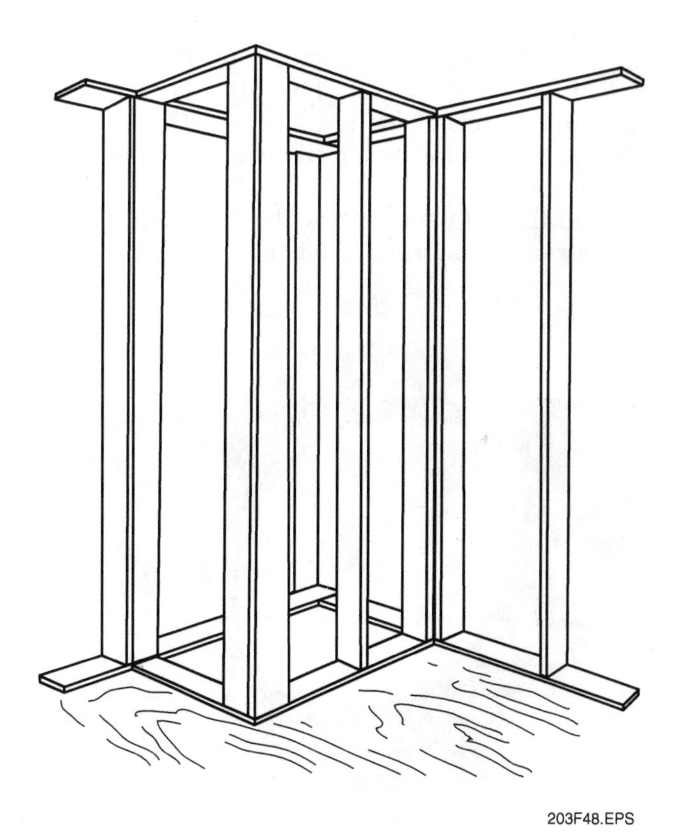

Figure 48 ◆ Chase.

Figure 49 ◆ Intumescent material in block form.

Many systems use an **intumescent** material, which expands when exposed to flame. Like water on a dry sponge, fire causes properly installed intumescent material to swell and completely seal the gap between the pipe and the structural opening. This prevents the passage of flame, smoke, or harmful gases through the opening. Intumescent material is available in preformed shapes such as blocks, foil-backed strips, and plugs, (see *Figure 49*) as well as in spray, foam, mortar, and putty form. Other fire-stopping systems use silicone foam or sealant that expands while curing to provide a complete seal when dry.

Manufacturers provide accessories to act as forming and backing support for the fire-stopping materials. For example, steel collars lined with preformed intumescent blocks can be screwed or bolted to the structural member around the penetration (see *Figure 50*). The pipe is then run through the collar. Cast-in-place adapters (see *Figure 51*) are used for penetrations in fire-rated concrete floors.

In addition to preventing the spread of flame, fire-stopping materials must be smoke-, gas-, and watertight. Use only approved fire-stopping materials and sealants, such as those recognized by Underwriters Laboratories, Inc. (UL) or the American Society for Testing and Materials (ASTM). Select fire-stopping materials that have the same fire rating as the structural member.

Ensure that the fire-stopping materials you select will not degrade over time from exposure to moisture or climate. Consult the local applicable fire code for approved methods and materials before cutting, drilling, or notching structural members. Your fire code may require a qualified or experienced professional to install fire-stopping materials. Always follow the manufacturer's instructions when applying fire-stopping materials.

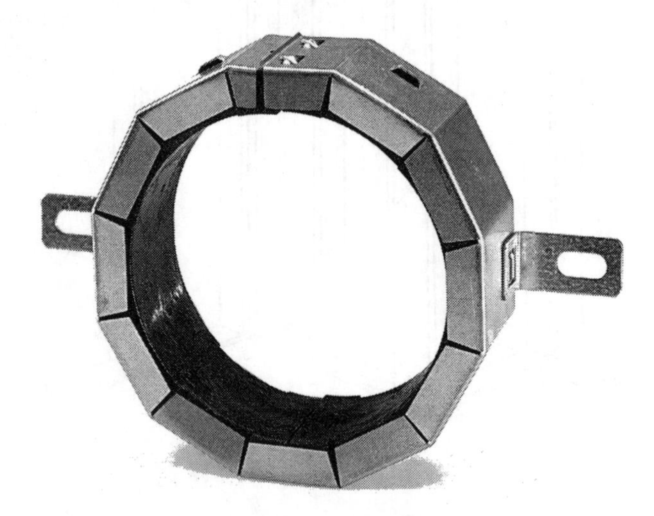

203F50.EPS

Figure 50 ◆ Fire-stopping collar.

203F51.EPS

Figure 51 ◆ Cast-in-place fire stop.

Review Questions

Sections 3.0.0–4.0.0

1. The largest hole that could be drilled in the center of a 2 × 10 floor joist would be _____ inches in diameter.
   a. 2
   b. $3\frac{5}{16}$
   c. $3\frac{1}{2}$
   d. $4\frac{3}{16}$

2. A hole drilled in a floor joist to allow for a run of pipe must be _____.
   a. no more than one-third the depth of the joist
   b. within 4 inches from the end of the joist
   c. at least 1 inch from the nearest edge of the joist
   d. at least 1 inch larger than the pipe

3. Notches in a structural member can be strengthened by using _____.
   a. boxing
   b. frames
   c. braces
   d. chases

4. When notching a structural member, the depth of the notch *cannot* exceed _____ of the depth of the structural member.
   a. one fourth
   b. one third
   c. one half
   d. two thirds

5. Floor joists may be fastened to the header using _____.
   a. strap hangers
   b. suspension hangers
   c. U-bolts
   d. universal pipe clamps

6. _____ is recommended when a floor joist interferes with installation of the closet flange in water supply piping.
   a. Furring
   b. Notching
   c. Drilling
   d. Boxing

7. When adding furring strips, locate the fixtures from the original wall.
   a. True
   b. False

8. Because a chase is usually permanently sealed when completed, you should make sure that _____ are installed for ease of access.
   a. access doors
   b. traps
   c. cleanouts
   d. valves

9. When exposed to fire, an intumescent material will _____.
   a. melt
   b. expand
   c. solidify
   d. contract

10. Silicone foams and sealants expand _____ to completely seal a structural penetration.
    a. when mixed with a reactant
    b. when wet
    c. immediately on contact
    d. while curing

Summary

This module introduced you to the methods and materials that can be used to hang and support pipe, modify structural members, and install fire-stopping systems. All of these activities require you to attach items to or modify the components of a building's structure. When performing this work, always follow your local applicable code and the requirements identified in the project plans and specifications. Always use the proper materials and carefully follow the manufacturer's instructions. This will protect the structural integrity of the building while also preventing damage to the water supply and DWV systems that you install.

Hangers and supports provide pipe with adequate support and prevent sagging. They are commonly divided into three categories: pipe attachments, connectors, and structural attachments. Use hangers and supports that are compatible with the type of pipes you are installing.

Structural members in walls and floors can be drilled, notched, or boxed to provide passage for water supply and DWV pipes. Furring strips and chases can be built to provide sufficient space for wider pipes.

Fire codes require penetrations through fire-rated structural members, walls, floors, and ceilings to be fitted with code-approved fire-stopping material. Fire-stopping systems completely seal structural penetrations in the event of a fire. They do this by expanding to completely seal the penetration. Intumescent materials expand on contact with fire. Other materials expand while curing. Use only approved fire-stopping materials, and be sure to install them according to the manufacturer's instructions. When attaching things to, modifying, or protecting structural members, remember that the health, safety, and comfort of the building's occupants should be your first and last consideration.

Notes

Trade Terms Introduced in This Module

Fire stopping: A system of materials and fittings used to prevent the passage of flame, smoke, and/or toxic byproducts of fire through structural penetrations.

I-beam: A rolled or extruded structural metal beam having a cross-section that looks like the capital letter I.

Intumescent: Having the ability to expand when exposed to flame.

Structural penetration: A modification to a structural member in the form of notching, drilling, or boxing.

Resources & Acknowledgments

Additional Resources

This module is intended to be a thorough resource for task training. The following reference works are suggested for further study. These are optional materials for continued education rather than for task training.

Handbook of Materials Selection. 2002. Myer Kutz, ed. New York: J. Wiley.

NFPA 1, Uniform Fire Code, Latest Edition. Quincy, MA: National Fire Protection Association.

Practical Plumbing Engineering. 1991. Cyril M. Harris, ed. New York: McGraw-Hill.

References

Dictionary of Architecture and Construction, Third Edition. 2000. Cyril M. Harris, ed. New York: McGraw-Hill.

Figure Credits

Ivey Mechanical Company	Module divider
Anvil International, Inc.	203F01, 203F02 (clevis hanger, swivel rings), 203F05, 203F07, 203F08, 203F10–203F14, 203F21–203F23, 203F24 (concrete plates), 203F25–203F28, 203F31
NIBCO International	203F02 (J-hanger), 203F03 (tin strap, U-hook)
Sioux Chief Manufacturing Co., Inc.	203F03 (multipurpose clips)
Reprinted with permission from ERICO International Corporation	203F06, 203F09, 203F41
Hilti, Inc.	203F49–203F51
J. Blanco Associates, Inc.	203F24 (steel washer plate)
International Code Council, Inc. *2003 International Plumbing Code. Copyright 2003. Falls Church, Virginia: International Code Council, Inc. Reproduced with permission. All rights reserved.*	203SA02

CONTREN® LEARNING SERIES — USER UPDATE

The NCCER makes every effort to keep these textbooks up-to-date and free of technical errors. We appreciate your help in this process. If you have an idea for improving this textbook, or if you find an error, a typographical mistake, or an inaccuracy in NCCER's Contren® textbooks, please write us, using this form or a photocopy. Be sure to include the exact module number, page number, a detailed description, and the correction, if applicable. Your input will be brought to the attention of the Technical Review Committee. Thank you for your assistance.

Instructors – If you found that additional materials were necessary in order to teach this module effectively, please let us know so that we may include them in the Equipment/Materials list in the Annotated Instructor's Guide.

Write: Product Development and Revision
National Center for Construction Education and Research
P.O. Box 141104, Gainesville, FL 32614-1104

Fax: 352-334-0932

E-mail: curriculum@nccer.org

Craft ______ Module Name ______

Copyright Date ______ Module Number ______ Page Number(s) ______

Description

______

______

______

______

(Optional) Correction

______

______

______

(Optional) Your Name and Address

______

______

______

02204-05

Installing and Testing DWV Piping

02204-05

Installing and Testing DWV Piping

Topics to be presented in this module include:

Overview

Drain, waste, and vent (DWV) systems remove liquid and solid wastes from buildings. Because efficient waste removal is essential to maintain public health, plumbers must ensure that DWV systems are installed correctly. Plumbers work according to construction plans. When plumbing a structure, they use the floor plan to locate the building fixtures. In residential and light commercial structures, basic framing usually accommodates plumbing installations without any unusual problems. Plumbers prefabricate fittings when possible to save time.

Because DWV piping systems rely on gravity to move solid and liquid wastes, they are installed at a slope toward the point of disposal. This slope is called the grade. Information about proper grade is located in plumbing codes, job specifications, or construction plans. Plumbers must install DWV piping systems at the specified, constant grade. To ensure accuracy when laying out and measuring grade, plumbers use leveling tools, including builder's levels that measure elevations and horizontal angles.

Plumbers work with the carpenters to determine how they will lay out the structure. Plumbers also are responsible for locating fixtures, installing the stack, calculating pipe grade, testing the installation, and arranging for the inspection. Completing these jobs safely and accurately ensures that the DWV piping system safely moves wastes away from the structure without letting in harmful gases and disease.

Focus Statement

The goal of the plumber is to protect the health, safety, and comfort of the nation job by job.

Code Note

Codes vary among jurisdictions. Because of the variations in code, consult the applicable code whenever regulations are in question. Referring to an incorrect set of codes can cause as much trouble as failing to reference codes altogether. Obtain, review, and familiarize yourself with your local adopted code.

Objectives

When you have completed this module, you will be able to do the following:

1. Develop a material takeoff from a given set of plans.
2. Use plans and fixture rough-in sheets to determine location of fixtures and route of the plumbing.
3. Install a building sewer and a building drain.
4. Locate the stack within the structure.
5. Install a DWV system using appropriate hangers and correct grade or slope.
6. Test a DWV system.

Trade Terms

Accessibility requirements	Rebar
Batter board	Sanitary increaser
Bench mark	Sewer tap
Blocking	Sheathing
Building cleanout	Sheetrock®
Carrier fittings	Shoring
Cure	Sizing
Daily log	Slab-on-grade
Invert elevation	Sleeve
Load factor	Slope
Percentage of grade	Soleplate
Perineal bath	Time-critical material
Pre-construction plan	Top plate
Prefabricated	Vernier

Required Trainee Materials

1. Appropriate personal protective equipment
2. Pencil and paper
3. Copy of local applicable code

Prerequisites

Before you begin this module, it is recommended that you successfully complete *Core Curriculum; Plumbing Level One; Plumbing Level Two,* Modules 02201-05 through 02203-05.

This course map shows all of the modules in the second level of the *Plumbing* curriculum. The suggested training order begins at the bottom and proceeds up. Skill levels increase as you advance on the course map. The local Training Program Sponsor may adjust the training order.

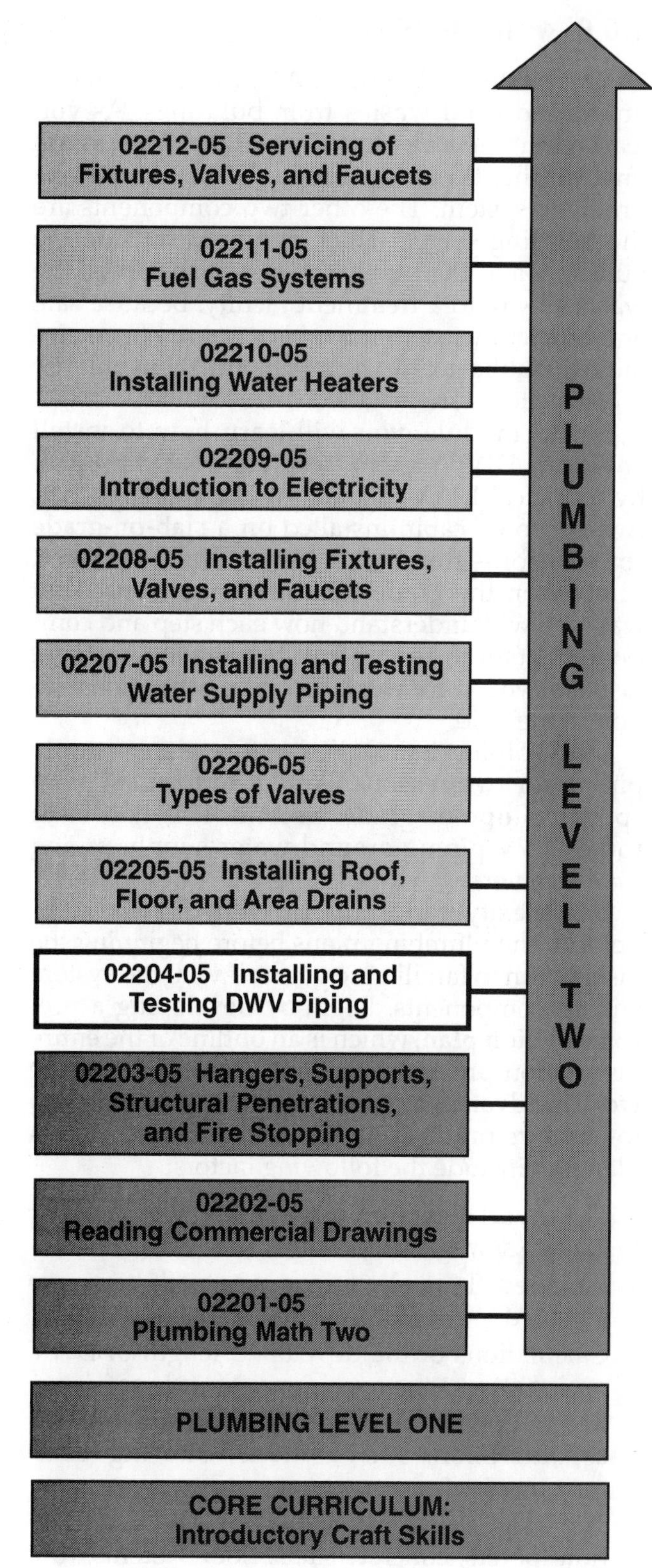

204CMAP.EPS

1.0.0 ◆ INTRODUCTION

Drain, waste, and vent (DWV) systems remove liquid and solid wastes from buildings. As you learned in *Plumbing Level One*, DWV systems are one of the three components of a sanitary drainage system. The other two components are the building sewer, which is buried outside the building, and the public sewer, which carries the wastes away to a treatment facility. Because safe and efficient waste removal is essential for maintaining public health, plumbers must ensure that DWV systems are installed correctly.

In this module, you will learn how to install and test a DWV system. You will work through a hypothetical DWV installation for a single-story, two-bedroom cabin installed on a **slab-on-grade** foundation—that is, the base slab is placed directly on the grade without a basement. That way you will understand how each step and component fits into the overall installation process, just as if you were working on an actual installation in the field.

DWV piping is installed before water supply piping for two reasons. First, it must maintain a specific **slope** or grade. Second, it is harder to route DWV piping around water supply piping than vice versa.

Complexity varies among DWV installations. Refer to the plumbing plans before beginning the installation to familiarize yourself with the system and its components. Begin by developing a **pre-construction plan**, which is an outline of the entire installation process. By allowing you to plan the work in advance, a pre-construction plan will help to improve productivity. A good pre-construction plan will include the following factors:

- The **invert elevation** of the inlet to the sewer or private sewage system
- Finished floor elevations where fixtures are installed
- Calculations of the developed length of DWV piping versus the requirement for slope
- Requirements for excavation, including safety considerations such as trench **shoring** (reinforcement)
- The main structural elements such as foundation footers, concrete slabs, floor systems, roof systems, and stairs
- Calculations of drain and vent **sizing** based on the load requirements
- Identification of the plans and specifications to be used
- Types of fixtures and fixture supports and their locations within the building
- Coordination of measurements and timing with other trades
- Delivery of **time-critical material**, or material that must be available before the installation can proceed
- Delivery and storage of materials
- Arrangements for inspection and testing equipment
- **Accessibility requirements**, which are requirements outlined in your local code or other standards that ensure access by physically challenged people

The construction method being used for the building will determine the way you install the DWV system. For example, if the building is being erected using slab-on-grade construction, like the example in this module, you would have to start your work before the walls are constructed. If the building is completely framed up and lacks only the basement slab and the roof, you would be able to install the complete system at one time. The construction method, as you can see, affects how you order materials, use equipment, and schedule inspections. Your pre-construction plan should reflect this.

2.0.0 ◆ PLANS

For economy and efficiency, plumbers work according to construction plans. Plumbing plans are typically included with the architectural plans for most industrial, commercial, and multiresidential structures. The plumbing plans usually allow some leeway for changes caused by unforeseen circumstances. In many cases, particularly in light residential construction, plumbing plans are not included with the set of construction drawings. In these cases, the plumber uses the floor plan to locate and size the runs of pipe, and to locate the various plumbing fixtures (see *Figure 1*).

The system's **load factors** will determine the sizing of the DWV system (the process of calculating the correct sizes for the system's drains, stacks, sewer lines, and vents). A load factor is a specified percentage of the total flow from connected fixtures that is likely to occur at any point along the DWV system. Your local applicable code will include tables and other information that you can use to calculate load factors. You will learn more about how to size DWV systems in *Plumbing Level Three*.

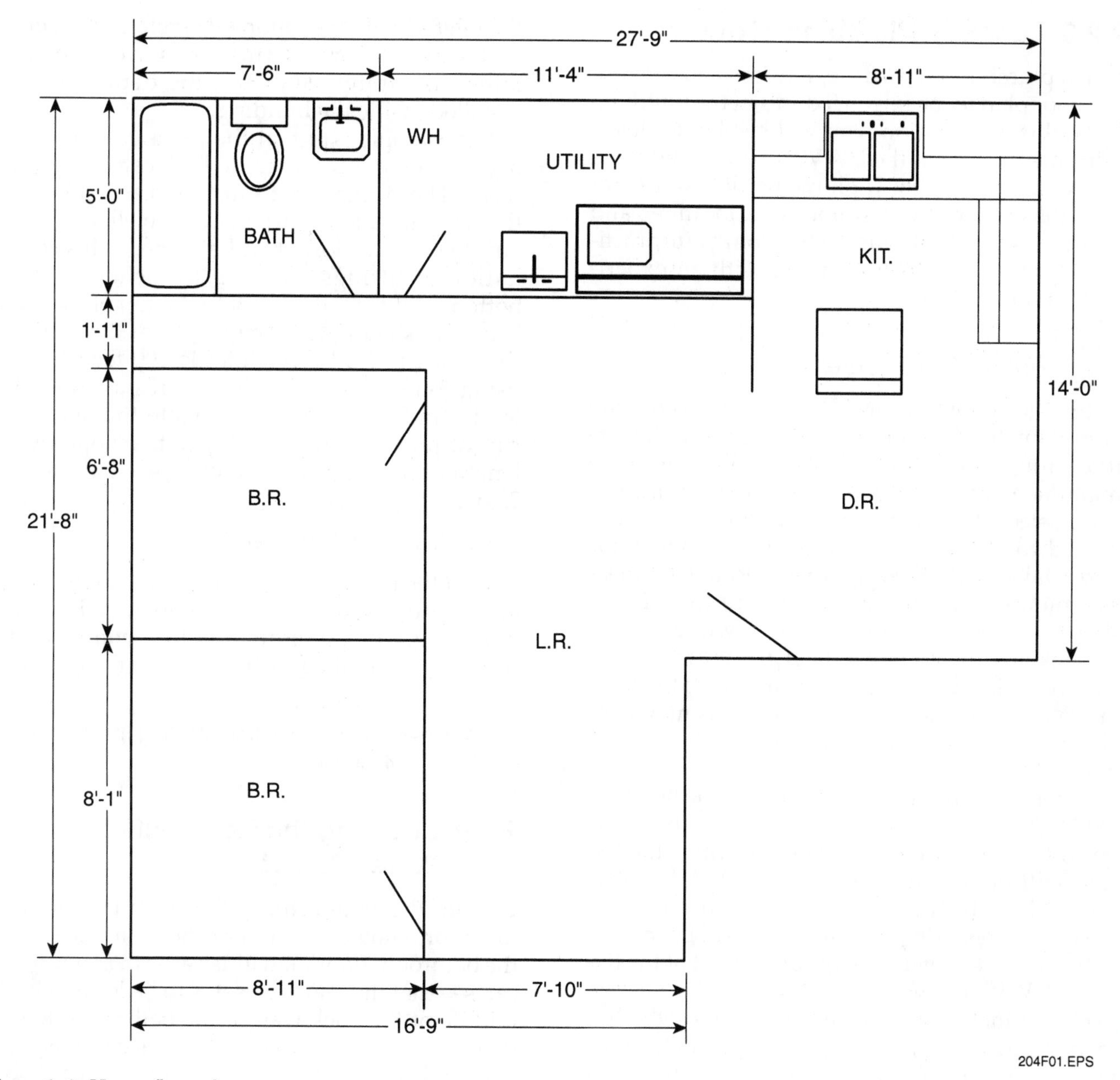

Figure 1 ◆ House floor plan.

2.1.0 Material Takeoffs

A material takeoff is a list of the types and quantities of materials required for a given job. An estimator who is knowledgeable about plumbing practices and materials may work from a set of drawings to do the takeoff. There is no standard form used for material takeoffs. A typical form is shown in *Figure 2*.

MATERIALS ESTIMATE

PROJECT Joe Doe – Residence — ESTIMATE NO. 107
LOCATION Anywhere, U.S.A. — SHEET NO. 1
ARCHITECT ENGINEER John Smith — DATE 10-2-01
SUMMARY BY E.K.B. — PRICES BY P.J. — CHECKED BY M.B.

QUANTITY	DESCRIPTION	UNIT PRICE	TOTAL ESTIMATED MATERIAL COST	UNIT PRICE	TOTAL ESTIMATED LABOR COST	TOTAL
2	19" Round Lavs.	21.95	43.90	18/hr.	90.00	133.90
2	Closets	51.00	102.00	"	72.00	174.00
1	Tub - Shower Value	22.86	22.86	"	12.00	34.86
1	Kitchen Sink	56.10	56.10	"	54.00	110.10
20'	4" ABS Pipe	1.23/FT.	24.60			24.60
80'	1½" ABS Pipe	.31/FT.	24.80			24.80
4	4" TY	4.23	16.92			16.92
8	2" 90° Els	.30/EA.	2.40			2.40
2	3" x 1½" Tees	.96/EA.	1.92			1.92
2	3" Flashings	8.65	17.30			17.30

204F02.TIF

Figure 2 ◆ Material takeoff with estimate.

2.2.0 Locating Plumbing Fixtures

When plumbing a structure, the first step is to use the floor plan to locate the fixtures in the building. From this information, you will be able to determine the location of the DWV stacks and the route of the fixture branch. If plumbing details are not available, you can determine stack locations and branch line routes to the stacks from your practical experience or by consulting with an experienced plumber.

2.2.1 Rough-In Measurements

After using the floor plan to determine the approximate location of a fixture, study the fixture's rough-in measurements, available from the manufacturer (see *Figure* 3). Once a fixture has been selected, rough-in measurements are submitted to the architect, engineer, or owner for approval and marked approved. Rough-in measurements tell you exactly where the piping should exit the walls or floors. You will use these dimensions for reference later to run the water supply piping and to attach the fixture to the floor or wall. Notice that critical dimensions may vary among styles of the same brand (refer to A and B in *Figure* 3).

When determining where to locate the holes for the DWV piping, remember to factor in the thickness of the finished wall or floor to which the fixture will be attached (see *Figure* 4). Add these dimensions to those of the rough-in measurements. The resulting measurement will be the one you will use to determine the placement of the fixture's DWV piping.

Before installing fixtures, determine whether **carrier fittings** or other fixture support elements will cause conflicts with the piping. A carrier fitting is a mounting apparatus that attaches bathroom fixtures to a wall. The rough-in sheet will specify the type, dimensions, and locations of fixture supports.

For example, say you are installing a lavatory carrier that requires a 3½-inch stud wall. Will you be able to run a trap arm behind the vertical portions of this carrier? The width of the vertical portion plus the outside diameter of the trap arm will have to be less than 3½ inches. If the width is greater than this, you will have to install an additional stack directly behind the lavatory to avoid a conflict.

2.3.0 Locating and Verifying the Depth of the Building Drain and Sewer

You must verify the location and the depth of either the sewage disposal system's **sewer tap** or the inlet's invert elevation (also referred to simply as the invert). This information is critical to determine the piping route and the depth that it is installed inside the building.

For example, say that the plans call for a half-bath (toilet and lavatory) to be located in a basement. The fittings that must be stacked beneath the floor are a long-turn 90-degree ell and a sanitary tee with a closet bend attached. This combination of fittings requires 16 inches from the bottom of the 90-degree ell to the top of the closet bend (finished floor). The length of the run from this stack to the sewer inlet is 100 feet. You are using 3-inch pipe, which requires a slope of at least ¼ inch per foot. To calculate the depth this run of pipe requires, multiply the slope by the length of the run. This run of pipe would require 25 inches of depth:

¼ inch per foot × 100 feet = 25 inches

Add the required amount of slope to the height of the fittings used below the lowest floor with plumbing. The minimum depth required for the sewer tap from the finished floor of the basement is 41 inches:

16 inches (the combination of fittings) + 25 inches (depth) = 41 inches

2.4.0 Locating the Residential Water Closet

Consult the architectural floor plan to determine the approximate location of the water closet. On the bathroom floor plan illustrated in *Figure 5*, you can see that the water closet is to be located in the middle of the back wall of the bathroom. Ensure that the water closet is centered accurately from front to back and from side to side. Consult your local applicable plumbing or building code to determine the side-to-side measurement and the minimum permissible distances between the center lines of the plumbing fixtures. For this hypothetical installation, assume that the code specifies the minimum permissible distance from the finished wall (on each side) to the center line of the water closet as 15 inches.

Remember to consider the thickness of the wall-finishing materials when locating the water closet. As an example, say that the finished wall will consist of ¼-inch ceramic tile over ½-inch drywall. Therefore, you will need to add ¾ inch to the minimum permissible distance. The center line of the waste outlet from the rough framing will be 15¾ inches. Mark the wall parallel to the sidewall at that location (refer to *Figure 1*).

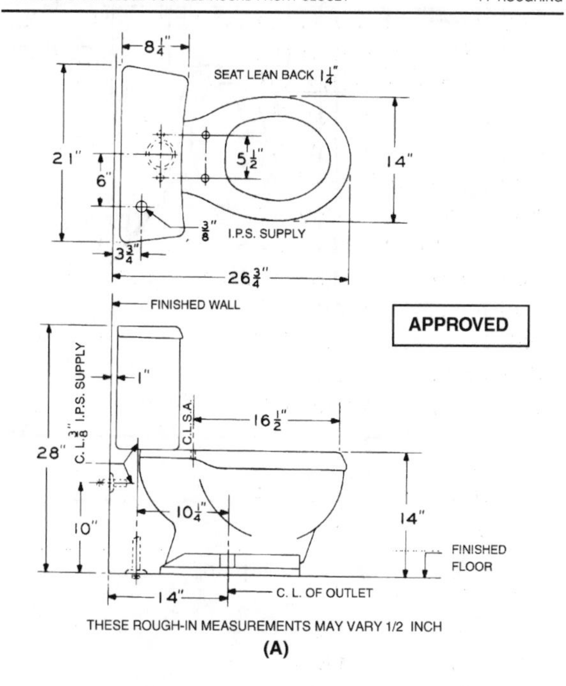

(A)

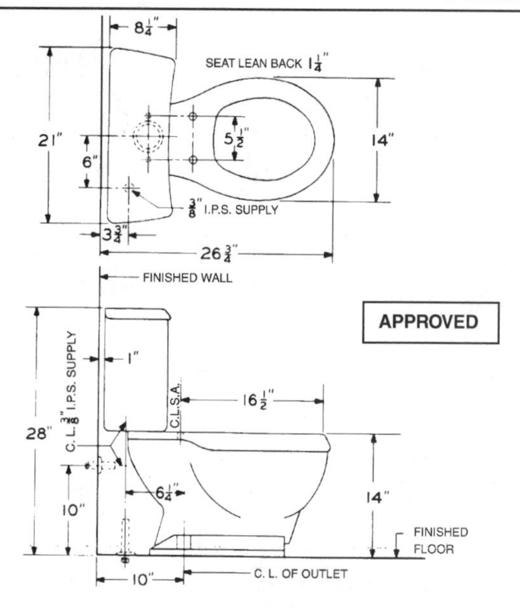

(B)

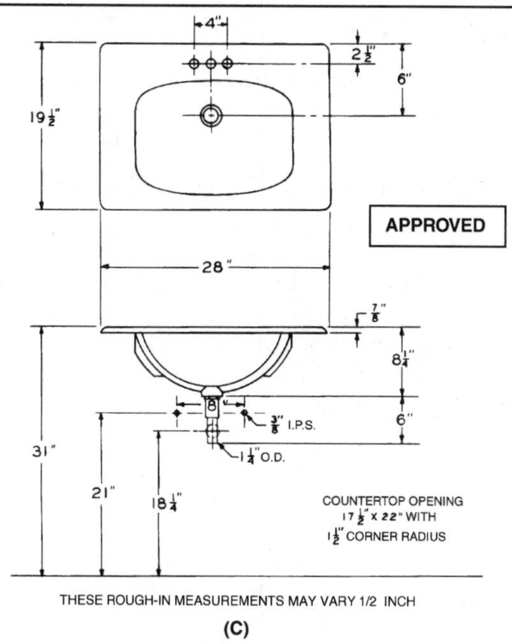

(C)

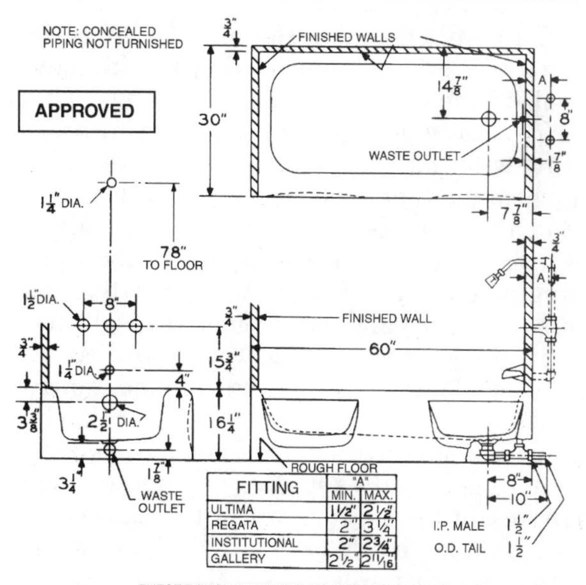

FITTING	"A" MIN.	"A" MAX.
ULTIMA	1 1/2"	2 1/2"
REGATA	2"	3 1/4"
INSTITUTIONAL	2"	2 3/4"
GALLERY	2 1/2"	2 11/16"

(D)

204F03.TIF

Figure 3 ◆ Rough-in measurements.

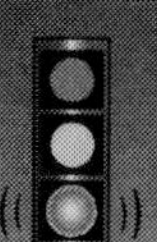

NOTE

In some cases, you may need to increase the wall size to fit the DWV pipe or its chase. In such cases, discuss the change with the general contractor before undertaking any changes.

DID YOU KNOW?

A Museum for Architecture

Your local code specifies the permitted ways to address conflicts with structural members such as joists. Codes will specify approved methods and measurements. Be sure to follow your local code requirements carefully when modifying structural members.

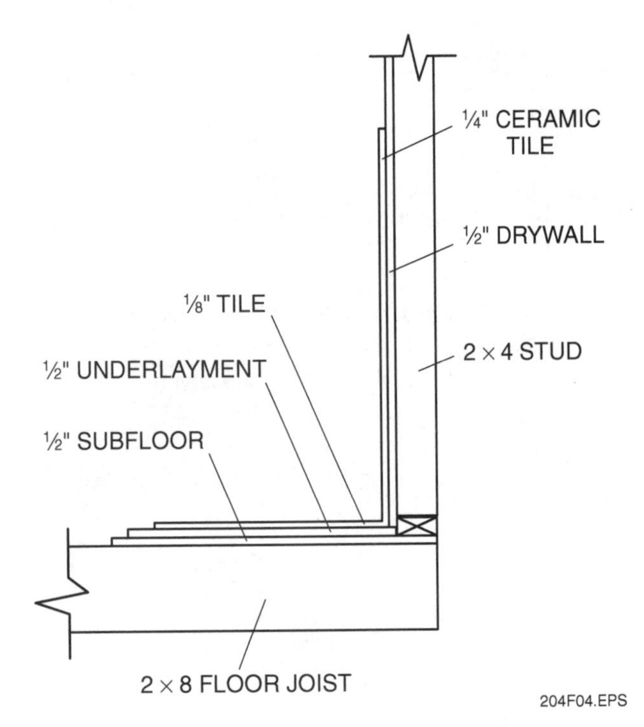

Figure 4 ◆ Thickness of wall-finishing materials.

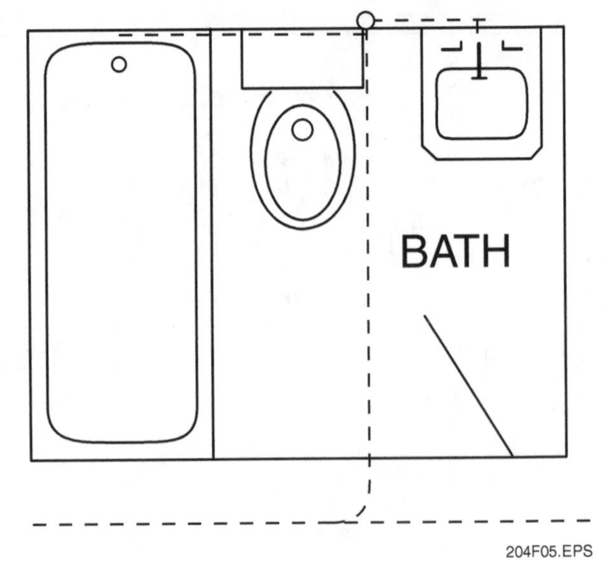

Figure 5 ◆ Bathroom floor plan.

Now, consult the manufacturer's rough-in measurements to determine the distance from the rear wall to the center of the closet flange. This will center the water closet from front to back. Referring to *Figure 3A*, you'll see that this distance is 14 inches, and that it includes a 1-inch clearance space at the rear. Because the wall is rough-framed, you must add the thickness of the wall-finishing material to this dimension. Say that the finishing material consists of ½-inch drywall covered with ¼-inch ceramic tile. You would add ¾ inch, making a total of 14¾ inches from the framed rear wall to the center line of the closet flange. Mark this measurement on the floor, perpendicular to the line already laid out. The two crossed lines represent the exact center of the location of the closet flange (see *Figure 6*). Cut the hole for the closet flange with a drill and scroll saw.

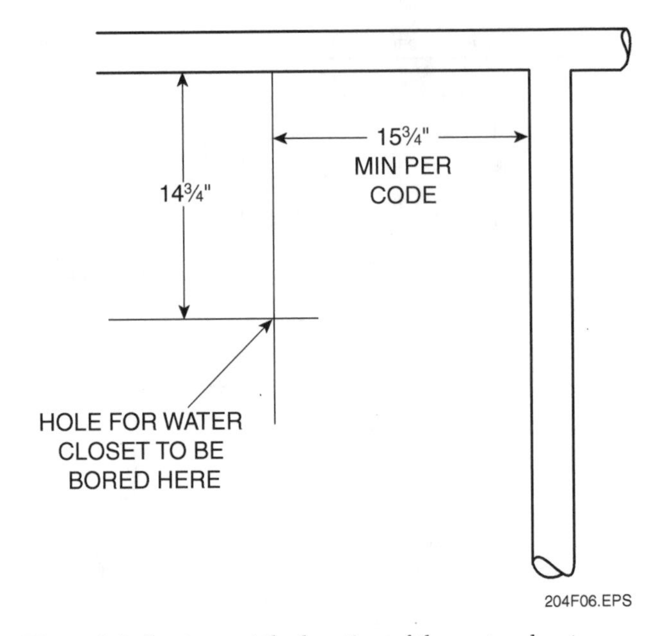

Figure 6 ◆ Laying out the location of the water closet.

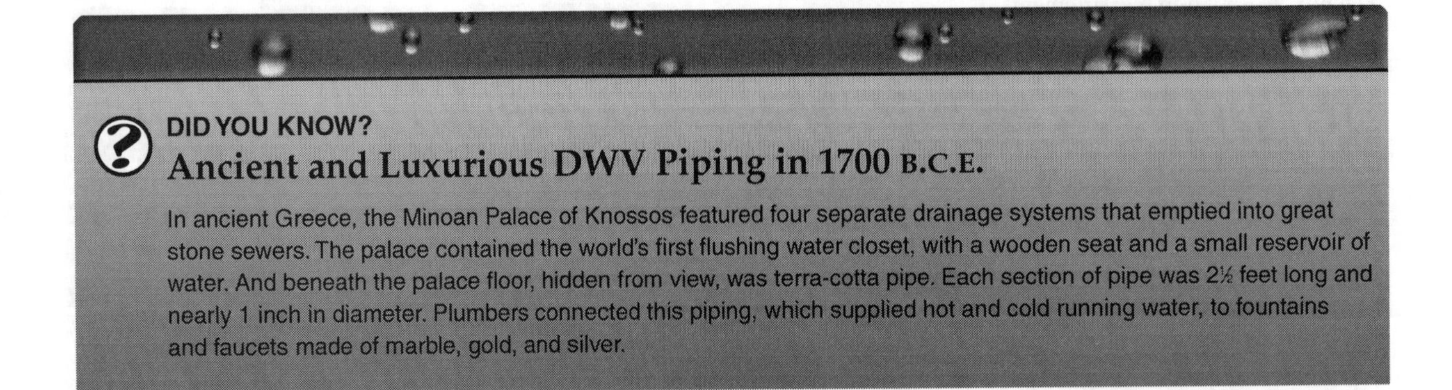

DID YOU KNOW?
Ancient and Luxurious DWV Piping in 1700 B.C.E.

In ancient Greece, the Minoan Palace of Knossos featured four separate drainage systems that emptied into great stone sewers. The palace contained the world's first flushing water closet, with a wooden seat and a small reservoir of water. And beneath the palace floor, hidden from view, was terra-cotta pipe. Each section of pipe was 2½ feet long and nearly 1 inch in diameter. Plumbers connected this piping, which supplied hot and cold running water, to fountains and faucets made of marble, gold, and silver.

2.5.0 Locating the Commercial Water Closet

Commercial water closets are often mounted on walls, or wall-hung, using special carrier fittings (see *Figure 7*) and special DWV piping. As a result, wall-hung water closets can be more difficult and more expensive to install. On the other hand, they allow easier access to the floor beneath the seat (see *Figure 8*). This makes them easier to clean and maintain than floor-mounted fixtures.

Because wall-hung water closets drain through the wall, you will not need to bore holes through the floor at each closet location. This feature can save time and labor when installing fixtures in buildings with concrete floors. In addition, wall-hung water closets do not require holes for the vent or false ceilings to conceal pipes. Carrier fittings for wall-hung water closets are available with either horizontal or vertical outlets, allowing you to select the type best suited for the installation.

Figure 7 ◆ Water closet carrier detail.

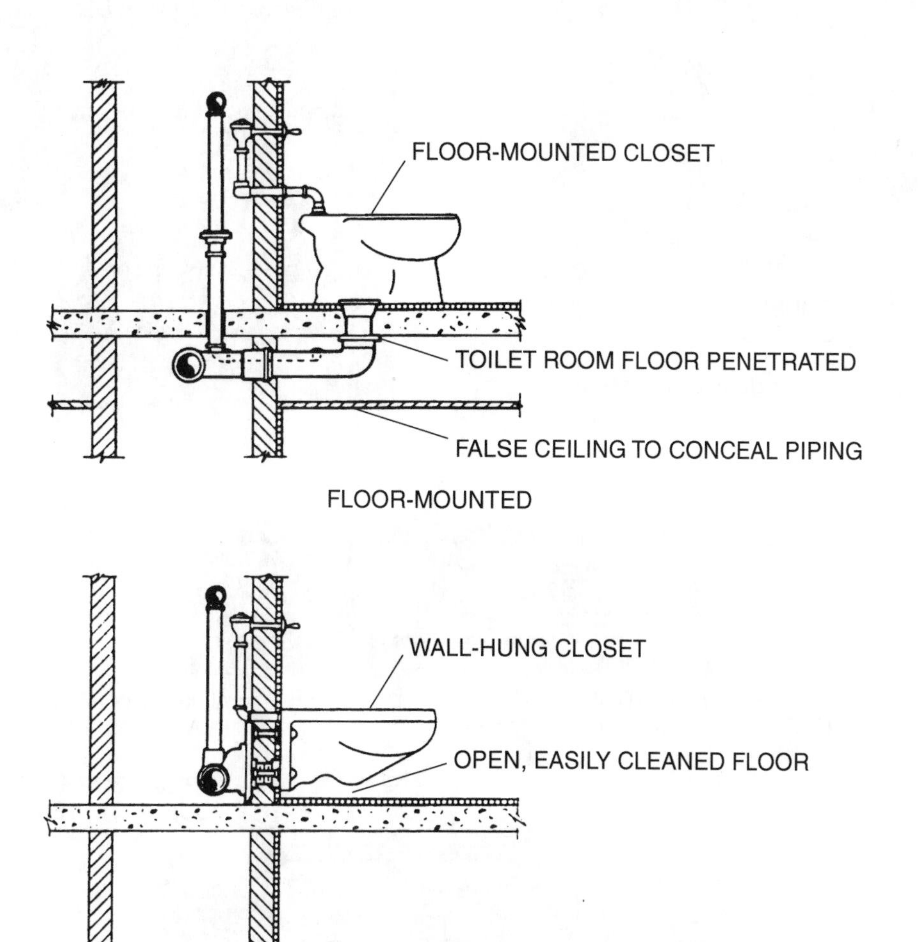

Figure 8 ◆ Comparison of floor-mounted and wall-hung water closets.

Commercial restrooms are often built back to back with a plumbing access area between them (see *Figures 9* and *10*). You can see from this figure how the components of the carrier assembly function together as a unit. Double-system units for back-to-back installations are used along with lateral pipe connections to span the width of the access area between restrooms. Auxiliary faceplates are applied to the lateral members to hook up the fixtures in the adjacent restrooms. Carriers are often **prefabricated** before delivery to the project site, which saves time and labor during the installation process.

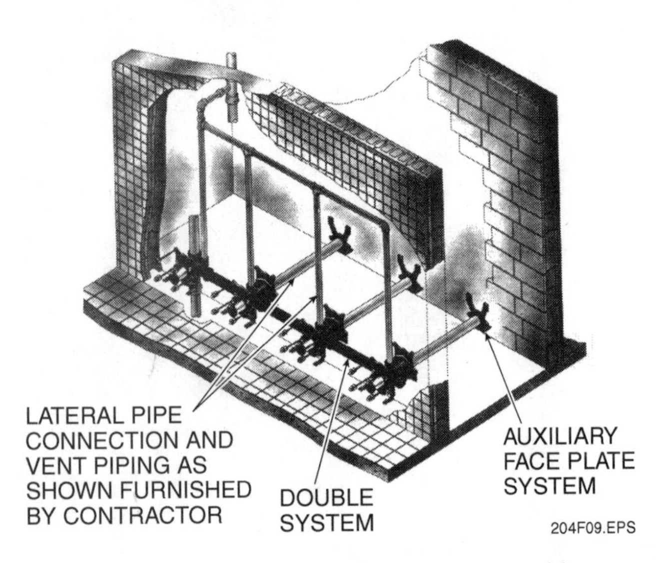

Figure 9 ◆ Back-to-back water closet carrier assembly.

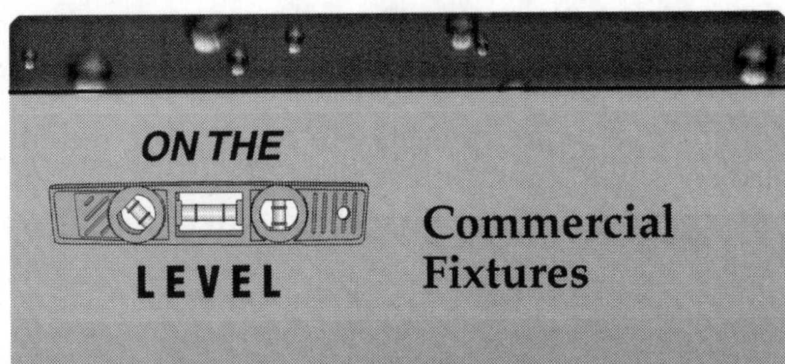

Commercial Fixtures

When roughing-in commercial fixtures, you must consider the possibility that fixtures may be abused or vandalized. To protect fixtures from this, therefore, you will need to provide additional structural strength for their mountings. One common method for protecting fixtures is to use carrier fittings.

The following factors will also play a part in selecting, locating, and installing commercial fixtures:

- Ease of maintenance
- Ease of installation
- Cost
- Appearance

Specific installation requirements and rough-in dimensions depend entirely on the make and model of the carriers that are being used. Just as with residential fixtures, you should refer to the manufacturers' specification sheets before attempting to lay out the piping to these fixtures.

Prefabricated Carriers

Fixture carrier fittings for commercial applications are often prefabricated prior to installation. Prefabrication is a more efficient method than assembling components for each installation on the project site. Because of the significant time and money savings that result from using them, prefabricated carriers are widely used on commercial projects. (See *Figure S-1*.)

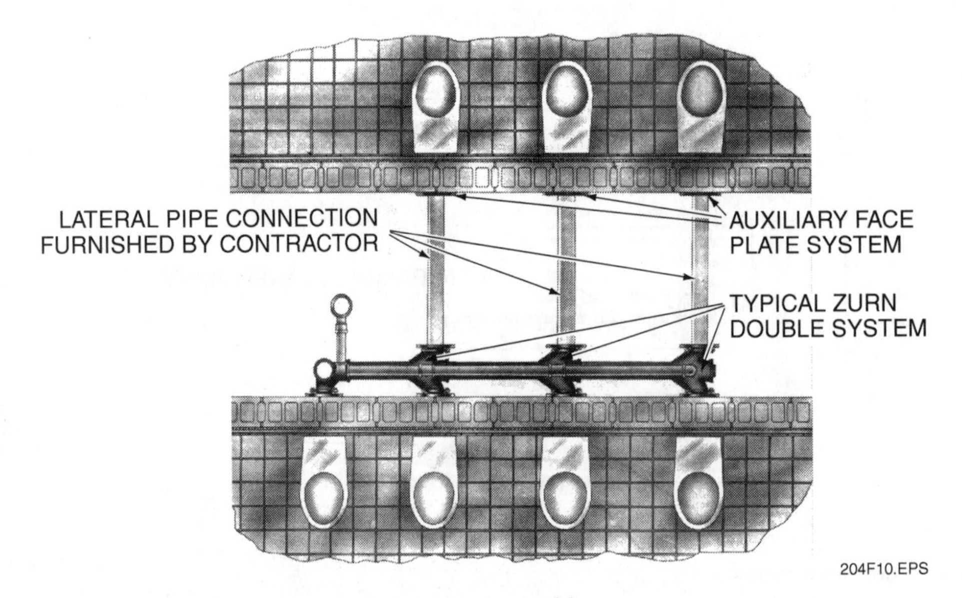

Figure 10 ◆ Plan view of a back-to-back water closet carrier assembly.

Prefabricated Carriers (continued)

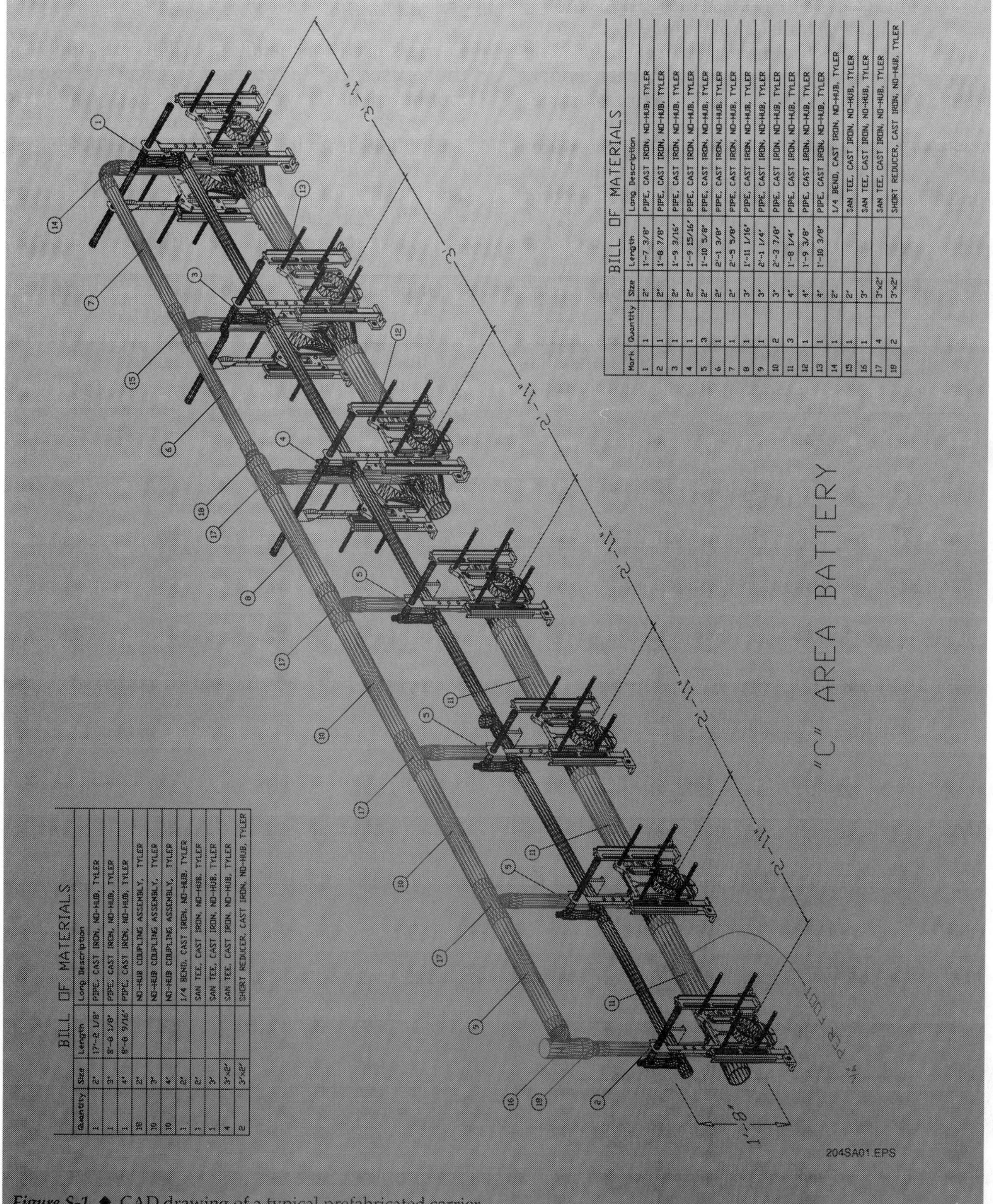

BILL OF MATERIALS

Quantity	Size	Length	Long Description
1	2"	17'-2 1/8"	PIPE, CAST IRON, NO-HUB, TYLER
1	3"	8'-8 1/8"	PIPE, CAST IRON, NO-HUB, TYLER
1	4"	8'-8 9/16"	PIPE, CAST IRON, NO-HUB, TYLER
18	2"		NO-HUB COUPLING ASSEMBLY, TYLER
10	3"		NO-HUB COUPLING ASSEMBLY, TYLER
10	4"		NO-HUB COUPLING ASSEMBLY, TYLER
1	2"		1/4 BEND, CAST IRON, NO-HUB, TYLER
1	2"		SAN TEE, CAST IRON, NO-HUB, TYLER
1	3"		SAN TEE, CAST IRON, NO-HUB, TYLER
4	3"x2"		SAN TEE, CAST IRON, NO-HUB, TYLER
2	3"x2"		SHORT REDUCER, CAST IRON, NO-HUB, TYLER

BILL OF MATERIALS

Mark	Quantity	Size	Length	Long Description
1	1	2"	1'-7 3/8"	PIPE, CAST IRON, NO-HUB, TYLER
2	1	2"	1'-8 7/8"	PIPE, CAST IRON, NO-HUB, TYLER
3	1	2"	1'-9 3/16"	PIPE, CAST IRON, NO-HUB, TYLER
4	1	2"	1'-9 15/16"	PIPE, CAST IRON, NO-HUB, TYLER
5	3	2"	1'-10 5/8"	PIPE, CAST IRON, NO-HUB, TYLER
6	1	2"	2'-1 3/8"	PIPE, CAST IRON, NO-HUB, TYLER
7	1	2"	2'-5 5/8"	PIPE, CAST IRON, NO-HUB, TYLER
8	1	3"	1'-11 1/16"	PIPE, CAST IRON, NO-HUB, TYLER
9	1	3"	2'-1 1/4"	PIPE, CAST IRON, NO-HUB, TYLER
10	2	3"	2'-3 7/8"	PIPE, CAST IRON, NO-HUB, TYLER
11	3	4"	1'-8 1/4"	PIPE, CAST IRON, NO-HUB, TYLER
12	1	4"	1'-9 3/8"	PIPE, CAST IRON, NO-HUB, TYLER
13	1	4"	1'-10 3/8"	PIPE, CAST IRON, NO-HUB, TYLER
14	1	2"		1/4 BEND, CAST IRON, NO-HUB, TYLER
15	1	2"		SAN TEE, CAST IRON, NO-HUB, TYLER
16	1	3"		SAN TEE, CAST IRON, NO-HUB, TYLER
17	4	3"x2"		SAN TEE, CAST IRON, NO-HUB, TYLER
18	2	3"x2"		SHORT REDUCER, CAST IRON, NO-HUB, TYLER

Figure S-1 ◆ CAD drawing of a typical prefabricated carrier

The dimensions of the chase between two restrooms will vary. The wall thickness, on the other hand, will be determined by the back-to-back water closet carrier that you are using. The manufacturer's specifications will provide these dimensions (see *Figure 11*). The dimensions for the model illustrated, for example, range from a minimum of 2¼ inches to a maximum of 6¼ inches. The ranges will vary depending on the carrier model.

Some carrier models feature invertible faceplates (see *Figure 12*). These faceplates are designed to ensure level placement of the water closet while maintaining a pitch in the drainage line. The maximum difference in the elevation of the fixture port in the model shown is 4¼ inches.

The project engineer will provide you with the necessary specifications for plumbing installations that involve a battery of wall-hung water closets. Carefully review the working drawings and the manufacturer's specifications before installing these items. Careful and correct installation practices will ensure that the assembly functions properly.

2.6.0 Locating Urinals and Other Fixtures

When locating other fixtures, such as urinals, water coolers, and corner sinks, note that the rough-in dimensions will vary depending on the fixture model being used. You will have little room for error when you rough-in these types of fixtures. You should therefore refer to the manufacturer's specifications when installing them.

2.6.1 Urinals

Carriers used for installing urinals are similar to those used for lavatories. Typically, a urinal mount will have two cross members or plates (see *Figure 13*). The upper cross member is designed to accept bolts from the fixture itself. The lower cross member is used to mount the base of the urinal. The design of the cross members will vary depending on the model of urinal being installed. Carriers also are available with adjustable couplings (see *Figure 14*).

Urinal carriers may also include a mount for the water supply piping. This mount ensures that the water supply piping is correctly located so that the flush valve can be installed properly. It also holds the water supply piping firmly in place to prevent damage to the supply piping or the valve.

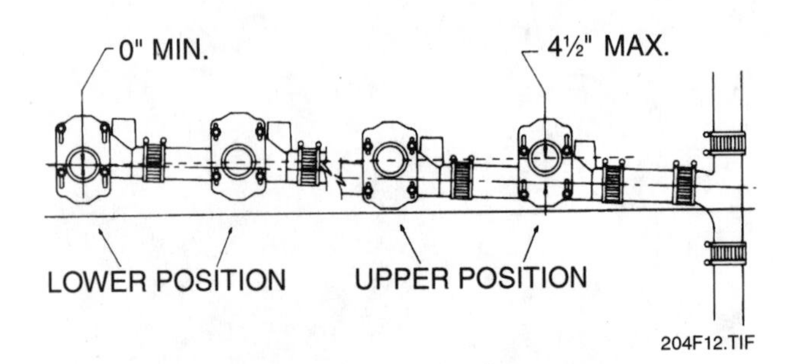

Figure 12 ◆ Carrier with invertible faceplates.

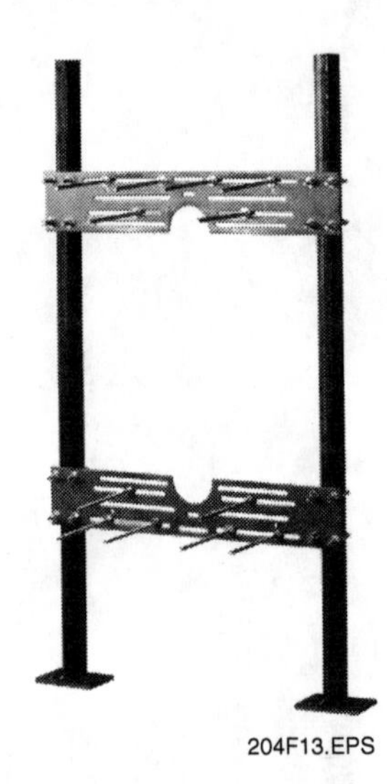

Figure 13 ◆ Double-plate, floor-mounted urinal carrier.

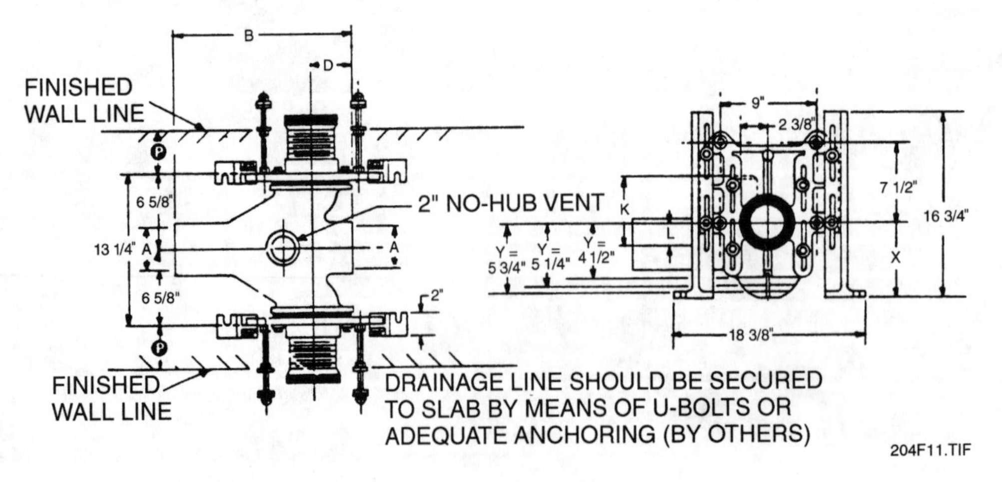

Figure 11 ◆ Manufacturer's specifications for a back-to-back water closet carrier.

204F14.EPS

Figure 14 ◆ Urinal carrier with adjustable coupling.

2.6.2 Perineal Baths

Perineal baths, also called sitz baths, are therapeutic bathtubs used in the treatment of injuries and diseases in the groin area (see *Figure 15*) They are found primarily in hospitals and other therapeutic settings. Wall-hung perineal baths require a special carrier (see *Figure 16*). These carriers are floor-mounted and are equipped with a second hanger for the top portion of the fixture.

2.6.3 Water Coolers

Wall-mounted water coolers (see *Figure 17*) also require carriers. The carriers are available in a variety of dimensions to accommodate the wide range of cooler sizes. Use two hangers to properly secure water coolers to a wall. Side-by-side and back-to-back carriers are also available.

2.6.4 Corner Sinks

Corner sinks require secure structural support. Most commercial corner sinks can be wall mounted to provide easy cleaning and maintenance. Ensure that the carriers you order are compatible with the fixtures planned for the installation.

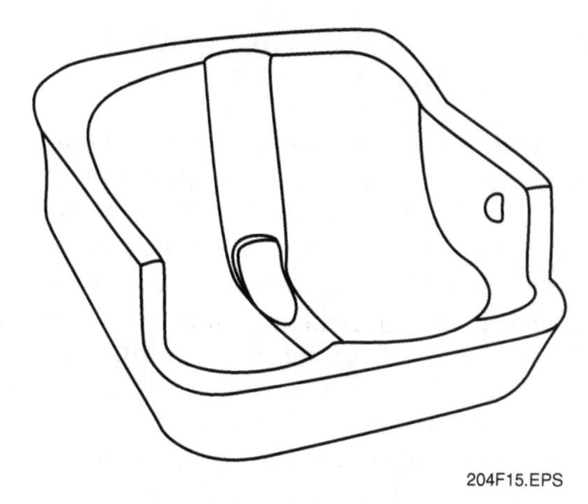

204F15.EPS

Figure 15 ◆ Perineal bath.

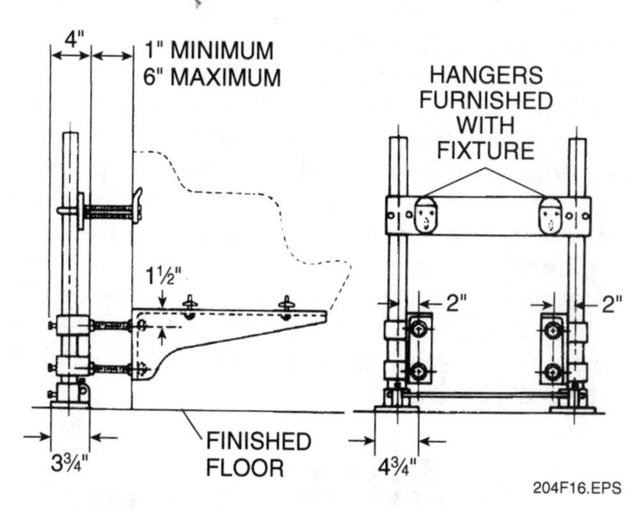

204F16.EPS

Figure 16 ◆ Perineal bath carrier.

204F17.EPS

Figure 17 ◆ Wall-mounted water cooler.

2.7.0 Locating the Residential Lavatory

Refer to the bathroom floor plan in *Figure 18* to determine the approximate location of the lavatory. Notice that the lavatory will be located in the right rear corner of the bathroom. If the lavatory will be inserted into a bathroom countertop, you will need to determine the location of the lavatory within the countertop before installing it.

Begin by determining the overall width and length of the lavatory. You can find this information on the rough-in measurements (refer to *Figure 3C*). Ideally, the lavatory should be installed with adequate front, back, and side clearance. This information is rarely provided on the construction plans. Plumbers generally rely on experience to determine proper placement of the lavatory.

Next, mark the exit point of the drainpipe. The rough-in measurements indicate that the drain is centered in the lavatory. That center is the point to be marked. However, because at this point the structure has only been roughed-in, plumbers usually place a mark on the **soleplate** to indicate the center of the lavatory drain to show where it will pass inside the framed wall. A soleplate is a horizontal framing member along the floor that acts as the base for wall studs.

3.0.0 ◆ BASIC FRAMING FOR LAVATORIES AND SINKS

In most residential and light commercial structures, basic framing usually accommodates plumbing installations without any unusual problems. You can construct a chase to house piping, or install various forms of **blocking** to anchor the screws that will hold lavatory brackets when installed (see *Figure 19*). Blocking is pieces of wood placed between and connected to studs.

The fixture rough-in procedures described in this section are the most basic and most often used. Refer to your local applicable code to determine if there are additional requirements in your area. Always follow the manufacturer's specifications closely. Prefabricate fittings wherever possible to save time.

3.1.0 Locating Commercial Lavatories and Sinks

Commercial wall lavatories and sinks usually require metal carriers that are located behind a wall (see *Figure 20*). Lavatory and sink carriers are adjustable. The arms may be concealed or exposed, depending on the model.

Most commercial carriers come with leveling screws, which allow you to ensure the fixture is level when installed on an uneven floor. Some lavatory carriers are designed to be mounted directly to wall surfaces. These carriers do not use vertical uprights. Instead, they are equipped with metal plates for mounting on wood or metal stud walls or masonry walls (see *Figure 21*). Though these carriers are convenient, they have poor structural strength.

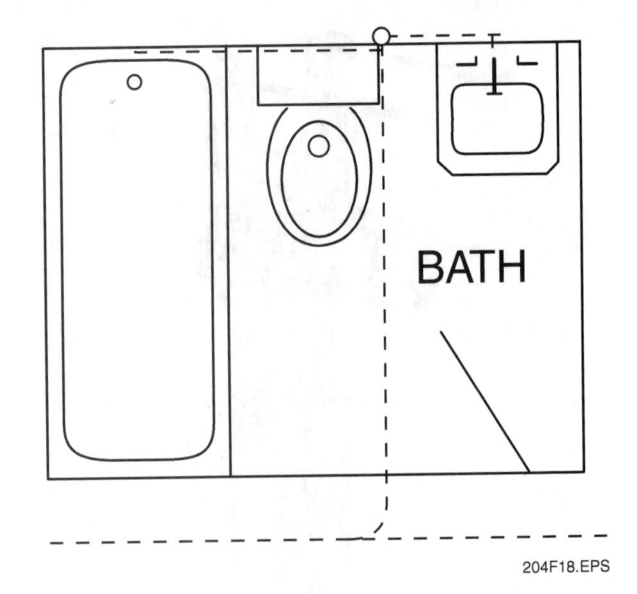

Figure 18 ◆ Bathroom floor plan.

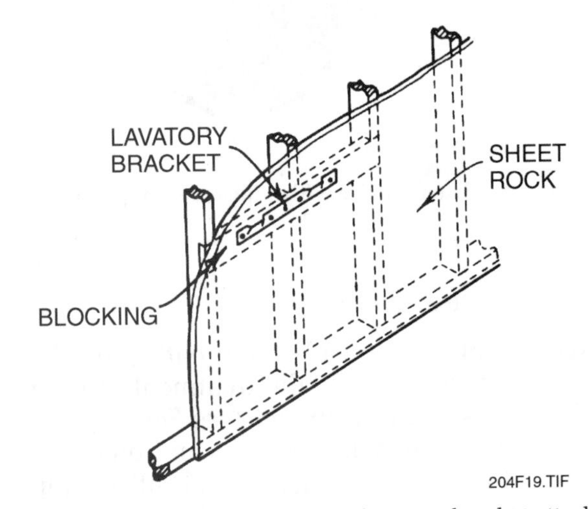

Figure 19 ◆ Wall section showing lavatory bracket attached to blocking.

Lavatory carriers are also available for back-to-back installations (see *Figure 22*). Because back-to-back lavatory installations share the same hardware, you can realize significant savings in costs and time. The headers (arm mounts) on each side of the carrier are designed to mesh so that lavatory heights can be equal in both rooms.

Sink carriers are available in a wide variety of designs to accommodate different sink models and manufacturers. The primary difference between lavatory carriers and sink carriers is their structural strength. Because sinks are generally

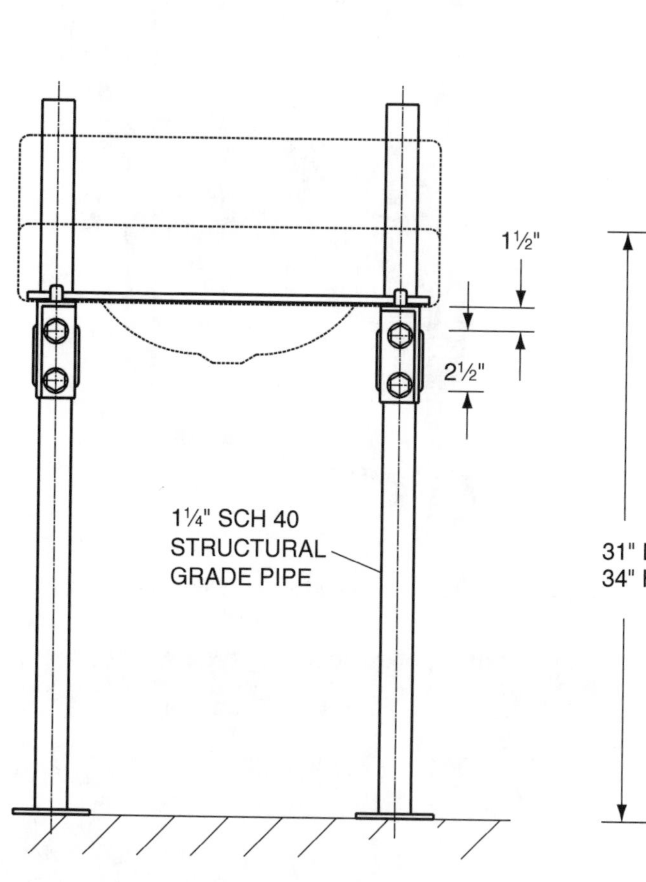

Figure 20 ◆ Lavatory carrier.

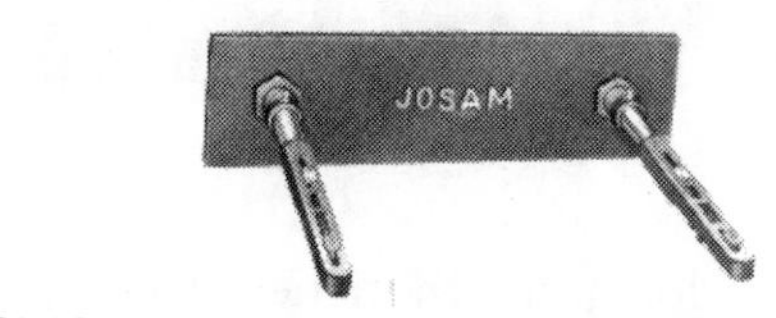

SINGLE LAVATORY CARRIER FOR STUD WALL

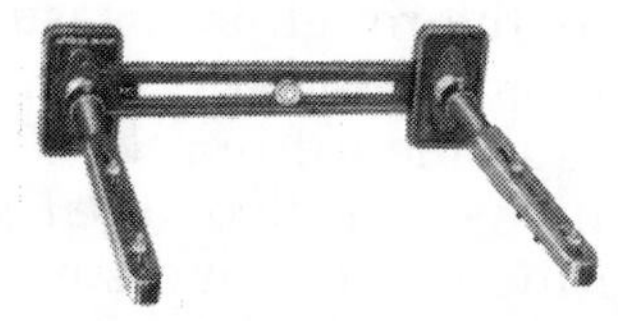

SINGLE LAVATORY CARRIER FOR MASONRY WALL

204F21.EPS

Figure 21 ◆ Single lavatory carriers for stud and masonry walls.

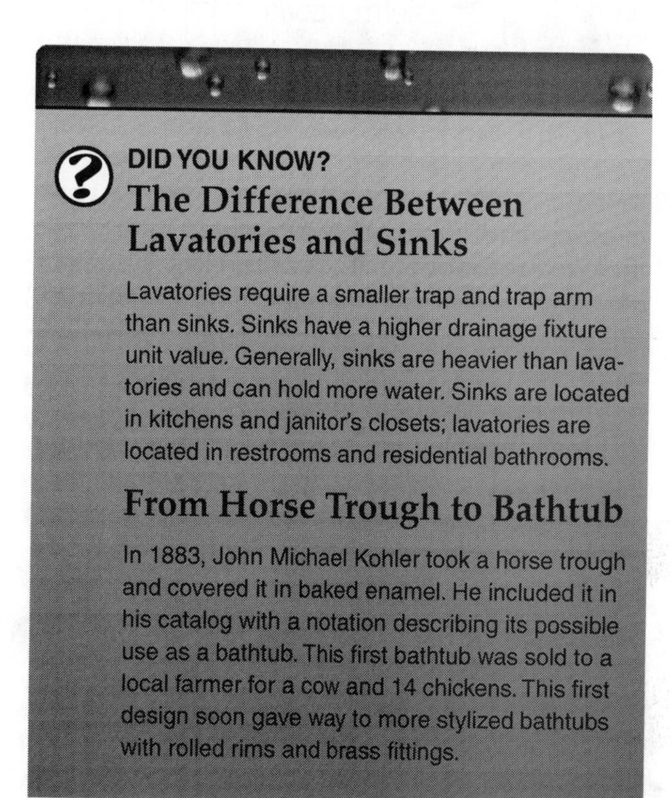

DID YOU KNOW?

The Difference Between Lavatories and Sinks

Lavatories require a smaller trap and trap arm than sinks. Sinks have a higher drainage fixture unit value. Generally, sinks are heavier than lavatories and can hold more water. Sinks are located in kitchens and janitor's closets; lavatories are located in restrooms and residential bathrooms.

From Horse Trough to Bathtub

In 1883, John Michael Kohler took a horse trough and covered it in baked enamel. He included it in his catalog with a notation describing its possible use as a bathtub. This first bathtub was sold to a local farmer for a cow and 14 chickens. This first design soon gave way to more stylized bathtubs with rolled rims and brass fittings.

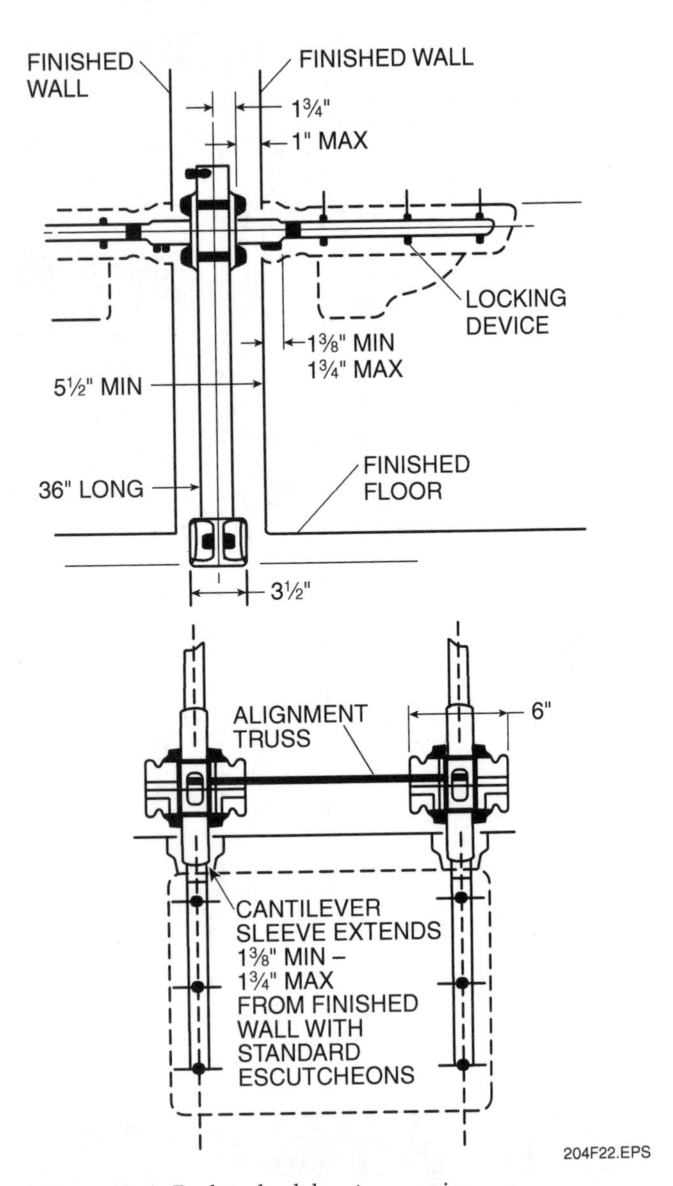

Figure 22 ◆ Back-to-back lavatory carrier.

heavier than lavatories and can hold more water, they require additional structural support. Floor-mounted carriers offer a greater load-bearing capacity because they transfer the weight of the sink directly to the floor.

Some floor-mounted sink carriers include a waste line bracket as part of the carrier assembly (see *Figure 23*). These carriers provide support for the sink as well as for the drain line. With a waste line bracket, you do not need to construct additional supports for the DWV piping.

3.2.0 Locating and Roughing-In Bathtubs and Showers

Refer back to the bathroom floor plan in *Figure 18*. Notice that the bathtub is located on the wall opposite the water closet. Refer to the rough-in measurements in *Figure 3D*. You'll see that the bathtub drain is located 14⅞ inches from the side wall. Because of their size, bathtubs are fitted into place before the finished walls are up. The finished wall will rest on the upper rim of the bathtub. Therefore, the rough-in measurements are made from the rough-framed wall. Lay out the 14⅞-inch measurement from the side wall and mark the position on the floor (see *Figure 24*).

Consulting the rough-in measurements again, you'll see that the bathtub drain is located 7⅞ inches from the rough-framed wall at the head of the bathtub. Lay out this measurement on the floor as well. The crossed lines where the two measurements intersect represent the center of the bathtub drain.

Figure 23 ◆ Floor-mounted sink carrier with waste fitting.

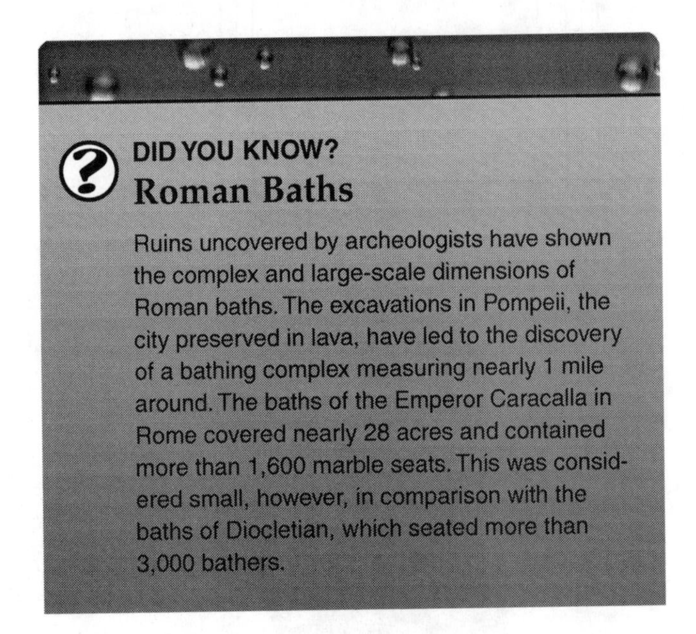

DID YOU KNOW?

Roman Baths

Ruins uncovered by archeologists have shown the complex and large-scale dimensions of Roman baths. The excavations in Pompeii, the city preserved in lava, have led to the discovery of a bathing complex measuring nearly 1 mile around. The baths of the Emperor Caracalla in Rome covered nearly 28 acres and contained more than 1,600 marble seats. This was considered small, however, in comparison with the baths of Diocletian, which seated more than 3,000 bathers.

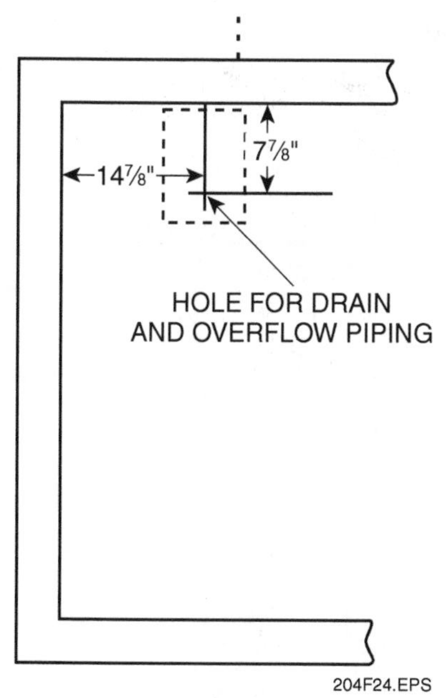

Figure 24 ◆ Locating the bathtub drain.

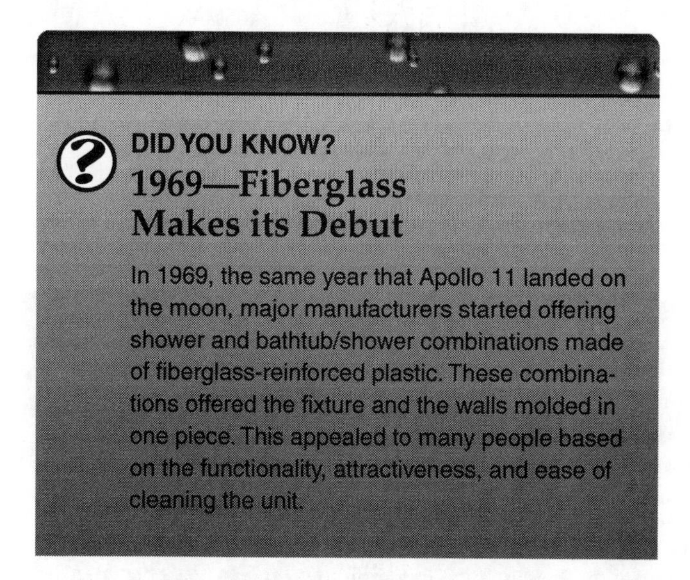

DID YOU KNOW?
1969—Fiberglass Makes its Debut
In 1969, the same year that Apollo 11 landed on the moon, major manufacturers started offering shower and bathtub/shower combinations made of fiberglass-reinforced plastic. These combinations offered the fixture and the walls molded in one piece. This appealed to many people based on the functionality, attractiveness, and ease of cleaning the unit.

Before cutting the drain hole in the floor, consider the distance from the bathtub overflow pipe to the bathtub drain. The hole must be large enough to provide clearance for both of these. Also, you will need to provide clearance so that you'll have enough room to use pipe wrenches to make up the joints. Typically, the hole will need to be approximately 12 inches long and 8 to 12 inches wide to accommodate these requirements.

One-piece fiberglass tub and shower fixtures require special attention. These fixtures are too large to be installed after framing is completed. They must be in place during construction. Ensure that the opening for these fixtures is properly framed, because the dimensions are critical. Fixture rough-in measurements do not include interior wall coverings such as gypsum core wallboard, commonly referred to by its trade name, **Sheetrock®**, and these measurements must be exact from stud wall to stud wall.

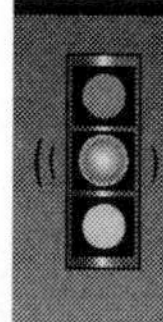

CAUTION

Fiberglass tub and shower enclosures are susceptible to damage from exposure to sunlight and weather, and from accidental impacts. Use the fixture's shipping carton to protect the item during construction.

Often, additional 2 × 4 framing is required to support the flexible upper portions of the shower enclosure. Provide the additional framing before the Sheetrock® is applied to the opposite side of the wall. Although the carpentry contractor will perform most of this work, a good plumber will oversee the rough-in work to be sure that the job is done correctly.

3.3.0 Locating Fixtures for the Physically Challenged

Special considerations apply when roughing-in fixture installations for physically challenged individuals. For example, the locations for the opening and mounting bolts on wall-mounted water closets are different. Toilets in horizontal battery installations must be at the upper end of the piping run, so that the higher rough-in required for the special fixture can be accommodated.

Figure 25 illustrates a typical bathroom for the physically challenged. Note that the room's dimensions include a 5-foot turning circle so that a wheelchair can turn easily. The final dimensions required are illustrated for both fixtures. You must adjust fixture rough-ins so that they comply with these dimensions.

Lavatories and water cooler carriers are available for wheelchair access. The fixtures themselves have shallow features, and the support arms from the carriers are hidden and extend out farther to give support to the fixture.

Be sure to refer to the American National Standards Institute (ANSI) code and your local code for provisions that apply to physically challenged individuals.

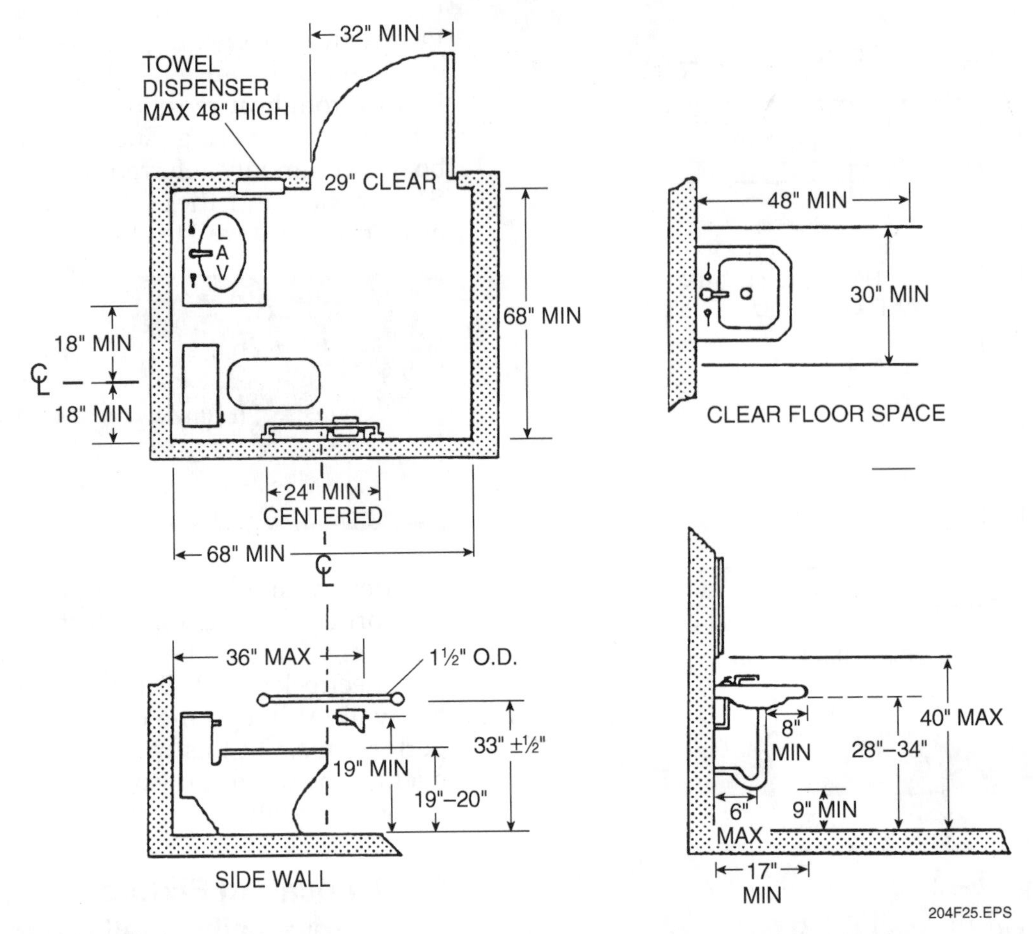

Figure 25 ◆ Fixture installation for the physically challenged.

ANSI/ICC A117.1, Accessible and Usable Buildings and Facilities

ANSI based its standard *A117.1, Accessible and Usable Buildings and Facilities*, on the Americans with Disabilities Act (ADA). The International Code Council (ICC) has also adopted this standard for its codes. In the construction industry, *ANSI A117.1* and the ADA are often referred to interchangeably.

ANSI/ICC A117.1, Accessible and Usable Buildings and Facilities

1 Purpose and Application

1.1 Purpose

The specifications in this standard make buildings and facilities accessible to and usable by people with such physical disabilities as: the inability to walk, difficulty walking, reliance on walking aids, blindness and visual impairment, deafness and hearing impairment, incoordination, reaching and manipulation disabilities, lack of stamina, difficulty interpreting and reacting to sensory information, and extremes of physical size based generally upon adult dimensions. Accessibility and usability allow a person with a physical disability to independently get to, enter, and use a building or facility.

This standard provides specifications for elements that are used in making functional spaces accessible. For example, it specifies technical requirements for making doors, routes, seating, plumbing fixtures, and other elements accessible. These accessible elements are used to design accessible functional spaces such as classrooms, bathrooms, hotel rooms, lobbies, or offices.

This standard is for adoption by government agencies and by organizations setting model codes to achieve uniformity in the technical design criteria in building codes and other regulations. This standard is also used by nongovernmental parties as technical design guidelines or requirements to make buildings and facilities accessible to and usable by persons with physical disabilities.

Review Questions

Sections 1.0.0–3.0.0

1. An estimator lists the type and quantity of plumbing supplies needed for a job on the _____.
   a. architect's plans
   b. material takeoff
   c. elevation drawing
   d. building plan

2. The approximate location of fixtures usually is determined by the _____.
   a. number of fixtures in the building
   b. manufacturer's catalog
   c. local building code
   d. floor plan

3. A residence with a basement half-bath requires a 90-degree ell and a sanitary tee that need 16 inches of depth below the finished floor. The sewer inlet is 60 feet from the stack, and a slope of at least ¼ inch per foot is required. The minimum depth of the sewer tap below the finished basement floor is _____ inches.
   a. 15
   b. 16
   c. 31
   d. 41

4. You must accurately center a water closet _____.
   a. from front to back
   b. from side to side
   c. from front to back and side to side
   d. along the center line of the main building stack

5. Wall-hung water closets require vent holes.
   a. True
   b. False

6. On a urinal mount, the _____ cross member accepts bolts from the fixture itself and the _____ cross member is used to mount the base of the urinal.
   a. lower; upper
   b. left; right
   c. upper; lower
   d. right; left

7. Wall-hung perineal baths require a special _____ carrier.
   a. floor-mounted
   b. wall-hung
   c. stand-alone
   d. three-armed

8. To properly secure a water cooler to a wall, use _____ hanger(s).
   a. one
   b. two
   c. three
   d. four

9. Place a mark on the _____ to indicate where the center of the lavatory drain will pass inside the framed wall.
   a. chase
   b. footer
   c. construction drawings
   d. soleplate

10. Be sure to refer to the _____ for provisions that apply to physically challenged individuals.
    a. ANSI standards and the local code
    b. ANSI and OSHA standards
    c. Uniform Plumbing Code
    d. manufacturer's rough-in sheets

4.0.0 ◆ GRADE

DWV piping systems rely on gravity to move solid and liquid wastes, so they must be installed at a slope toward the point of disposal. As you have already learned, this slope is called the grade. DWV systems are designed with the grade engineered into the system. Architects and engineers who design a piping system determine the grade. However, in some cases, such as in residential plumbing, the plumber specifies the grade.

4.1.0 The Importance of Grade

In a system with the proper grade, the velocity (speed) of the flowing liquid wastes will scour the insides of the pipe and carry solids away. If the grade is too steep, the liquid wastes may flow too fast, leaving the solids behind. If the grade is too shallow, the liquid wastes will not flow fast enough to scour the pipe and remove the solid wastes. In either case, solid wastes will soon obstruct the pipe.

Unnecessary changes to the grade of a line of pipe will cause the pipe to suffer blockages in use. For this reason, once the proper grade for a DWV system has been established, you must ensure that it is held constant while the system is being built.

4.2.0 Sources of Grade Information

You can obtain information about the proper grade for the job from local plumbing codes or job specifications, or directly from the construction plans. Local plumbing codes specify the required grade for various pipe sizes and plumbing applications. Codes may also require the grade to be held constant throughout the plumbing job. If the grade is not specified on the construction drawings or the project specifications or in your local code, contact the local plumbing inspector to get this information.

4.2.1 Velocity Tables

Local codes may require that the flow through a piping system maintain a minimum velocity, usually 2 feet per second (fps). Usually, the code includes a velocity table (see *Figure 26*) that you can use to determine the approximate velocity of flow through a pipe system of a given size installed with a given fall. Notice that a 2-inch pipe requires a fall of ¼ inch per foot to obtain a velocity of 2 fps while an 8-inch pipe requires a fall of only ¹⁄₁₆ inch per foot to obtain the same velocity.

4.2.2 Specifications

Grade information is provided in the project specifications. You learned about specifications in *Plumbing Level One*. To determine the grade from the specifications, refer to the section related to plumbing. Under that should be a section that covers either the storm water system or the sanitary sewer system, depending on the project. Look for a section titled *Grades and Elevations*. This section will specify the grades and elevations you must use for the job, along with other important information that you should also read and follow.

4.2.3 Construction Plans

Grade information is often included in the project's architectural plans. Plot plans, which are included in the set of prints, show the location of the structure on the lot as well as dominant features and the elevation at various points on the lot (see *Figure 27*). On the plot plan you can find the finished elevation of the sewer manhole cover, the invert elevation, and the slope of the surrounding terrain. The invert elevation is the lowest elevation on the project site through which liquid will flow, such as the lowest inside surface of a drainage channel, sewer pipe, or drain.

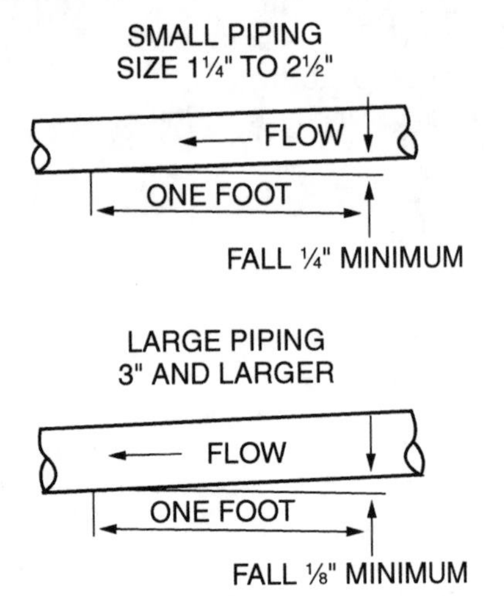

APPROXIMATE VELOCITY IN FT/SEC
AT INDICATED FALL/FT

Pipe Size	¹⁄₁₆ in/ft	⅛ in/ft	¼ in/ft	½ in/ft
2 in	1.02	1.44	2.03	2.88
3 in	1.24	1.76	2.49	3.53
4 in	1.44	2.03	2.88	4.07
5 in	1.61	2.28	3.53	4.56
6 in	1.76	2.49	4.07	5.00
8 in	2.03	2.88	4.23	5.75
10 in	2.28	3.23	4.56	6.44

204F26.EPS

Figure 26 ◆ Relationship between grade and velocity.

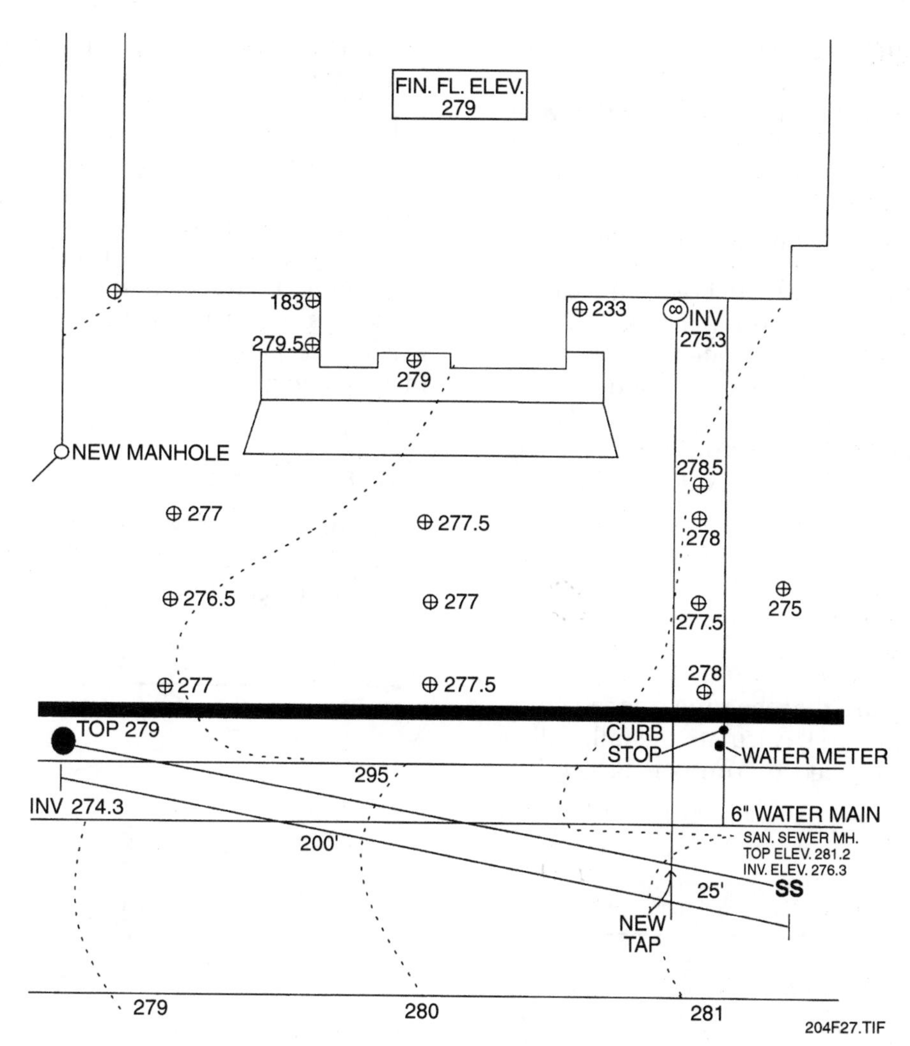

Figure 27 ◆ Typical plot plan.

4.2.4 Elevations

To determine the elevation, locate a **bench mark**, or reference point, at the project site. Bench marks are permanent markers set at specific elevations. You can mark the elevation directly on top of the bench mark. You and others on the job site will use the bench mark as a reference. In situations where a bench mark cannot be located, some other point of known elevation must be located. That point is often located from construction drawings.

To make reading and referring to local elevations easier, your county engineer or surveyor has already established local bench marks, which are usually set at 0' or 100'. In addition, project architects often specify a print bench mark. The print bench mark is referenced to either a local or national bench mark. This information can greatly simplify the task of reading a set of construction drawings.

4.3.0 Calculating Grade

Grade is the slope or fall of a line of pipe in reference to a horizontal plane. Plumbers use grades from ⅛ inch per foot to ½ inch per foot. Designers specify grades such as $\frac{1}{16}$ inch per foot and $\frac{1}{32}$ inch per foot for large pipelines and other special applications. Grade is expressed as vertical fall in fractions of an inch per foot of horizontal run. For example, a grade of ¼ inch means that the line of pipe will fall ¼ inch for every foot of run. The fall is the total change in elevation for the length of pipe. The run is the horizontal length of the run of pipe.

4.3.1 Relationship of Grade, Fall, and Run

When you work with grade (G), fall (F), and run (R), remember the following:

- Grade is expressed in inches per foot.
- Fall is expressed in inches.
- Run is expressed in feet.

Depending upon which of the components you need to calculate, use one of the following formulas:

- Grade = Fall ÷ Run (G = F/R)
- Fall = Grade × Run (F = G × R)
- Run = Fall ÷ Grade (R = F/G)

Consider the following example (see *Figure 28*). The horizontal run of a line of pipe is 40 feet. Its fall is 10 inches. What is the grade for this run of pipe?

G = F/R

G = 10"/40' = 0.250 or ¼" per foot

Therefore, the grade for a 40-foot horizontal run of pipe with a 10-inch fall is ¼ inch per foot.

Note that the answer first appears in decimal form. You must convert this to a fraction. You could use a conversion table (see *Table 1*). However, your job will be easier if you memorize the common fractional decimal equivalents. They are as follows:

1/16 = 0.0625 ⅛ = 0.125

¼ = 0.250 ⅜ = 0.375

½ = 0.500

Use grade to ensure that the pipe trench is dug at the correct depth. First, calculate the fall. Fall is equal to grade multiplied by run (see *Figure 29*). The line of pipe runs a horizontal distance of 16 feet at a grade of ¼ inch per foot.

F = G × R

F = ¼" per foot × 16' = 4"

Therefore, the fall is 4 inches.

Once you know both grade and fall, you can determine the horizontal run of the pipe without actually measuring. Run equals fall divided by grade (see *Figure 30*). The fall in this figure is 8 inches. The grade is ¼ inch per foot.

R = F/G

R= 8" ÷ ¼" per foot = 32'

Therefore, the run is 32 feet. You can also convert the grade fraction (¼) to a decimal (0.250) to perform this calculation.

Table 1 Decimal Equivalents of Fractions

1/64	0.015625	33/64	0.515625
1/32	0.03125	17/32	0.53125
3/64	0.046875	35/64	0.546875
1/16	0.0625	9/16	0.5625
5/64	0.078125	37/64	0.578125
3/32	0.09375	19/32	0.59375
7/64	0.109375	39/64	0.609375
⅛	0.125	⅝	0.625
9/64	0.140625	41/64	0.640625
5/32	0.15625	21/32	0.65625
11/64	0.171875	43/64	0.671875
3/16	0.1875	11/16	0.6875
13/64	0.203125	45/64	0.703125
7/32	0.21875	23/32	0.71875
15/64	0.234375	47/64	0.734375
¼	0.25	¾	0.75
17/64	0.265625	49/64	0.765625
9/32	0.28125	25/32	0.78125
19/64	0.296875	51/64	0.796875
5/16	0.3125	13/16	0.8125
21/64	0.328125	53/64	0.828125
11/32	0.34375	27/32	0.84375
23/64	0.359375	55/64	0.859375
⅜	0.375	⅞	0.875
25/64	0.390625	57/64	0.890625
13/32	0.40625	29/32	0.90625
27/64	0.421875	59/64	0.921875
7/16	0.4375	15/16	0.9375
29/64	0.453125	61/64	0.953125
15/32	0.46875	31/32	0.96875
31/64	0.484375	63/64	0.984375
½	0.5	1	1.0

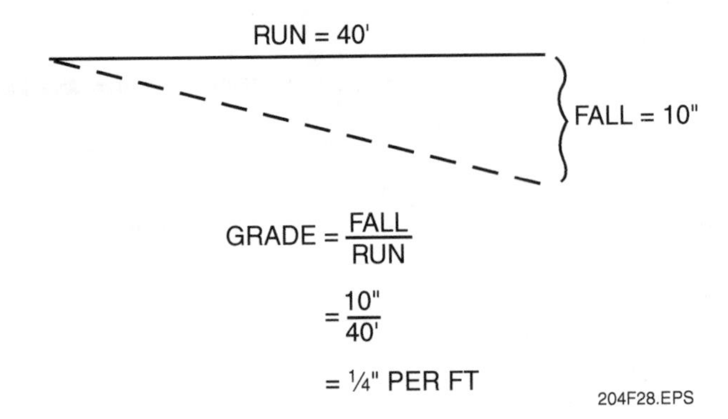

Figure 28 ◆ Calculating grade.

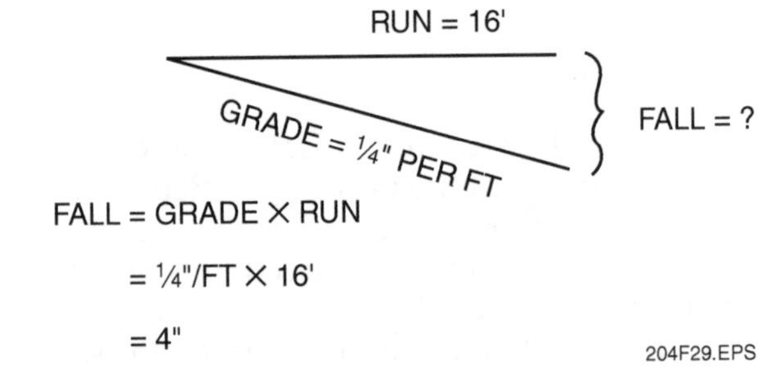

Figure 29 ◆ Calculating fall.

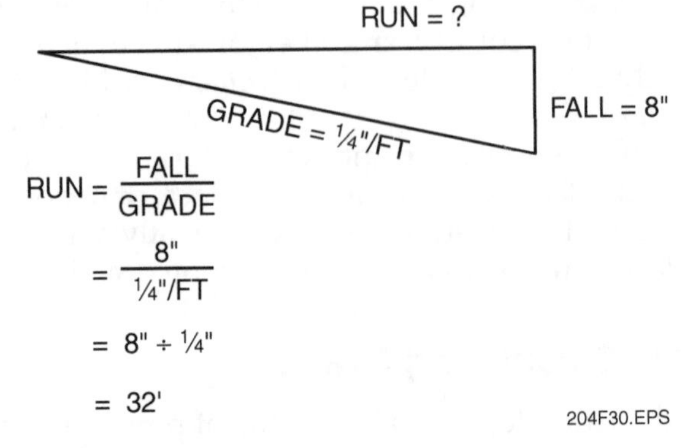

Figure 30 ◆ Calculating run.

Practice Exercises

4.3.2 Fall, Grade, and Run

Test your ability to calculate fall, grade, and run by completing the following study problems.

1. Calculate G where R = 28' and F = 7". G = _____
2. Calculate F where G = ¼" per foot and R = 50'. F = _____
3. Calculate R where G = ⅛" per foot and F = 13". R = _____

4.4.0 Calculating Percentage of Grade

Another way to describe the slope of a line of pipe is to calculate the **percentage of grade**. Percentage of grade (PG) is equal to the fall in feet, divided by the run in feet, multiplied by 100. This can be expressed mathematically as PG = 100(F/R). The multiplication of grade by 100 turns the result into a percentage figure. Percentage of grade calculations are used primarily for sewer and water mains and for large diameter pipelines. The procedure used to calculate the percentage of grade is similar to that used for calculating grade. When working with greater distances, you'll find that it's much easier to work with percentage of grade than with grade.

Consider the following example (see *Figure 31*). The fall is 10 feet. The run is 1,000 feet. What is the percentage of grade?

$$PG = 100(F/R)$$

$$PG = 100(10' \div 1{,}000')$$

$$PG = 100 \times 0.01$$

$$PG = 1$$

Therefore, the percentage of grade is 1 percent.

Refer to *Figure 31*. Fall and run can also be calculated with derived formulas, as shown.

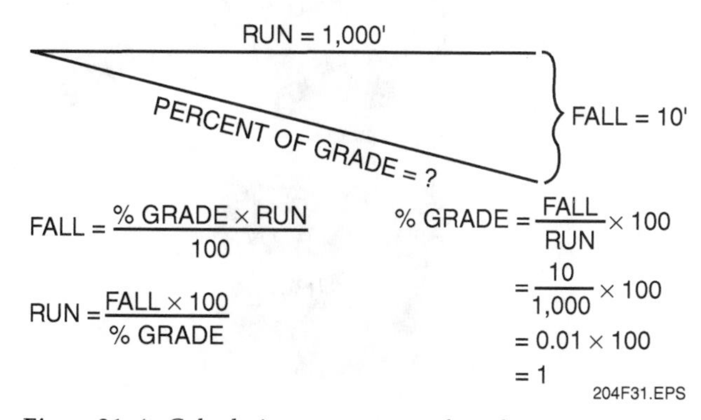

Figure 31 ◆ Calculating percentage of grade.

Practice Exercises

4.4.1 Percentage of Grade

Test your ability to calculate percentage of grade by completing the following study problems.

1. Calculate PG where F = 10' and R = 500'. PG = _____
2. Calculate PG where F = 30" and R = 83.33'. PG = _____ (hint: convert F to feet)

4.5.0 Measuring Grade Using a General-Purpose Level

You must install drainage and waste piping systems at a specified, constant grade. To ensure accuracy when laying out and measuring grade, use a leveling tool. You were introduced to different types of levels and their uses in *Plumbing Level One*.

The most common type of level, especially among plumbers and carpenters, is the general-purpose level (see *Figure 32*). This type of level has three vials. The two end vials measure vertical alignment, such as DWV stacks. The center vial is used to level horizontally, as in a run of pipe. General-purpose levels are not recommended for precise grade calculation, but they can be used when no other leveling tools are available.

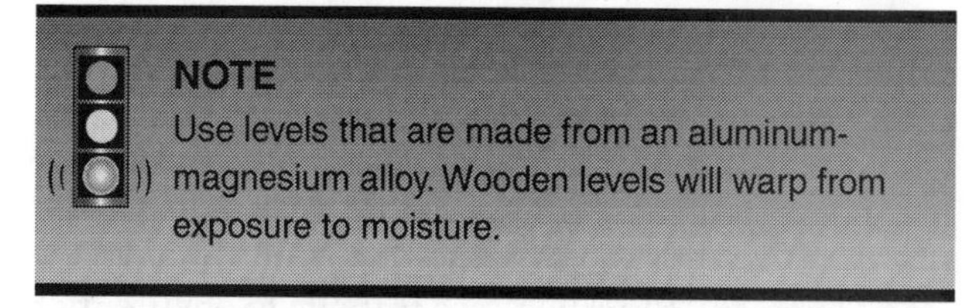

NOTE

Use levels that are made from an aluminum-magnesium alloy. Wooden levels will warp from exposure to moisture.

A general-purpose level will not indicate the precise grade. Use more accurate devices wherever possible. If no other options are available, you can use a general-purpose level to obtain a very rough approximation of grade. Use the following steps:

Step 1 Place the level on pipe that has already been laid with the correct grade.

Step 2 Look at the level's center vial. Notice how much of the bubble (say, ⅛, ¼, or ½) crosses the upper indicator mark.

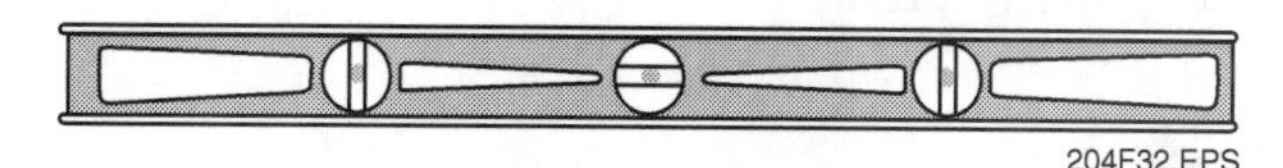

Figure 32 ◆ General-purpose level.

Step 3 Place the level on the pipe for which the grade needs to be determined.

Step 4 Holding the level on the pipe, tilt the pipe until the bubble reaches the same position. The pipe is now at the proper grade.

Step 5 Repeat this process for the remaining pipe lengths to be installed.

This method should only be used as a last resort. A more effective option is to modify the general-purpose level to make it more suitable for measuring grade. Again, use this method only in cases where more precise tools are unavailable.

Begin by determining the grade to be used in the system. Multiply the grade by the length of the level to determine the fall of the line of pipe. Place or tape a wood block that is the same height as the fall under one end of the level (see *Figure 33*). For example, assume that the desired grade is ¼ inch per foot and that your level is 36 inches (3 feet) long. Multiply the grade by the length of the level to determine the fall over the length of the level.

The line of the pipe will fall ¾ inch over the length of the level. Place the level on the pipe, with the modified end on the downward end of the pipe (see *Figure 34*). If the bubble appears level, the pipe is positioned at the proper grade. Note that the piece of wood must be located precisely at the end of the level. Otherwise, the grade will not be accurate.

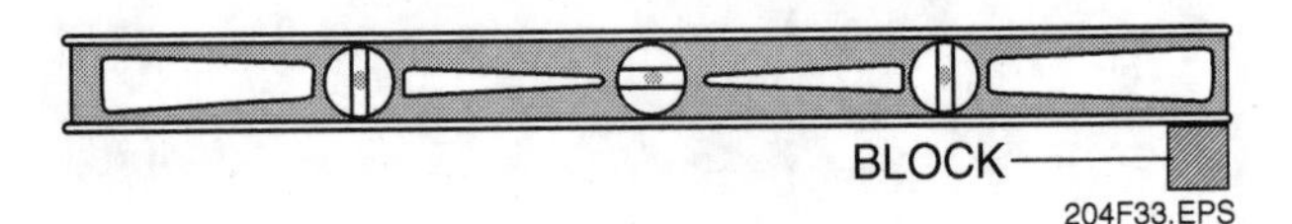

204F33.EPS

Figure 33 ◆ Modifying a general-purpose level to determine grade.

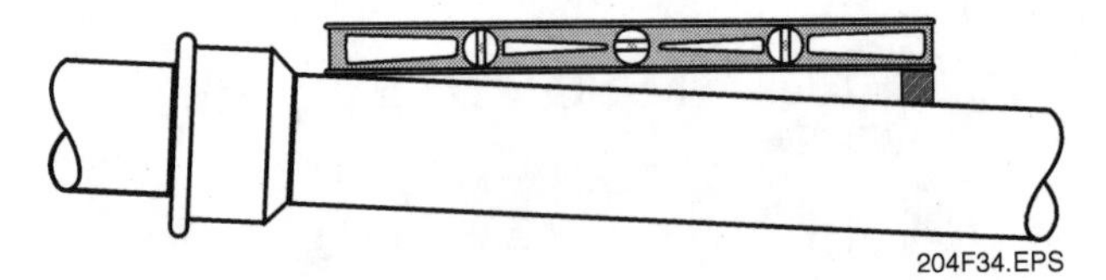

204F34.EPS

Figure 34 ◆ Using a general-purpose level to determine grade.

4.6.0 Measuring Grade Using Builder's Levels

Plumbers often use builder's levels (*Figure 35*) when laying pipe over long distances. Sometimes called a dumpy level, the builder's level measures elevations and horizontal angles. Unlike builder's levels, transit levels (also called transits) can be tilted upward to measure angles from the vertical (see *Figure 36*). Plumbers use transits to ensure that vertical walls and pipe runs are plumb. Because transits perform more functions, they are more expensive than builder's levels. Plumbers generally use the builder's level, because they are mainly concerned with elevations and grades. This section focuses on using the builder's level, although the same procedures also apply to transit levels.

The builder's level consists of seven basic parts:

- Telescope
- Leveling vial
- Leveling screws
- Protractor circle
- Index
- Tripod
- Plumb bob

204F35.EPS

Figure 35 ◆ Builder's level.

204F36.EPS

Figure 36 ◆ Transit level.

The telescope is the fundamental part of the builder's level. You can measure elevations and angles through the line of sight established by the telescope. The line of sight is weightless and continuous; unlike a string line, it won't sag and is therefore more accurate (see *Figure 37*).

The telescope magnifies the image so that you can see the object, or target, more clearly. A magnification of 20x (also referred to as "20 power") is common, though scopes with larger power are also available. As a general rule, use a greater magnification power for jobs involving greater working distances.

When you look through the telescope, you will see crosshairs overlaying the view. Crosshairs improve the telescope's accuracy by serving as reference marks (see *Figure 38*). The horizontal crosshair is used in reading elevations. The vertical crosshair is used to check the plumb of objects.

A highly accurate leveling vial and a set of leveling screws allow you to adjust the telescope to level before you use it. The vial is attached directly to the telescope. Turn the leveling screws to center the bubble in the vial (see *Figure 39*). The proper use of the leveling screws will be discussed in detail later in this section.

The protractor allows you to adjust the telescope accurately and to measure horizontal angles (see *Figure 40*). The protractor rotates above a circle attached to the stationary base. The circle is divided into 360 degrees. The protractor indicator, called the index, rotates over these markings from a center pivot. This feature makes it easy to take 90-degree and 45-degree views and measurements. A **vernier** indicator, which is a finer scale, allows more precise alignment.

The builder's level is mounted on a tripod (see *Figure 41*). You can adjust each of the tripod's three legs individually so that you can set up the level on soft or irregular ground. Spades are attached to the ends of the tripod legs to allow for firm placement in soil.

Often the builder's level must be set up over a specific point, such as a monument, or a stake at the corner of a lot. To center the builder's level over the desired point, suspend a plumb bob from a special hanger at the base of the tripod (see *Figure 42*). Move the legs of the tripod around until the plumb bob is over the desired point. Some builder's levels have a fine adjusting screw that may be used to accurately center the instrument over the desired point.

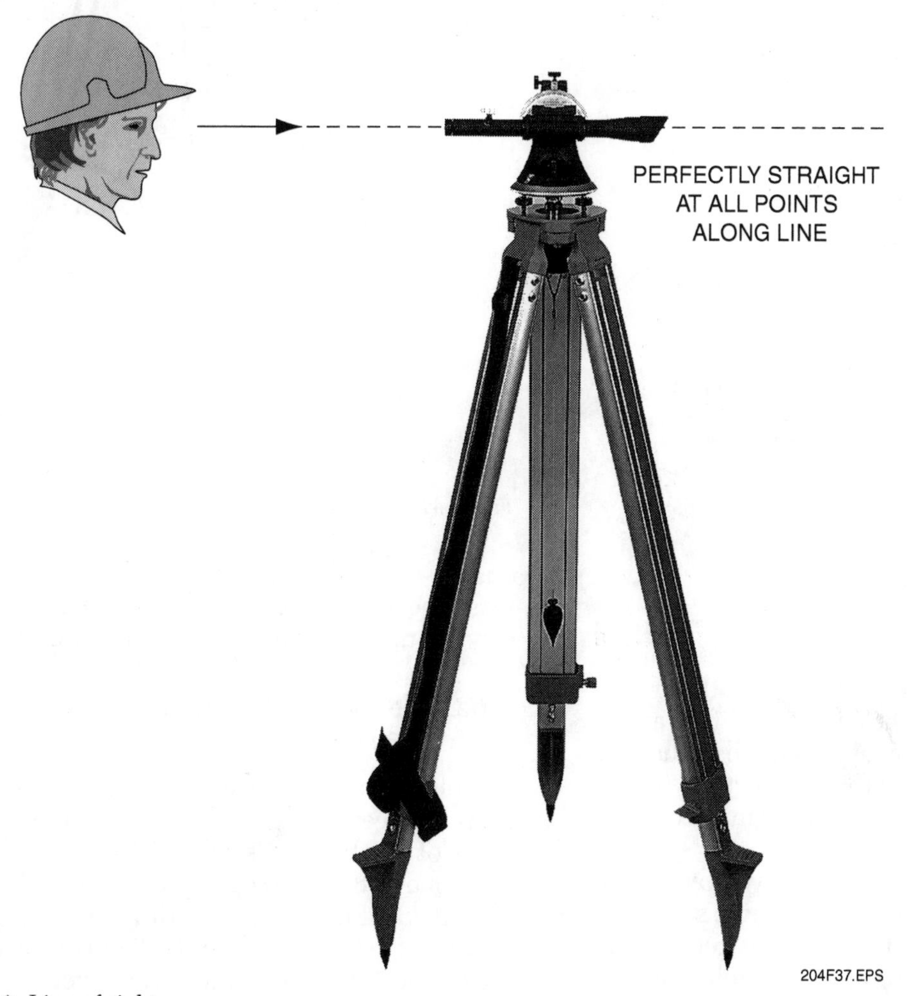

Figure 37 ◆ Line of sight.

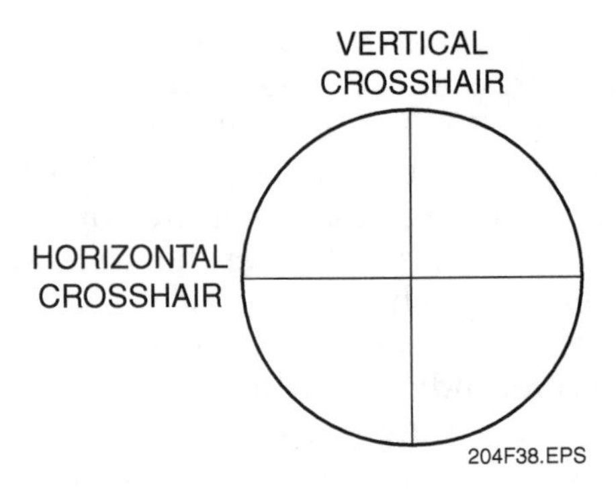

Figure 38 ◆ Crosshairs.

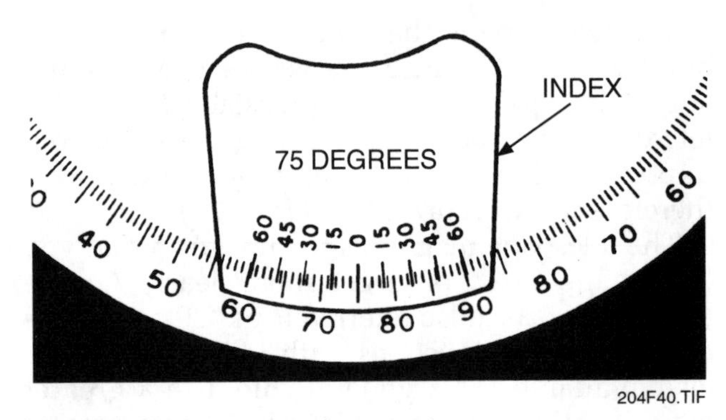

Figure 40 ◆ Protractor circle.

FOR A FINAL LEVEL CHECK, rotate the telescope over each of four leveling points to be sure the bubble remains centered.

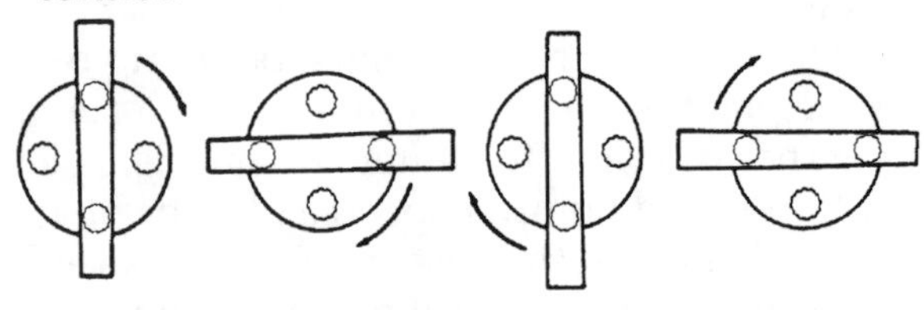

The Golden Rule for quick and simple leveling is THUMBS IN, THUMBS OUT. Turn BOTH screws equally and simultaneously. Practice will help you get the feel of the screws and the movement of the bubble. It will also help to remember that the direction your left thumb moves is the direction the bubble will move.

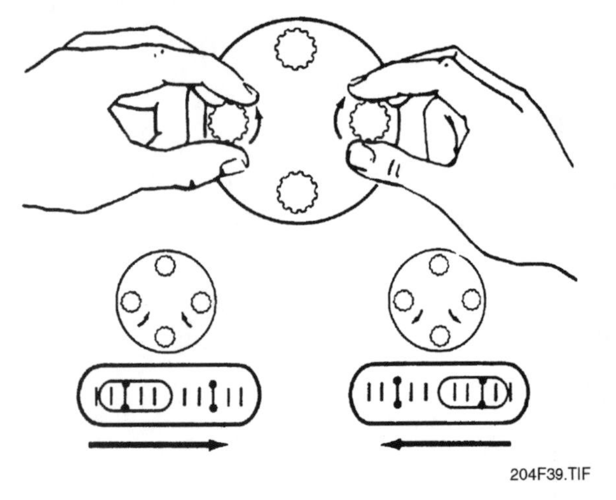

Figure 39 ◆ Adjusting the leveling screws.

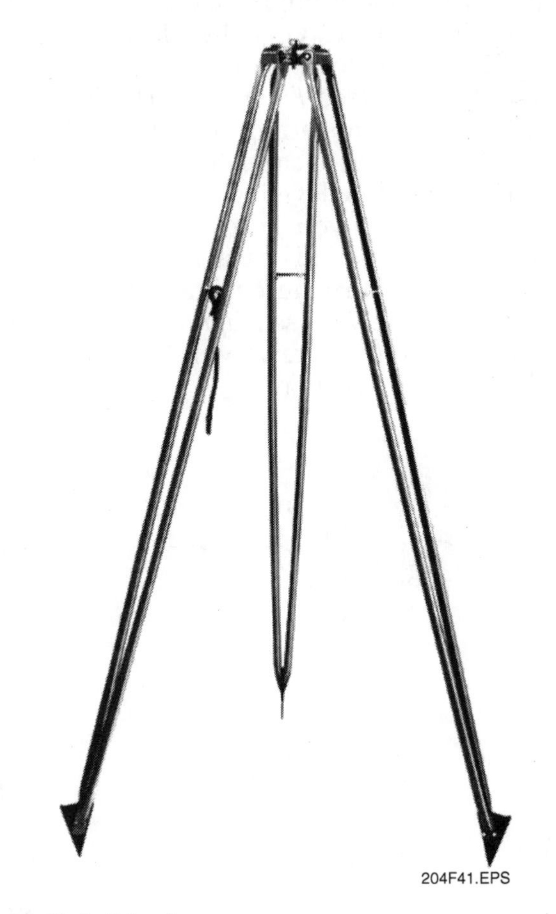

Figure 41 ◆ Tripod.

4.6.1 Setting Up the Builder's Level

To set up a builder's level, begin by determining the proper location. Once you have determined where the tripod is to be placed, force the tripod legs into the ground, approximately 3 feet apart, over that spot. Use the leveling vial to adjust the tripod legs until the unit is roughly level. With the tripod legs firmly positioned, you can then accurately level the instrument.

Rotate the telescope until it lays over two opposing leveling screws. Move the two screws in opposite directions until the bubble is centered in the leveling vial. After rotating the telescope 90 degrees, adjust the two remaining screws. To ensure accuracy, repeat the leveling procedure at least one more time. With the builder's level properly set up, you are ready to measure grade.

4.6.2 Measuring Grade with the Builder's Level

For plumbing jobs, you will often measure elevations between two points that are below the line of sight of the builder's level. This is a two-person job—one at the telescope, one at the stadia rod (see *Figure 43*). Position the builder's level midway between the two points, and set it up as discussed in the previous section. A movable target on the stadia rod provides accuracy in reading the measurements.

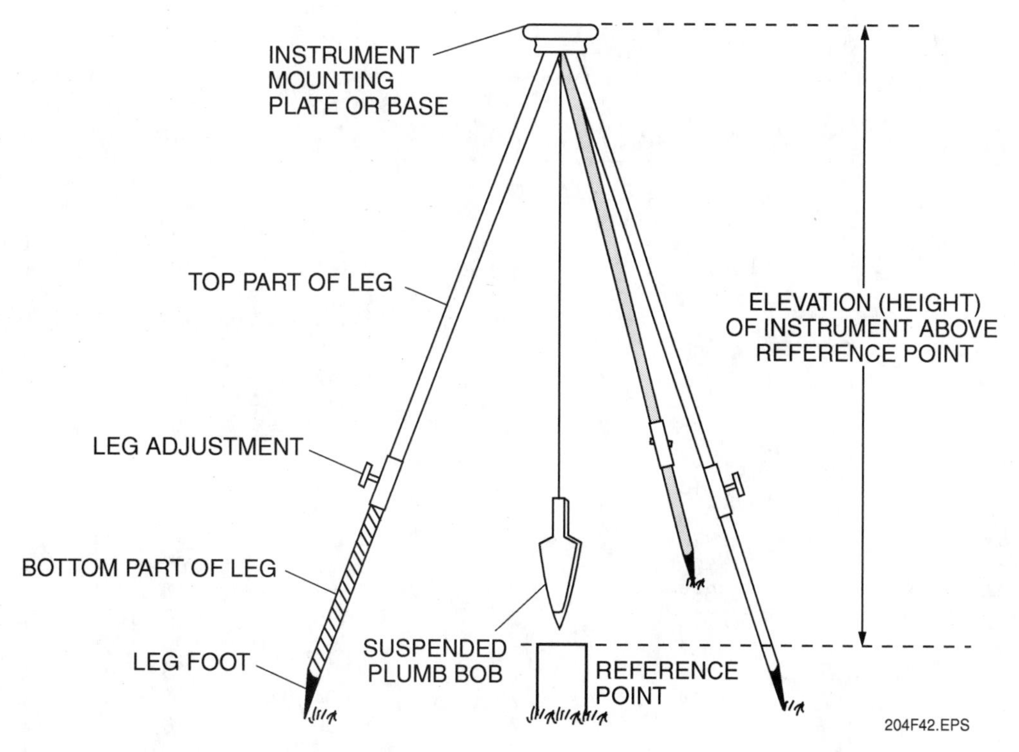

Figure 42 ◆ Positioning a tripod using a plumb bob.

The stadia rod may be marked off in feet, feet and inches, tenths of a foot, or hundredths of a foot. You may substitute a folding rule for the stadia rod.

Tenths of a foot are also called decimals of a foot. You learned in *Plumbing Level One* that a decimal of a foot is a decimal fraction in which the denominator is 12, instead of the usual 10. Use a conversion table, such as the one shown in *Table 2*, to convert decimals of a foot to the equivalent inches. Conversions are shown for measurements ranging from 1⁄16 inch to 12 inches.

Now you are ready to measure the elevations. Follow these steps, referring to *Figure 44*:

Step 1 Hold the stadia rod on top of the first point.

Step 2 Look through the builder's level and focus on the stadia rod.

Step 3 Signal the person holding the stadia rod to move the target up or down until it is centered in the crosshairs of the telescope. Remember, the vertical crosshair is used to make sure the stadia rod is vertical.

Step 4 Read the numbers at the horizontal crosshairs to measure the elevation (*Figure 45*).

Step 5 Repeat this procedure to obtain the relative elevation of the second point.

Step 6 Once you have the relative elevation and the distance between both points, use the grade formula to calculate the grade.

Figure 43 ◆ Stadia rod.

Table 2 Inches Converted to Decimals of a Foot

Inches	Decimals of a Foot	Inches	Decimals of a Foot	Inches	Decimals of a Foot	Inches	Decimals of a Foot
1/16	0.005	3 1/16	0.255	6 1/16	0.505	9 1/16	0.755
1/8	0.010	3 1/8	0.260	6 1/8	0.510	9 1/8	0.760
3/16	0.016	3 3/16	0.266	6 3/16	0.516	9 3/16	0.766
1/4	0.021	3 1/4	0.271	6 1/4	0.521	9 1/4	0.771
5/16	0.026	3 5/16	0.276	6 5/16	0.526	9 5/16	0.776
3/8	0.031	3 3/8	0.281	6 3/8	0.531	9 3/8	0.781
7/16	0.036	3 7/16	0.286	6 7/16	0.536	9 7/16	0.786
1/2	0.042	3 1/2	0.292	6 1/2	0.542	9 1/2	0.792
9/16	0.047	3 9/16	0.297	6 9/16	0.547	9 9/16	0.797
5/8	0.052	3 5/8	0.302	6 5/8	0.552	9 5/8	0.802
11/16	0.057	3 11/16	0.307	6 11/16	0.557	9 11/16	0.807
3/4	0.063	3 3/4	0.313	6 3/4	0.563	9 3/4	0.813
13/16	0.068	3 13/16	0.318	6 13/16	0.568	9 13/16	0.818
7/8	0.073	3 7/8	0.323	6 7/8	0.573	9 7/8	0.823
15/16	0.078	3 15/16	0.328	6 15/16	0.578	9 15/16	0.828
1	0.083	4	0.333	7	0.583	10	0.833
1 1/16	0.089	4 1/16	0.339	7 1/16	0.589	10 1/16	0.839
1 1/8	0.094	4 1/8	0.344	7 1/8	0.594	10 1/8	0.844
1 3/16	0.099	4 3/16	0.349	7 3/16	0.599	10 3/16	0.849
1 1/4	0.104	4 1/4	0.354	7 1/4	0.604	10 1/4	0.854
1 5/16	0.109	4 5/16	0.359	7 5/16	0.609	10 5/16	0.859
1 3/8	0.115	4 3/8	0.365	7 3/8	0.615	10 3/8	0.865
1 7/16	0.120	4 7/16	0.370	7 7/16	0.620	10 7/16	0.870
1 1/2	0.125	4 1/2	0.374	7 1/2	0.625	10 1/2	0.875
1 9/16	0.130	4 9/16	0.380	7 9/16	0.630	10 9/16	0.880
1 5/8	0.135	4 5/8	0.385	7 5/8	0.635	10 5/8	0.885
1 11/16	0.141	4 11/16	0.391	7 11/16	0.641	10 11/16	0.891
1 3/4	0.146	4 3/4	0.396	7 3/4	0.646	10 3/4	0.896
1 13/16	0.151	4 13/16	0.401	7 13/16	0.651	10 13/16	0.901
1 7/8	0.156	4 7/8	0.406	7 7/8	0.656	10 7/8	0.906
1 15/16	0.161	4 15/16	0.411	7 15/16	0.661	10 15/16	0.911
2	0.167	5	0.417	8	0.667	11	0.917
2 1/16	0.172	5 1/16	0.422	8 1/16	0.672	11 1/16	0.922
2 1/8	0.177	5 1/8	0.427	8 1/8	0.677	11 1/8	0.927
2 3/16	0.182	5 3/16	0.432	8 3/16	0.682	11 3/16	0.932
2 1/4	0.188	5 1/4	0.438	8 1/4	0.688	11 1/4	0.938
2 5/16	0.193	5 5/16	0.443	8 5/16	0.693	11 5/16	0.943
2 3/8	0.198	5 3/8	0.448	8 3/8	0.698	11 3/8	0.948
2 7/16	0.203	5 7/16	0.453	8 7/16	0.703	11 7/16	0.953
2 1/2	0.208	5 1/2	0.458	8 1/2	0.708	11 1/2	0.958
2 9/16	0.214	5 9/16	0.464	8 9/16	0.714	11 9/16	0.964
2 5/8	0.219	5 5/8	0.469	8 5/8	0.719	11 5/8	0.969
2 11/16	0.224	5 11/16	0.474	8 11/16	0.724	11 11/16	0.974
2 3/4	0.229	5 3/4	0.479	8 3/4	0.729	11 3/4	0.979
2 13/16	0.234	5 13/16	0.484	8 13/16	0.734	11 13/16	0.984
2 7/8	0.240	5 7/8	0.490	8 7/8	0.740	11 7/8	0.990
2 15/16	0.245	5 15/16	0.495	8 15/16	0.745	11 15/16	0.995
3	0.250	6	0.500	9	0.750	12	1.000

ON THE

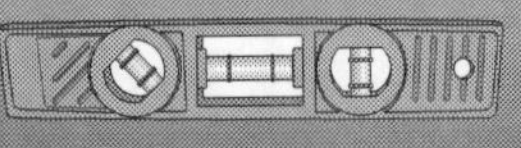

LEVEL

Converting Decimals Using a Calculator

You can use a calculator to convert decimals of a foot to the equivalent inches. Simply multiply by 12. For example, if you want to convert 0.0833 decimals of a foot into inches, multiply by 12. The answer is 0.99 or 1. On the other hand, if you want to convert inches into decimals of a foot, divide by 12 (1 divided by 12 is 0.0833). Try this yourself. Convert 2 inches into decimals of a foot using your calculator. (The answer is 0.1667.)

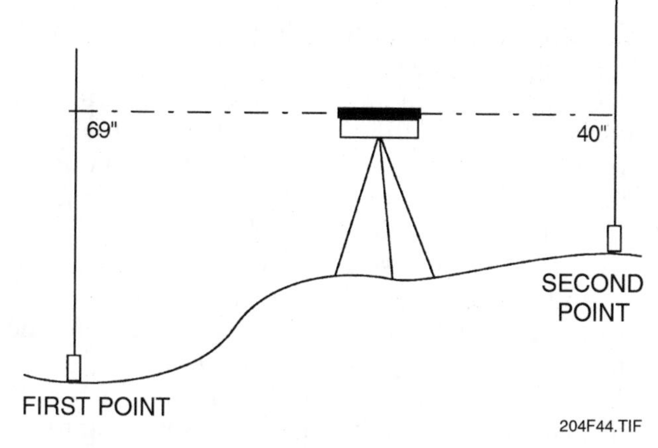

Figure 44 ◆ Measuring elevations.

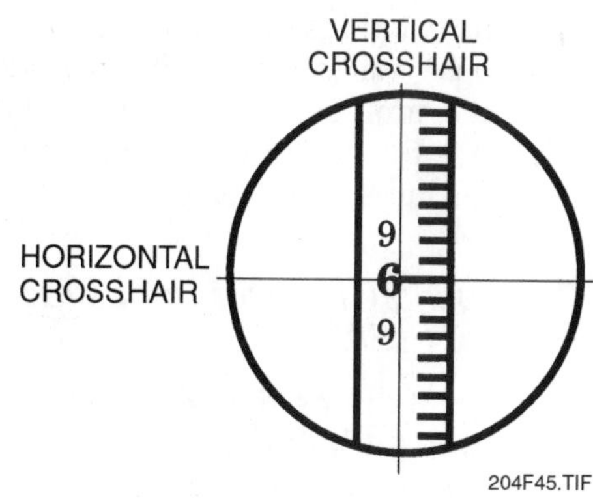

Figure 45 ◆ Reading the elevation.

Occasionally, you will have to determine the elevation for lines of pipe that are to be run above the height of the builder's level (see *Figure 46*). This is also a two-person job. To measure this elevation, follow these steps:

Step 1 Position the builder's level between the two unknown points, and set it up.

Step 2 Have your partner hold the stadia rod upside down against the desired point.

Step 3 Sight through the telescope, and take a reading.

Step 4 Repeat this procedure to obtain the elevation of the second point.The cold beam laser has recently come into wide use in the plumbing industry (see *Figure 47*). It projects a fine beam of infrared light that is perfectly straight and will not sag. The beam does not expand appreciably. Because of these characteristics, cold beam lasers allow grade to be set with a high degree of accuracy. Although the laser is considered harmless, never look directly into the beam of light.

You may position the laser inside the pipe, outside the pipe, or even at grade level above the pipe (see *Figure 48*). Use a tilting telescope to determine the height of the beam below ground. To adjust the laser unit, shoot a reading on a stadia rod and transfer it down to the laser unit.

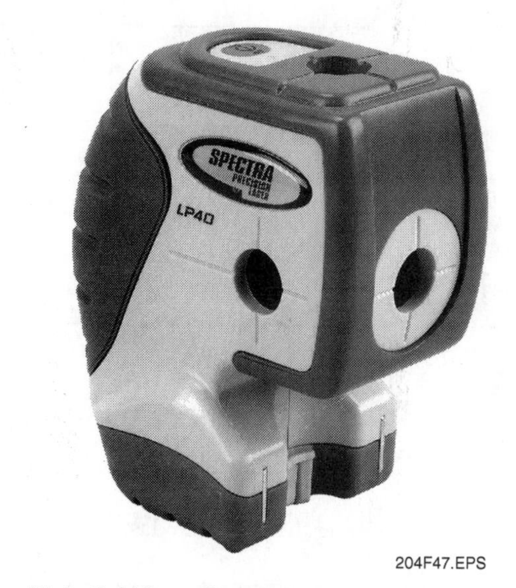

Figure 47 ◆ Cold beam laser.

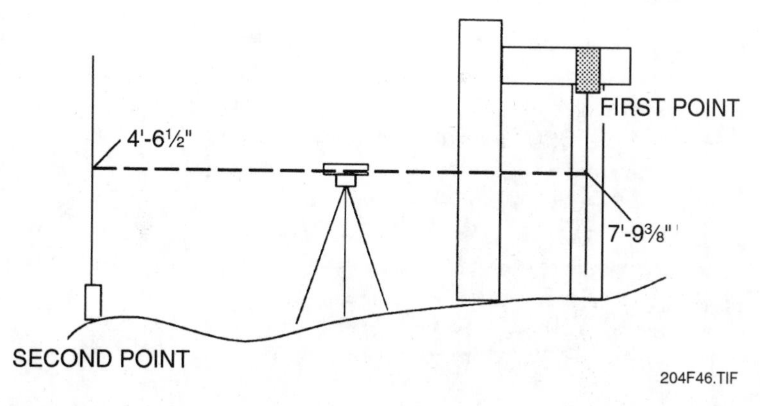

Figure 46 ◆ Measuring elevations above the instrument.

The target for a cold beam laser must be placed along the center line of the pipe. The target can be temporarily mounted inside the pipe or hung from the top of the pipe wall (see *Figure 49*). Begin by turning on the cold beam laser. To set the pipe at the correct grade, move the end of the pipe until the laser beam strikes the target in its center (see *Figure 50*).

4.6.3 Computing Grade for Unobstructed Runs of Piping

The grade for roof drains, fixture branch lines, and occasionally even the building drain and the sewer main itself can be determined simply by using the grade formula. Consider the following hypothetical example of a building drain that runs from the building to the sewer main (see *Figure 51*).

Using the stadia rod and builder's level, say that your reading at the first elevation is 9'-6". This means that the top of the building branch outlet is 9'-6" below the line of sight established through the telescope on the builder's level. Repeat the sighting procedure for the second elevation. Say that the second elevation is 10'-6". With both points established, you can compute the grade. The difference in elevation, or fall, is 12 inches (10'-6" – 9'-6" = 12"). Remember that for the grade formula (G = F/R), fall is always expressed in inches. Referring to *Figure 51*, you can see that the run is 32 feet. Use the formula to determine the grade:

$$G = F/R$$

$$G = 12''/32'$$

$$G = 0.375 = \tfrac{3}{8}'' \text{ per foot}$$

Therefore, the grade is ⅜ inch per foot.

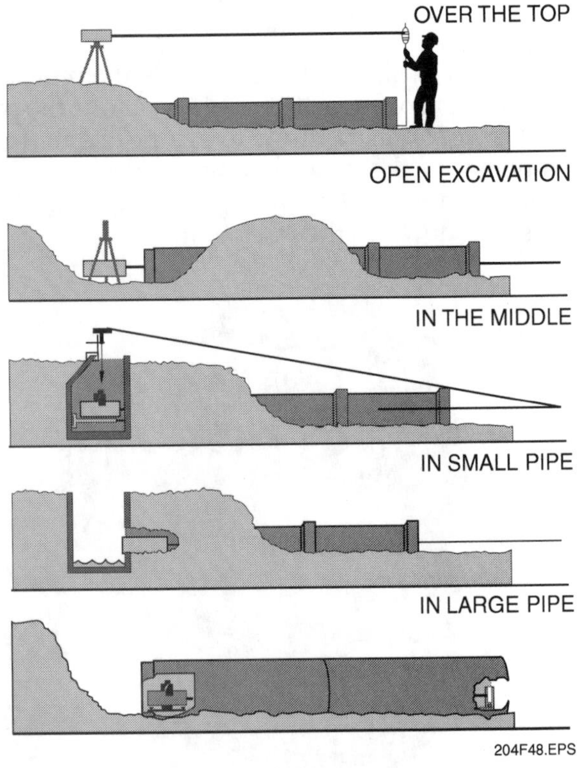

Figure 48 ◆ Positioning the cold beam laser.

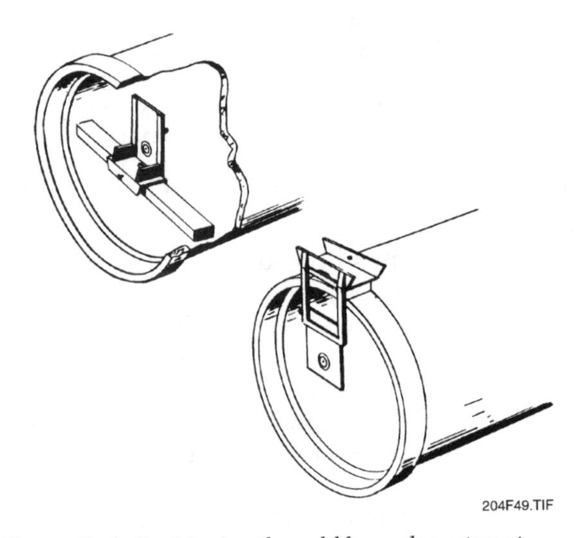

Figure 49 ◆ Positioning the cold beam laser target.

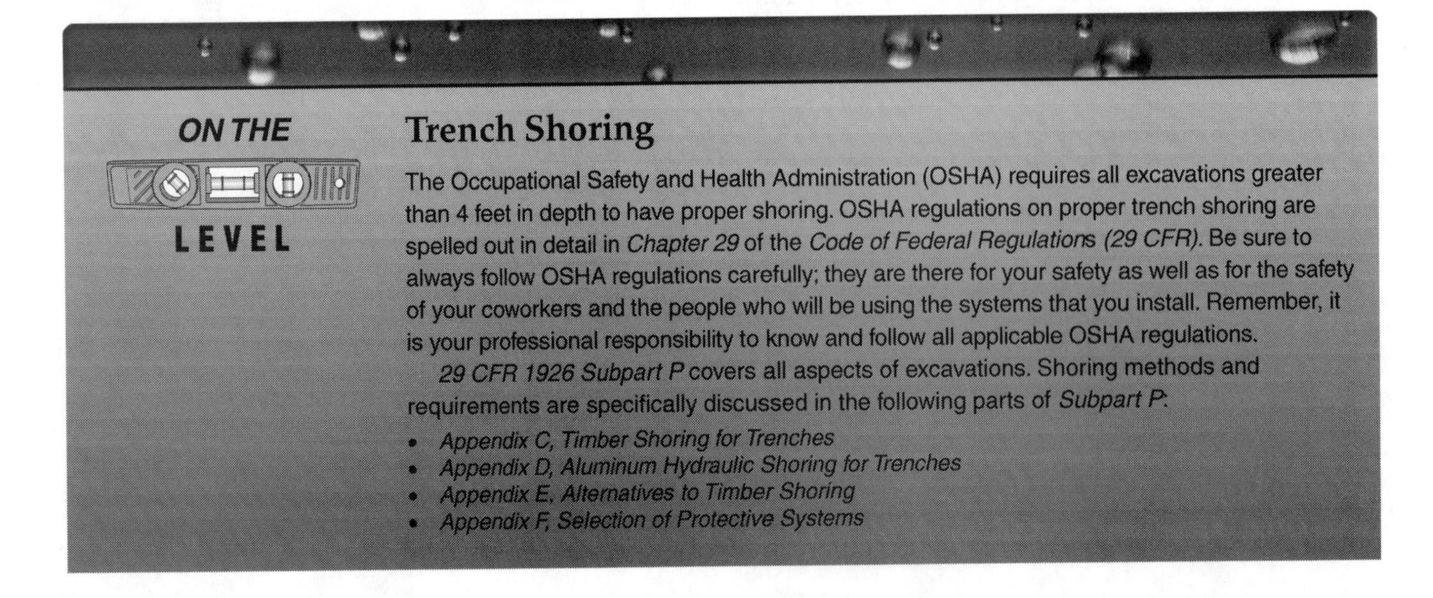

Trench Shoring

The Occupational Safety and Health Administration (OSHA) requires all excavations greater than 4 feet in depth to have proper shoring. OSHA regulations on proper trench shoring are spelled out in detail in *Chapter 29* of the *Code of Federal Regulations (29 CFR)*. Be sure to always follow OSHA regulations carefully; they are there for your safety as well as for the safety of your coworkers and the people who will be using the systems that you install. Remember, it is your professional responsibility to know and follow all applicable OSHA regulations.

29 CFR 1926 Subpart P covers all aspects of excavations. Shoring methods and requirements are specifically discussed in the following parts of *Subpart P*:

- *Appendix C, Timber Shoring for Trenches*
- *Appendix D, Aluminum Hydraulic Shoring for Trenches*
- *Appendix E, Alternatives to Timber Shoring*
- *Appendix F, Selection of Protective Systems*

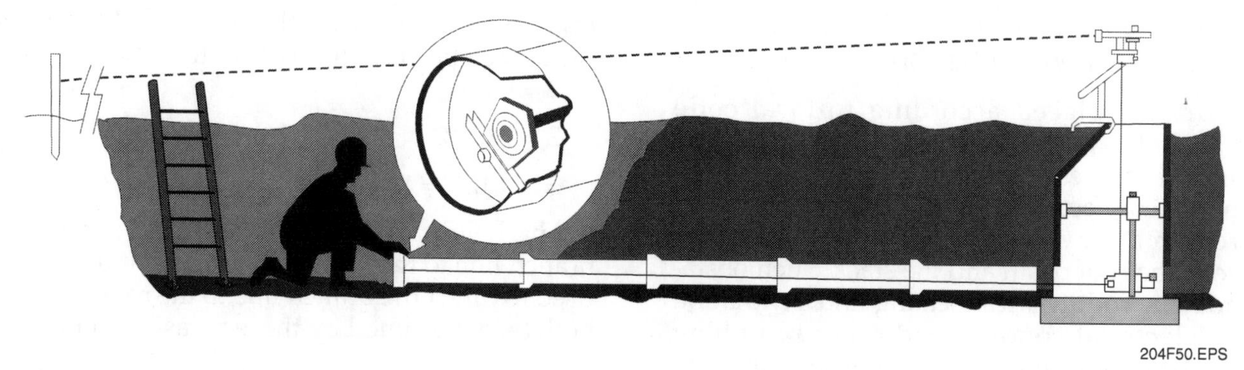

Figure 50 ◆ Positioning the pipe using a cold beam laser.

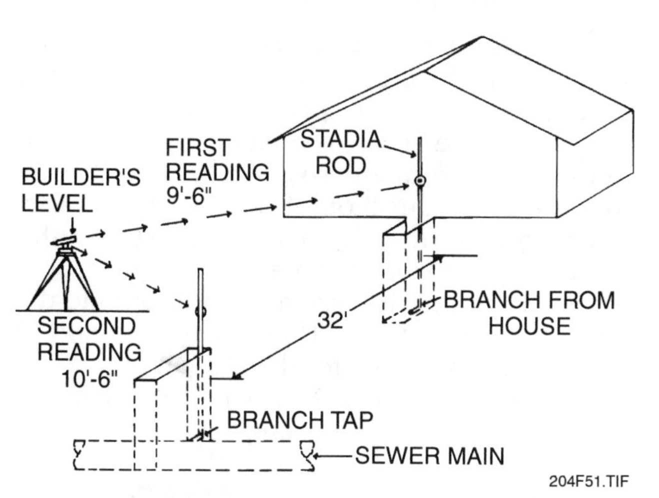

Figure 51 ◆ Computing grade for a building drain.

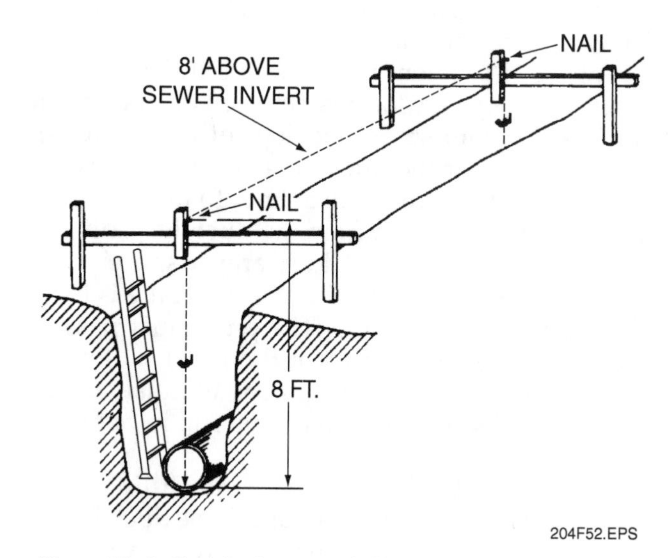

Figure 52 ◆ Positioning a grade line using batter boards.

4.6.4 Computing Grade Using a Grade Line

You can also use the builder's level to accurately position grade lines (see *Figure 52*). A grade line, as its name suggests, is a length of taut string or other line material that is run along the planned piping route at the desired grade. The line is fastened at regular intervals to **batter boards**, which are horizontal boards that are nailed at right angles to each other to three posts set beyond the corners of a building elevation. To use a grade line with batter boards to determine the grade of a line of pipe, follow these steps:

Step 1 Drive a stake on one side of the excavation trench.

Step 2 Drive a second stake on the opposite side of the trench, in line with the first stake. If a length of pipe has already been positioned in the trench, drive the stakes beside the hub end of the pipe.

Step 3 Place the batter board in position, and level it using a builder's level.

Step 4 For ease in calculating grade, locate the batter board at a convenient height above the pipe. General-purpose levels are suitable for this task, though the results will not be as accurate as those obtained by using a builder's level.

Step 5 Once the batter board is correctly positioned, use C-clamps or nails to attach it to the stakes.

Step 6 Use a plumb bob or level to transfer the center line of the run of pipe to the batter board.

Step 7 Drive a nail into the batter board at the height established in Step 4.

Step 8 Repeat this procedure every 25 to 50 feet. Make sure that the line of batter boards is level throughout the run.

Step 9 Stretch a string line (which becomes the line of reference) tightly between the nails on the batter boards. This is the grade line.

Step 10 Measure from the string down to the pipe as you lay the pipe.

4.6.5 Computing Grade for a Roof Drain within the Ceiling

Roof drains, placed according to local code requirements, are located at various points on flat-roofed structures to remove rainwater. These drains are connected to a storm drainage stack.

Problems may arise when obstacles occur between the roof drain and the stack. Such obstacles may be roof trusses, heating and ventilation ducts, electrical conduit, and other plumbing lines. These obstacles require you to carefully determine the grade of the branch line so that its grade can be constant.

A typical problem is illustrated in *Figure 53*. A duct and a pipe are located between the roof drain and the drainage stack. The slope of a line that will pass both under the duct and over the pipe is the desired grade for the branch line. To measure that grade, measure down from the duct the radius of the pipe. Remember to add clearance space.

For example, assume that a 4-inch pipe is to be used for the branch line. The pipe must pass at least 2 inches (the radius of the pipe) under the duct. The thickness of the pipe wall must also be figured in. Assume also that a minimum of 1-inch clearance is required.

Therefore, the center of the branch line must pass approximately 3 inches under the duct. Using the bottom of the ceiling cavity as a reference, measure up to a point within 3 inches of the duct. Say this distance is 15 inches. You must determine the relative elevation of the second critical point. That is the point at which the branch line will pass above the obstacle pipe. In the same way as before, you determine that the center line of the branch line must pass at a relative elevation of 3 inches from the obstacle pipe plus 13.5 inches above the bottom of the ceiling cavity. The elevations are relative to the bottom of the ceiling cavity.

To calculate the grade for the line, measure the distance between the two points. Assume that this distance is 12 feet. Use the grade formula (G = F/R) to calculate the grade. In this example, the grade for the branch line is ⅛ inch per foot.

G = F/R

G = 1.5"/12'

G = 0.125" = ⅛" per foot

The height of the fitting used to connect the branch line to the stack and the height of the bend used to direct the branch line to the roof drain will both be determined by these measurements.

4.6.6 Computing Grade for Hanging Pipe Trapezes

Pipe run near the ceiling of a structure is generally supported with a pipe trapeze (see *Figure 54*). Local codes specify how far apart trapezes may be set. When drainage and waste pipe is involved, you must install the trapezes so that they allow the pipe to slope in the direction of flow. Therefore, each trapeze must be installed at a lower height (farther from the ceiling) than the previous trapeze.

Install the first trapeze at an appropriate distance from the ceiling. That distance should be enough to allow room for the pipe and for you to work. After the trapeze is installed, level it using a torpedo level or other leveling device.

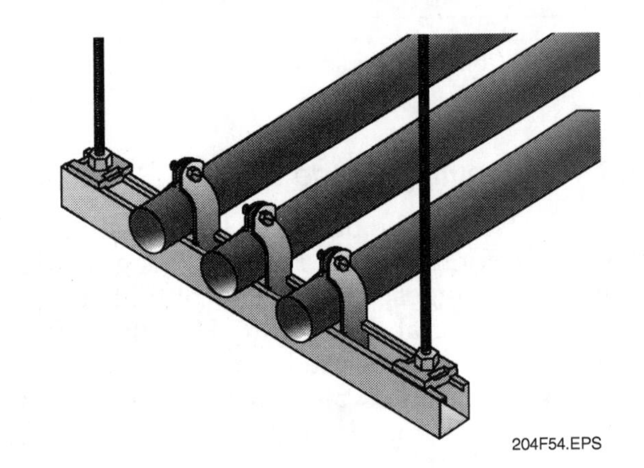

Figure 54 ◆ Pipe trapeze.

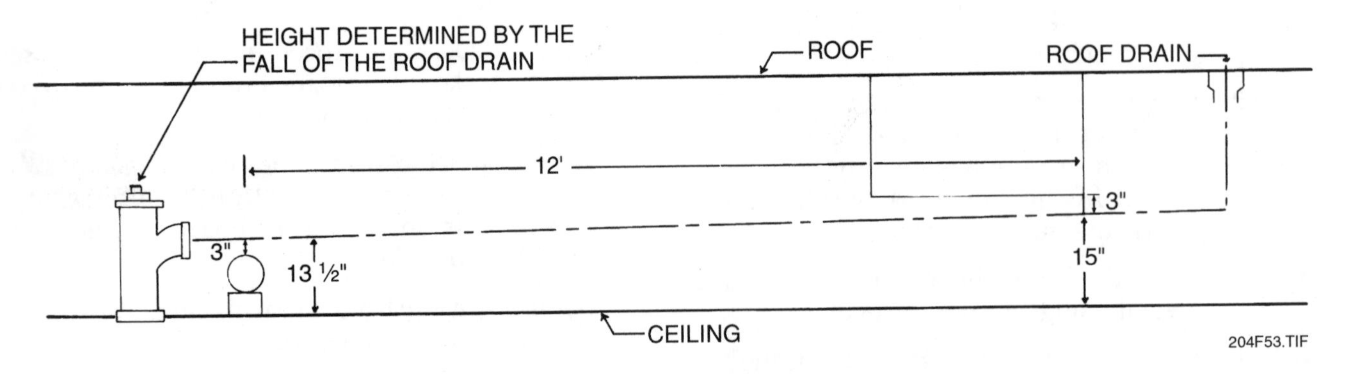

Figure 53 ◆ Figuring grade.

Locate and set up a builder's level beside the desired path of the pipe (see *Figure 55*). Place an extended folding rule upside down against the installed trapeze. Sight through the builder's level and take a reading. Let's assume the reading is 8'-2". This reading is relative and tells you nothing in itself. However, it does provide you with a starting point that is a level line.

For example, assume that you want a grade of ¼ inch per foot. Also assume the trapezes will be positioned 12 feet apart. Use the grade formula to determine the fall of the line of pipe between the two adjacent trapezes.

F = G × R

F = ¼" × 12'

F = 3"

Therefore, the line of pipe will fall 3 inches over the 12-foot distance between the trapezes.

By knowing this fall and the relative elevation of the first trapeze, you can determine the elevation of, and adjust, the second trapeze. If the first trapeze was located 8'-2" from the ceiling and the fall of the pipe to the next trapeze is 3 inches, then the second trapeze must be set at a relative elevation of 7'-11" (8'-2" – 3" = 7'-11"). You will note that as the trapezes fall further downward, the subsequent readings will be ever smaller.

Once you know its proper relative elevation, you can install the second trapeze. Hold the bottom of the folding rule against the trapeze. Adjust the trapeze until you get a reading of 7'-11". Then, secure the trapeze in position. Repeat this procedure to install the rest of the system.

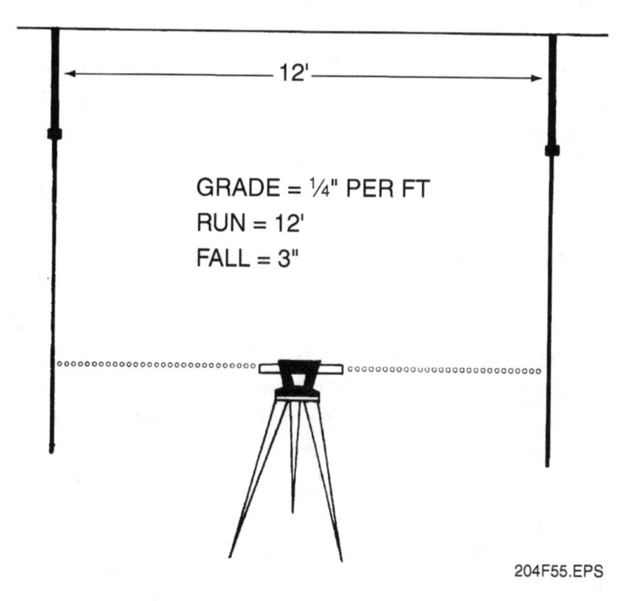

Figure 55 ◆ Determining grade.

5.0.0 ◆ DWV ROUGH

Consult with the carpenters to determine how they will lay out the structure before you attempt to lay out the plumbing system. Often, the north and east walls are designated as the walls from which all the trades will measure. This method prevents small variations in the existing structure from affecting progress. It will ensure that the pipes are in correct relationship to the walls that will be built later. The pipe installer has the authority to decide the most appropriate routing for the pipes.

In structures built on concrete slabs, such as the hypothetical installation in this module (refer to *Figure 1*), the building drain and the sweep at the base of the stack are placed before the slab is poured. There is no need to bore holes in the floor because the DWV pipes already have been extended above the surface of the slab. The carpenters will bore the holes in the soleplates as they erect the walls over the piping. From the slab on up, the plumbing is a matter of locating the fixtures, running the stacks, and connecting the fixtures. Buildings with basements and crawl spaces have different requirements.

Some pipes will pass through concrete structural elements, such as footings, stem walls, and grouted masonry. Place **sleeves** around pipes where they will pass through such structural elements. Sleeves prevent the pipe from being damaged by contact with concrete. They are made from plastic pipe that is one size larger than the DWV pipe being installed. Secure the sleeve in place before the concrete is poured. Sleeves must be sloped so that the pipe passing through is not pinched.

Secure the piping installation to prevent movement caused by backfilling and soil compaction around the pipes. **Rebar**, or ribbed reinforcing metal bar, is often used for this purpose. Drive the rebar into the soil adjacent to the pipe, at a 45-degree angle. Then, clamp the pipe to the rebar using tie wire. Two of these bars placed perpendicular to each other—that is, driven into the ground at opposing 45-degree angles to create a 90-degree V into which the pipe is placed—provide good protection against movement.

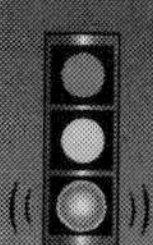

NOTE

Use string to mark the front edge of plumbing walls; this enables you to stretch a tape measure parallel to the plumbing wall to measure for stacks and fixture locations. Make sure dirt is not piled where a string line might be needed.

Before you begin the installation of a DWV system, ensure that the following conditions have been met:

- The plans have been approved
- The plans are available for referral during the installation
- All appropriate permits have been pulled
- All fixtures have been sized

Once these conditions have been met, you are ready to begin installing the DWV system. Refer to the floor plan to locate the point where the DWV piping will penetrate the floor. Do this by using a steel tape measure to lay out the measurements from a corner of the building or from a wall. Make sure that excavated dirt does not block access to areas where you need to measure.

5.1.0 Verifying the Layout

Begin by locating all underground site utility lines. Otherwise, you might cut into an existing utility as you dig trenches for the below-grade piping. This information can be obtained from drawings or by determining the end points of a buried line. In some cases you may have to use pipe locators if drawings and end points are unavailable. You will learn more about how to locate buried pipe later in the *Plumbing* curriculum.

When all underground utilities have been identified, verify the system's layout. Use grade lines attached to batter boards to establish the locations of the drains. You have already learned how to establish a grade line. Chalk lines may also be used instead of grade lines; refer to your local codes and established company guidelines for the correct procedure. Confirm the elevation of the building sewer, and use the plans to determine the route of the building drain, proceeding from right to left on the drawing (see *Figure 56*).

When you have calculated the depth of the building drain, locate the major points along the building drain and calculate the slope and depth of the pipe. Determine the depth at which the building drain will pass under the foundation footing. This information can usually be found on the plumbing plans.

If the depth is not indicated on the plans, dig a hole beside the foundation wall and measure the depth to the underside of the footing. To this measurement, add the desired distance between the center line of the pipe and the underside of the foundation. For example, suppose the foundation footing is 22 inches below the finished floor. Also suppose that you want the center line of the pipe to pass 8 inches under the footing. The depth of the center line of the building drain will be 30 inches below the finished floor (22" + 8" = 30"). Your local applicable code will also include information to help you determine proper drain and vent placement.

At this point, you can determine the fixture locations. Review the section in this module that discusses how to locate lavatories, sinks, baths, and showers. Remember to use the rough-in dimensions provided by the manufacturer when locating a fixture. When you finish, create an isometric sketch of the complete DWV system.

5.2.0 Installing Below-Grade Piping

When you have verified the layout of the DWV system, determine the following:

- Invert elevations
- Footer
- Branches
- Farthest point of the system
- Fall

When you have established these positions, you can then begin to excavate the pipe trenches along the chalk or string lines that you established earlier. The process of installing DWV pipe below grade can be broken down into distinct phases. Each phase is treated separately to show you how they relate to each other and to the system as a whole.

5.2.1 Locating and Installing the Building Drain

After locating the main stack, begin work on the building drain. The NSPC defines a building drain as "that part of the lowest piping of a drainage system which receives the discharge from soil, waste, and other drainage pipes inside the walls of the building and conveys it to the building sewer beginning 3 feet outside the building wall."

By running the building drain before installing the stack, you will ensure that the sweep at the base of the stack is located at the proper elevation. If the sweep is not accurately located, liquid and solid wastes will not flow properly from the stack to the building drain.

Run the building drain according to the building's plumbing plans. In cases where plumbing plans are not provided, you must determine the route of the building drain. When determining the drain's path, consider the location of the floor drains, stacks, fixture drains, and the invert elevation of the building sewer.

In some installations, the building drain may change direction to connect with the floor drain and the secondary stack. In such cases, you will need to install a sanitary wye to the outlet of the sweep to permit this. Temporarily join the sanitary

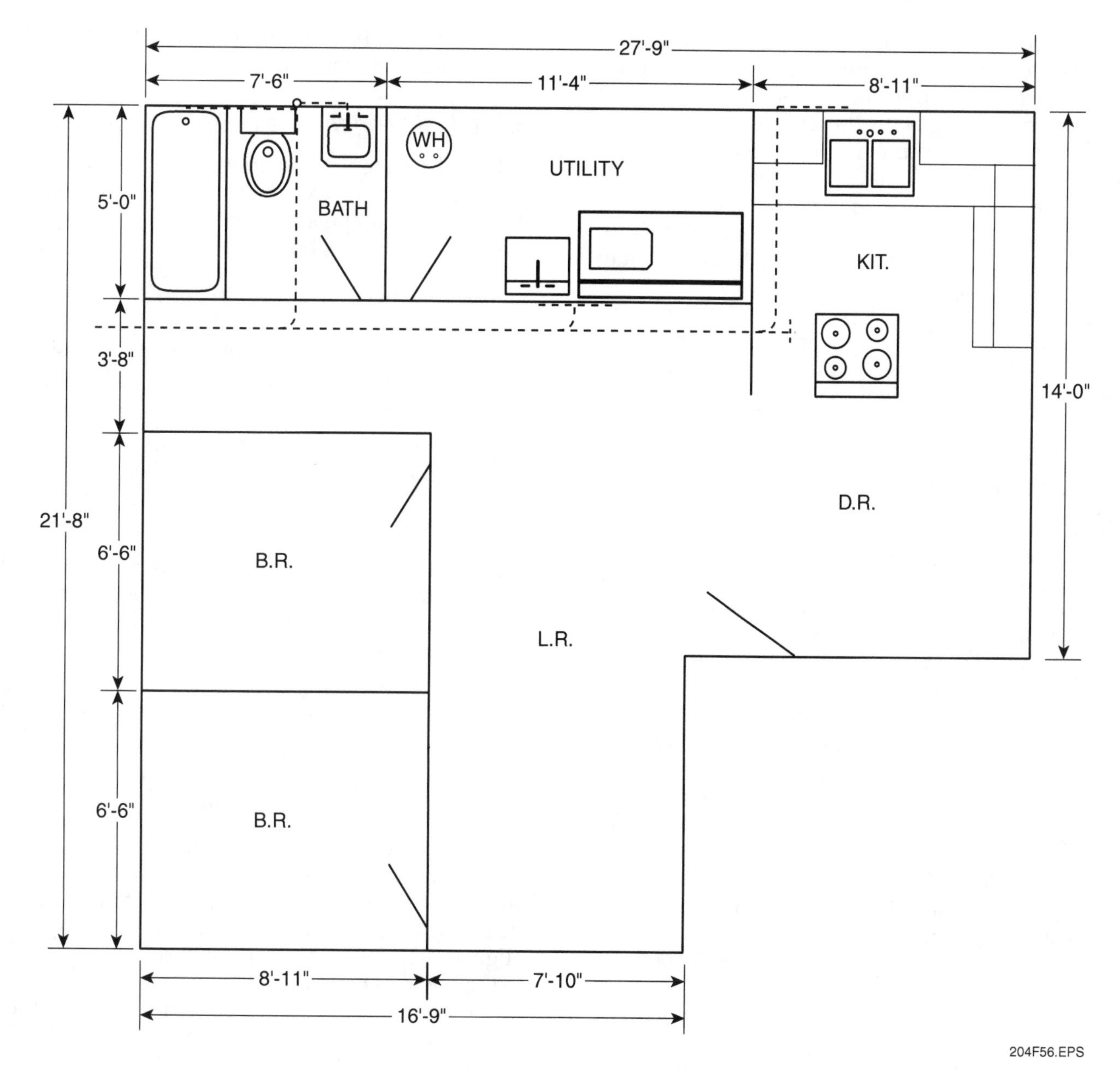

Figure 56 ◆ Route of the building drain shown on the house floor plan.

wye to the exit of the sweep. Then, string a line from the secondary stack to the branch inlet of the sanitary wye. Use this line to make sure the branch inlet of the sanitary wye is pointing in the proper direction. The line also indicates the path of the trench to the secondary stack. After checking for accuracy, join the sanitary wye to the sweep.

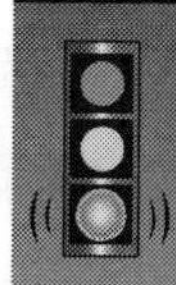

NOTE

The building drain is only one type of drain that you will encounter in a DWV system. Other types of drains include drains for indirect waste and special wastes. You will learn more about locating, sizing, and installing these drains later in the *Plumbing* curriculum.

If the plans call for the size of the building drain to be reduced at the wye, the trench for the next leg of the building drain must be reduced in depth accordingly. A 1-inch reduction of the pipe diameter means you will need to reduce the depth of the trench by ½ inch, or the radius from the center of the pipe to its outer wall.

Note that the method discussed in this section is only one of several methods available to you. Your local applicable code will specify the method that is approved for use in your area.

5.2.2 Extending the Building Drain to the Building Sewer

After locating the center line of the main stack, you can determine where the building drain will pass under the foundation footing and connect to the building sewer. The building drain must extend from the building for a specified distance before connecting with the building sewer. This distance can be found in your local applicable code. For example, the NSPC specifies that the building drain should extend 3 feet outside the building wall. For economy, the building drain and building sewer usually run to the sewer main in a straight line.

Best practice for laying out the center line of the main stack is usually perpendicular to the foundation wall. This is the building drain's path to the building sewer and then to the sewer main.

5.2.3 Calculating the Elevation of the Main Stack Sweep

After calculating the depth of the building drain at the foundation wall, you can determine its rise to the sweep at the base of the main stack. The drain must rise for the stack to drain into it. This information may be provided on the plans. If not, you can obtain it by direct measurement.

Say that the sweep is 6 feet from the foundation wall. At a slope of ¼ inch per foot, the building drain would rise 1½ inches (6 × ¼"=1½"). Next, calculate the elevation of the sweep. In this example, you know that the building drain center line will pass under the foundation wall 30 inches below the finished floor level. You also know that the drain will rise 1½ inches at the center line of the sweep. Subtracting the rise from the depth of the building drain center line, the center line of the sweep will be located 28½ inches below the level of the finished floor (30" – 1½" = 28½").

These measurements are made to the center line of the pipe. Remember to consider the pipe's diameter when calculating the elevation. In this example, this leg of the building drain will be constructed with 4-inch pipe. Add half its diameter, or 2 inches, to the depth of the trench. This is the radius from the pipe's center line to its outer wall. The end of the pipe at the foundation footing will be at a depth of 32 inches (30" + 2" = 32") and rise to 30½ inches (28½" + 2" = 30½") at the sweep.

Once you have determined these measurements, dig the trench from the foundation wall to the sweep. Stretch a string beside the selected path to aid in keeping the trench on course. Ensure that the trench is dug at the proper slope so that wastes will flow properly. Use a spirit level to maintain the correct grade. Dig an enlarged area at the location of the sweep to provide enough space for its installation.

If you dig too deeply, you may need to backfill the trench. This can cause problems because backfill may settle over time, causing the pipe to sag. In cases where backfilling cannot be avoided, use wet sand as fill material.

5.2.4 Installing the Main Stack Sweep

Follow these steps to position and install the main stack sweep in the trench. These steps will ensure that the sweep stays on center and supports the weight of the pipe above it.

Step 1 Suspend your plumb bob from the center of the sanitary tee attached to the closet bend. Mark the point where the plumb bob comes to rest. This is the position of the sweep.

Step 2 Move the sweep around until the plumb bob points to its center. Ensure that the exit of the sweep is at the proper slope and that it points straight down the trench.

Step 3 Brace and support the sweep to prevent it from moving. Refer to your local applicable code for approved bracing and supporting techniques.

Step 4 Pour concrete around the sweep to secure it in position.

5.2.5 Calculating the Elevation of the Secondary Stack Sweep

Before proceeding with the installation, you must determine the vertical center of the secondary stack sweep. This will ensure that the building drain and the secondary stack are properly aligned. The procedure discussed here is similar to the one you used to locate the sweep at the base of the main stack, though there are several other ways to calculate the secondary stack sweep elevation. Refer to your local applicable code to determine the approved method for your area.

Suspend the plumb bob through the center of the hole cut for the stack in the first floor soleplate. The spot at which the plumb bob comes to rest indicates the center of the stack at ground level. With the center of the secondary stack identified, determine the elevation of the sweep. In this case, you can base your calculations on the fitting that is used to connect the floor drain.

When reducing the size of the pipe from 3 inches to 2 inches, remember to adjust the depth of the trench accordingly. In this example, the trench will be ½ inch shallower following the reduction. Subtract this ½ inch from the 2-inch rise. The building drain will therefore rise 1½ inches from the floor drain fitting to the secondary stack sweep. Dig the trench to the proper depth, and extend it to the center line of the secondary stack, which is indicated by the suspended plumb bob.

5.2.6 *Connecting the Secondary Stack Sweep to the Building Drain*

After extending the trench, position the sweep using the plumb bob and string. With the plumb bob still suspended, place the sweep in the trench. Move the sweep around until the plumb bob points to the center of its vertical end. Pour concrete around the sweep to enable it to support the weight of the stack. With the sweep accurately located, positioned, and supported, you can now install the building drain between the floor drain and the sweep.

5.2.7 *Connecting the Building Drain to the Floor Drain*

You can refer to the plumbing plan to locate the floor drain, but as you gain experience you will be able to locate the floor drain yourself. For economy, the floor drain usually is placed near the building drain. After locating the drain, mark the path of the trench. Then, dig a trench at the proper slope and install the pipe. Join a trap to the end of the pipe to turn the drain upward to the floor drain.

The installation of the floor drain is critical. It must be accurately positioned before the concrete floor is poured. Plumb the floor drain above the unfinished floor, and support it in position. As the concrete is being poured, concrete finishers will work the concrete around the drain with a trowel. This will ensure that the floor will slope toward the drain.

5.2.8 *Connecting the Building Drain to the Building Cleanout*

In buildings with basements, most local codes require a cleanout in the building drain just before it exits the building. This is called the **building cleanout**. A building cleanout may be one of the following fittings:

- A cleanout and test tee
- A sanitary tee
- A sanitary wye
- A combination tee-wye

Begin by temporarily placing the cleanout fitting in the trench. Position the fitting so that the branch inlet of the fitting is pointing upward. Ensure that the fitting is an appropriate distance inside the front wall. This distance should be long enough to allow the next length of pipe to be joined to the cleanout fitting without difficulty.

With the cleanout fitting located, you can then install the pipe to connect the sanitary wye at the main stack to the cleanout fitting. Attach a vertical section of pipe to the branch inlet of the cleanout. This will ensure that it is accessible at surface level inside the structure.

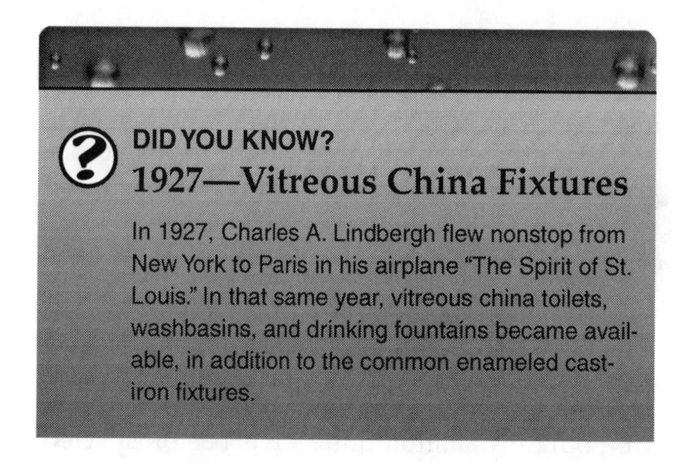

DID YOU KNOW?

1927—Vitreous China Fixtures

In 1927, Charles A. Lindbergh flew nonstop from New York to Paris in his airplane "The Spirit of St. Louis." In that same year, vitreous china toilets, washbasins, and drinking fountains became available, in addition to the common enameled cast-iron fixtures.

5.2.9 *Connecting the Building Drain and Building Sewer to the Sewer Main*

From the cleanout, run the building drain and building sewer to the sewer main. First, establish the path of the drain with a tightly drawn string. Dig the trench with the aid of a backhoe or a trenching machine. Then, extend the building drain from the cleanout to the building sewer. Do not make the connection to the sewer main until you have taken steps to protect the pipe, for example by installing a plug. This prevents sewer gases from entering the structure.

In some jurisdictions, backwater valves must be installed on the building drain. Refer to your local applicable code. Install the valve according to the manufacturer's instructions.

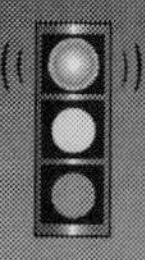

WARNING!

Be aware that toxic and flammable vapors may be present when a sewer tie-in is made. The public sewer is also a biohazard. Use the appropriate personal protective safety equipment and wear hand and eye protection as necessary.

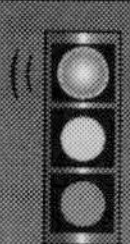

WARNING!

The trench from the building to the sewer is usually the deepest trench you will dig on the job site. Cave-ins and falls are possible, so trench safety is extremely important. You or others may be seriously injured or even killed in a cave-in or fall.

You must use proper shoring to support the sides of the trench to prevent a cave-in. If shoring is not required, you may be required to terrace (step or shelve) the sides of the ditch. Consult the OSHA requirements for restrictions on trench depth without shoring and follow OSHA guidelines on safely shoring or terracing your trench.)

5.2.10 First Rough

After the below-grade piping has been completely installed, it is inspected and tested. This inspection and test is called first rough. In a slab-on-grade installation, the first rough must take place before the slab is laid. The inspection ensures compliance with code requirements. It also ensures that the installation and construction of the finished system are the same as described in the approved plans. The inspection and testing process is covered in detail later in this module.

Never backfill any part of the roughed-in plumbing installation until it has been inspected and approved. Codes prohibit backfilling before a test. It usually carries a stiff penalty.

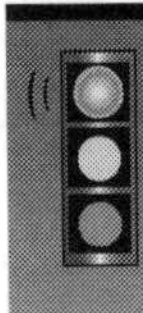

WARNING!

Incorrect DWV installations—such as leaks, breaks, defective installations, poorly located or improper venting, or improper joining—can spread diseases such as typhoid, scarlet fever, infectious hepatitis, poliomyelitis, and various types of dysentery.

5.3.0 Installing Above-Grade Piping

When the plumbing under the slab has been tested successfully, the slab is poured and allowed to **cure.** Carpenters then rough-frame the structure. After the floor has cured, the walls have been erected, and the roof has been put on, plumbers are again called to the site to complete the DWV piping installation. In this section, you will learn how to install the main DWV stack, run the stack vent through the roof, and then connect the stack to the various fixture drains. The procedures described below for the installation of the main stack can be used to install the secondary stack as well.

In a residential installation, the main stack is located within one of the bathroom walls. This is because the bathroom fixture group creates the greatest amount of discharge. When determining the exact location of the stack, consider the following:

- The direction and path of pipe from the fixtures
- Ease of installation
- Economy

Refer to the position of the DWV stack in the floor plan (refer to *Figure 18*). Mark the location of the stack on the center of the soleplate. Remember to center the stack in the wall. This will prevent nails and other fasteners from damaging the pipe. The location also allows more flexibility when placing fittings within the wall.

5.3.1 Locating the Stack Run within the Wall Frame

Now you will be able to determine how to transfer the center of the stack from the basement up through the roof. The stack must be precisely vertical. To ensure this, use a plumb bob and string (see *Figure 57*).

Suspend the plumb bob from the center of the hole that you cut in the floor for the stack. After the plumb bob has stopped moving, it will indicate the location of the center of the sweep at the base of the stack (see *Figure 58*).

Use the following steps to determine the location of the exit for the stack through the roof:

Step 1 Hold the string against the top plate and move it around until the plumb bob comes to rest in the center of the hole cut in the soleplate (see *Figure 59*). Once the plumb bob stops moving, the point where the string touches the top plate marks the center of the path the pipe will take through the top plate.

Step 2 Cut a hole in the **top plate.** The top plate is the topmost horizontal frame member, to which the studs or rafters are fastened.

Step 3 Using the center of the pipe path, bore a hole of the appropriate size for the stack to pass through.

Step 4 Now hold the plumb bob string against the roof **sheathing,** which is the wooden boards or other materials placed over the exterior studs or rafters. Allow the plumb bob to hang through the hole in the top plate to the hole in the soleplate. Move the string around until the plumb bob comes to rest in the center of the hole cut in the soleplate (see *Figure 60*). The point where the string touches the roof marks the exit of the stack through the roof.

Step 5 Mark this location, and cut an appropriately sized hole in the roof.

5.3.2 Locating the Center Line of the Main Stack

Use your plumb bob to locate the center of the sweep at the base of the main stack. Suspend the plumb bob through the center of the hole that you bored through the soleplate for the stack. Lower the plumb bob until it is approximately ¼ inch off the ground. This point represents the location of the center of the main stack at the level of the unfinished floor.

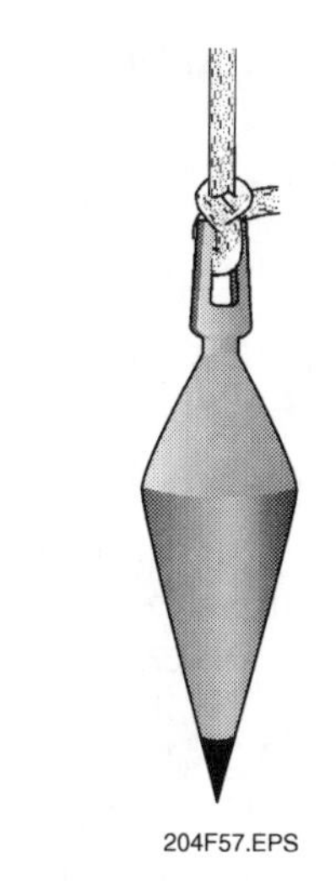

204F57.EPS

Figure 57 ◆ Plumb bob.

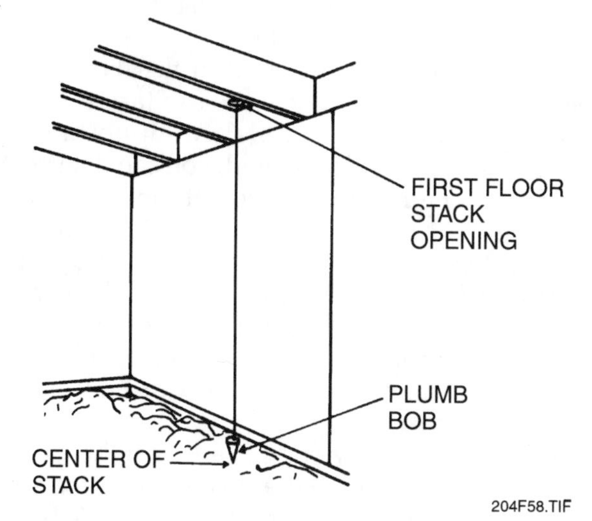

204F58.TIF

Figure 58 ◆ Transferring the center of the stack.

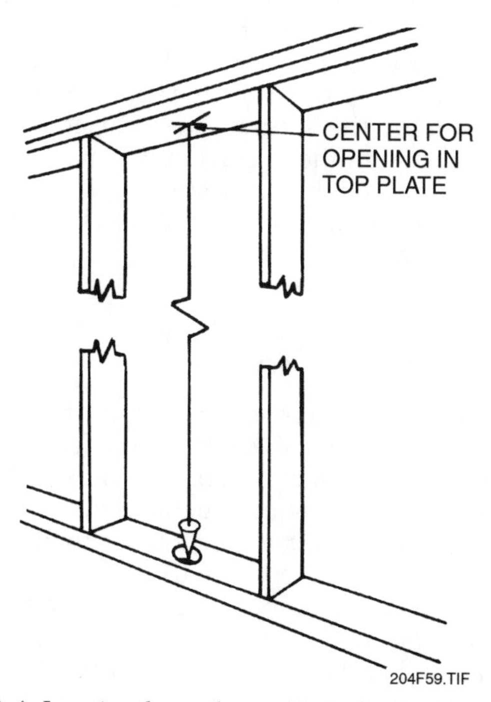

204F59.TIF

Figure 59 ◆ Locating the stack opening in the top plate.

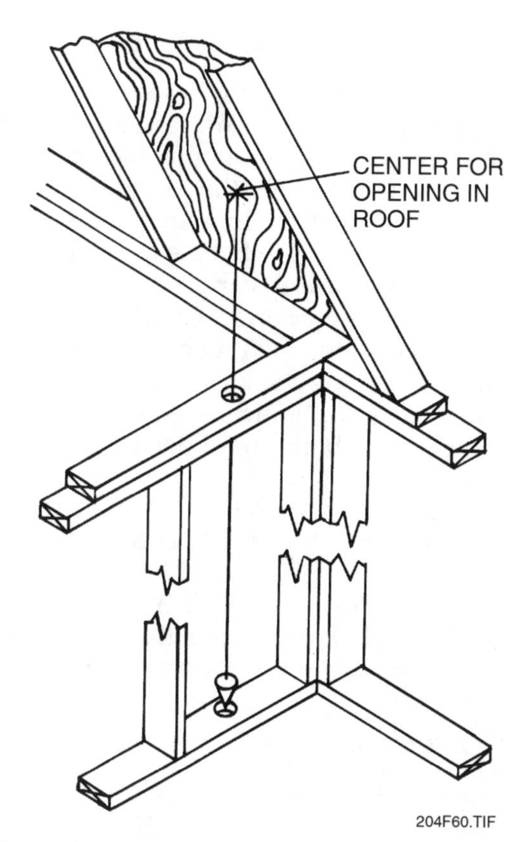

204F60.TIF

Figure 60 ◆ Locating the stack exit through the roof.

5.3.3 Locating the Center Line of the Secondary Stack

Next, determine the location of the secondary stack. Suspend the plumb bob through the hole that you cut for the stack. Lower the plumb bob until it is suspended approximately ¼ inch above the ground. Mark that point on the ground. As with the main stack, this point represents the location of the center of the secondary stack at the level of the unfinished floor.

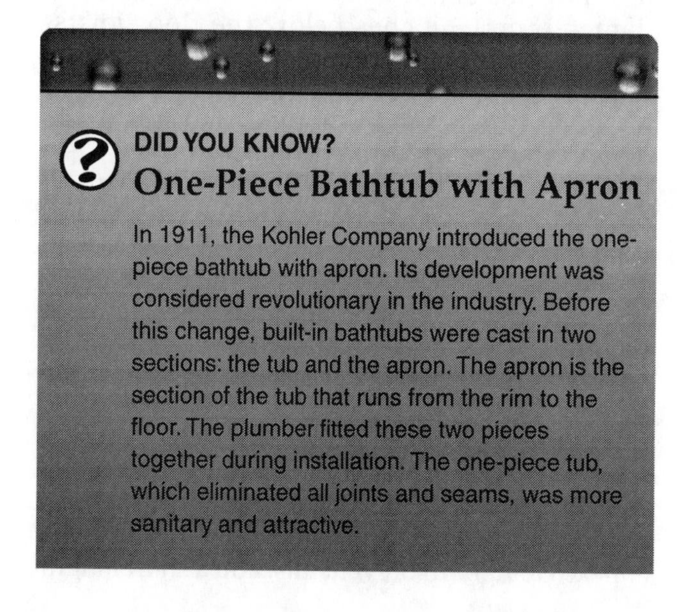

DID YOU KNOW?

One-Piece Bathtub with Apron

In 1911, the Kohler Company introduced the one-piece bathtub with apron. Its development was considered revolutionary in the industry. Before this change, built-in bathtubs were cast in two sections: the tub and the apron. The apron is the section of the tub that runs from the rim to the floor. The plumber fitted these two pieces together during installation. The one-piece tub, which eliminated all joints and seams, was more sanitary and attractive.

5.3.4 Selecting the Closet Fitting

The fitting that connects the water closet to the stack is extremely important. It accurately positions the sweep at the base of the stack in relation to the building drain. Furthermore, its installed height determines the slope of fixture branch lines. You should ensure that the location, height, and type of fitting are all correct.

When selecting the correct closet fitting for the installation, consider the number of fixtures and the routes of their drains. This will help you determine the number of branches required by the fitting. Refer to the bathroom floor plan shown in *Figure 5*.

Refer to guidelines in your local applicable code to select an approved fitting. After consulting your local code, say that you select a 3" × 3" × 3" sanitary tee with a side inlet. Use a 3" × 3" × 1 ½" sanitary wye to join the sanitary tee. Join the water closet at the end and the bathtub at the branch inlet.

5.3.5 Determining the Closet Fitting Height

With the proper closet fitting selected, you can now determine its height in the DWV stack. Generally, you would measure its height by the slope of the branch line from the water closet. The water closet drainpipe is larger than the lavatory and bathtub drain pipes; given an equal slope, it needs more space than smaller drain lines. To determine the fall of the water closet branch line, measure the distance between the water closet and the stack and multiply that distance by the slope. Say that the distance in this example is 4 feet. At a slope of ¼ inch per foot, the line would fall 1 inch (4 × ¼" = 1").

Let's assume that the start of the run of pipe will begin with the pipe touching the lower surface of the floor joists. Using 3-inch pipe, this means that the center line of the pipe would be one-half of its diameter, or 1½ inches below the floor joists. Remember, measurements are made to the pipe's center line. This is important because the height of the fitting is determined by measurements made to the center line of the branch inlet.

The height of the sanitary tee is in relation to the lower surface of the floor joists. The center line of drain pipe begins 1½ inches below the floor joists and falls another 1 inch before joining the sanitary tee. Therefore, the center line of the branch inlet of the sanitary tee must be a minimum of 2½ inches below the floor joists. This accurately locates the height of the fitting.

You will have to adjust to problems that may occur. For example, where working clearance is a problem, you may have to place the fitting at a lower height. However, you cannot allow the slope to change. To prevent this, add a short length of pipe to the vertical drain from each fixture. The assembled length of the pipe will correspond to the distance the fitting was lowered to obtain working clearance.

5.3.6 Connecting the Fitting to the Closet Flange

With the position of the sanitary tee known, you can determine the best method to couple it to the closet flange. Either a closet bend or a quarter bend and a short length of pipe can be used to connect the closet flange to the sanitary tee. Let's assume that you select a closet bend. Place the closet bend under its hole in the bathroom floor. Place blocking equal to the thickness of the finished floor under the flange. Secure the flange in position to prevent it from moving. You can use a weight or temporarily fasten the flange to the floor.

Next, insert the closet bend in the flange from below the floor level. Place the sanitary tee in its correct position. By temporarily supporting the sanitary tee and the closet bend, you can determine the correct length of the horizontal run of the closet bend. Measure the distance between the two.

If the closet bend is too long, cut it to the proper length. If it's too short, add a length of pipe between the bend and the sanitary tee. Now temporarily assemble the bend and the tee, place the assembly in its correct position, and support it from beneath.

Next, determine the correct length of the vertical run of the closet bend. If the closet bend is too long, as it will be in most cases, mark it to indicate where it should accurately join the flange and cut it to the proper length. Sometimes, the vertical run of the closet bend will be too short, so you'll have to measure the distance between the bend and flange and cut a length of pipe to join the bend and flange.

Assemble the sanitary tee, closet bend, and closet flange in their correct position. Then, raise the entire assembly and support it in position. Make sure the sanitary tee and the closet bend are centered correctly.

5.3.7 Connecting the Sweep to the Sanitary Tee

With the sweep temporarily supported in place, you can install part of the stack (see *Figure 61*). Place the cleanout on the sweep, making sure that the branch of the cleanout is pointed in an accessible position. Then, join the cleanout tee to the sweep.

Next, connect the cleanout to the closet fitting with a length of pipe. Be sure to maintain the proper slope of the closet bend by accurately measuring and cutting this length of pipe. First, measure the distance between the stack and sanitary

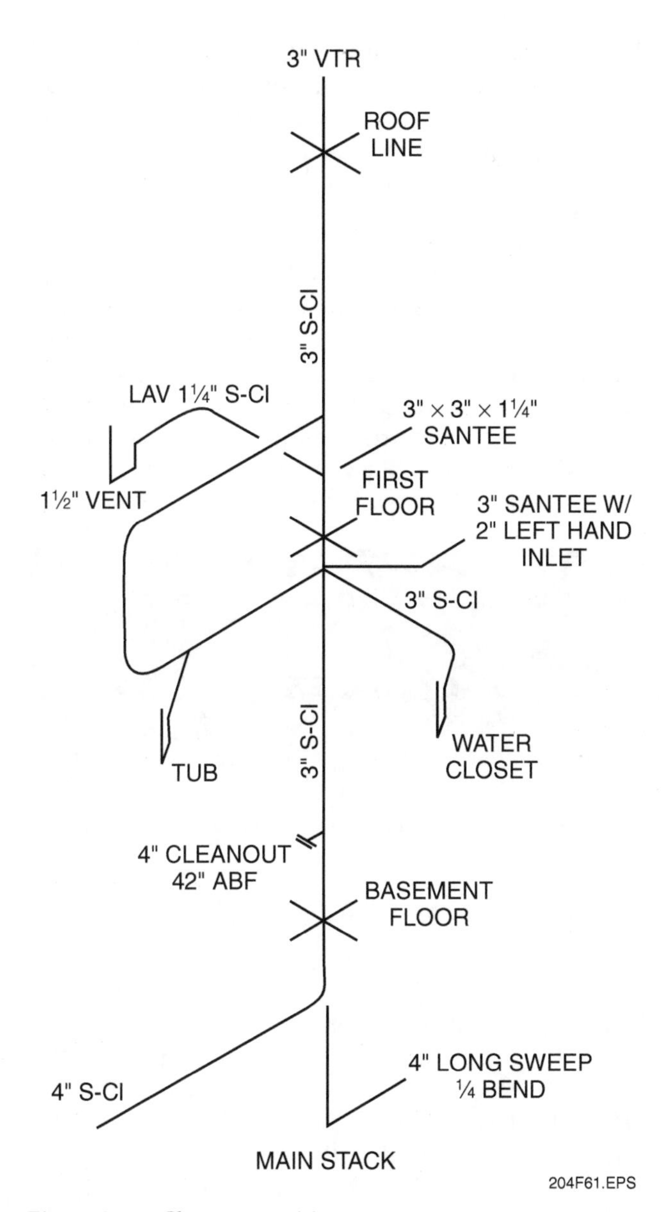

Figure 61 ◆ Cleanout position.

tee. Let's say this distance is 70 inches. Use direct measurements or manufacturer's tables to determine the distance the pipe and the sanitary tee spigot will slip down into the hub of the pipe. For a 3-inch diameter pipe, this is roughly 2½ inches per joint. You have two joints, so add the distance plus the joints (70" + 2½" + 2 ½"). You should cut a pipe that is 75 inches long.

A simpler method of determining this length is to hold a length of pipe next to the sanitary tee (see *Figure 62*). Begin by placing the hub next to the sanitary tee at the height it would be after the joint is made. Next, make a mark on the pipe where it would bottom in the hub of the cleanout fitting. Cut the pipe to that length. Lift the sanitary tee and closet flange off their supports to install the section of pipe. After ensuring that the fit is correct, make up the joints.

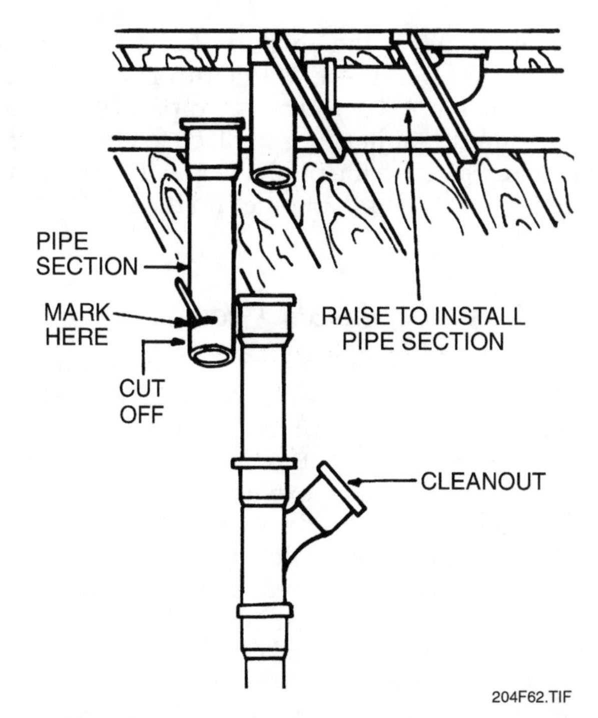

Figure 62 ◆ Determining the correct length of pipe.

5.3.8 Installing the Lavatory Drainage Fitting

Before you can run the stack through the roof, you must determine the elevation of the fitting joining the lavatory branch line. This is necessary to ensure proper slope of the lavatory drain.

Begin by consulting the manufacturer's rough-in measurements for the lavatory. Refer to *Figure 3C*. You can see that the center line of the drain enters the wall 18¼ inches above the floor. Remember that you must add the thickness of the finished floor to this measurement. Let's assume that the finished floor will be 1⅛ inches thick. So the center line of the lavatory drain is 19⅜ inches off the unfinished floor (18¼" + 1½"). Because the walls are only rough-framed at this stage of construction, mark this dimension on the nearest stud. To run the lavatory drain simply and economically, run it horizontally to the stack.

Next, determine the type of fitting you will use to connect the lavatory branch line to the stack. In this case, a 3" × 3" × 1¼" sanitary tee would be best. Now, determine the height of the fitting on the DWV stack. Start back at the lavatory drain and calculate the slope to the stack. You know that the lavatory drain is located 19⅜ inches above the surface of the roughed-in floor. From that point, the branch line will run approximately 1 foot to the DWV stack. With a slope of ¼ inch per foot, the branch line will drop only ¼ inch. This also means that the center line of the inlet to the sanitary tee will be 19⅛ inches above the surface of the floor.

Now you can measure the length of pipe that will run from the sanitary tee at the closet bend to the sanitary tee at the lavatory drain. Use direct measurement or the manufacturer's rough-in measurements. After you cut the pipe, dry fit (trial fit) it with the sanitary tee in place to ensure that you've maintained the proper slope. If the slope is correct, join the length of pipe and the sanitary tee to the stack.

5.3.9 Extending the Stack Vent

With all fittings used to connect branch lines in place, you can now extend the DWV stack through the roof. The procedure is simple, consisting of measuring, cutting, and joining lengths of pipe until the stack extends a minimum of 6 inches above the high side of the roof. Check your local applicable code for details.

In some areas, you may reduce the diameter of the stack above the last branch line. This is mainly an economic decision. The smaller size is less expensive and easier to install, so labor costs are lower. Always check the local code to determine if and to what extent the stack diameter may be reduced.

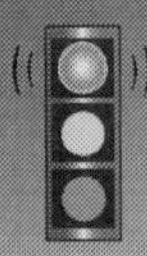

WARNING!
Never work unprotected on a roof. Fall protection is required when working at any height over 6 feet.

In colder climates, local codes may require the addition of increasers to the stack. **Sanitary increasers** (see *Figure 63*) are necessary in cold climates to prevent condensing water vapor from freezing and gradually closing the vent opening. The loss of the vent will cause a pressure differential aspiration in the system that could result in the siphoning of the trap seals. Loss of the trap seals would allow sewer gas to enter the structure. Refer to your local code requirements to determine whether an increaser is required.

5.3.10 Flashing the Stack

The final step in the process of running the stack is to install vent flashing. Depending on the location, the project, and the contract requirements, this may be a job for the plumber, the carpenter, or the roofer. Regardless of who installs the flashing, however, the plumber must ensure that it is installed correctly. Check with your supervisor, the job foreman, or the construction supervisor to determine whether your job includes the installation of vent flashing.

Vent flashing prevents rainwater from running down the outside edges of the pipe and entering the building. It is commonly made of lead and plastic. Two types of vent flashings are in common use (see *Figure 64*). The first type totally covers the exposed portion of the stack. The upper portion of the flashing is tucked into the top of the stack and the lower portion is nailed to the roof. In the second type of flashing the upper portion extends only partway up the exposed pipe. The hole in the flashing is smaller than the pipe's diameter, so the flashing fits tightly against the pipe, thus preventing any leaks. Caulking placed around the seal further strengthens the seal and guards against leaks. The height at which the flashing joins the stack is the main difference between the two types.

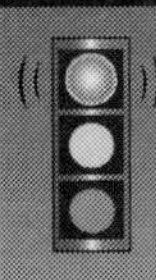

WARNING!
Incorrectly installed vent flashing can allow sewer gases to re-enter a building, causing illness or even death. Always ensure that vent flashing is installed according to your local code. Also ensure that the vent opening has been sized according to your local code.

5.3.11 Connecting the Fixture Drains

After you have finished installing the stack, you can run the various fixture drain lines. Much of this work has already been done. You've already marked the location of the fixture drains on the walls and floors. You've correctly positioned the fitting that will connect the branch line to the stack. You've calculated the slope of each branch line to determine the correct placement of the fitting. Therefore, to join the fixture drains, you only need to measure, cut, join, and brace pipe.

For the water closet, all you will need to do is make up the joints at the stack fitting, sanitary wye, closet bend, and closet flange. For the bathtub, measure and cut a length of drainpipe to run from the branch inlet of the sanitary wye to the bathtub drain. Then, calculate the length of pipe needed to run vertically from the horizontal drainpipe to the bathtub waste outlet.

Note the back outlet, waste, and overflow. To accurately determine these measurements, refer to the manufacturer's rough-in measurements (refer to *Figure 3D*). You'll find that the waste outlet of the bathtub is 1⅞ inches above the surface of the rough floor. Measure the distance from the horizontal run of pipe to an imaginary point 1⅞ inches above the rough floor level. Note that the length of both runs of pipe is measured in reference to the 90-degree ell that will join them. After cutting the pipe to the correct length, assemble the branch line starting from the sanitary tee. After

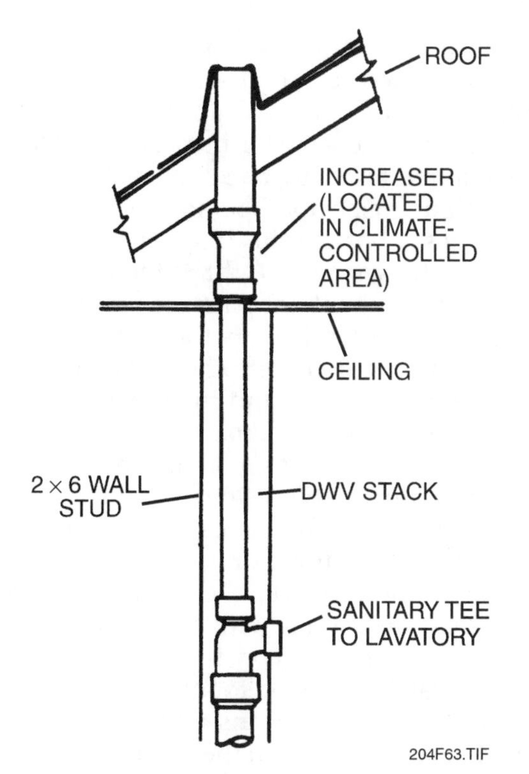

Figure 63 ◆ Sanitary increaser.

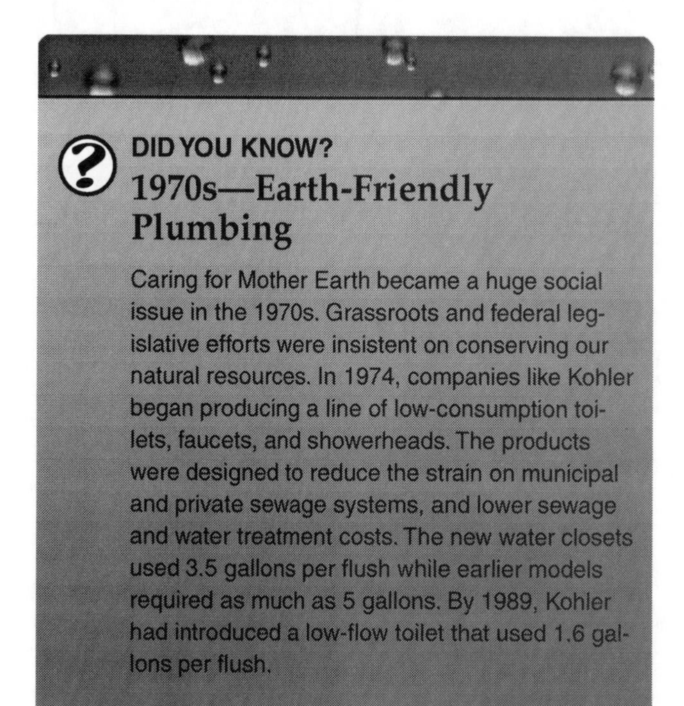

DID YOU KNOW?

1970s—Earth-Friendly Plumbing

Caring for Mother Earth became a huge social issue in the 1970s. Grassroots and federal legislative efforts were insistent on conserving our natural resources. In 1974, companies like Kohler began producing a line of low-consumption toilets, faucets, and showerheads. The products were designed to reduce the strain on municipal and private sewage systems, and lower sewage and water treatment costs. The new water closets used 3.5 gallons per flush while earlier models required as much as 5 gallons. By 1989, Kohler had introduced a low-flow toilet that used 1.6 gallons per flush.

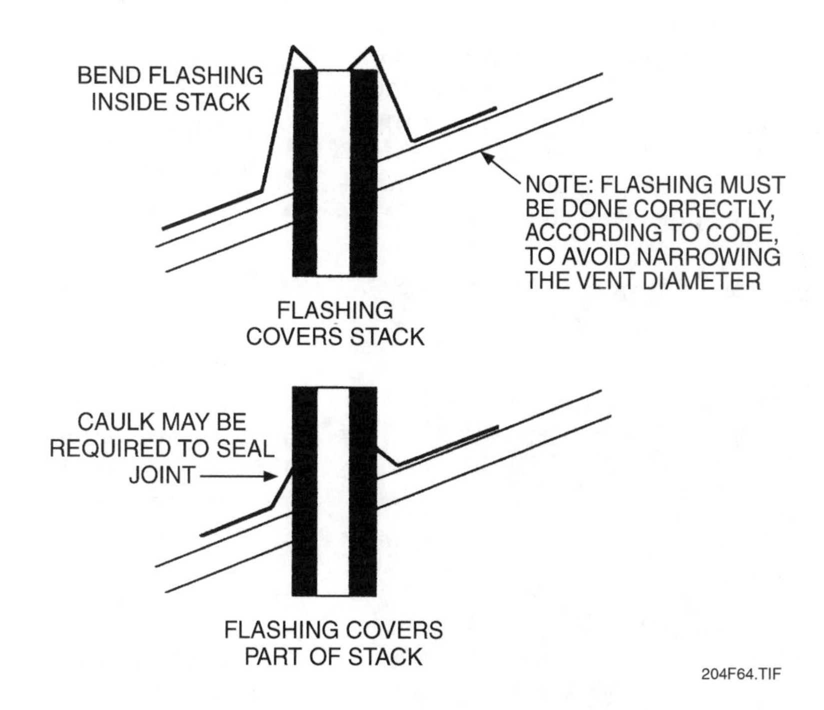

Figure 64 ◆ Vent flashing.

final assembly, check to see that the slope is correct and that the vertical length of pipe rises $1\frac{7}{8}$ inches off the rough floor.

Next, install the lavatory drain line. Because the wall is rough-framed, you will have to mark the height of the center line of the lavatory drain on the nearest stud. You will find this information on the manufacturer's rough-in measurements. When measuring the length of the horizontal run of pipe, measure to the true point of intersection with the lavatory drain. Cut a length of pipe and join it to the stack. Check the accuracy of your work.

5.3.12 Second Rough

As with below-grade DWV piping, the above-grade DWV installation is also thoroughly inspected and tested when it has been completed. This round of inspection and testing is called second rough, or the in-wall inspection and test. The inspection and test are conducted after the fixtures are set. That way, the test will reveal the location of leaky traps or other unsatisfactory conditions. The inspection and testing procedure is discussed in detail in the next section.

5.4.0 First and Second Roughs

The process of conducting the inspections and tests in the first and second roughs is explained in your local applicable code. The information in this section serves as an introduction to the inspection and testing process. It also explains the plumber's role and responsibilities during the first and second roughs.

The person doing the work authorized by the work permit must notify the plumbing inspector that the work is ready for inspection. The inspector may require that every request for inspection be filed a certain number of working days before the inspection is wanted. These requests can be made in writing or by telephone, though a formal written request is preferred. Be sure to note the request and the scheduling information in the **daily log**. The daily log, as its name suggests, is the daily description of all work completed on the project. The site supervisor maintains the daily log.

> **DID YOU KNOW?**
>
> **A Sinking Bathtub Spurs a White House Renovation**
>
> In 1948, Harry S. Truman was elected president. Under his administration, the White House underwent nearly $5 million in renovations. Several experts were in favor of tearing down the building rather than trying to correct the tangle of pipes, wires, and inefficiencies that had developed over the years. The president was spurred into action because he noticed that his bathtub was sinking into the floor.
>
> This same year witnessed the increased use of a powder room or guest bathroom in residential building design. It was advertised as a benefit by "saving traffic through the house, making toilet training easier, and a welcome convenience for guests."

If you request the tests, you must provide access to the work and the means for proper inspection. You must furnish all equipment, material, and labor necessary for inspections or tests.

The required tests must be done in the presence of a plumbing inspector. You should conduct the pressure tests before the inspector arrives to ensure that the system is tight. If the plumbing is not completed or will not pass the required tests, you will have to reschedule the inspection. The best method is to complete the installation, perform a preliminary test, and then work on something else until the inspector arrives to witness the formal test. In many areas, you may be allowed to keep the pressure in the system until the inspector arrives. This procedure saves time for both the inspector and the contractor.

The inspector approves the plumbing system, certifies that it complies with the local code, and posts the certificate in a conspicuous place. These certificates can be revoked for any violation of the prevailing code. This means that it is illegal for anyone to make changes in the installed and certified piping system that would violate the code. Even in areas where no plumbing code exists, an inspection may be required to satisfy the agency that finances the construction.

If tests show defects, the defective work or material must be replaced and the test repeated. In all cases the inspector will designate the points at which the pressure is relieved or drawn off.

Tests are not required after:

- Repairing or replacing old fixtures, fittings, faucets, or valves
- Forcing out stoppage
- Repairing leaks
- Relieving frozen pipes and fittings

These four cases are the most common exceptions to the requirement for testing and inspection. In addition, no tests or inspections are required where building or drainage systems are set up solely for exhibition purposes.

Repairs are limited to replacing defective components in a piping system. Installing new vertical or horizontal lines of soil, waste, vent, or interior leader pipes is considered new work. Even the change of relative location of these pipes qualifies as new work. In cases of buildings condemned because of unsanitary conditions in the plumbing, the code for new plumbing must be followed. Four different kinds of tests are used with DWV piping:

- Water test
- Air test
- Smoke test
- Peppermint test

The water and air tests are the most frequently used. The other tests are used mainly to locate leaks that are difficult to find.

5.4.1 Testing Tools and Equipment

The tools and equipment you will use to conduct the test include test plugs. Two types of test plugs are illustrated in *Figure 65*. Mechanical test plugs block fitting openings by expanding a rubber gasket inside the fitting.

Test plugs are inflated with air to close the opening (see *Figure 66*). Note that you can use the extension hose to inflate the test plugs, which are placed inside the pipe.

The water removal plate allows a test plug to be deflated so the water can flow out of the piping system (see *Figure 67*). The plate is installed as shown to prevent the test plug from going down the pipe. The water removal plate also retains most of the water within the pipe.

The test gauge assembly is connected between the pump and the piping system (see *Figure 68*). It allows the pressure to be held in the pipe and the pump to be removed. The test pump enables the plumber to inflate the test plugs and to apply the required pressure to the piping system. The pump is equipped with a check valve and a gauge.

5.4.2 The Water Test

In the water test, all openings in the drainage pipes are closed except those at the tops of the stacks. The openings can be closed by test plugs or by pipe fittings such as caps or plugs.

Where test pressures are less than about 5 pounds per square inch (psi) and the pipes are large (for example, concrete or vitrified-clay pipes), other types of plugs are used. A bag of sand or clay around which oakum or similar material is lightly caulked is one alternative. A wooden diaphragm surrounded by an inflated inner tube and supported by a sandbag is a second alternative. Some companies produce a special rubber diaphragm with flaring edges to seal large pipes. Quick-setting plaster of Paris and Portland cement formed in the pipe can serve as a plug. When a test is to be made on a waste pipe that is already connected to the sewer main, the house drain can be plugged by inserting a wye-test plug through a cleanout opening. If no other alternative is available, a section of the pipe can be broken out to permit installation of a plug.

When all openings (except at the tops of the stacks) are closed, water is run into the pipes until it overflows from the top of a stack. The water level may drop immediately after the initial filling

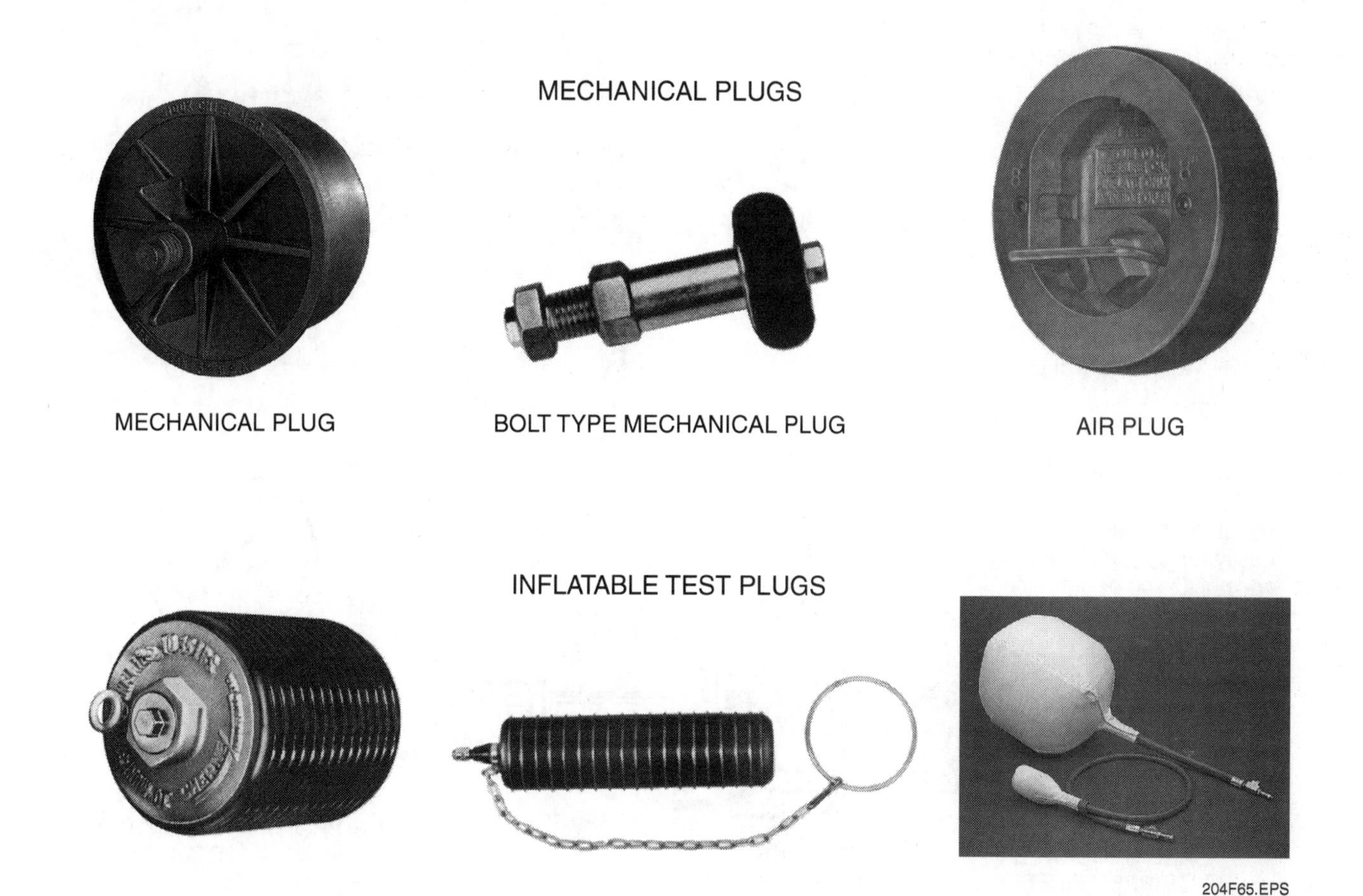

Figure 65 ◆ Mechanical and inflatable test plugs.

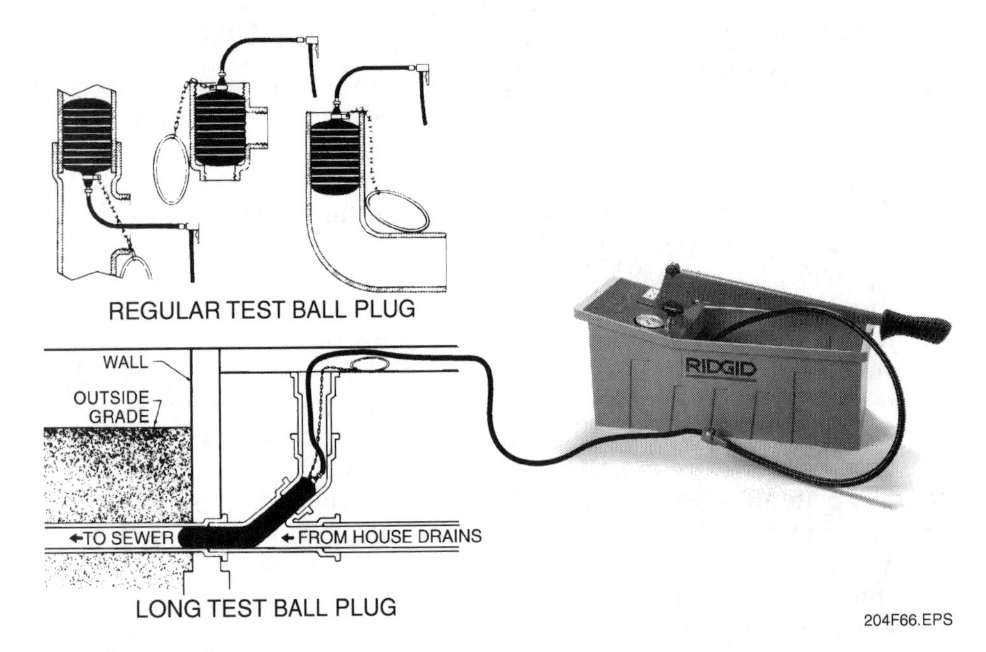

Figure 66 ◆ Test plug installation and test pump.

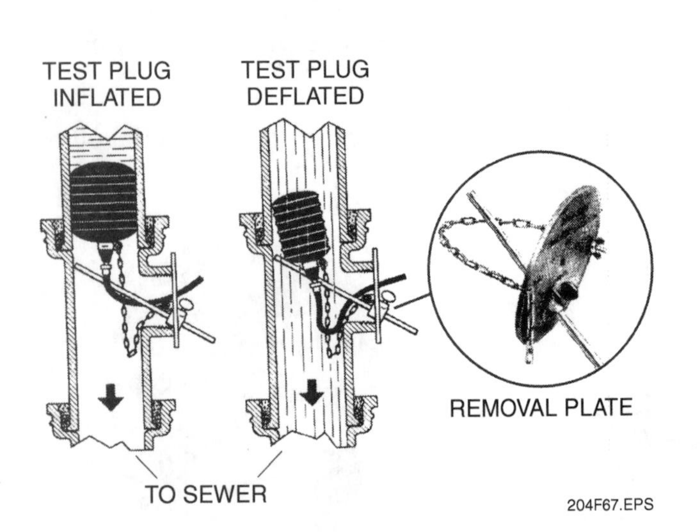

Figure 67 ◆ Water removal plate.

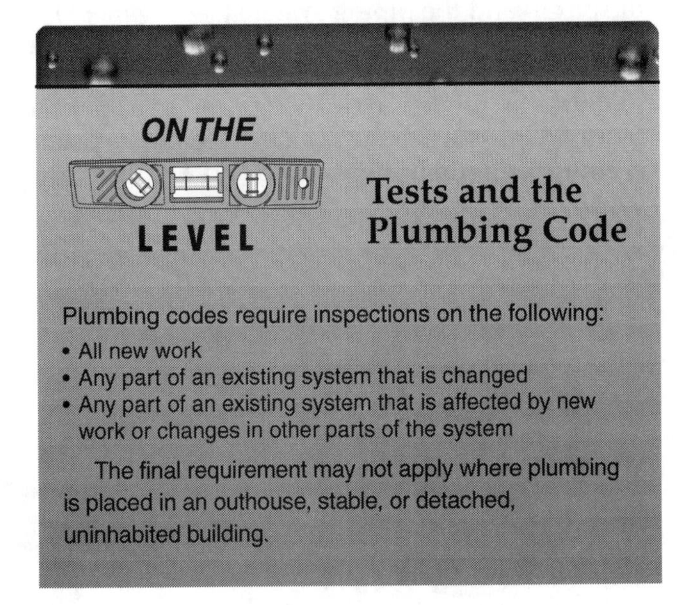

Tests and the Plumbing Code

Plumbing codes require inspections on the following:

- All new work
- Any part of an existing system that is changed
- Any part of an existing system that is affected by new work or changes in other parts of the system

The final requirement may not apply where plumbing is placed in an outhouse, stable, or detached, uninhabited building.

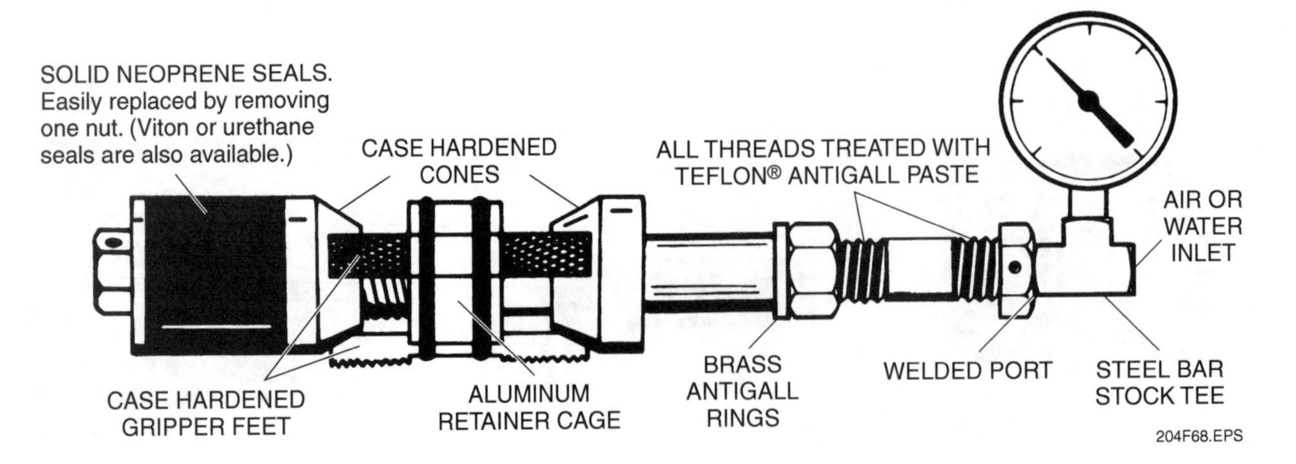

Figure 68 ◆ Test gauge assembly.

because of air pockets in the piping system. This does not indicate a leak. Simply refill the stack until the air pockets are eliminated. The water can be introduced in one of three ways:

- Into the top of a stack with a hose
- Through a connection temporarily installed in a cleanout or other convenient opening
- Through the hollow handle of a mechanical test plug

If the water level drops after the pipes are full, there is a leak. During a satisfactory test, the water level in the pipe should remain stationary for at least 15 minutes. All parts of the system should be subjected to a pressure of at least 10 feet of water (4.34 pounds per square inch gauge [psig] pressure). It is not desirable to use pressures over 30 to 40 feet of water. Higher stacks should be tested in sections of 10 to 40 feet. In high-rise buildings this piping is tested in sections, one story at a time. Sectional testing of DWV piping is frequently required for horizontal piping installed in commercial buildings. In these cases, the horizontal piping is sectioned off by adding test tees. The inspector will be required to witness the test of each individual section of the piping system.

5.4.3 The Air Test

The air test is not recommended for testing plastic or cast-iron pipe. In the air test, all openings are closed and an air pressure of at least 5 psi is exerted in the pipes for at least 15 minutes. Falling pressure, as shown by a sensitive gauge attached to the pipes, indicates a leak. A large air leak can usually be detected by sound. Where little or no sound is made and you suspect a leak, you can find its location by applying a smoke or odor test to the piping system. Air tests are useful in cold weather when a water test could freeze and damage the pipes. The air test also allows you to maintain uniform pressure throughout the section being tested. However, the air test is not as simple to apply as the water test, and it is sometimes more difficult to discover leaks with the air test.

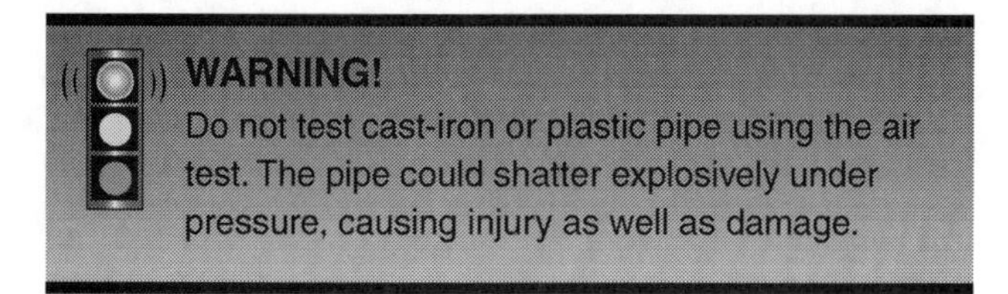

WARNING!

Do not test cast-iron or plastic pipe using the air test. The pipe could shatter explosively under pressure, causing injury as well as damage.

The test pressure for an air test plug using a manometer or test can is about 1 inch of water column (see *Figure 69*). If, after all openings have been closed, this pressure can be maintained for 15 minutes without additional pumping, the system can be considered airtight. You may have some difficulty in interpreting the pressure readings if the pressure falls slowly because the air has cooled and not because of a leak. When in doubt, restore the air pressure to the original level without releasing air already in the pipes. If the rate of decrease of pressure is slower, the loss of pressure is probably caused by temperature change and not a leak. If the test shows that the pipes are not airtight, you must locate the leak.

To locate a leak, spray or squeeze a solution of liquid soap onto the joints in question. If a steady flow of air bubbles up through the soap solution, you have found the leak.

5.4.4 The Smoke Test

In the smoke test, oily waste, tar paper, or similar materials are burned to produce a thick smoke that is blown into the DWV piping (see *Figure 70*). Smoke bombs, which look like large firecrackers, can be dropped inside the stack to produce a large volume of brightly colored smoke. This procedure is much simpler than using a smoke generator. Do not create smoke by mixing chemicals such as ammonia and muriatic acid. This method produces toxic smoke and dangerous particles that are hard to see.

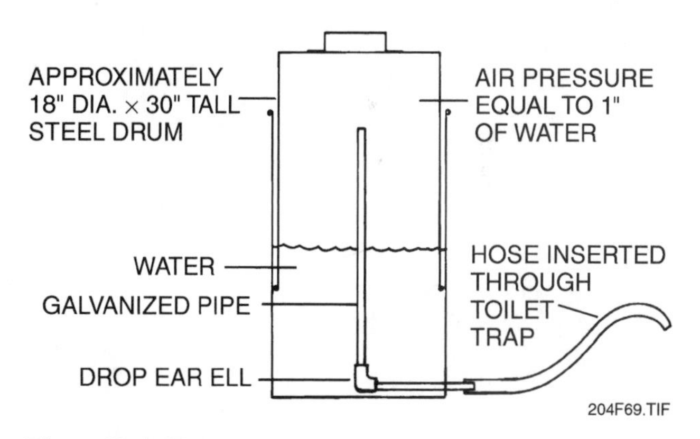

Figure 69 ◆ Test can.

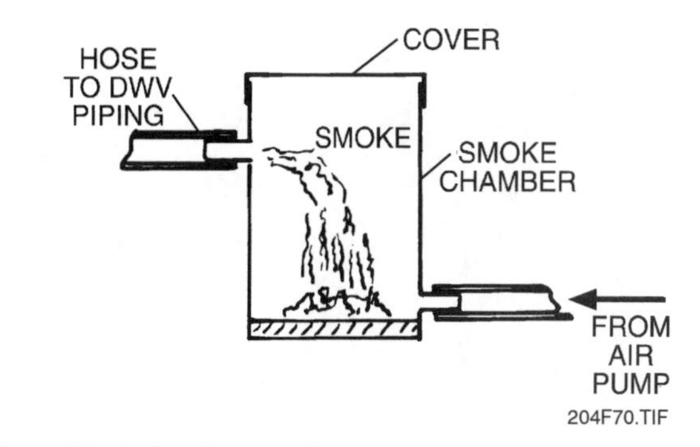

Figure 70 ◆ Smoke chamber.

When the smoke begins to escape from the top of the stack, close the stack with a test plug and increase the pressure to 1 inch of water. This pressure should be maintained until the end of the test. If the pressure cannot be maintained, a leak is indicated. To find it, look for the escaping smoke. Because the leaks are sometimes too small for the escaping smoke to be seen, close windows and doors to retain the odor of the escaping smoke and locate the leak by scent.

5.4.5 The Peppermint Test

You can use an odor test if other tests fail to locate the leak. Use oil of peppermint to create the odor (ether is sometimes used instead). Close the outlet end of the drainage system and all vent openings except the top of one stack. Empty about 1 ounce of oil of peppermint for every 25 feet of stack (but not less than 2 ounces) down the stack. Pour a gallon or more of warm water into the stack and immediately close the top of the stack.

Add air pressure to force the odor through the leak. Search for the leak using your sense of smell. Anyone who has recently handled peppermint should not enter the building until the search is complete. The odor test is simpler than the smoke test, but results are not always satisfactory because there may have been insufficient pressure in the pipes to force the odor through a leak. Also, it may be difficult to locate the leak after you've detected the odor.

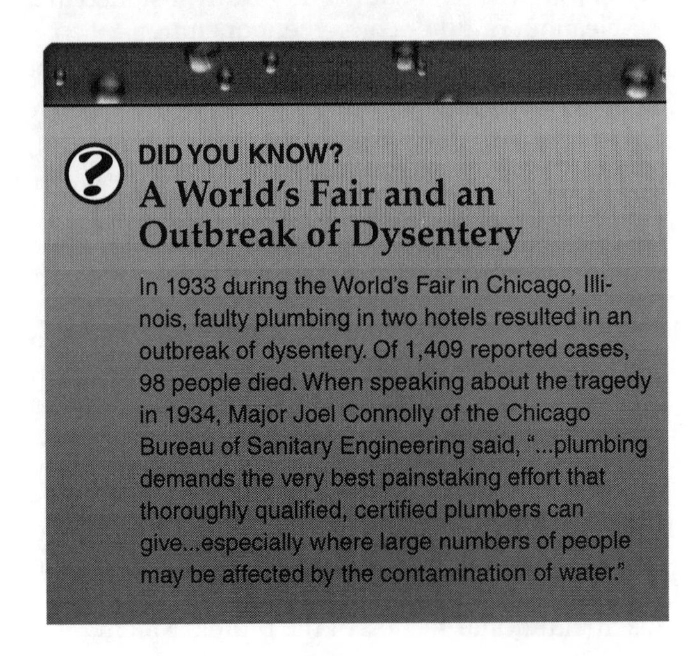

DID YOU KNOW?
A World's Fair and an Outbreak of Dysentery
In 1933 during the World's Fair in Chicago, Illinois, faulty plumbing in two hotels resulted in an outbreak of dysentery. Of 1,409 reported cases, 98 people died. When speaking about the tragedy in 1934, Major Joel Connolly of the Chicago Bureau of Sanitary Engineering said, "...plumbing demands the very best painstaking effort that thoroughly qualified, certified plumbers can give...especially where large numbers of people may be affected by the contamination of water."

Review Questions

Sections 4.0.0–5.0.0

1. The formula for calculating the grade in a DWV piping system is _____.
   a. Grade = Fall ÷ Run
   b. Grade = Run ÷ Fall
   c. Grade = Fall × Run
   d. Grade = Elevation × Fall ÷Run

2. Percentage of grade can be expressed mathematically as _____.
   a. PG = 100(F/R)
   b. PG = 100(G/R)
   c. PG = F(R/100)
   d. PG = 10(F/R)

3. When using a builder's level to determine the elevation for lines of pipe that are to be run above the height of the builder's level, position the stadia rod _____.
   a. upside down
   b. below grade
   c. at the center line of the pipe
   d. above the line of sight

4. To secure DWV piping to prevent movement from backfilling and soil compaction, _____.
   a. drive two rebars into the ground on either side of the pipe, perpendicular to the pipe and parallel to each other
   b. run the pipes through protective sleeves
   c. drive two rebars into the ground on either side of the pipe at opposing 45-degree angles to create a 90-degree V
   d. drive two rebars into the ground on either side of the pipe and use pipe clamps to secure the pipe to the rebars

5. You are installing a DWV system in a building with slab-on-grade foundation. The footing is 17 inches below the finished floor. The center line of the pipe must pass 9 inches under the footing. The depth of the center line of the building drain will therefore be _____ inches below the finished floor.
   a. 8
   b. 26
   c. 30
   d. 153

6. In installations where the building drain changes direction to connect with the floor drain and the secondary stack, install a _____ to the outlet of the sweep.
   a. sanitary reducer
   b. sanitary tee
   c. check valve
   d. sanitary wye

7. The NSPC specifies that the building drain should extend _____ feet outside the building wall.
   a. 1
   b. 2
   c. 3
   d. 4

8. When installing the main stack sweep, the position of the sweep is _____.
   a. where the plumb bob, suspended from the sanitary tee, comes to rest
   b. where the building drain and the secondary stack are properly aligned
   c. determined by the depth of the foundation
   d. determined by the distance between the stack and sanitary tee

9. Each of the following may be used as a building cleanout *except* a(n)_____.
   a. combination tee-wye
   b. sanitary tee
   c. sanitary reducer
   d. cleanout and test tee

10. You can measure the height of the closet fitting by the _____.
    a. diameter of the branch line from the water closet
    b. dimensions of the water closet
    c. flow rate of the water closet
    d. slope of the branch line from the water closet

Review Questions

11. You are connecting the cleanout to the closet fitting. The distance is 40 inches and the pipe diameter is 3 inches. You have three joints. You should cut a pipe that is _____ inches long.
    a. 37.5
    b. 40
    c. 45
    d. 47.5

12. Codes for areas with cold climates often require sanitary increasers on stacks extending through the roof.
    a. True
    b. False

Match the type of test with its description.

13. _____ Air test

14. _____ Odor test

15. _____ Water test

16. _____ Smoke test

    a. In this test, all openings in the drainage pipes are closed except those at the tops of the stacks.
    b. Some versions of this test use a device that looks like a large firecracker.
    c. This test is useful in cold weather because it will not freeze and damage the pipes.
    d. This test should only be conducted on DWV systems installed in commercial buildings.
    e. This test uses a mixture of oil of peppermint and warm water.

17. For a DWV system to pass inspection, tests must be performed in the presence of a _____.
    a. site supervisor
    b. plumbing inspector
    c. utility representative
    d. safety inspector

18. Air tests should *not* be performed on _____ or _____ pipe.
    a. copper; plastic
    b. copper; cast-iron
    c. plastic; cast-iron
    d. plastic; stainless steel

Summary

Experienced plumbers develop an action plan for a DWV installation, covering the entire location and installation process from start to finish. A good plan includes the following elements:

- Ordering fixtures and materials, ensuring their delivery, and storing them properly
- Developing plans, specifications, and schedules
- Performing slope and grade calculations
- Testing the system and preparing it for inspection
- Ensuring that all safety and accessibility requirements are met

When installing DWV systems, you must pay attention to both on-the-job safety and the health and safety of the people who will eventually use these systems. While plumbers can find installation requirements in plans, prints, or local codes, they also rely on their experience and skill to complete the installation and solve problems on site.

Plumbers are responsible for locating fixtures, installing the stack, calculating pipe grade, testing the installation, and arranging for the inspection. Plumbers do all of these jobs safely and accurately to ensure that the DWV piping system moves wastes away from the structure while also keeping harmful gases and disease out.

Notes

Trade Terms Introduced in This Module

Accessibility requirements: The requirements outlined in some building codes and by ANSI (the American National Standards Institute) that affect physically challenged people and their access to public buildings and facilities.

Batter board: One of a pair of horizontal boards that are nailed at right angles to each other to three posts set beyond the corners of a building elevation. Strings fastened to these boards indicate the exact corner of a building.

Bench mark: In surveying, a marked reference point on a permanent, fixed object such as a metal disk set in concrete, the elevation of which is known, and from which the elevation of other points may be obtained.

Blocking: Pieces of wood used to secure, join, or reinforce members, or to fill spaces between them.

Building cleanout: A cleanout located in a building drain immediately prior to its exit from a building, usually required by code in buildings that have basements.

Carrier fittings: The support apparatus for wall-hung bathroom fixtures.

Cure: A natural process in which excess moisture in concrete evaporates and the concrete hardens.

Daily log: A description of all work accomplished during the day on a project, maintained by the site supervisor.

Invert elevation: In plumbing, the lowest point or the lowest inside surface of a channel, conduit, drain, pipe, or sewer pipe. It is the level at which fluid flows.

Load factor: A specified percentage of total flow from connected fixtures that is likely to occur at any point along the DWV system, usually provided in the local applicable code.

Percentage of grade: The slope of a line of pipe expressed as the fall in feet, divided by the run in feet, multiplied by 100.

Perineal bath: A therapeutic bathtub used to treat diseases and injuries of the groin.

Pre-construction plan: A thorough outline of an entire installation process for a plumbing system.

Prefabricated: Assembled prior to delivery to the project site.

Rebar: A ribbed steel bar that provides a good bond when used as a reinforcing bar in concrete.

Sanitary increaser: An enlargement of a stack vent used in cold climates to prevent water vapor from condensing and freezing inside the vent opening.

Sewer tap: The inlet of a private sewage disposal system.

Sheathing: The covering (usually over wood boards, plywood, or wallboard) that is placed over the exterior framing or rafters of a building. It provides a base for the application of exterior cladding.

Sheetrock®: A proprietary name for gypsum board. Also called wallboard.

Shoring: The act of using timbers set diagonally to temporarily hold up a wall. In excavations, the use of such timbers to stabilize the ditch sides is employed to prevent cave-ins.

Sizing: The process of calculating the proper sizes for the drains, stacks, sewer lines, and vents in a DWV system.

Slab-on-grade: A term describing a building in which the base slab is placed directly on grade without a basement.

Sleeve: A piece of piping through which water supply or drain, waste, and vent piping is inserted when that piping penetrates a building's structural elements such as concrete footings or floors. The sleeve protects the piping from being damaged by concrete.

Slope: The ground elevation or level planned for or existing at the outside walls of a building or elsewhere on the building site.

Soleplate: A horizontal timber that serves as a base for the studs in a stud partition.

Time-critical material: Material that must arrive at a job site at a specified time and that must be installed before any other work can proceed.

Top plate: The top horizontal member of a frame building to which the rafters are fastened.

Vernier: An auxiliary scale that slides against, and is used in reading, a primary scale. The scale makes it possible to read a primary scale much closer than one division of that scale.

Additional Resources

This module is intended to be a thorough resource for task training. The following reference works are suggested for further study. These are optional materials for continued education rather than for task training.

Code Check Plumbing: A Field Guide to the Plumbing Codes. 2000. Redwood Kardon. Newtown, CT: Taunton Press.

Handbook of Materials Selection. 2002. Myer Kutz, ed. New York: J. Wiley.

Plumber's and Pipe Fitter's Calculations Manual. 1999. R. Dodge Woodson. New York: McGraw-Hill.

References

2003 International Plumbing Code, 2003. Falls Church, VA: International Code Council.

2003 National Standard Plumbing Code, 2003. Falls Church, VA: Plumbing-Heating-Cooling Contractors—National Association.

Dictionary of Architecture and Construction, Third Edition, 2000. Cyril M. Harris, ed. New York: McGraw-Hill.

Figure Credits

LeDuc and Dexter	Module divider
Jonathan Byrd	204F01, 204F05, 204F18, 204F56
Eljer Plumbingware, Inc. www.eljer.com	204F03
Watts Regulator Company	204F07, 204F13
Zurn Plumbing Products Group	204F09–204F12, 204F22, 204F25
Jay R. Smith Mfg. Co.	204F14
American Standard	204F15
Halsey Taylor	204F17
Wade Drains	204F20
Josam Company	204F21
Courtesy of Ridge Tool Company	204F66 (photo)
Courtesy of The L.S. Starrett Co.	204F35, 204F37
DeWALT Inc.	204F36, 204F43
CST/berger David White	204F39–204F41
Trimble Navigation Limited *©1999 Trimble Navigation Limited. All Rights Reserved.*	204F47
Kohler Co.	204F54
Sioux Chief Manufacturing Co., Inc.	204F65 (A)
Expansion Seal Technologies	204F65 (B, C, F), 204F68
Cherne Industries	204F65 (D, E), 204F66 (line drawings), 204F67
Universal Plumbing and Heating Company	204SA01

CONTREN® LEARNING SERIES — USER UPDATE

The NCCER makes every effort to keep these textbooks up-to-date and free of technical errors. We appreciate your help in this process. If you have an idea for improving this textbook, or if you find an error, a typographical mistake, or an inaccuracy in NCCER's Contren® textbooks, please write us, using this form or a photocopy. Be sure to include the exact module number, page number, a detailed description, and the correction, if applicable. Your input will be brought to the attention of the Technical Review Committee. Thank you for your assistance.

Instructors – If you found that additional materials were necessary in order to teach this module effectively, please let us know so that we may include them in the Equipment/Materials list in the Annotated Instructor's Guide.

Write: Product Development and Revision
National Center for Construction Education and Research
P.O. Box 141104, Gainesville, FL 32614-1104

Fax: 352-334-0932

E-mail: curriculum@nccer.org

Craft ______________________ Module Name ______________________

Copyright Date ____________ Module Number ____________ Page Number(s) ____________

Description

____

____

____

____

(Optional) Correction

____

____

____

(Optional) Your Name and Address

____

____

____

02205-05

Installing Roof, Floor, and Area Drains

02205-05
Installing Roof, Floor, and Area Drains

Topics to be presented in this module include:

Overview

Roof, floor, and area drains are similar in design and installation. These drains collect storm water and direct it to the storm drainage system. Manufacturers offer roof, floor, and area drains in several basic parts and in a variety of styles. Specifications or plumbing drawings provide information about the type of drains to install and where to install them.

The basic parts of a drain include the body, grate, deck clamp, clamping ring or flashing clamp, drain receiver, and extensions. Plumbers interchange these parts to create a variety of drains. Some structures require a high degree of sanitation in their drains. Other installations require drains that can be washed, flushed, or primed. To determine what types of roof, floor, and area drains to install, plumbers refer to the local code or plans and specifications. When high levels of sanitation are required, plumbers also consult the health department.

To install roof drains, plumbers lay out and cut the opening, attach the drain body, connect it to the piping system, install expansion joints, install the waterproof membrane or flashing, and check the drains. Installing floor drains is similar to installing roof drains. While the pipe-joining techniques are identical, to install floor and area drains, plumbers also locate, support, and prevent damage to the drains. To avoid project conflicts, plumbers coordinate their work with the other trades.

Focus Statement

The goal of the plumber is to protect the health, safety, and comfort of the nation job by job.

Code Note

Codes vary among jurisdictions. Because of the variations in code, consult the applicable code whenever regulations are in question. Referring to an incorrect set of codes can cause as much trouble as failing to reference codes altogether. Obtain, review, and familiarize yourself with your local adopted code.

Objectives

Objectives

When you have completed this module, you will be able to do the following:

1. Use a surveyor's level or transit level to set the elevation of a floor or area drain.
2. Install a roof drain, a floor drain, and an area drain.
3. Install waterproof membranes and flashing.

Trade Terms

Box-out	Extension
Can wash drain	Flashing clamp
Clamping ring	Floor sink
Deck clamp	Flushing floor drain
Dome	Grate
Drain body	Sediment bucket
Drain receiver	Trap primer
Expansion joint	Trap primer valve

Required Trainee Materials

1. Appropriate personal protective equipment
2. Pencil and paper
3. Copy of local applicable code

Prerequisites

Before you begin this module, it is recommended that you successfully complete *Core Curriculum: Plumbing Level One; Plumbing Level Two,* Modules 02201-05 through 02204-05.

This course map shows all of the modules in the second level of the Plumbing curriculum. The suggested training order begins at the bottom and proceeds up. Skill levels increase as you advance on the course map. The local Training Program Sponsor may adjust the training order.

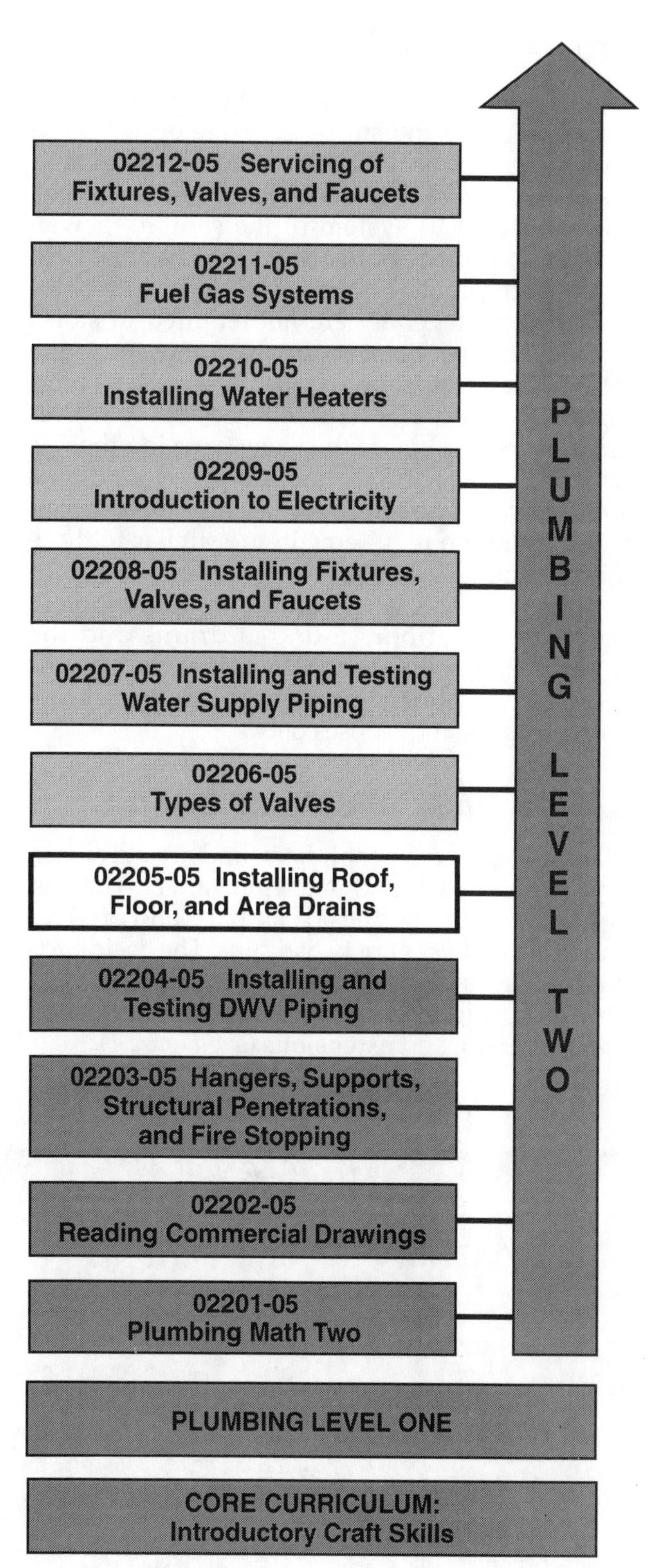

205CMAP.EPS

1.0.0 ◆ INTRODUCTION

Roof, floor, and area drains collect storm water and direct it to the storm drainage system. The piping that connects the roof and area drains to the storm sewer is separate from the drain, waste, and vent (DWV) system of the sanitary sewer. However, plumbers use the same materials to install both piping systems. The major difference is that traps generally are not required in storm drainage piping. Floor drains are connected to the sanitary sewer system because the wastes entering them are usually contaminated. These wastes must be treated before the water carrying them is returned to the natural water supply. Floor drains are included in this module because their design and installation requirements are similar to those for roof and area drains.

In this module, you'll learn about the different types of roof, floor, and area drains and the specifics of installing each. Note that dimensions shown in the figures included in this module are for instructional purposes only.

2.0.0 ◆ BASIC PARTS OF DRAINS

Manufacturers offer roof, floor, and area drains in several basic parts and in a wide variety of styles. You can interchange these basic parts and produce an almost endless variety of drains. The basic parts are the **drain body, grate (dome), deck clamp, clamping ring or flashing clamp, drain receiver (bearing pan),** and **extensions** (see *Figure 1*).

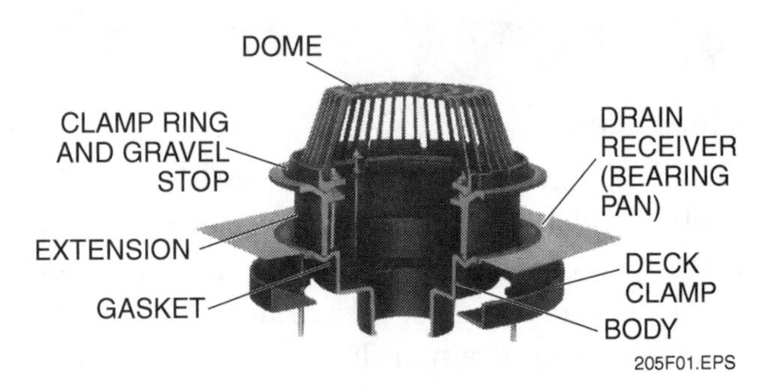

Figure 1 ◆ Basic parts of a drain.

2.1.0 Drain Body

The basic component of all drains, the drain body, funnels the water into the piping system. Usually made of cast iron or plastic, it is designed to connect to the roof, floor, or foundation. Special installations may require stainless steel or porcelain-enameled cast iron. You can see different types of roof drains in *Figure 2* and floor drains in *Figure 3*.

Floor and area drains often contain a separate **sediment bucket** to prevent solids from entering the piping (see *Figure 4*). The drain body connects to the piping system in one of several ways. Inside-caulk drain bodies are compatible with hub-and-spigot cast-iron pipe. A threaded drain body can receive a standard male pipe thread. Drain bodies can be joined with no-hub gaskets and clamps.

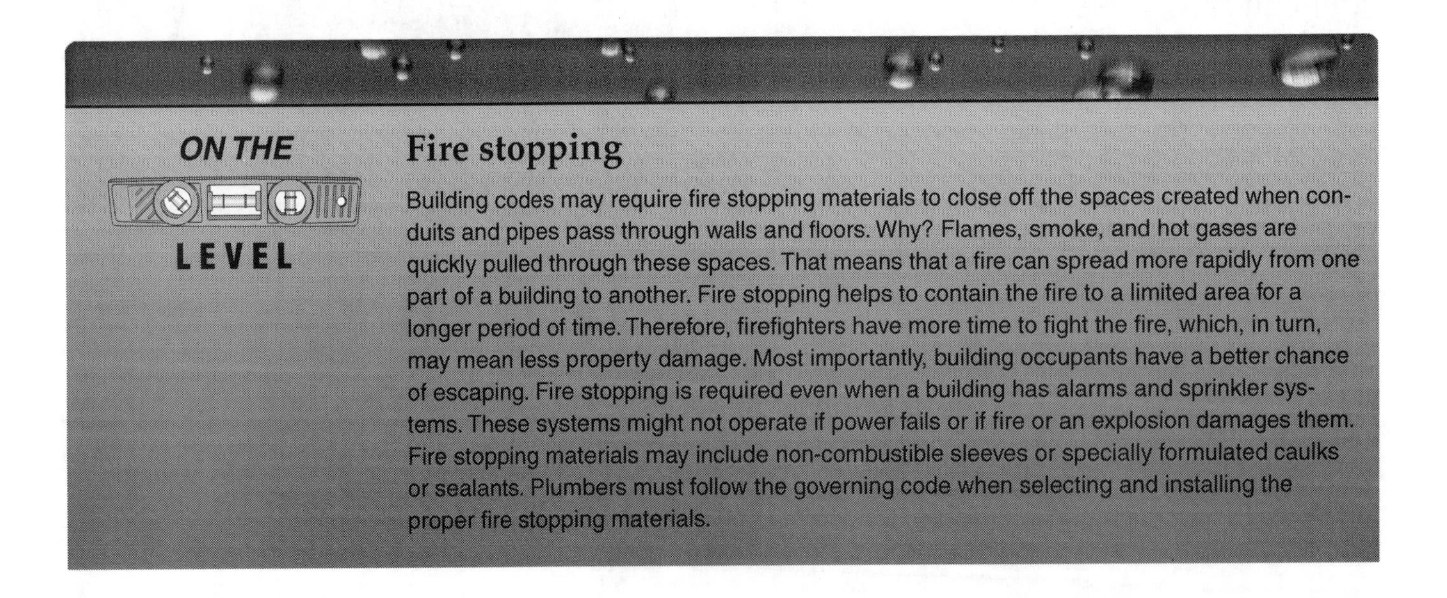

Fire stopping

Building codes may require fire stopping materials to close off the spaces created when conduits and pipes pass through walls and floors. Why? Flames, smoke, and hot gases are quickly pulled through these spaces. That means that a fire can spread more rapidly from one part of a building to another. Fire stopping helps to contain the fire to a limited area for a longer period of time. Therefore, firefighters have more time to fight the fire, which, in turn, may mean less property damage. Most importantly, building occupants have a better chance of escaping. Fire stopping is required even when a building has alarms and sprinkler systems. These systems might not operate if power fails or if fire or an explosion damages them. Fire stopping materials may include non-combustible sleeves or specially formulated caulks or sealants. Plumbers must follow the governing code when selecting and installing the proper fire stopping materials.

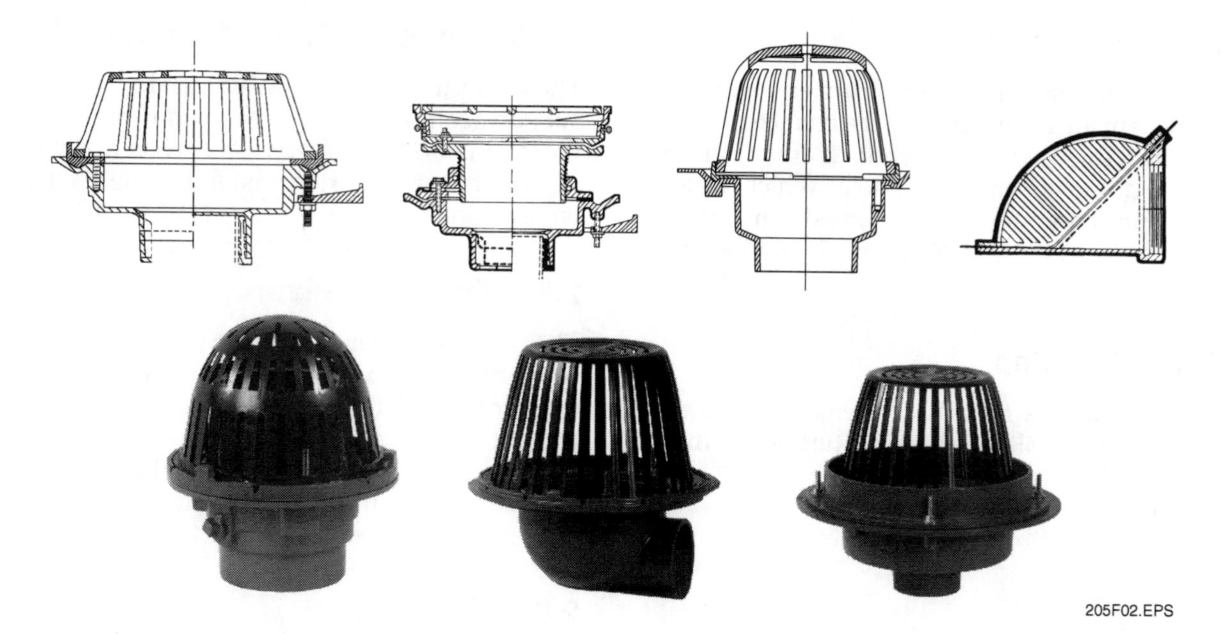

Figure 2 ◆ Roof drains.

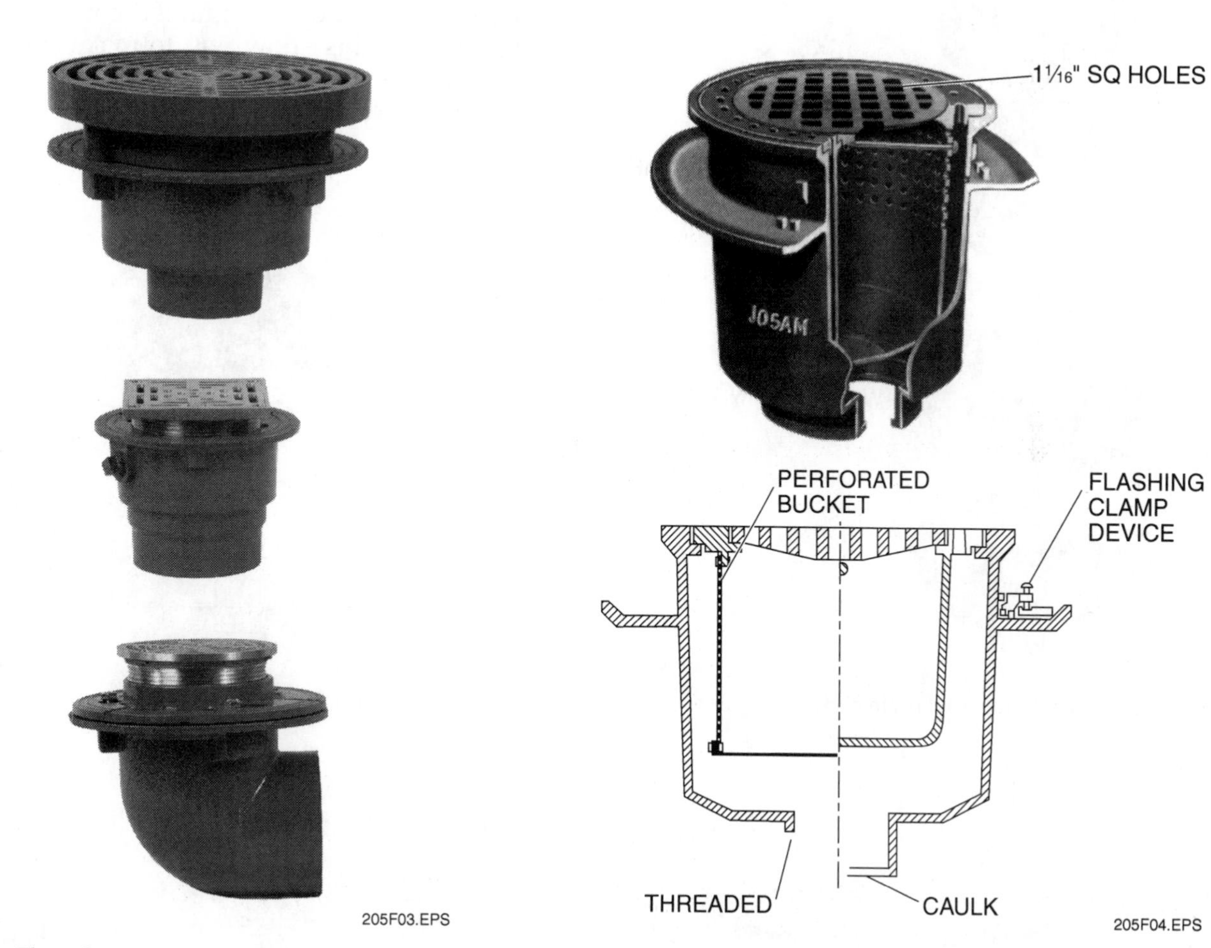

Figure 3 ◆ Floor drains.

Figure 4 ◆ Floor drain with sediment bucket.

2.2.0 Dome Grate

The dome grate prevents debris from entering the piping system and helps reduce the chance of pipe clogs (refer to *Figure 2*). A roof drain with an adjustable dome controls the rate at which water enters the piping system in a designed controlled-flow system (see *Figure 5*). You can adjust this type of drain on site.

2.3.0 Deck Clamp

The deck clamp is a ring that fastens to the drain body with several bolts. Tightening the bolts clamps the drain securely to the roof deck (see *Figure 6*).

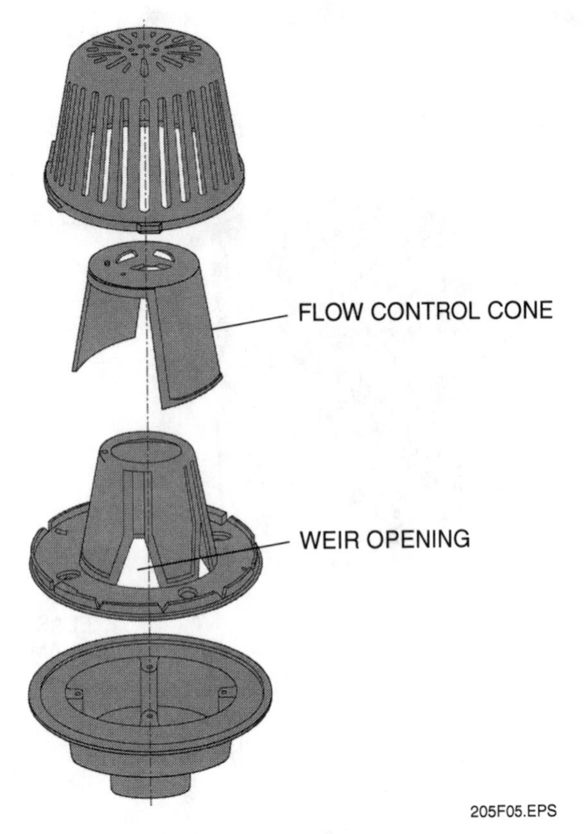

Figure 5 ◆ Adjustable roof drain dome.

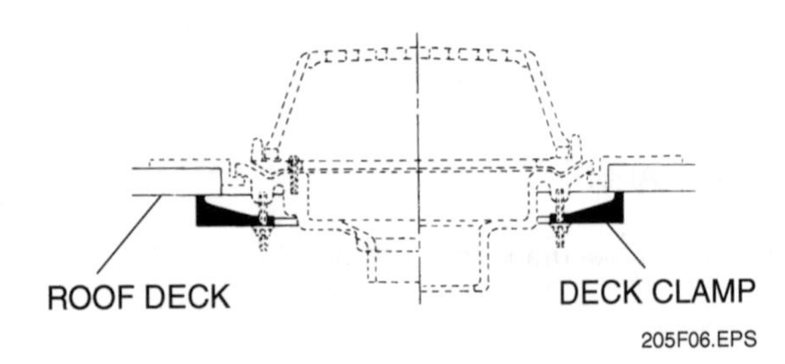

Figure 6 ◆ Deck clamp.

2.4.0 Clamping Ring (Flashing Clamp)

The clamping ring, or flashing clamp, secures the subsurface flashing or waterproof membrane to the drain body (see *Figure 7*). A roofer or plumber installs this ring when installing the flashing or waterproof membrane.

2.5.0 Drain Receiver

The drain receiver, or bearing pan, (*Figure 8*) helps support a roof drain body. It serves two important functions. First, it distributes the roof drain's weight over a large area, which is particularly important when the drain is installed over insulation. Second, it supports a drain in an oversized or off-center roof opening.

2.6.0 Extension

A drain should be set at the proper grade within the roof deck or the concrete floor. Use an extension to ensure that the drain's grade is correct (see *Figure 9*). Floor drains allow you to make some adjustments without extensions because the strainers are threaded (see *Figure 10*).

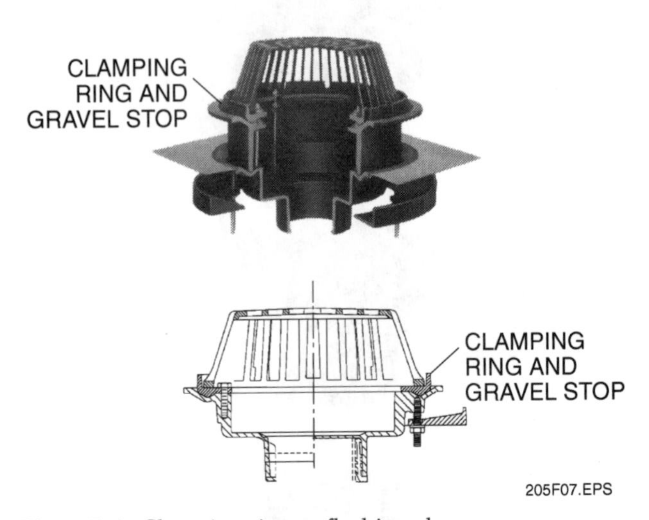

Figure 7 ◆ Clamping ring or flashing clamp.

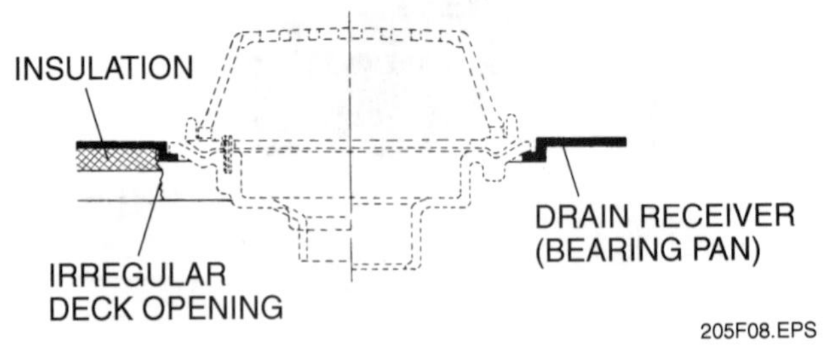

Figure 8 ◆ Drain receiver.

Figure 9 ◆ Extension.

Figure 10 ◆ Threaded strainer.

3.0.0 ◆ TYPES OF DRAINS

Some structures, such as hospitals, laboratories, cafeterias, and restaurants, require a high degree of sanitation in their drains. Some installations require drains that can be washed, flushed, or primed. Special drains meet these needs.

3.1.0 Can Wash Drain and Flushing Floor Drain

A **can wash drain** is designed to flush solids from a drain (see *Figure 11*). A sediment bucket is a part of this type of drain. Can wash drains come in several sizes with either side or bottom outlets.

Figure 11 ◆ Can wash drain.

You must connect a water supply pipe and remote valve to the can wash drain. Codes require the addition of a reduced pressure-zone principle (RPZ) backflow preventer assembly to the line on which a can wash drain is installed.

In a **flushing floor drain** installation, an inlet is provided in the trap. Connect a water supply pipe to this inlet to flush the trap. Be sure to fit the water supply with a valve and vacuum breaker to prevent the possibility of backflow. Some codes may require an RPZ backflow preventer assembly in high-hazard applications.

Examples of areas where can wash drains and flushing floor drains are used include packing plants, areas where food is processed, hospitals, and any area where solids get washed into floor drains but remain in the trap. The flushing action allows the solids to pass through the trap. In non-flushing drains, solids sit, creating foul odors.

3.2.0 Floor Sinks

Floor sinks are used in installations that require a high degree of sanitation, such as hospitals, laboratories, cafeterias, and restaurants. The basic difference between a floor drain and a floor sink is that the floor sink is coated with a porcelain or other easily cleaned finish. Floor sinks reduce the amount of contaminant that can collect in the grate and drain body (see *Figure 12*). They come in a variety of sizes and in round, square, or rectangular shapes. Your local applicable code may require the installation of an RPZ backflow preventer assembly on the same line as the floor sink.

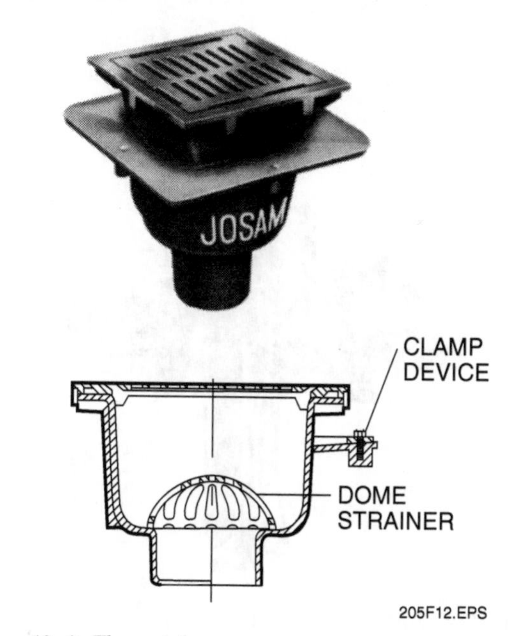

Figure 12 ◆ Floor sink.

3.3.0 Trap Primer

Traps that receive limited use, or that are installed in floors in a moist environment, may lose their seal. To prevent this, install a **trap primer**, which periodically adds a small amount of water to the trap. This keeps the trap seal from failing as a result of evaporation, for example on a relief valve line or condensate piping. A deep seal trap with a trap primer connection is shown in *Figure 13*. The water supply pipe connects at the same point as the flushing connection. Some trap primers are designed to work in conjunction with a waste valve. Many codes prohibit cleanouts with traps in certain applications. Refer to your local code for restrictions in your area.

Manufacturers offer a variety of **trap primer valves** (see *Figure 14*). Some are installed in conjunction with the water supply valve to a fixture. Each time the valve operates, a small amount of water enters the trap. Other trap primer valves meter water into the trap at regular intervals.

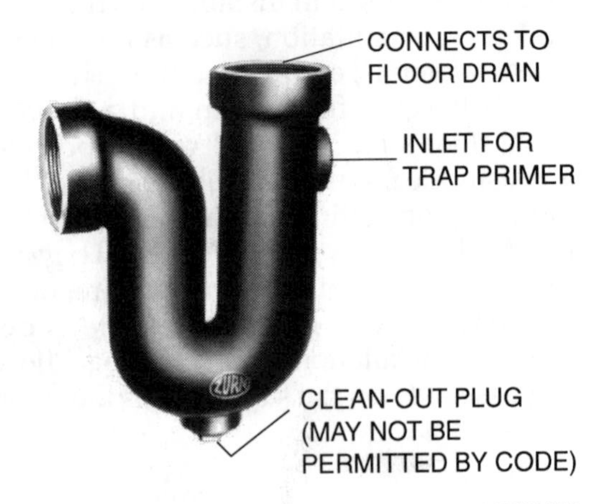

Figure 13 ◆ Deep seal trap with primer connection.

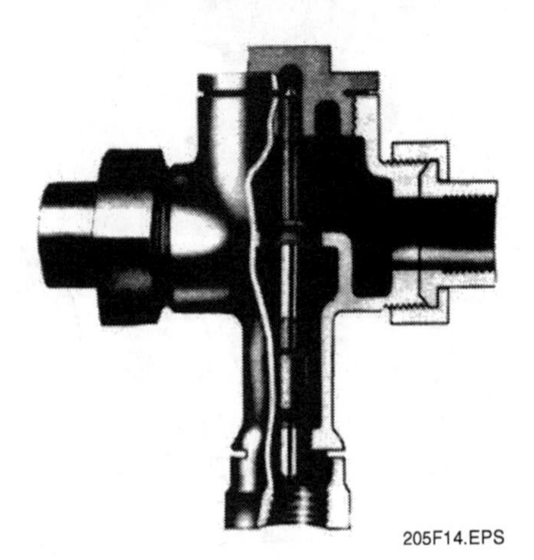

Figure 14 ◆ Trap primer valve.

Always follow the manufacturer's specifications and your local applicable code when installing trap primers and trap primer valves.

3.4.0 Graded Drain

Graded drains are concrete storm drainage pits that are installed in multi-story parking lots, car garages, and other similar facilities. They are designed systems, and local codes specify their application, location, and sizing. Graded drains are designed to be tapped by trench drains.

When graded drains are used for storm drainage in parking garages where vehicle washing or floor rinsing will take place, the authority having jurisdiction may require the installation of sand interceptors and/or oil separators. Consult your state and local authorities having jurisdiction for requirements.

4.0.0 ◆ DETERMINING REQUIREMENTS FOR FLOOR DRAINS

To determine what types of roof, floor, and area drains to install, refer to the local building code or plans and specifications. Installations that require a high degree of sanitation have been designed to meet stringent health requirements. Consult and follow the design specifications and your local applicable code. If you are uncertain about any aspect of the installation, consult the design engineer or architect.

4.1.0 Codes

Plumbing codes govern the installation of floor drains. Refer to the sizing tables in your local code to determine the appropriate sizes for drains (refer to *Table 1*). Acceptable drain dimensions may vary depending on the local code. While codes specify the areas where drains are to be installed, they do not specify exact drain location. You will find this information in the plans and specifications. If necessary, refer to the manufacturer's specifications to determine the capacity of a particular drain. Codes will specify changes required for heating, ventilation, and air conditioning (HVAC) systems.

4.2.0 Plans and Specifications

The plans and specifications state what types of drains are used in construction projects such as commercial buildings, hospitals, manufacturing facilities, homes, etc. These buildings have so many unique features that it is impossible to write a code that covers all the possible alternatives. Engineers custom design the piping systems to meet specific needs and then write the plans and specifications. (Refer to the partial set of specifications covering drains in *Table 1*.) Generally, you'll find this information in Division 15 of the specifications. As you have already learned, the bigger the project, the more detailed the plans and specifications are.

Table 1 Specifications for Drains

P-5 Roof Drain
Shall be J.R. Smith Figure No. 1010 if no insulation above roof slab and Figure No. 1015 if roof insulation is above roof slab. Provide Figure No. 1710 expansion joint if no offset in rainwater leader line is required. Size as shown on drawings.

P-6 Floor Drain (Regular)
Shall be J.R. Smith Figure No. 2010-T, cast-iron floor drain with sediment bucket and polished nickel bronze and adjustable strainer with flashing clamp device. Size as shown on drawings. (Flashing clamp required on drains installed above first floor, slab on grade.)

P-11 Floor Drain (Equipment Rooms)
Floor drain shall be J.R. Smith Figure No. 2233 round cast-iron 12-inch diameter area drain, with sediment bucket, 4-inch pipe connections, and polished bronze top. Furnish complete with a flange. A clamp device is required if the drain is installed above first floor, slab on grade.

Review Questions

Sections 2.0.0–4.3.0

1. The basic component of all drains, the _____ funnels water into the piping system.
   a. drain body
   b. deck clamp
   c. extension
   d. drain receiver

2. The part of a floor or area drain that is designed to prevent solids from entering the piping is called the _____.
   a. extension
   b. deck clamp
   c. drain receiver
   d. sediment bucket

3. The _____ prevents debris from entering the piping system and helps reduce the chance of pipe clogs.
   a. flashing
   b. drain receiver
   c. grate or dome
   d. trap or trap inlet

4. The _____ secures the waterproof membrane to the drain body in a roof drain.
   a. clamping ring
   b. deck clamp
   c. extension
   d. grate

5. A drain that receives limited use should be installed _____ to protect against loss of the water seal.
   a. with a floor sink
   b. with a trap primer
   c. with a funnel
   d. without a trap

5.0.0 ◆ INSTALLING FLOOR AND AREA DRAINS

Installing floor drains is similar in many ways to installing roof drains. The pipe-joining techniques are identical, correct positioning of the floor drains is critical, and you may have to attach a waterproof membrane. The steps for installing floor and area drains include locating the drains, supporting the drains, preventing damage to the drains, and installing waterproof membranes.

5.1.0 Locating Floor and Area Drains

Often, the floor drain is set before the concrete floor is poured. Make sure that the elevation of the floor drain is correct. Once the concrete has set, correcting a poor elevation is time-consuming and costly. Locate the drain at the lowest point in the floor or area. Refer to the floor plan, plumbing drawings, or specifications for the drain's position, the horizontal distances from walls, and the elevation of the finished floor drain. Coordinate your work with the concrete contractor to ensure accuracy and to protect the drain during the pouring of concrete.

It's best to take the elevation measurements with a surveyor's level or transit (see *Figure 15*). In small areas you can use a level and a straight board to locate the top of the drain with respect to a known elevation (see *Figure 16*). You can also stretch a string line across the area that indicates the highest level of the floor. Measure down from the line to determine the floor drain elevation (see *Figure 17*).

5.2.0 Supporting the Floor Drain

Supporting the drain at the proper elevation is often difficult. The area below the drain is excavated to install the pipe, so the fill is probably not stable. Use concrete blocks, bricks, or tamped gravel to support the drain body.

Sometimes you'll have to drive three or four lengths of pipe or reinforcing rod into the ground and place the drain on top of these supports. Never use wood blocks. They expand when concrete is poured around them and can cause the concrete to fail.

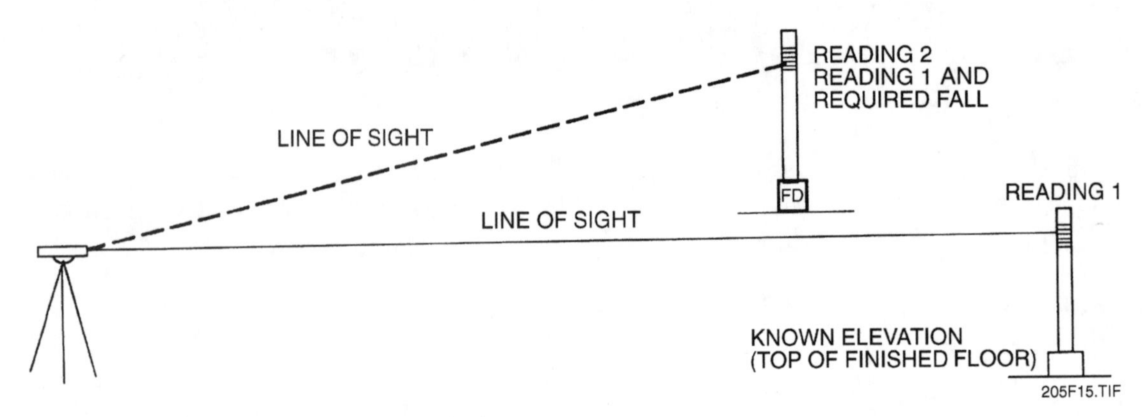

Figure 15 ◆ Measuring the elevation with a surveyor's level.

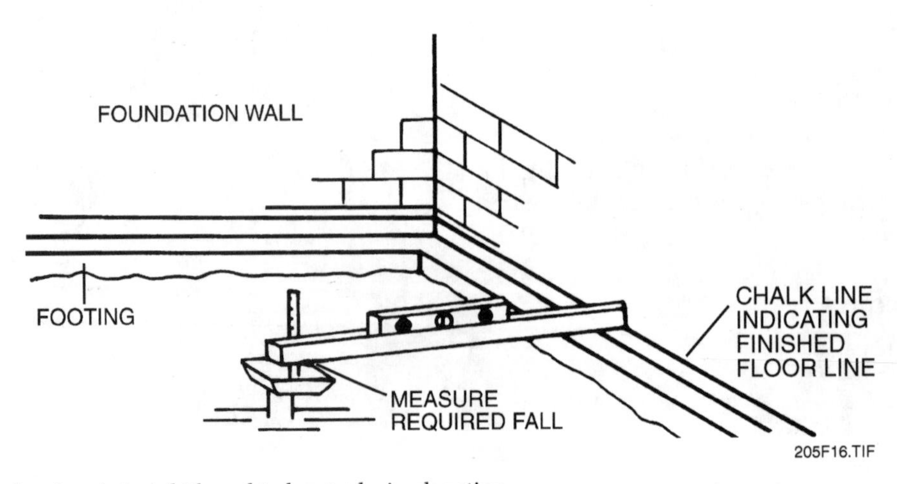

Figure 16 ◆ Using a level and straight board to locate drain elevation.

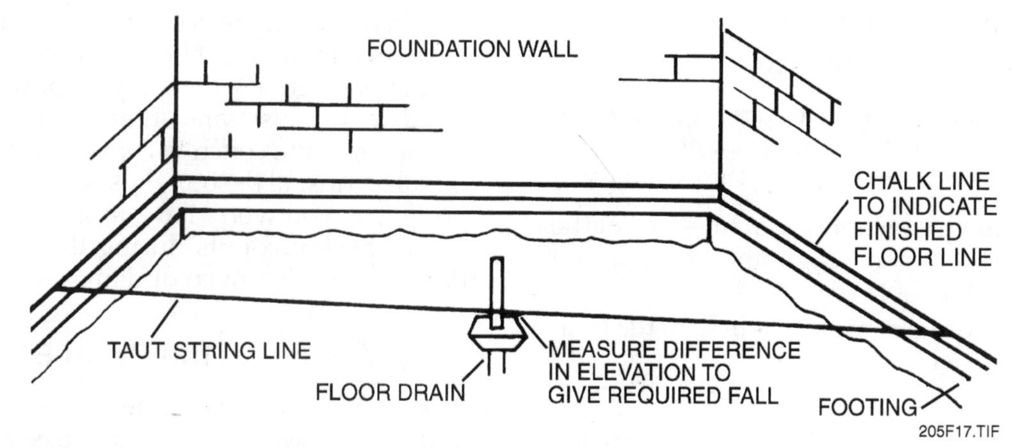

Figure 17 ◆ Stretching a string line to measure elevation of a floor drain.

5.3.0 Preventing Damage to the Floor Drain

When concrete is poured, some of it may enter the drain. To prevent this, use test plugs to temporarily block the drain. Temporarily block small drains with duct tape.

Drain grates with a special finish or polish can be damaged when cement is poured. You can remove the grate until the concrete work is completed, but be sure to protect the inside of the drain body. To do this, tape a piece of plywood or metal to the grate.

5.4.0 Installing Floor Drains with Waterproof Membranes

Some installations require a waterproof membrane between the structural concrete and the topping slab (see *Figure 18*). This type of installation is necessary for floor drains above ground level, such as showers. The membrane prevents water that may come through the topping slab from penetrating the structural concrete. The drain body is designed with weep holes so that water from the top of the membrane can enter the drain near the clamp. You must initially install the drain body without the strainer in place. After the structural concrete cures, put the waterproof membrane in place and secure it with the membrane clamp. Finally, set the strainer at the correct elevation.

If the drain is not located in the basement, you must add a vinyl or sheet lead material between the drain and the floor (see *Figure 19*). This material protects against leaks if the floor cracks.

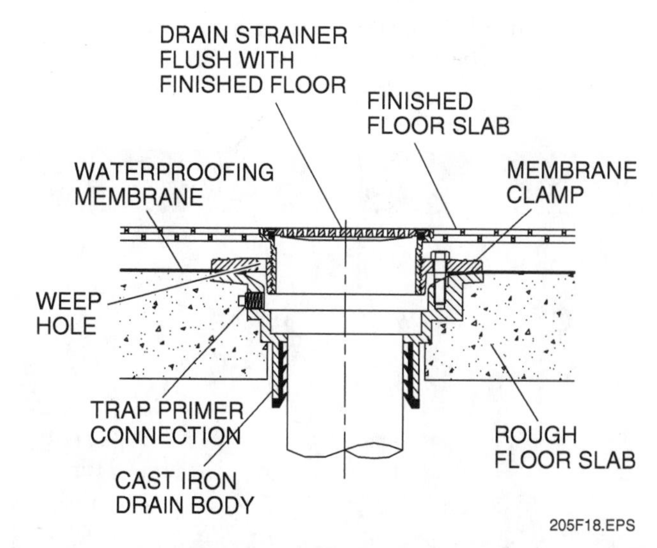

Figure 18 ◆ Floor drain with waterproof membrane clamping device.

Figure 19 ◆ Drain with vinyl or sheet lead material in

6.0.0 ◆ INSTALLING ROOF DRAINS

Installing roof drains is fairly easy, but, depending on the application and the type of drain, you will have several options to consider. The steps for installing roof drains include laying out and cutting the opening, attaching the drain body, connecting to the piping system, installing **expansion joints,** installing the waterproof membrane or flashing, and checking the drains. As pointed out before, it is important to protect the drain opening.

6.1.0 Laying Out and Cutting the Opening

Use the dimensions given on the drawings to locate the roof drain position. Two important questions must be considered before cutting the openings:

- Is the position reasonable given the slope of the roof? Stated another way, is the location in a low point of the roof?
- Does the building frame interfere with the location you have chosen? If it does, make a slight change in the location to solve this problem.

Once you have checked the drain location, lay out the shape of the opening and cut it. The tools you use to make the cut depend upon the type of roof deck. For example, on a metal roof or decking, you might use tin snips, a Sawzall, a jigsaw, or a cutting torch. On wood decks, you might use a Sawzall, jigsaw, or keyhole saw.

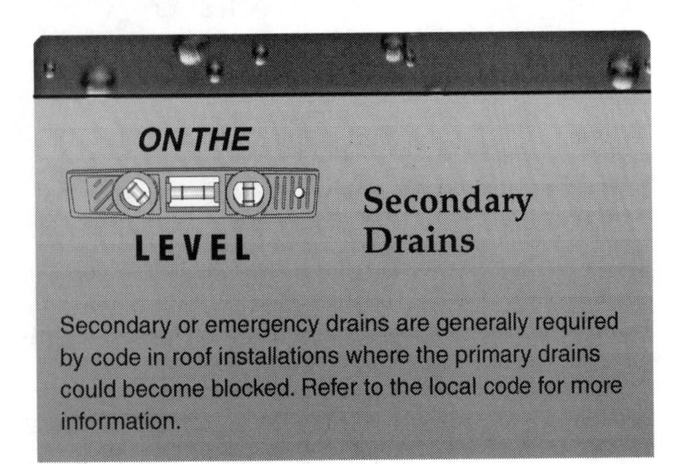

Secondary Drains

Secondary or emergency drains are generally required by code in roof installations where the primary drains could become blocked. Refer to the local code for more information.

When working on a concrete roof deck that will be poured in place, set **box-outs** on the deck forms. Box-outs are short pieces of pipe or wooden boxes, filled with moist sand, that preserve roof drain openings as the concrete is poured (see *Figure 20*). Be sure to install the box-outs before the concrete is poured so you won't have to drill these openings later. Check box-outs during the pour to ensure that they aren't moved or damaged.

6.2.0 Attaching the Drain Body

Insert the drain body through the opening you cut, and secure it with a deck clamp (see *Figure 21*). For installations over rigid insulation or in cases where the opening is oversized, install a drain receiver before putting the drain body in position. The drain receiver provides a wide flange to support the drain.

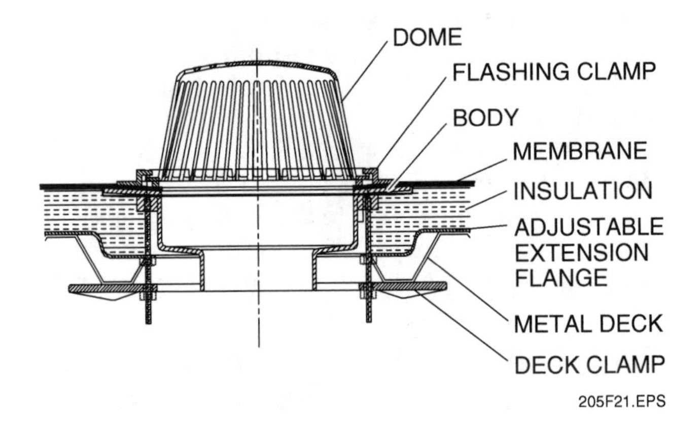

Figure 21 ◆ Deck clamp.

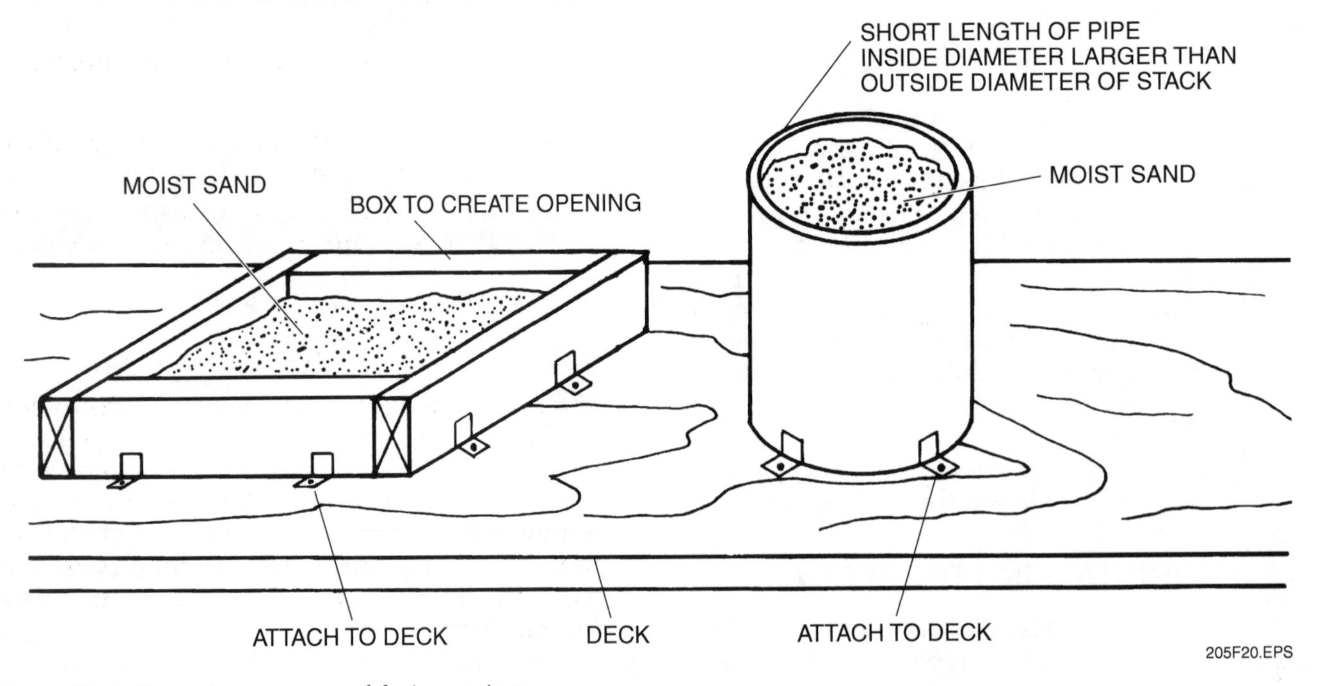

Figure 20 ◆ Box-outs preserve roof drain openings.

6.3.0 Connecting to the Piping System

You can make the joint between the drain body and the storm water piping system in several ways. If the piping material is cast iron, the joint may be caulked, no-hub, compression, or threaded. The procedures for making these joints are identical to those used in joining cast-iron pipe (see *Figure 22*). You can join certain types of plastic roof drain bodies to the piping system with solvent welding or threaded joints.

6.4.0 Installing Expansion Joints

To deal with potential movement between the roof deck and the rest of the structure, install an expansion joint (see *Figure 23*). This joint is installed in the vertical pipe between the drain body and the horizontal piping. Ensure that there is proper spacing between expansion joints. Refer to the manufacturer's instructions for the correct spacing.

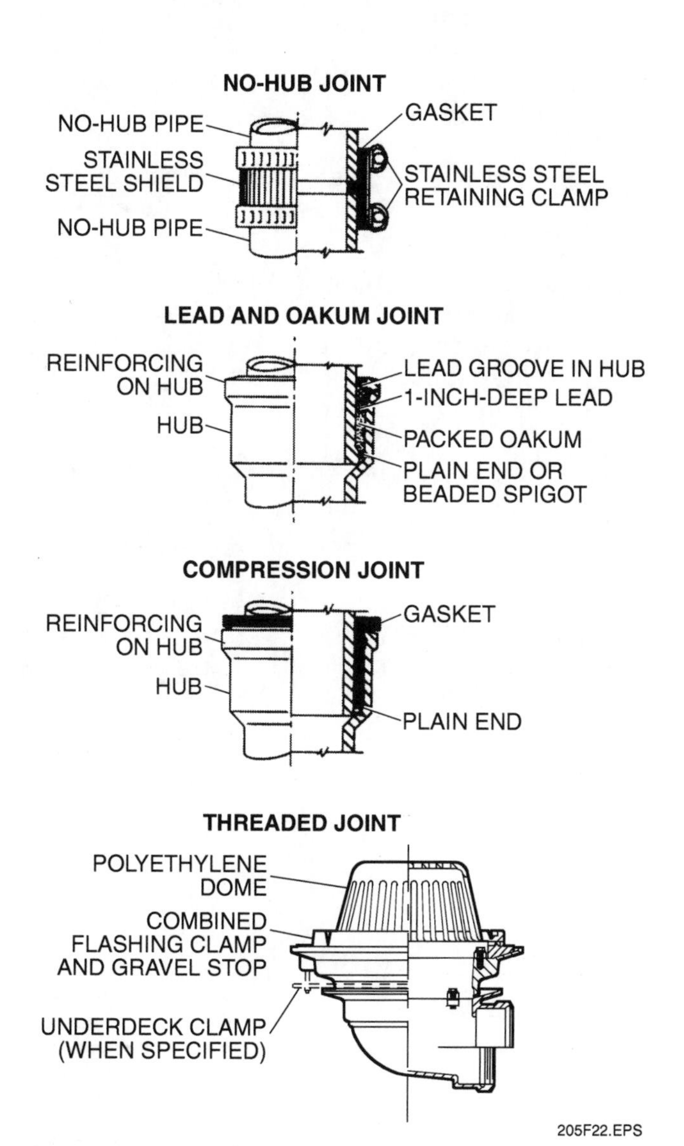

Figure 22 ◆ Joints for cast-iron roof drains.

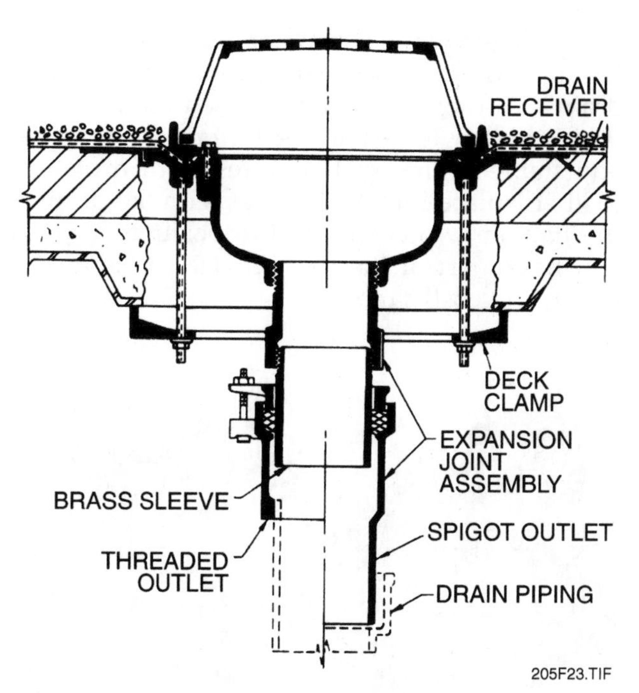

Figure 23 ◆ Expansion joint installation.

6.5.0 Installing Flashing or a Waterproof Membrane

Specifications may require installation of a waterproof membrane or flashing. A roofer may do this, or it may be your responsibility. If you are responsible for installing the membrane or flashing, follow these steps (refer to *Figure 21*):

Step 1 Place the flashing or waterproof membrane around the drain after installing the roof drain.

Step 2 Make an opening in the membrane or flashing the size of the drain.

Step 3 Secure the membrane or flashing with the flashing clamp.

Depending on the roof design, you may place the waterproof membrane either above or below the insulation.

6.6.0 Checking the Roof Drains

You are responsible for ensuring that the roofers' work, as it relates to your job, is completed satisfactorily. Once the roof is completed, check the roof drains to make sure that they are not clogged with roofing material. Also, make sure that the roof drain domes are properly secured. Make sure that adjustable flow roof drains are at the correct setting (refer to *Figure 5*).

A problem you'll encounter frequently is maintaining the correct elevation on adjustable roof drains. The compression joint used for some roof drains is easy to install (see *Figure 24*). However, adjustable drains and compression fittings are easily displaced if someone steps on the drain. If this happens, you will have to reposition the drain and have part of the roof replaced, requiring a major, and costly, repair.

Figure 24 ◆ Compression joint.

Review Questions

Sections 5.0.0–6.6.0

1. The best time to set the elevation of a floor drain in a concrete floor is _____.
   a. after the concrete floor is poured
   b. after finished flooring is installed
   c. before the concrete floor is poured
   d. before the building foundation is finished

2. The elevation of a floor drain should be measured with a(n) _____.
   a. T-square and a steel ruler
   b. surveyor's level or transit
   c. architect's ruler
   d. folding rule

3. The best way to support a drain at the proper elevation in a concrete floor before the floor is poured is with _____.
   a. tamped fill
   b. loose gravel
   c. wooden blocks
   d. concrete blocks or bricks

4. A waterproof membrane is needed in a floor drain that is not located in a _____.
   a. bathroom
   b. basement
   c. kitchen
   d. tub or shower

5. The best way to create the opening for roof drains that will be installed in a concrete roof deck is to _____.
   a. drill each opening after the roof deck has cured
   b. frame the openings in pipe or wood before the roof deck is poured
   c. ask the concrete crew to create the openings
   d. mark where the roof drains are needed with cans after the deck is poured

Summary

Roof, floor, and area drains are similar in design and installation. In each case, refer to specifications or plumbing drawings for information on the type of drains to install and where to install them. Always consult local codes, and in areas that require high levels of sanitation, check with the local health department. Drain installation work will take place at the same time that roofers, concrete workers, and other trades will be doing their jobs. Be sure to coordinate with the other trades to avoid conflicts.

The precise location of drains is vital because poorly located drains will not function properly. Once you've positioned the drains at the proper elevation, protect them from movement, damage, and clogging during the construction and hanging process.

You will use conventional pipe-joining techniques to connect roof, area, and floor drain piping. You'll use many different sizes of pipes and a variety of pipe materials. Consult the local codes and job specifications.

Notes

Trade Terms Introduced in This Module

Box-out: A short length of pipe or a wooden box attached to a roof deck before concrete is poured to preserve roof drain openings.

Can wash drain: A can that is placed upside down over a drain to allow water to squirt upward, washing the inside of the container.

Clamping ring: A clamp that secures subsurface flashing or a waterproof membrane to the drain body. Also called a flashing clamp.

Deck clamp: A ring that fastens to a drain body with several bolts.

Dome: A ventilated drain cover that prevents debris from entering the piping system and helps reduce the chance of pipe clogs. Also called a grate or strainer.

Drain body: The basic element of all drains, the drain body funnels water into the piping system.

Drain receiver: A type of flange that distributes the roof drain's weight over a large area. It also supports a drain in an oversized or off-center roof opening.

Expansion joint: A joint or gap between adjacent parts of a building that permits the parts to move as a result of temperature changes or other conditions without damage to the structure.

Extension: A part that helps set a fixture, such as a drain, at the proper grade.

Flashing clamp: Another term for clamping ring.

Floor sink: A type of floor drain that is coated with a porcelain or easy-to-clean finish. Most often used in areas requiring a high level of sanitation.

Flushing floor drain: A floor drain that is equipped with an integral water supply connection, enabling flushing of the drain receptor and trap.

Grate: Another term for dome.

Sediment bucket: A removable device inside a drain body that traps small solids that pass through the grate, or dome, to keep the solids out of the piping. Also called a sediment trap.

Trap primer: A device or system of piping to maintain a water seal in a trap.

Trap primer valve: A valve that allows a small amount of water into a trap when the valve is operated.

Resources & Acknowledgments

Additional Resources

This module is intended to be a thorough resource for task training. The following reference works are suggested for further study. These are optional materials for continued education rather than for task training.

Controlled Storm Water Drainage, 1979. Louis Blendermann. New York: Industrial Press.

Water and Plumbing, 2000. Ifte Choudhury and J. Trost. Upper Saddle River, N.J.: Prentice Hall.

References

2003 National Standard Plumbing Code, 2003. Falls Church, VA: PHCC

Dictionary of Architecture and Construction, Third Edition. 2000. Cyril M. Harris, ed. New York: McGraw-Hill.

Pipefitters Handbook. 1967. Forrest R. Lindsey. New York: Industrial Press Inc.

Figure Credits

Ivey Mechanical Company	Module divider
Josam Company	205F01, 205F02 (line drawings), 205F04, 205F06, 205F07, 205F08-205F10, 205F12, 205F23
Watts Regulator Company	205F02 (photos), 205F03, 205F11, 205F18, 205F21
Jay R. Smith Mfg. Co.	205F05, 205F22 (threaded joint)
Zurn Plumbing Products Group	205F13, 205F14, 205F19

CONTREN® LEARNING SERIES — USER UPDATE

The NCCER makes every effort to keep these textbooks up-to-date and free of technical errors. We appreciate your help in this process. If you have an idea for improving this textbook, or if you find an error, a typographical mistake, or an inaccuracy in NCCER's Contren® textbooks, please write us, using this form or a photocopy. Be sure to include the exact module number, page number, a detailed description, and the correction, if applicable. Your input will be brought to the attention of the Technical Review Committee. Thank you for your assistance.

Instructors – If you found that additional materials were necessary in order to teach this module effectively, please let us know so that we may include them in the Equipment/Materials list in the Annotated Instructor's Guide.

Write: Product Development and Revision
National Center for Construction Education and Research
P.O. Box 141104, Gainesville, FL 32614-1104

Fax: 352-334-0932

E-mail: curriculum@nccer.org

Craft ______ Module Name ______

Copyright Date ______ Module Number ______ Page Number(s) ______

Description ______

______

______

______

(Optional) Correction ______

______

______

(Optional) Your Name and Address ______

______

______

Plumbing Level Two

02206-05

Types of Valves

02206-05
Types of Valves

Topics to be presented in this module include:

Overview

Valves regulate flow in water supply systems; drain, waste, and vent (DWV) systems; and other plumbing systems. They may provide on/off service, act as a throttling device, or prevent flow reversal through a line. Valves regulate flow according to their design. The most common types of valves are gate, globe, angle, ball, butterfly, check, flushometer, plug, and temperature and pressure (T/P) safety and relief.

Valves consist of stems, packing, bonnets, and end connections. Packing material provides a tight seal around the stem. Plumbers join valves in several ways. The four basic types of end connections are internal threads, external threads, solder ends, and flanged ends. Many factors contribute to proper valve selection, including its application, the valve material, its size, and the rating on the valve body. Plumbers must select the valve that is appropriate for each particular type of plumbing system and always refer to local code.

For repair purposes, valves are divided into five categories: globe valves, gate valves, flushometers, flush valves, and float-controlled valves/ball cocks. To service or repair valves, plumbers identify the problem and troubleshoot the cause. With experience, plumbers develop the knowledge and skills to select and use the right tools and efficiently repair or replace parts. When repairing valves, plumbers always take the necessary precautions to protect themselves and others as well as to prevent property damage.

Focus Statement

The goal of the plumber is to protect the health, safety, and comfort of the nation job by job.

Code Note

Codes vary among jurisdictions. Because of the variations in code, consult the applicable code whenever regulations are in question. Referring to an incorrect set of codes can cause as much trouble as failing to reference codes altogether. Obtain, review, and familiarize yourself with your local adopted code.

Objectives

When you have completed this module, you will be able to do the following:

1. Identify the basic types of valves.
2. Describe the differences in pressure ratings for valves.
3. Demonstrate the ability to service various types of valves.

Trade Terms

- Angle valve
- ASME Boiler and Pressure Vessel Code
- Ball-check valve
- Blowdown
- Bolted bonnet
- Bonnet
- Butterfly valve
- External thread
- Flanged end
- Float-controlled valve/ball cock
- Flush valve
- Huddling chamber
- Internal thread
- Lever-and-weight swing-check valve
- Lift-check valve
- Nonrising stem
- Outside screw and yoke stem
- Packing
- Plug valve
- Pressure regulator valve
- Rising stem
- Screwed bonnet
- Slurries
- Solder end
- Solid-wedge gate valve
- Split-wedge gate valve
- Supply stop valve
- Swing-check valve
- Trim
- Union bonnet
- Wire drawing

Required Trainee Materials

1. Appropriate personal protective equipment
2. Pencil and paper
3. Copy of local applicable code

Prerequisites

Before you begin this module, it is recommended that you successfully complete *Core Curriculum; Plumbing Level One; Plumbing Level Two,* Modules 02201-05 to 02205-05.

This course map shows all of the modules in the second level of the Plumbing curriculum. The suggested training order begins at the bottom and proceeds up. Skill levels increase as you advance on the course map. The local Training Program Sponsor may adjust the training order.

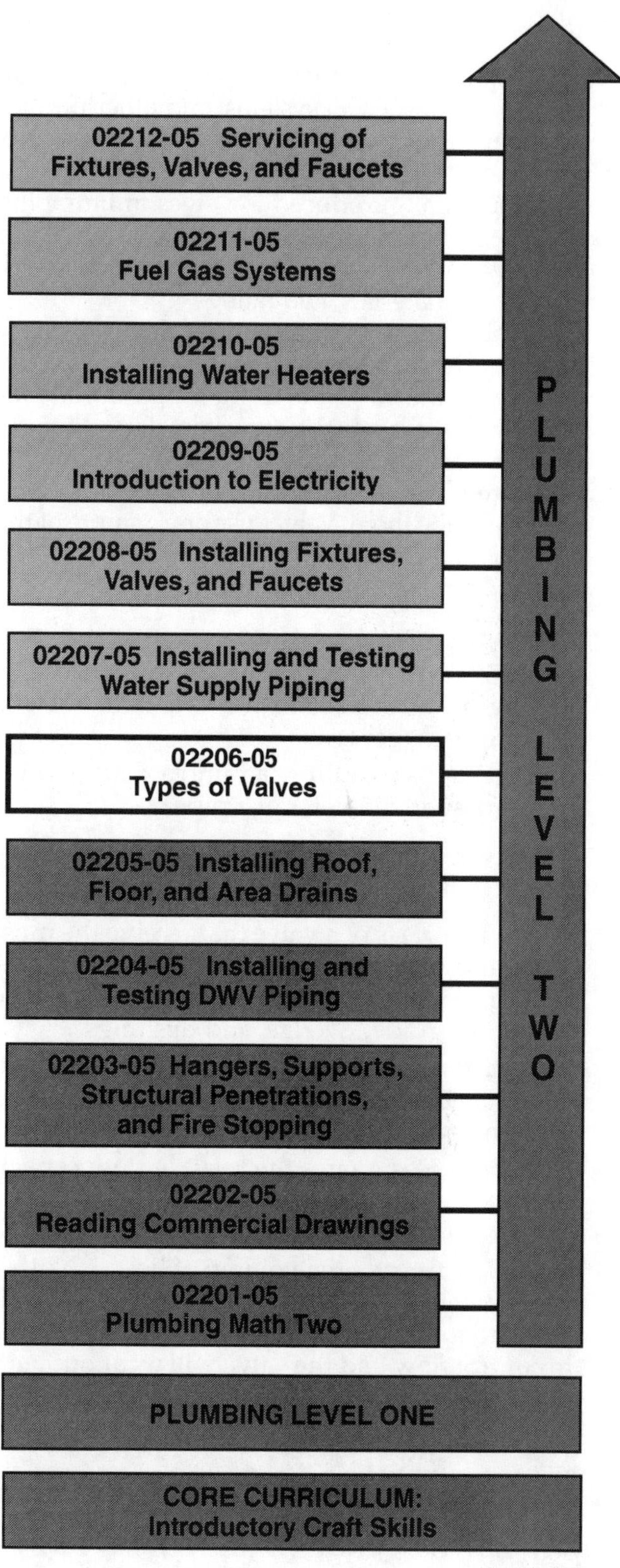

206CMAP.EPS

1.0.0 ◆ INTRODUCTION

Valves regulate flow in water supply systems, drain, waste, and vent (DWV) systems, and other plumbing systems. They may provide on/off service, act as a throttling device, or prevent flow reversal through a line. You were introduced to valves in Plumbing Level One. In this module, you will learn more about how valves work, and about their components, applications, materials, and ratings.

The way that a valve regulates flow will vary with the design of the valve. A plumber must know how to select the valve that is most appropriate for a particular type of plumbing system. First, though, you must learn the ways that basic valves operate and become familiar with some of the terminology used to describe them. Valves use one or more of the following methods to control flow through a piping system:

- Moving a disc or plug into or against a passageway.
- Sliding a flat cylindrical or spherical surface across a passageway.
- Rotating a disc or ellipse around a shaft extending across the diameter of a pipe.

Study the following valve-related terms. They will help you understand how valves operate:

- ***Trim*** – The parts of a valve that receive the most wear and tear and, consequently, are replaceable. Trim includes the stem, disc, seat ring, disc holder (or guide), wedge, and bushings.
- *Straight-through flow* – Describes an unrestricted flow. The element that closes the valve retracts until the passage is clear.
- *Full flow* – Designates the relative flow capacities of various valves.
- *Throttled flow* – A valve designed to partially control or throttle the volume of liquid. Not all types of valves are suitable for throttled flow. If you choose the wrong valve, it will be difficult to control flow and the valve will wear out very quickly.

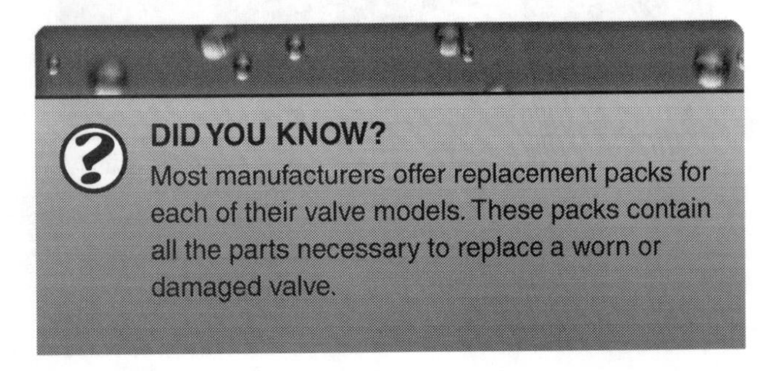

DID YOU KNOW?
Most manufacturers offer replacement packs for each of their valve models. These packs contain all the parts necessary to replace a worn or damaged valve.

- *Valve connections* – The various methods used to join valves to the piping system, including threaded, solvent welded, soldered, and flanged. When purchasing valves, be sure to specify the material type, valve pattern, required pressure rating, and end connections.

2.0.0 ◆ TYPES OF VALVES

The following are the valves you will work with when installing and maintaining water supply and DWV systems:

- Gate valves
- Globe valves
- **Angle valves**
- Ball valves
- **Butterfly valves**
- Check valves
- Flushometer valves
- **Flush valves**
- **Plug valves**
- Temperature and pressure (T/P) safety and relief valves
- **Pressure regulator valves**
- **Supply stop valves**
- **Float-controlled valves (ball cocks)**
- Backwater valves

2.1.0 Gate Valves

A gate valve controls flow with a gate, or disc, that slides in machined grooves at right angles to the flow. The action of the threaded stem on the control handle moves the gate. With the gate fully opened, the valve provides an unobstructed passageway. Gate valves are used for terminations on lines containing steam, water, gas, oil, and air. Gate valves can be of either the **rising-stem** or **nonrising-stem** variety. You will learn more about rising and nonrising stems elsewhere in this module.

Gate valves provide good on-off service. They cause little pressure drop and permit full two-way flow through a pipe. This makes them good choices for lines requiring unrestricted flow, such as pump lines, main supply lines, and lines that require stop-valve service. They do not provide throttled flow control. Gate valves with metal-to-metal seating are not well suited for frequent operation because the gate will wear down with use. Gate valves are difficult to automate.

Although gate valves can be completely cleared from the line of flow, materials can lodge between the wedge and the seat. This can damage the valve so that it won't close properly. To avoid this problem, close the valve slowly just before the wedge meets the seat. This creates a greater flow between the wedge and the seat that washes away any trapped solid materials.

In a gate valve, approximately 80 percent of full flow occurs in the first 20 percent of opening. The sudden release of pressure from a quickly opened gate valve may cause water hammer, which can damage the pipes. Always open and close a gate valve slowly and completely.

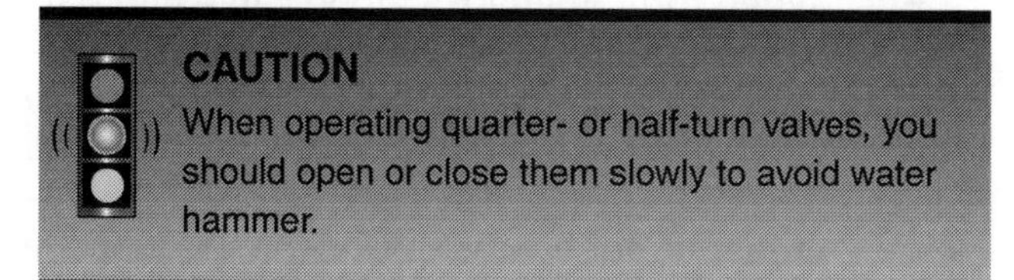

CAUTION

When operating quarter- or half-turn valves, you should open or close them slowly to avoid water hammer.

Liquids can remain trapped in a gate valve after the line has been drained. If liquid freezes inside the valve, it could expand and destroy the valve. To avoid this, remember to drain the valves as well as the lines.

Gate valves contain either a solid wedge or a split wedge. Each of these types is described in more detail below. Manufacturers are gradually phasing out split-wedge gate valves, but you are likely to still see them in construction specifications and existing installations. Be sure to follow the project specifications that call for one or the other type of gate valve.

2.1.1 Solid-Wedge Gate Valves

Use **solid-wedge gate valves** for steam, hot water, and other services where shock waves are a factor (see *Figure 1*). The wedge is precision-machined at an angle that matches the body seat. This angle conforms to the taper of the body seat to ensure a tight seal. Solid-wedge gate valves can be installed at any angle.

2.1.2 Split-Wedge Gate Valves

Use **split-wedge gate valves** on lines where you need a more positive closure, such as on low-pressure lines and cold water lines (see *Figure 2*). Also use split-wedge valves on fluid lines such as gasoline, benzene, kerosene, and solvents, all of which are light, volatile, and difficult to handle. The wedge's ball-and-socket joint helps align both faces against the body seat. When the wedge lowers completely, the downward motion of the stem forces the wedged halves outward against the body seats, forming a tight closure. When the valve opens, it relieves pressure on the split wedge. The split wedge is then removed by the upward motion of the stem from contact with the seats. The split-wedge design retains better closure than the solid-wedge design in cases of sudden temperature changes or flexing in the line.

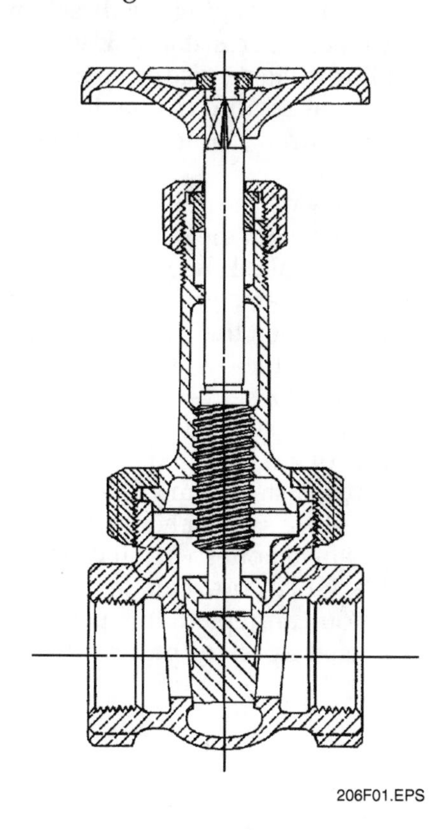

Figure 1 ◆ Solid-wedge gate valve.

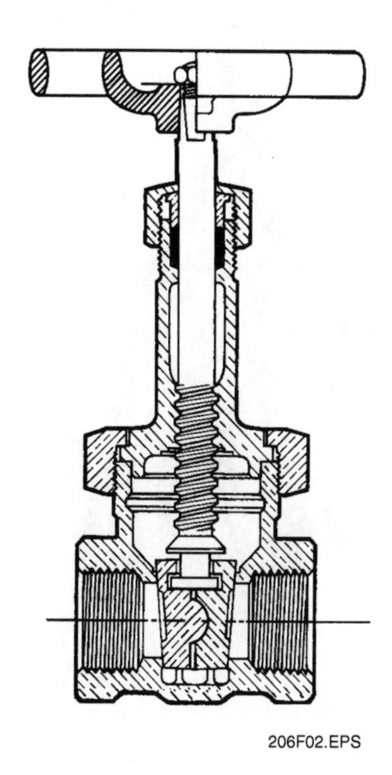

Figure 2 ◆ Split-wedge gate valve.

2.2.0 Globe Valves

A globe valve controls flow by moving a circular disc against a metal seat that surrounds the flow opening (see *Figure 3*). The screw action of the turning handle forces the disc onto the seat or draws it away. By changing the direction of flow, globe valves increase resistance. Use globe valves for general throttling service on steam, water, gas, and oil lines that require frequent operation and precise flow control. Do not use globe valves to balance a system.

A partition inside the valve body closes off the valve's inlet side from the outlet side, except for a circular opening called the valve seat. The upper side of the valve seat is ground smooth to ensure a complete seal when the valve is closed. To close the valve, turn the handwheel clockwise until the valve stem firmly seats the washer, or disc, between the valve stem and the valve seat. This stops the flow of gas or liquid.

Threads on the stem screw into corresponding threads in the valve's upper housing. The top of the housing is hollowed out so that it can hold a **packing** material, which prevents leaks through the valve stem. You can replace this packing if a leak occurs between the packing nut and the valve stem.

The critical parts of a globe valve—the washer, seat, and packing—are all replaceable. Globe valves provide accurate control of water flow. They are durable and won't wear out easily despite repeated use. Because of these features, plumbers frequently install globe valves in water supply lines. They are used, for example, to cut off the hot or cold water supply to a bathroom while a fixture is repaired. In less expensive models, the seat seal washer may require more frequent replacement to ensure a positive seal.

Even when fully opened, a globe valve will partially obstruct flow. Furthermore, it cannot be drained completely. Any liquid remaining in the valve could freeze, expand, and damage the valve. Refer to the manufacturer's specifications to determine whether a globe valve is appropriate for the installation.

If globe valves are specified in the plans, they are usually accounted for in the sizing of the water pipe. If you decide to use globe valves rather than gate valves, you should calculate the added pressure drop to check the pipe size as with any other added device or restricting valve. You will learn how to size water supply systems later in the *Plumbing* curriculum.

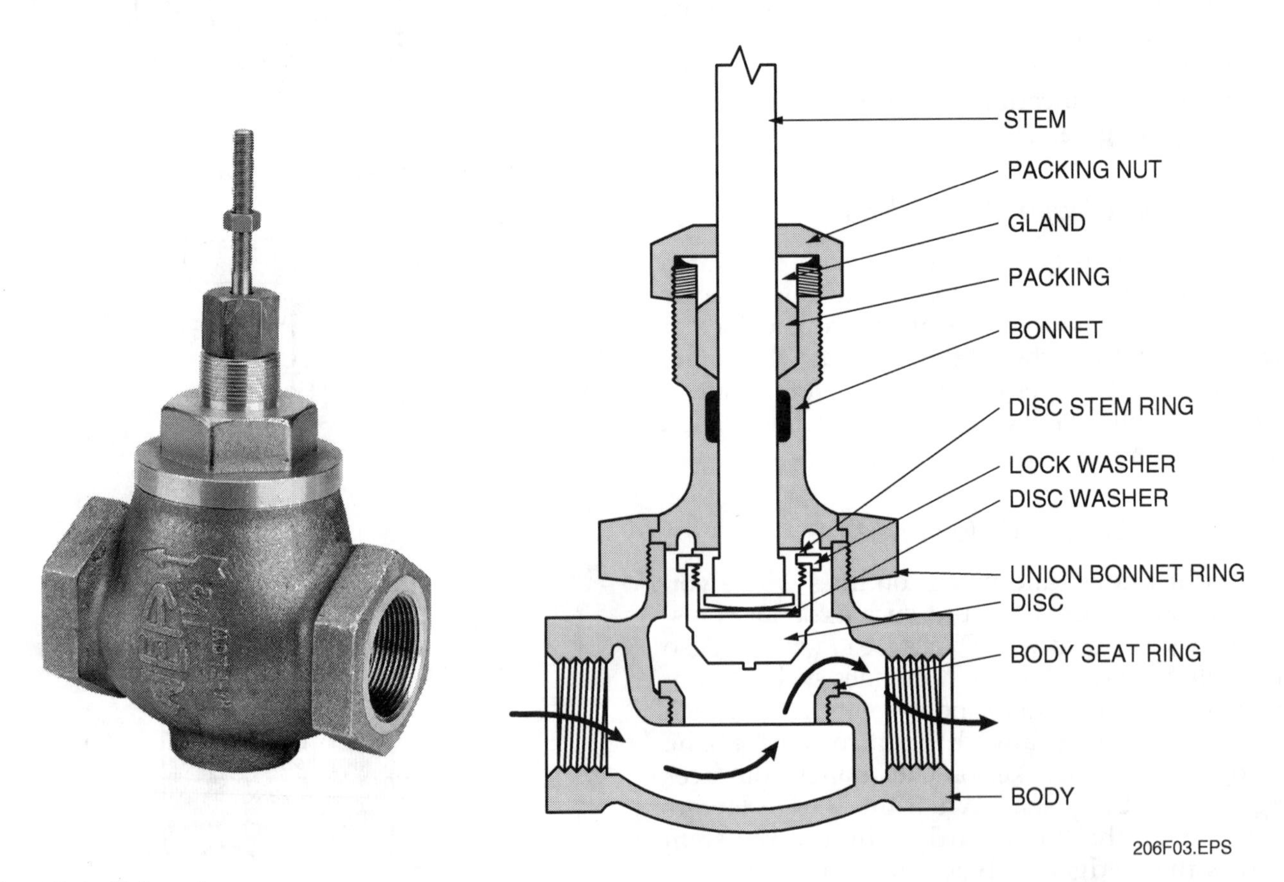

Figure 3 ◆ Globe valve.

Some globe valve models use a conical plug instead of a flat disc. The plug offers more precise flow control. Project specifications may specifically call for globe valves with conical plugs. Be sure to refer to the specifications before installing globe valves.

Globe valves must be installed in the correct flow direction. Flow must be up through the seat toward the disc or plug. Reverse flow through a globe valve will result in poor throttling control and chatter in the valve.

2.3.0 Angle Valves

The angle valve is similar to the globe valve but can serve as both a valve and a 90-degree elbow (see *Figure 4*). Because flow changes direction only twice through an angle valve, the valve is less resistant to flow than the globe valve, in which flow must change direction three times. Angle valves come with conventional, plug-type, or composition discs.

2.4.0 Ball Valves

Ball valves provide positive, quick flow control on piping systems (see *Figure 5*). A handle on the outside of the valve body rotates a spherical plug, or ball, into and out of the flow. Ball valves are best suited for main supply lines and pump lines, or where quick shutoff is desired. They can also be used for in-line maintenance or on lines in which liquids and gases are mixed. Unlike gate and globe valves, which require multiple turns of the handwheel to fully open or fully close, ball valves only require a quarter turn to operate. The valve seat material limits the temperature range within which a ball valve can be used. Consult the manufacturer's specifications to ensure that the valve is appropriate for the conditions on the line.

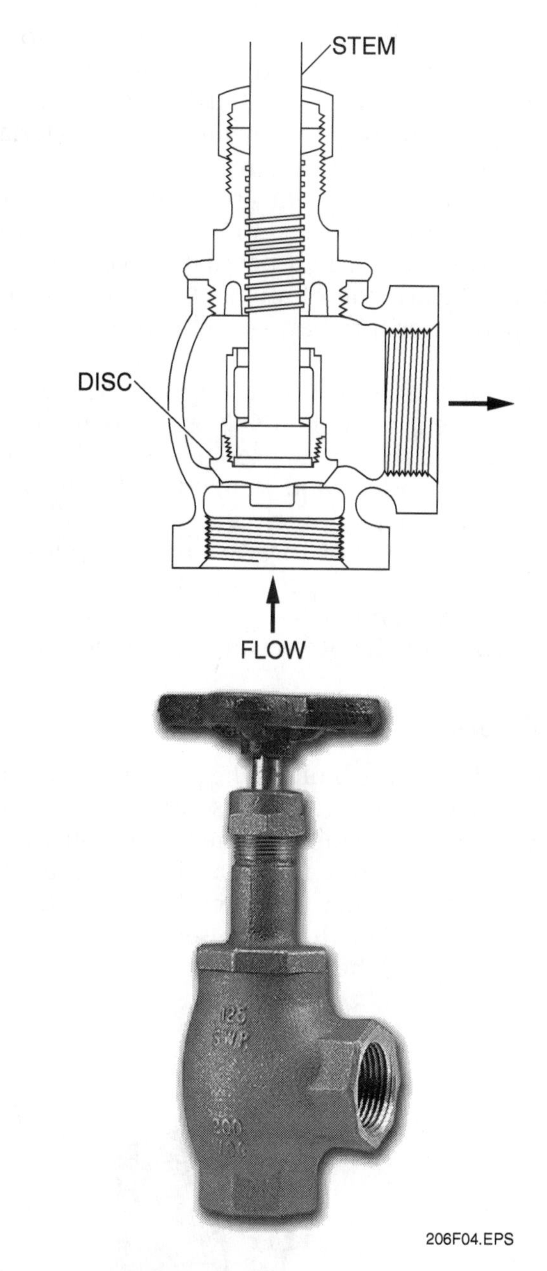

Figure 4 ◆ Angle valve.

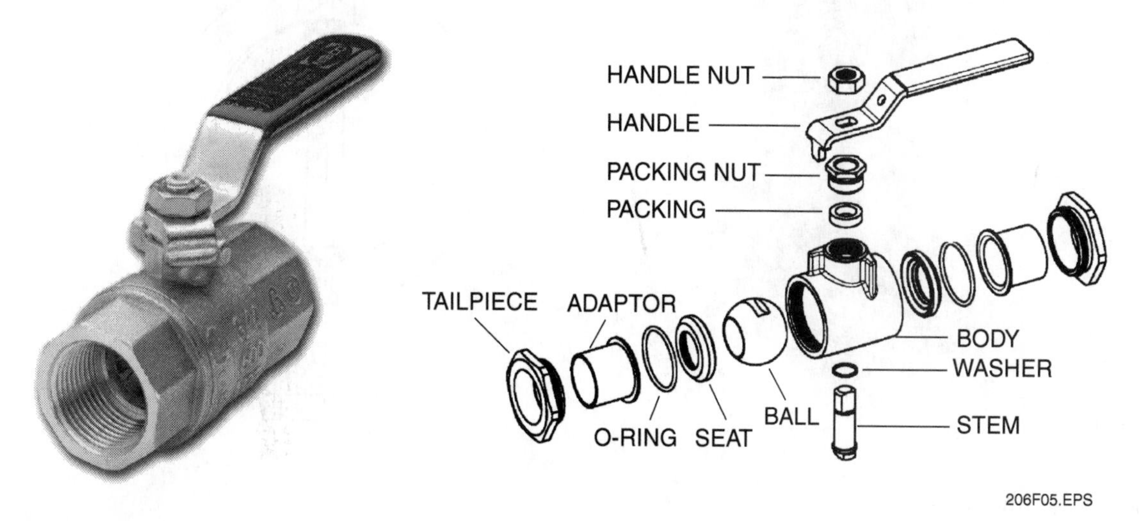

Figure 5 ◆ Ball valve.

Ball valves are fast becoming the workhorse valve of the plumbing industry. Ball valves are generally less expensive than other types of valves. They are available in almost any material, pressure classification, and end connection required. Furthermore, they offer good flow characteristics and can be used to throttle flow as well as for on/off service. Full-port ball valves offer the least flow restriction of any style valve. Because they are quarter-turn valves, ball valves are easily automated.

Remember to select the appropriate type of ball valve for the service. Ball valves are available in one-, two, and three-piece designs, as well as with soldered, threaded, and flanged ends. In high-maintenance areas, three-piece valves will save time and money, but they are more expensive. Refer to the project's specifications for the type of valve required, and then submit the selected valve for approval by the project architect or engineer.

2.5.0 Butterfly Valves

A butterfly valve is built on the principle of a pipe damper (see *Figure 6*). A disc about the same diameter as the inside diameter of the pipe controls the flow. The disc rotates and provides a tight seal against a seat machined on the inside diameter of the pipe. For throttling service, secure the disc in place with handle-locking devices.

Butterfly valves, especially the larger sizes, offer many advantages over the valves already discussed. They are lighter, take up less space, cost less, have fewer moving parts, and contain no pockets to trap fluids. They provide nearly full flow and offer little resistance. They are more compact than multi-turn valves, are easily automated, and have a long operational life. They should not be used on steam lines. If larger sized butterfly valves are operated too quickly, they can cause destructive water hammer on the line.

Use butterfly valves for either on/off or throttling service. They are suitable for installations requiring frequent operation. They do a good job of handling **slurries**, which are liquids that contain large amounts of suspended solids. In addition, butterfly valves offer a positive shutoff for gases and liquids. A positive shutoff valve forms a leak-proof seal when closed. To remember, use this memory aid: "A positive shutoff valve positively will not leak when shut off."

Butterfly valves with locking handles are widely used as balancing valves. Motor-driven butterfly valves can be used for flow control and as a remotely operated on/off valve. Butterfly valves can also be installed between two flanges or stud types.

Butterfly valves are available with flanged and groove-end connections, and with discs of either lug or wafer pattern. Codes require lug-pattern butterfly valves on lines providing dead-end service.

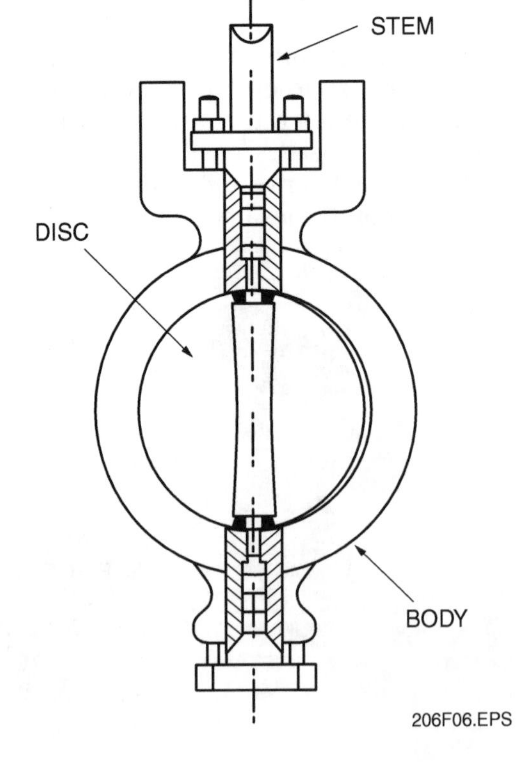

Figure 6 ◆ Butterfly valves.

2.6.0 Check Valves

Check valves prevent flow reversal in a piping system. Under normal operating conditions, flow in the line keeps the valve in an open position. In cases of flow reversal, the weight of the disc mechanism will automatically close the valve. Check valves are used on the steam side of water heating equipment. They allow air into the system as the system cools. This prevents damage to the coils in the heat exchanger. There are several types of check valves available, including the **ball-check valve**, the **swing-check valve**, and the **lift-check valve**.

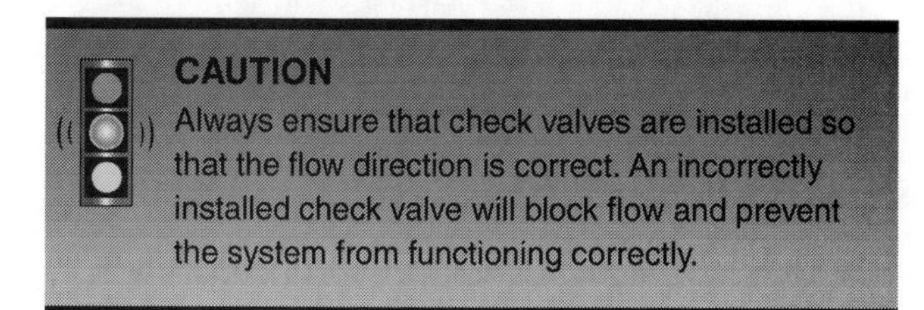

CAUTION
Always ensure that check valves are installed so that the flow direction is correct. An incorrectly installed check valve will block flow and prevent the system from functioning correctly.

2.6.1 Ball-Check Valve

The ball-check valve allows one-way flow in water supply or drainage lines. It is used in lines that have extremely low back pressure (see *Figure 7*).

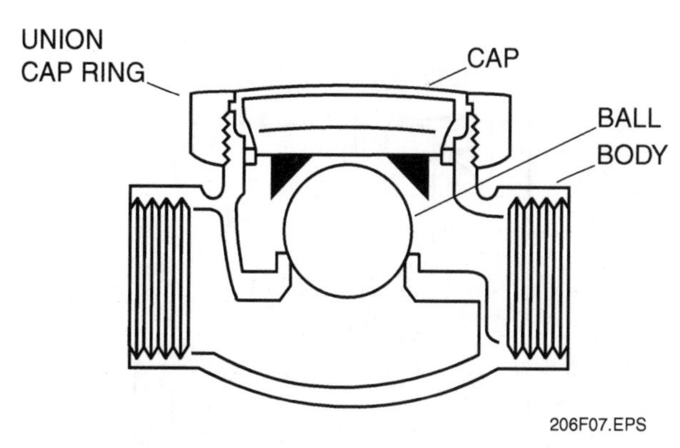

Figure 7 ◆ Ball-check valve.

2.6.2 Swing-Check Valve

Swing-check valves are well suited for lines containing liquids or gases with low to moderate pressures (see *Figure 8*). They offer lower flow resistance than ball-check and lift-check valves. Depending on the manufacturer, the swing-check valve comes in four different types: bronze mounted, all iron, rubber-faced disc, and **lever-and-weight swing-check**.

Plumbers use bronze-mounted swing-check valves most often. Because the bronze facing is soft and pliable, it offers a more positive sealing effect than facings made from harder materials. Use all-iron swing-check valves for services where bronze might erode. Any fluid or gas that contains corrosive chemicals will eventually erode the bronze valve.

Use a rubber-faced swing-check valve when noise in the check valve disc seating is a problem. This non-slam disc muffles the closing sound of the valve. It also helps protect the seat when high head pressures sharply reverse the flow.

Use the lever-and-weight swing-check valve when pulsation occurs in the lines. You can install these valves in either a horizontal or a vertical position and adjust them to prevent chatter from turbulence or pulsation in the line. The disc also adjusts to open at different amounts of pressure.

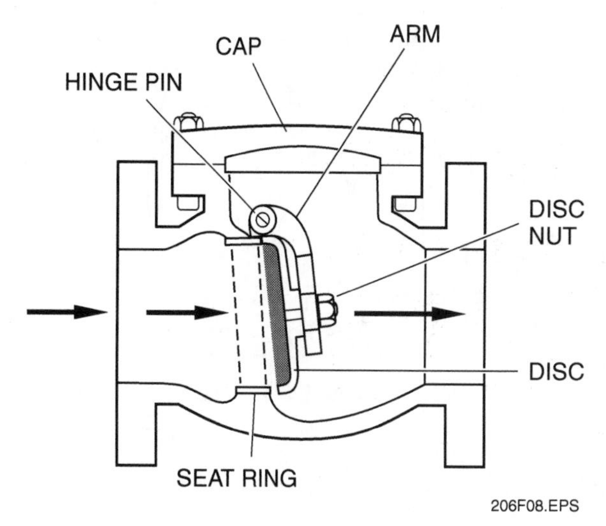

Figure 8 ◆ Swing-check valve.

2.6.3 Lift-Check Valve

Use the lift-check valve for gas, water, steam, or air, and for lines where frequent fluctuations in flow occur (see *Figure 9*). These valves come in horizontal and vertical styles. The integral construction of the horizontal type is similar to the globe valve. The vertical type allows a straight-through flow. Some in-line models are spring-actuated, which allows them to close immediately when flow stops. These are often called silent checks.

2.7.0 Flushometer Valves

Flushometer valves, also referred to simply as flushometers, are used for water closets or urinals (see *Figure 10*). They do not require a storage tank. Instead, a predetermined amount of water flows under pressure directly into the fixture. The scouring action of the water created when the flushometer opens generally cleans more effectively than the gravity flow from a storage tank. Flushometers

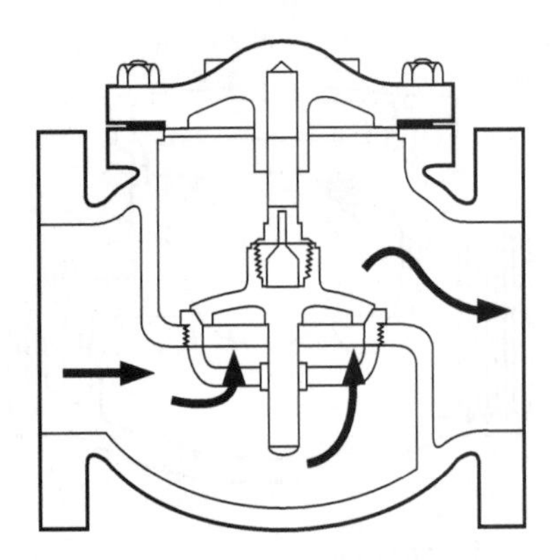

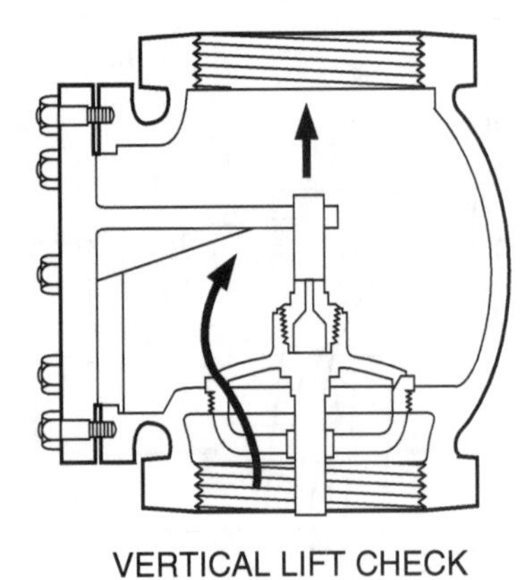

Figure 9 ◆ Vertical and horizontal lift-check valves.

are popular in commercial installations, institutions, and large assembly halls. Plumbers can connect them directly to the water supply pipe, and they can flush repeatedly in a short space of time without waiting for a storage tank to refill.

The flushometer in *Figure 10* is a manual type. The handle is depressed to trigger the flushing action. Electronic flushometers are widely used in heavy-use public restroom facilities. These are more sanitary because they do not require physical contact by the user. Electronic flushometers use an infrared device to trigger the flushing action (see *Figure 11*).

Flushometers are available in diaphragm and piston types. Both work the same way. In the diaphragm type, water flow stops when water pressure in the upper chamber forces the diaphragm against the valve seat (see *Figure 12*). Motion of the handle in any direction pushes the plunger against the auxiliary valve. Even the slightest tilt will cause

> **DID YOU KNOW?**
> ### Variety of Valves
> Valves come in many different sizes. Very small ones are used in automobile and bicycle tires. Huge valves control the flow of water in canal locks. No matter the size, all valves have two basic parts—an enclosed body containing a hole through which fluids or gases pass, and a device that opens and closes the hole.

Figure 10 ◆ Flushometer.

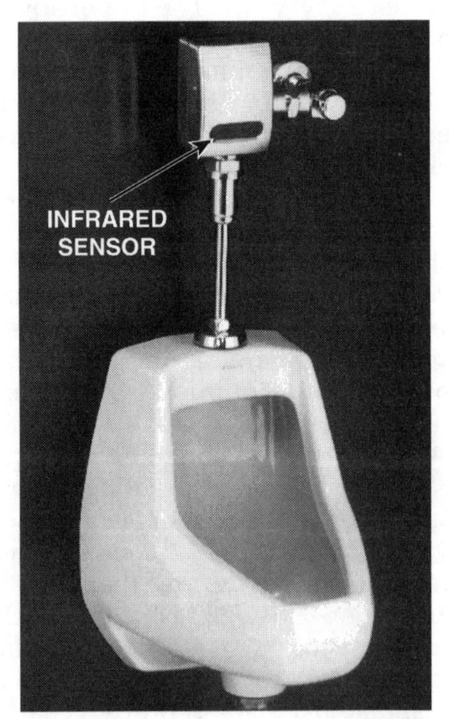

Figure 11 ◆ Electronic flushometer installed on a siphon-jet urinal.

water leakage from the upper chamber around the diaphragm.

When pressure lessens in the upper chamber, the diaphragm rises, allowing water to flow into the fixture. While the fixture is in flushing mode, a small amount of water flows through the bypass. As the water fills the upper chamber, the diaphragm reseats against the valve seat. This action shuts off the flow of water to the fixture.

2.8.0 Flush Valves

A flush valve is a valve located at the bottom of a tank for the purpose of flushing water closets and similar fixtures. They are mostly found in residential and small commercial buildings. Flush valves are installed in the bottom of water closet flush tanks and control the flow of water from the tank into the bowl (see *Figure 13*). When the tank lever is depressed, the valve lifts above the tank outlet and floats there. This allows the water in the tank to flow rapidly into the bowl. When the water level in the tank drops to the point where the flush valve no longer floats, the valve reseats on the tank drain and the tank—regulated by the float-controlled valve—refills. Flush valves are available in both manual and electronic models. *Figure 13* shows a manual type.

2.9.0 Plug Valves

Plug valves are widely used in supply, sanitary, and gas lines to provide both on/off and throttling control (see *Figure 14*). They are also widely used as balancing valves in water lines. However, ball valves and butterfly valves have taken their place on heating, ventilating, and air conditioning

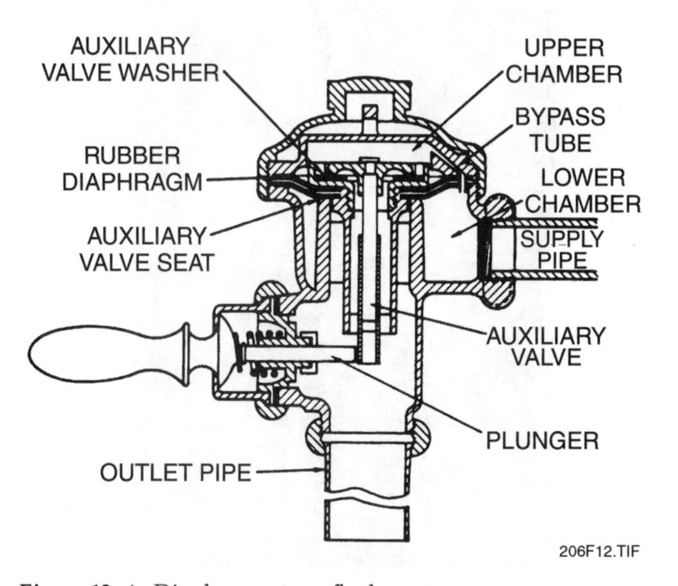

Figure 12 ◆ Diaphragm-type flushometer.

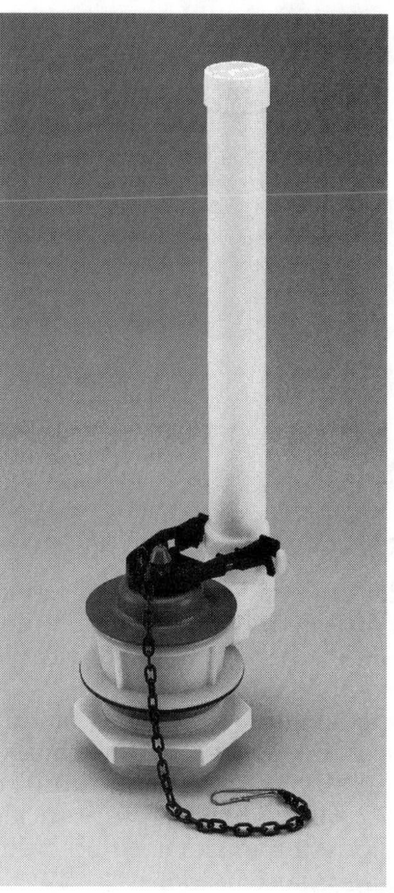

Figure 13 ◆ Manual flush valve.

(HVAC) water systems. Plug valves provide positive shutoff when closed.

Plug valves operate in one of two ways, depending on the model of valve. In some designs, rotation of the valve handle clockwise raises a cylindrical or cone-shaped plug out of its seat, increasing the flow. Rotating the valve handle counter-clockwise lowers the plug back into its seat, restricting flow. In other models, the cone-shaped plug has a horizontal opening through which flow is permitted until the opening is turned away from the flow. These types of plug valves are nonrising stem, quarter-turn valves.

Plug valves are widely used in industrial applications such as liquid and gas pipelines, boiler feed lines, and condensate drain lines. They are also used on lines that handle slurries.

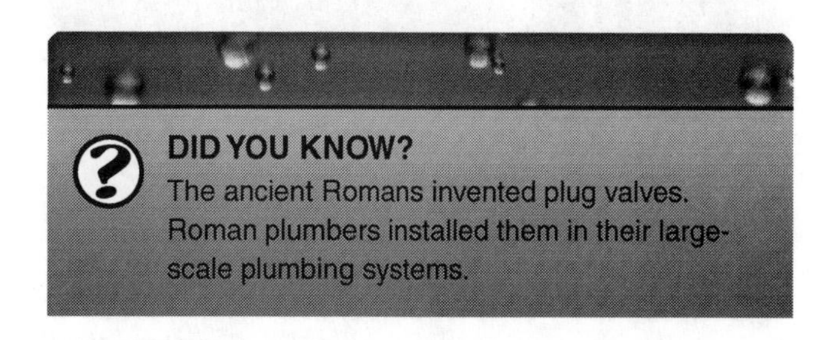

DID YOU KNOW?

The ancient Romans invented plug valves. Roman plumbers installed them in their large-scale plumbing systems.

206F14.EPS

Figure 14 ◆ Plug valve

2.9.1 Non-Lubricated Plug Valves

Some older plug valves use metal-to-metal seals without lubrication. Non-lubricated plug valves may feature mechanical lifting devices to reduce the torque required to turn them.

Some models use elastomeric, or rubber-like, sleeves or coatings that reduce friction when the plug is turned. Note that non-lubricated plug valves often suffer from chafing or sticking. They should be carefully maintained according to the manufacturer's specifications.

2.9.2 Lubricated Plug Valves

Lubricated plug valves, as their name suggests, use a special type of grease as a lubricant between the plug and the valve seat. The grease reduces wear and eliminates sticking. The lubricant also reduces leakage around the plug.

2.10.0 Temperature and Pressure (T/P) Safety and Relief Valves

The terms temperature and pressure (T/P) relief valve and T/P safety valve are often used interchangeably. Spring-loaded T/P safety valves and T/P relief valves look similar. Both limit fluid pressure by discharging some of the pressurized liquid or gas.

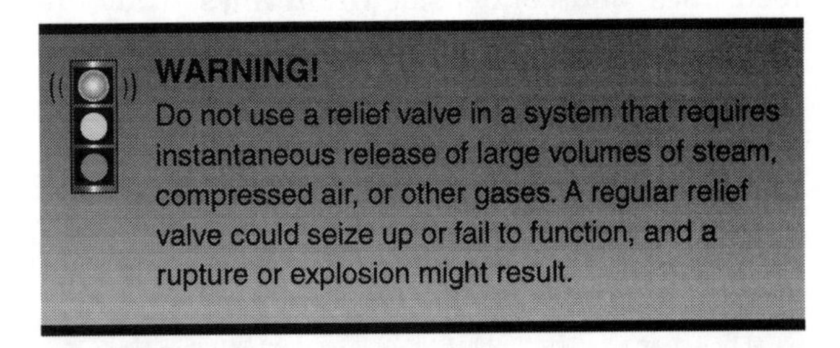

WARNING!

Do not use a relief valve in a system that requires instantaneous release of large volumes of steam, compressed air, or other gases. A regular relief valve could seize up or fail to function, and a rupture or explosion might result.

Codes usually require safety and relief valves to be installed in sections of a plumbing system that must have pressure protection. Safety and relief valves relieve or divert pressure away from the system, which may not be designed to handle excess pressure.

Use steam safety valves in lines with gases that carry air and steam (see *Figure 15*). The design includes a **huddling chamber** that harnesses the expansion forces of the pressurized gases to quickly open (pop) or close the valve. The difference between the opening and closing pressures is called **blowdown**. Blowdown limitations for steam safety valves are stated in the **American Society of Mechanical Engineers (ASME) Boiler and Pressure Vessel Code**.

Use relief valves for liquid service (see *Figure 16*). Ordinarily, relief valves do not have an accentuating huddling chamber or a regulator ring that varies or adjusts the blowdown. They therefore operate with a relatively lazy motion. As pressure increases they slowly open, and as pressure decreases they slowly close. In vessels or systems that don't require instantaneous release of large volumes, these valves provide sufficient protection. They are also appropriate for use in systems that provide sufficient leeway between design and operating pressures. Ensure that T/P safety and relief valves terminate at an approved point of disposal.

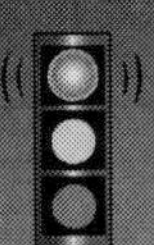

WARNING!

Codes require the vent lines for safety and relief valves to be terminated at a safe location. Improperly terminated vents can cause injury when the valve operates. Refer to your local applicable code for venting requirements for T/P valves.

206F15.EPS

Figure 15 ◆ Steam safety valve.

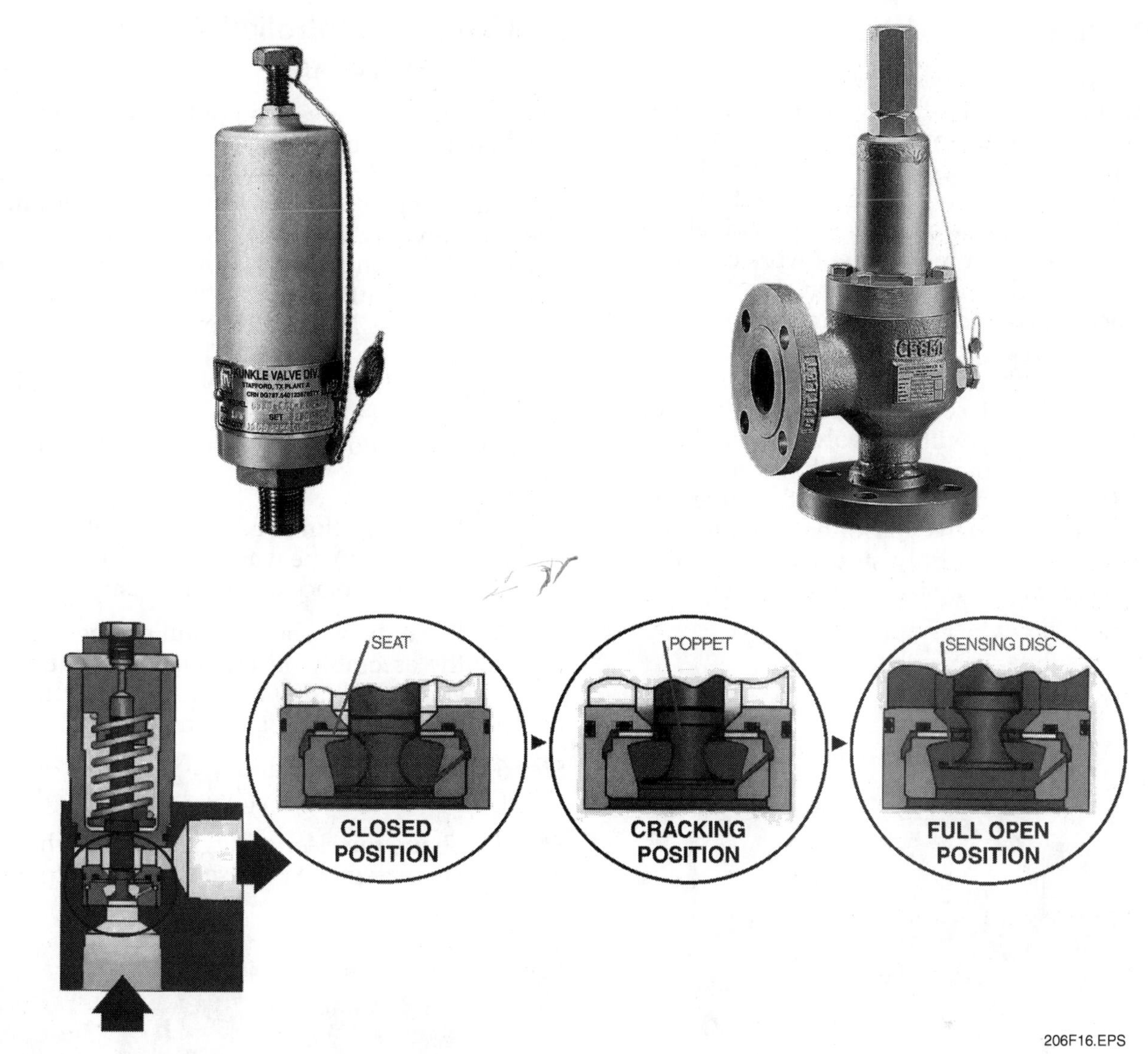

Figure 16 ◆ Relief valve.

2.11.0 Pressure Regulator Valves

Pressure regulator valves reduce water pressure in a water supply system (see *Figure 17*). Pressure changes in the system activate the valve. As the pressure changes, a spring located in the valve's dome acts on a diaphragm to move the valve up or down. The valve opens when it is pressed down away from the valve seat and remains open until the building's water pressure reaches a set level. The valve then closes and remains closed until the building's water pressure begins to drop.

Install pressure regulator valves on boilers, residential systems that are subject to excess pressure, and other mechanical devices as required by the local applicable code.

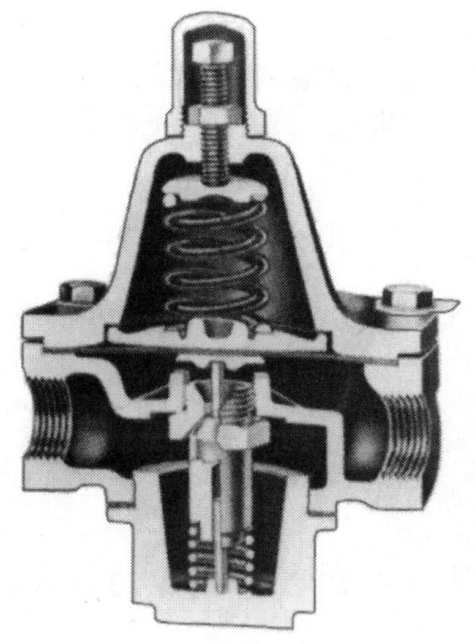

Figure 17 ◆ Pressure regulator valve.

2.12.0 Supply Stop Valves

Supply stop valves are installed on individual fixtures such as water closets, lavatories, and sinks. They are used to stop flow and permit the disconnection of the hot or cold water supply. These valves, also called supply valves, make it easy to control the water connection at an individual fixture, making repair or replacement work easier as well. They are chrome-plated for an attractive appearance, and they come in both right-angle and straight designs (see *Figure 18*). Supply stop valves are made with different end connections, such as compression, ridged, solder, or glued. The type of connection will vary with the valve material and the application. A compression valve is shown in *Figure 18A*. Note that the inlet and the outlet are different sizes. Such valves are often used on tank-type water closets. Supply stop valves are also available with the same size inlet and outlet.

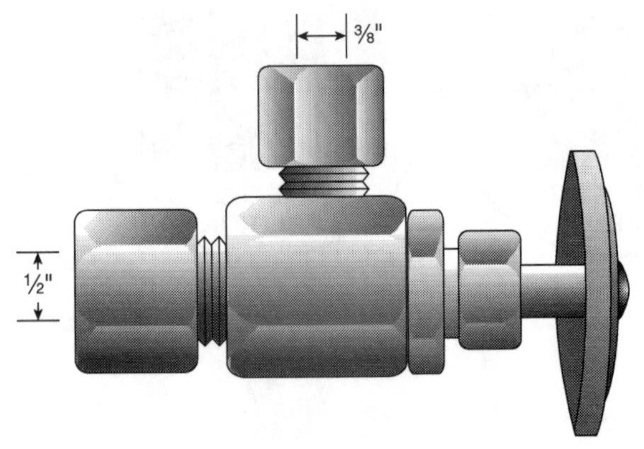

RIGHT-ANGLE SUPPLY STOP VALVE

STRAIGHT-DESIGN SUPPLY STOP VALVE

206F18.EPS

Figure 18 ◆ Right-angle and straight-design supply stop valves.

2.13.0 Float-Controlled Valves (Ball Cocks)

Float-controlled valves, also called ball cocks, are installed in water closet flush tanks (see *Figure 19*). They maintain a constant water level in the tank by controlling the water supply into the tank. Their design may vary, but all use the water level in the tank to control flow. When the water level drops, the valve lifts up and away from the seat as the arm lowers. The float arm then rises as the water level in the tank rises. This forces the valve down against the seat, which closes the valve and stops the flow of water.

To install the float valve assembly, follow these steps:

Step 1 Lower the threaded base through the bottom hole in the water closet tank with the gasket in place against the flange.

Step 2 Place the washer and nut on the base of the assembly on the outside of the tank and tighten. Do not overtighten the nut because the toilet tank may crack.

Step 3 Place the riser and coupling nut on the base of the float valve, and hand tighten.

Step 4 Shape the riser to align with the shutoff valve, remove the riser, and cut it to fit.

206F19.EPS

Figure 19 ◆ Float-controlled valve.

DID YOU KNOW?

Valve Component Terms

Terms for valve components vary widely depending on the manufacturer, region, and the size or type of valve. For example, followers are used in certain types of valves to put pressure on the packing when the packing nut is turned. The follower is also called a packing gland or simply a gland. Flanges that incorporate more than one nut to tighten followers are also called follower flanges, packing flanges, stem flanges, and gland flanges. Take the time to learn the valve terms used in your area.

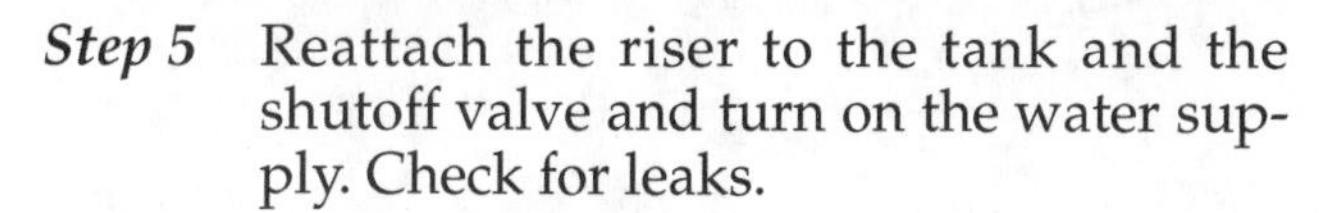

Step 5 Reattach the riser to the tank and the shutoff valve and turn on the water supply. Check for leaks.

Step 6 Adjust the float ball in the tank to achieve the water level shown on the tank.

Step 7 Adjust the float ball with the adjusting screw. Do not bend the float arm. The rod may work around a half-turn (180 degrees) and cause the tank to overflow.

2.14.0 Backwater Valves

Backwater valves are a type of check valve installed on DWV lines. They provide a mechanical seal against backflow from public sewer lines. They are manufactured in gate (see *Figure 20*), flapper (see *Figure 21*), and ball float (see *Figure 22*) models.

Codes require backwater valves to be installed in DWV systems to protect fixtures and drain inlets from contamination by backflow or flooding from the public sewer. A fixture or drain inlet typically requires protection if its flood level rim is below the elevation of the nearest upstream sewer manhole cover, or if it is located above the crown level of the public sewer at the point of connection but below curb level where the building sewer crosses under the curb at the property line.

Depending on your local applicable code, backwater valves can be installed in the following ways:

- In the building drain
- On a branch of the building drain
- On the horizontal branch that serves the fixtures needing protection

206F20.EPS

Figure 20 ◆ Gate-style backwater valve.

206F21.EPS

Figure 21 ◆ Flapper-style backwater valve.

206F22.EPS

Figure 22 ◆ Ball float-style backwater valve.

Ensure that the valve you select complies with the applicable ASME standards for backwater valves that are referenced in your local applicable code. The backwater valve must offer no flow resistance when installed. That is, the capacity of the valve when fully opened must not be less than the pipes on which they are installed.

Install the valve so that it is accessible for servicing and repair. Most codes require backwater valves to be inspected following installation by a person who is licensed to test and certify or repair them. Refer to your local applicable code.

Review Questions

Sections 1.0.0–2.0.0

1. The part(s) of a valve that receives the most wear and tear is the _____.
   a. trim
   b. seat
   c. disc
   d. packing

2. Water hammer may result when gate valves are opened too _____.
   a. quickly
   b. slowly
   c. far
   d. little

3. Use _____ valves in lines that carry hot water or steam.
   a. split-wedge gate
   b. solid-wedge gate
   c. swing-check
   d. ball-check

4. A(n) _____ is most commonly used to control bathroom water supply lines.
   a. angle valve
   b. float-controlled valve
   c. check valve
   d. globe valve

5. A butterfly valve is best only for on/off flow control in a piping system.
   a. True
   b. False

Match the type of check valve with its characteristics.

6. _____ Ball-check valve
7. _____ Swing-check valve
8. _____ Lift-check valve
   a. Used in lines that experience frequent fluctuations in flow
   b. Used in lines that handle high-temperature gases and liquids
   c. Used in lines that have extremely low back pressure
   d. Used in lines that have low to moderate pressures

9. A _____ delivers water under pressure to a water closet or urinal without the need for a storage tank.
   a. globe valve
   b. flushometer
   c. float-controlled valve
   d. pressure regulator valve

10. Use _____ to disconnect the hot or cold water supply to individual fixtures such as water closets, lavatories, and sinks.
    a. gate valves
    b. globe valves
    c. supply stop valves
    d. ball check valves

11. A _____ controls water supply to a water closet storage tank.
    a. supply stop valve
    b. flushometer
    c. T/P relief valve
    d. float-controlled valve

12. Each of the following is an acceptable way to install a backwater valve *except* _____.
    a. in the building drain
    b. on fixtures with flood rims above the upstream manhole cover
    c. on a branch of the building drain
    d. on the horizontal branch that serves the fixtures needing protection

3.0.0 ◆ VALVE COMPONENTS

As well as the valve itself—whether in the form of a disc, a ball, a hinged plate, or a plug—and the valve body that holds the valve and its seat, a valve consists of a valve stem, packing, a **bonnet**, and end connections. The following sections review each of these valve elements in detail.

3.1.0 Stems

The three basic types of stems used on valves are as follows:

- Rising stem
- Nonrising stem
- Outside screw and yoke stem

3.1.1 Rising Stem

In a rising stem unit, as the name suggests, both the handwheel and the stem rise as the handwheel is turned (see *Figure 23*). The height of the stem gives an approximate indication of how far the valve is open. You can install valves with rising stems only in areas with sufficient headroom. The stem itself comes into contact with the liquid in the line. If the stem is damaged, the valve will not operate properly.

3.1.2 Nonrising Stem

With a nonrising stem, neither the handwheel nor the stem rises when the valve opens (see *Figure 24*). Because only a stem inside the valve body turns when the handwheel turns, wear is kept to a minimum. This type of stem also makes contact with the liquid in the line. Nonrising stems are suitable for installations with limited access space.

3.1.3 Outside Screw and Yoke Stem

Often abbreviated OS&Y, the outside screw and yoke stem is suitable for use with corrosive liquids because the stem does not come in contact with the liquid in the line (see *Figure 25*). As the handwheel turns, the stem moves up through the wheel. The height of the stem above the handwheel gives an approximate indication of how far the valve is open.

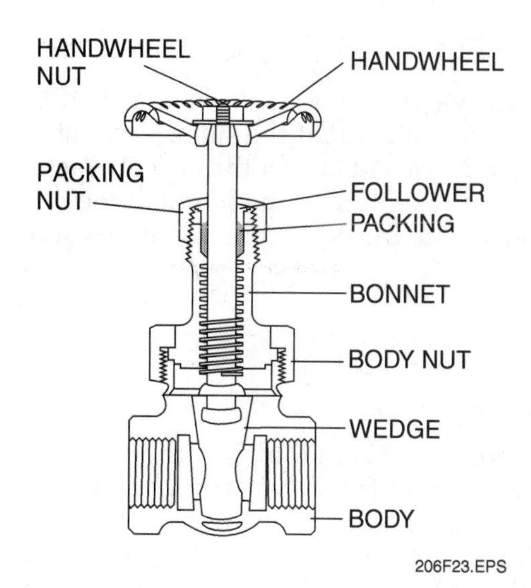

Figure 23 ◆ Rising stem.

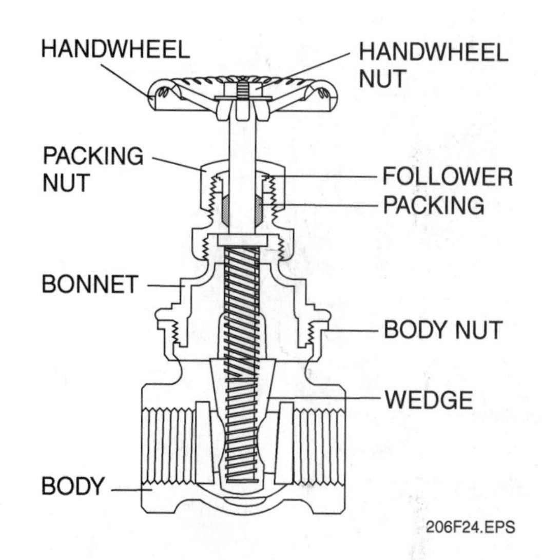

Figure 24 ◆ Nonrising stem.

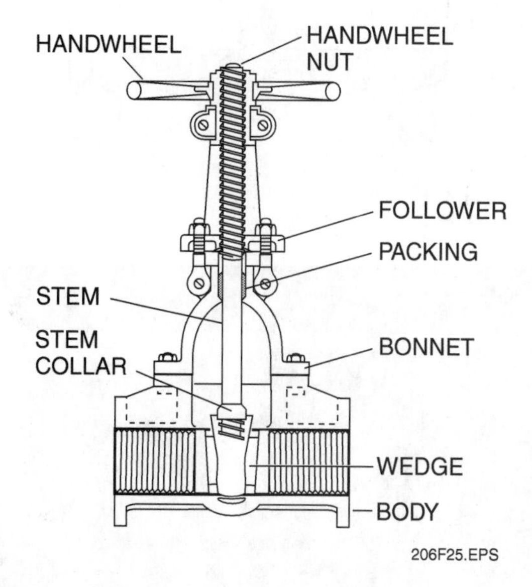

Figure 25 ◆ Outside screw and yoke stem.

3.2.0 Packing

Because valve stems move, it can be difficult to achieve a good seal at the stem. Packing seals the stem and prevents leaks through it. Packing also retains the pressure of the liquid in the valve. The following are the most common types of packing:

- Solid
- Braided
- Granulated fibers
- Grease

The packing material fills the stuffing box, a space between the valve stem and bonnet. A follower presses the packing against the stem in the stuffing box (see *Figure 26*). The stuffing box requires occasional tightening, especially if the valve has not been used for a while.

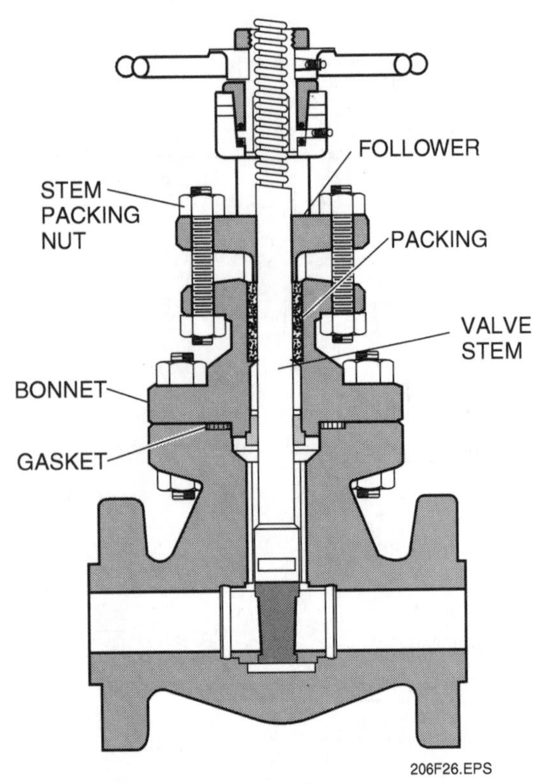

Figure 26 ◆ Packing.

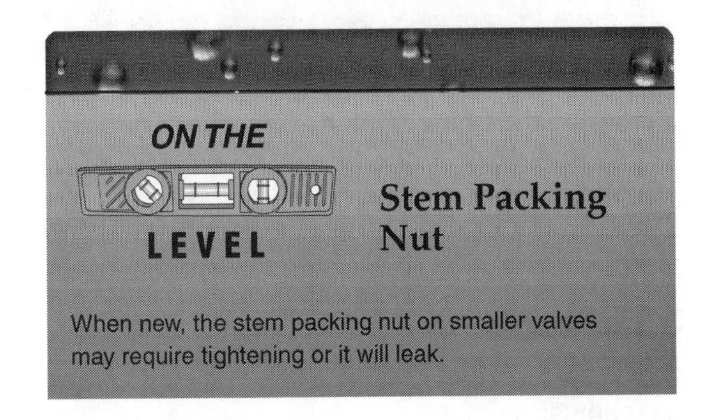

ON THE LEVEL

Stem Packing Nut

When new, the stem packing nut on smaller valves may require tightening or it will leak.

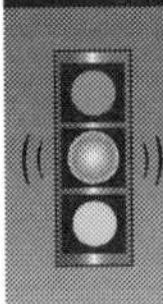

CAUTION

Be sure that the packing is suited for the material flowing through the line and the operating pressures and temperatures. Improper packing will fail and leak.

When choosing the correct packing material, consider the following factors:

- What materials will be flowing through the line?
- What are the anticipated operating pressures and temperatures in the line?
- What will be the minimum temperature of the piping system?
- What is the composition of the valve stem?

3.3.0 Bonnets

The valve bonnet is the cover that both guides and encloses the valve stem. The following are the three basic types of bonnets:

- Screwed bonnet
- Union bonnet
- Bolted bonnet

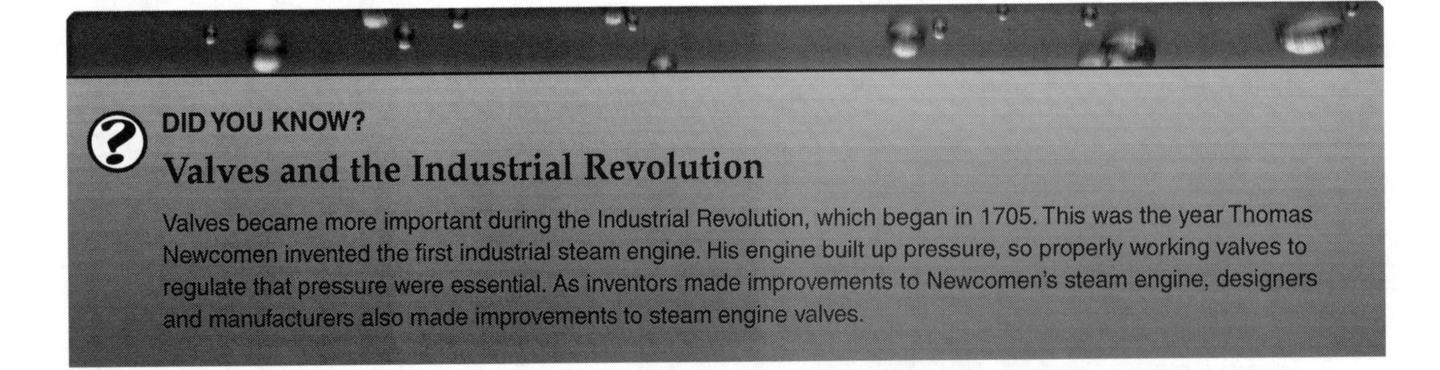

DID YOU KNOW?

Valves and the Industrial Revolution

Valves became more important during the Industrial Revolution, which began in 1705. This was the year Thomas Newcomen invented the first industrial steam engine. His engine built up pressure, so properly working valves to regulate that pressure were essential. As inventors made improvements to Newcomen's steam engine, designers and manufacturers also made improvements to steam engine valves.

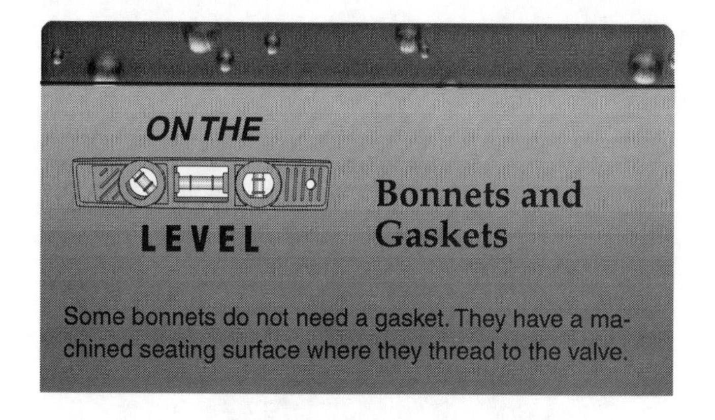

3.3.1 Screwed Bonnet

The screwed bonnet is the simplest and least expensive bonnet design (see *Figure 27*). It is a two-piece configuration, consisting of a threaded bonnet and a packing nut, which screws onto the bonnet. Screwed bonnets are frequently used in low-pressure applications and also in applications where periodic disassembly of the valve is not required. Screwed bonnets are often used on bronze gate, globe, and angle valves. They are also available with soldered connections.

3.3.2 Union Bonnet

The three-piece union bonnet is best suited for those applications where frequent disassembly of the valve is required (see *Figure 28*). A separate union ring holds the valve bonnet to the valve body, giving the body added strength against internal valve pressures. Union bonnets work well with smaller valves (¼ inch to 2 inches). However, they are impractical for larger valves (2 inches and up). Union bonnets are also available with soldered joints.

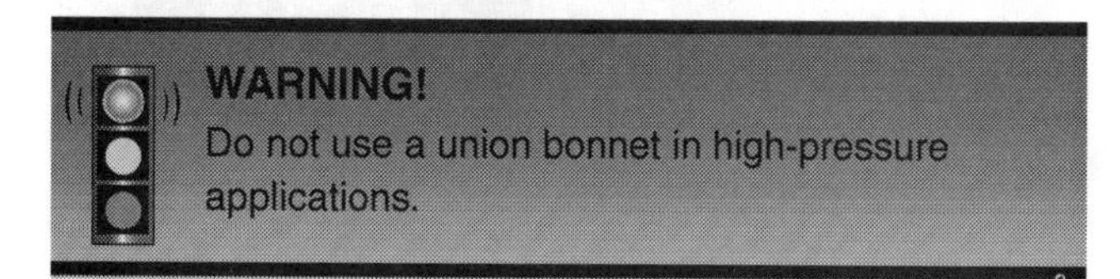

WARNING!
Do not use a union bonnet in high-pressure applications.

3.3.3 Bolted Bonnet

The bolted bonnet joint is used frequently on larger valves and in high-pressure applications (see *Figure 29*). The bonnet is attached to the valve body with a series of small-diameter bolts. This allows a uniform sealing pressure and maintenance with small wrenches.

3.4.0 End Connections

Valves that are commonly used in plumbing installations are manufactured with a range of end connections, including:

- Internal threads
- Solder ends
- Flanged ends
- External threads

Most metal-bodied valves are manufactured with internal threads. The threads allow the valve to be hooked up easily to threaded pipe (see *Figure 30*). Always refer to the manufacturer's specifications and to your local applicable code to ensure that the metals are compatible or that, if they are not, a proper dielectric connection is used.

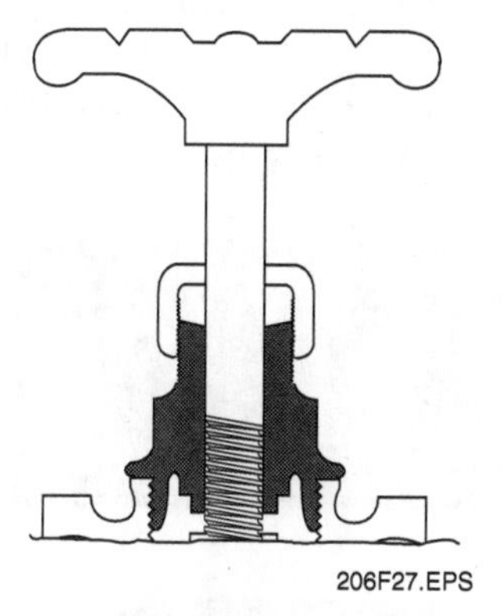

Figure 27 ◆ Screwed bonnet.

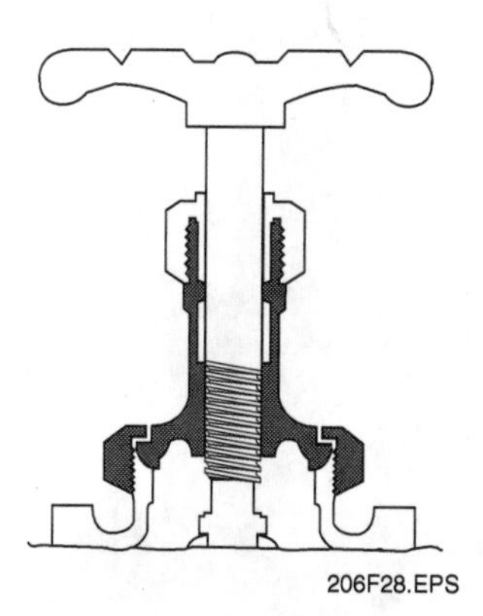

Figure 28 ◆ Union bonnet.

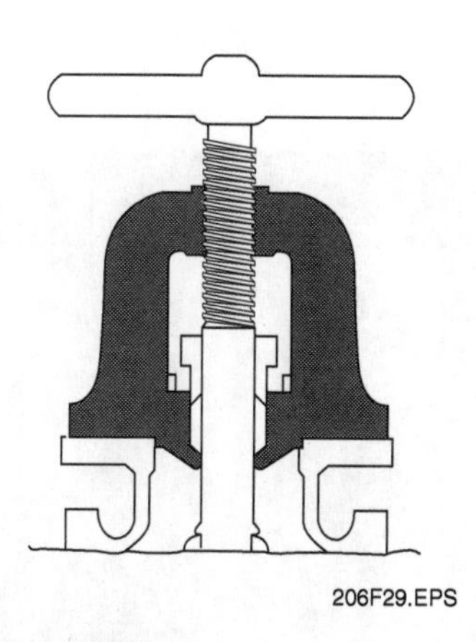

Figure 29 ◆ Bolted bonnet.

Use solder end valves to hook up to copper and brass pipes that have no internal or external threads. You can purchase these valves in bronze or bronze mounted and copper (see *Figure 31*).

In applications where valves must be changed out for repair or replacement, install valves with flanged ends (see *Figure 32*). Flanged end valves are more quickly and economically replaced than soldered valves. Flanged end valves are generally used on large pipes of at least 4 inches diameter, and in commercial and industrial installations.

The external thread design is frequently used on plastic valves. These valves require a nut in order to make a compression joint with the pipe. External thread valves are rarely used in water supply and DWV systems.

Figure 31 ◆ Solder end valve.

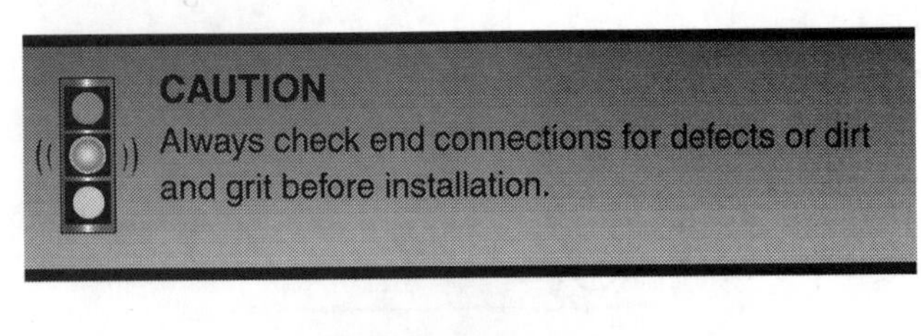

CAUTION

Always check end connections for defects or dirt and grit before installation.

Figure 30 ◆ Threaded connection.

Figure 32 ◆ Flanged end valves.

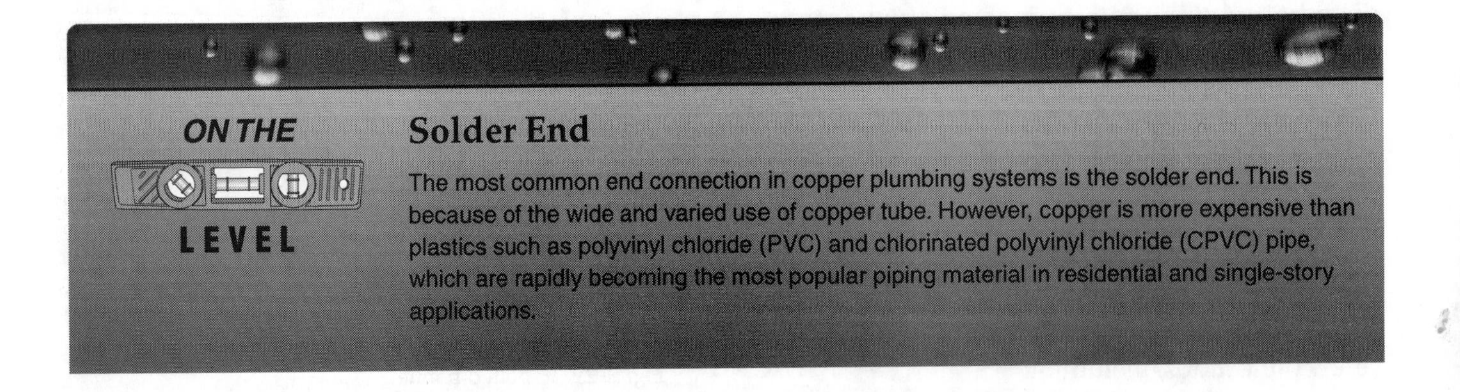

Solder End

The most common end connection in copper plumbing systems is the solder end. This is because of the wide and varied use of copper tube. However, copper is more expensive than plastics such as polyvinyl chloride (PVC) and chlorinated polyvinyl chloride (CPVC) pipe, which are rapidly becoming the most popular piping material in residential and single-story applications.

DID YOU KNOW?

During the Industrial Revolution, industries became major users of valves. These industries produced items such as textiles, chemicals, and processed food. With the birth of the petroleum industry came increased demand for high-performance valves. These valves had to handle the great pressure of oil and gas flowing from deep wells to the surface. As demand for higher-performance valves grew, manufacturers responded with improvements in engineering, materials, and design.

4.0.0 ◆ SELECTING VALVES FOR SPECIFIC APPLICATIONS

Many factors contribute to the proper selection of valves. For example, if you need a valve to provide on/off service, select a gate valve, butterfly valve, or ball valve. If you need a valve that will throttle and regulate flow, select a globe valve, butterfly valve, or ball valve.

Once you have determined the valve's application, analyze the materials that are used in the system. The following are some of the factors to consider:

- Is the material a liquid or a gas?
- How freely can the material move through the system?
- Is the material abrasive?
- Is the material corrosive?
- What is the temperature of the material?
- What is the pressure in the system?
- To what extent must the valve be leak-tight?
- What is the maximum pressure drop the system can tolerate?

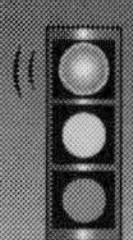

WARNING!

Be sure that the valve you use is suited for the intended system. You cannot use bronze or cast-iron materials in corrosive systems.

To select the correct valve that addresses all of these factors, you must consider the materials that the valve is made from, the size of the valve, and the valve's ratings. The following sections discuss these criteria in more detail. Remember to refer to the project specifications for the types of valve called for in the system.

4.1.0 Valve Materials

Valves are made from practically every material that can be machined. The most common materials used are iron, brass, carbon steel, bronze, aluminum, alloy steels, and polyvinyl chloride (PVC). Circumstances dictate which material to use in a particular application. Generally, more than one material will satisfy the requirements of a given situation.

For example, valves used in water service are made chiefly from bronze, brass, malleable iron, cast iron, copper, forged steel, stainless steel, thermoset plastic, and thermoplastic. Frequently, different materials make up the various parts of a single valve. The valve body may be formed from cast iron, the valve plug from cast bronze, the O-ring seal from rubber, and the compression washer from Teflon®.

Air and water service valves are made of bronze or iron with bronze trim. Valves used for low-pressure steam applications are made of iron. Valves that regulate noncorrosive products are made of steel.

DID YOU KNOW?

Pressure is the force exerted by one substance on another. For example, compressed gas exerts pressure on the walls of a pipe. Fluid pressure (liquid and gas) is measured in pounds per square inch (psi) in one of two ways: (1) pounds per square inch gauge (psig), which is measured with a pressure gauge, and (2) pounds per square inch absolute (psia), which is the atmospheric pressure (14.696 at sea level at 70°F).

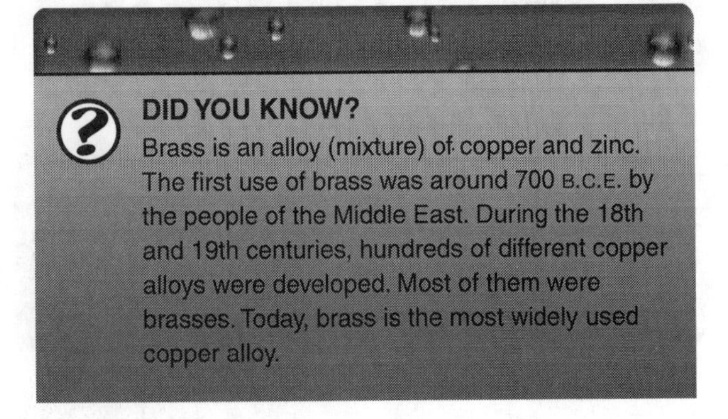

Although it is expensive, stainless steel is required for valves that regulate the flow of corrosive substances. Stainless steel is also used when it is necessary to prevent contamination of fluids flowing through a line. Most valves used in cryogenic (very low temperature) applications are made from stainless steel because it does not get as brittle as iron or carbon steel at low temperatures. Some chemicals—for example, chlorine and sulfuric acid—require specialized valve bodies. These are made from plastic, rubber, ceramics, or special alloys.

4.2.0 Valve Sizing

You may have some difficulty determining the correct size valve for a given application. The general rule of thumb is to use a valve with the same connection size as the pipeline in which it is installed. A valve's dimensions, however, are not always directly related to its capacity. As a result, relying solely on the rule of thumb may cause the system to operate inconsistently. Base valve sizing on valve capacity and known performance values. This can be found in the valve sizing tables in your local applicable code.

4.3.0 Valve Ratings

When you look at a valve body, you will see a combination of numbers and letters. Taken together, these are the valve's rating. Valves are rated according to the types and pressures of gases or liquids they are designed to handle. The number represents the safe operating pressure in pounds per square inch (psi). The letters represent the type of liquid or gas. The following letters are used:

- O – oil
- W – water
- S – steam
- G – gas
- L – liquid other than water
- SWP – steam working pressure

So for example, a valve marked 125 SWP with 200 WOG can operate safely at 125 psi of saturated steam or 200 psi of cold water, oil, or gas.

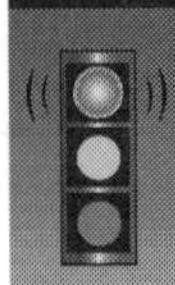

WARNING!

Using a valve in a system that exceeds its pressure rating can cause a serious accident.

5.0.0 ◆ REPAIRING VALVES

You now have a basic understanding of how valves operate. To service or repair valves, you need to identify the problem, determine what caused it, and efficiently make the repair. Experience will help you develop the knowledge and skills to select and use the right tools and efficiently repair or replace parts.

5.1.0 General Safety Guidelines

The severity of the problems you'll encounter when responding to a service call will vary greatly. For example, a small leak from a faucet spout into a kitchen sink is wasteful and annoying, but it is not an emergency. On the other hand, a valve leaking a lot of water above a suspended ceiling is an emergency. In a case like this, you must stop the flow of water, minimize the potential damage, and make the area safe before you continue with the repairs.

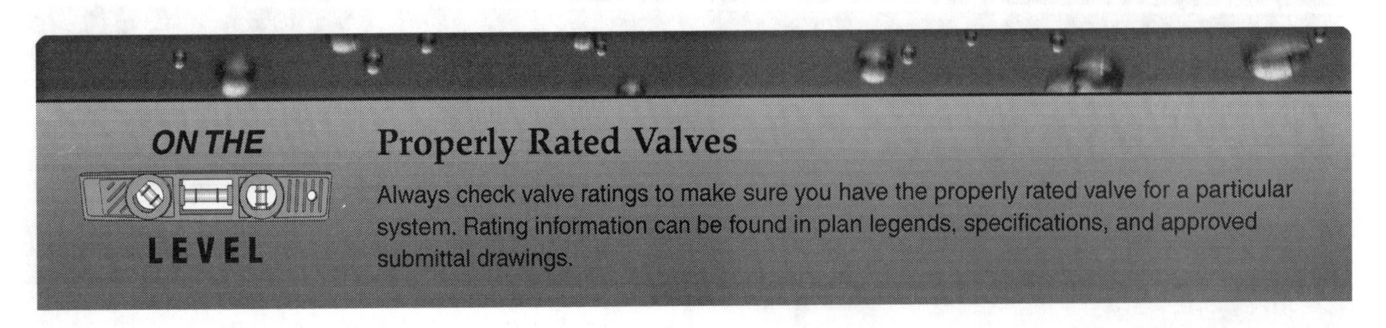

ON THE
LEVEL
Properly Rated Valves
Always check valve ratings to make sure you have the properly rated valve for a particular system. Rating information can be found in plan legends, specifications, and approved submittal drawings.

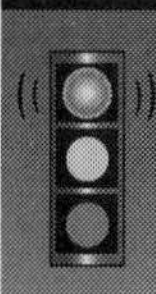

WARNING!

Always take safety precautions when repairing plumbing components. Do not assume that the electricity has been turned off. Always check for yourself. Follow the appropriate lockout/tagout procedures.

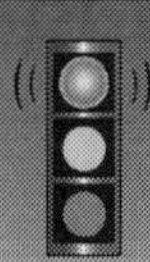

WARNING!

Never attempt to repair or repack a valve that is under pressure. Tightening of the packing nut to stop a leak around the stem is permitted.

The following are some general guidelines for repair work. As you progress in your career, add your experiences to these guidelines:

- Wear rubber-soled shoes or boots for protection from slipping and electric shock.
- Turn off electrical circuits, and tag-out.
- Shut off a valve upstream from the leak. If you think it will take you a while to locate and turn off this valve, direct the leak into a suitably sized container to minimize damage until you can turn the water off.
- Remove excess water.
- Move furniture, equipment, or other obstacles clear of the work area.
- Cover furniture, floors, and equipment to protect them from any damage that could occur as a result of your work on the valve.
- Place ladders carefully, bracing them if necessary.
- Turn the water back on to test your repair.
- Do your work neatly, and clean up when you finish.

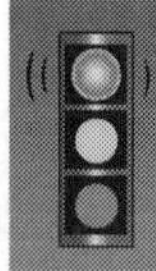

WARNING!

Never field repair a relief or safety valve. Instead, immediately replace them with a valve of identical characteristics and capacity. Safety and relief valves must never be plugged or capped to stop water or steam flow.

5.2.0 Identifying and Repairing Defects

After you've taken the necessary safety precautions, inspect the valve. The type of defect, and the method of repair, is related to the type of valve. Valves come in many styles, but for repair work they fit into the following five categories:

- Globe valves, including angle valves and compression faucets
- Gate valves
- Flushometers
- Float-controlled valves/ball cocks
- Flush valves

Note also that certain models of check valve, particularly larger models, can be repaired. Damaged or worn seats and discs can be replaced. Consult the manufacturer's specifications and parts catalogs.

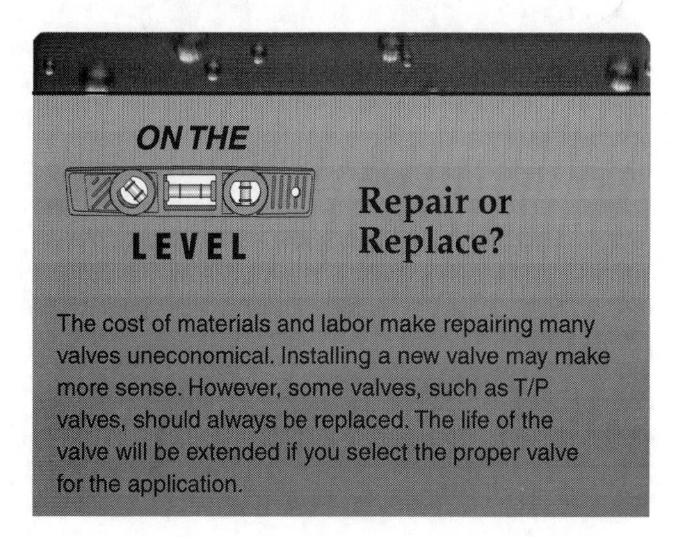

ON THE
LEVEL
Repair or Replace?
The cost of materials and labor make repairing many valves uneconomical. Installing a new valve may make more sense. However, some valves, such as T/P valves, should always be replaced. The life of the valve will be extended if you select the proper valve for the application.

5.2.1 Repairing Globe Valves

Internally, globe valves, angle valves, and compression faucets contain the same basic parts (see *Figure 33*). Therefore, the problems usually encountered and their solutions are similar (see *Table 1*).

To prevent damage to the finished surfaces of the valves, use the correct size wrenches. Don't use pliers or adjustable wrenches. To prevent scratching, make sure the wrench jaws are clean and smooth.

5.2.2 Repairing Gate Valves

Some of the problems common to gate valves are similar to those affecting globe valves (see *Figure 34*). Leaks around the stem are the result of either packing wear or wear on the stem.

Gate valves must operate either fully opened or fully closed. Never use them to throttle the volume of liquid flowing through the pipe. When a gate valve is only partly opened, the disc (gate) vibrates. This vibration erodes the disc edge, a defect commonly called **wire drawing**. A gate

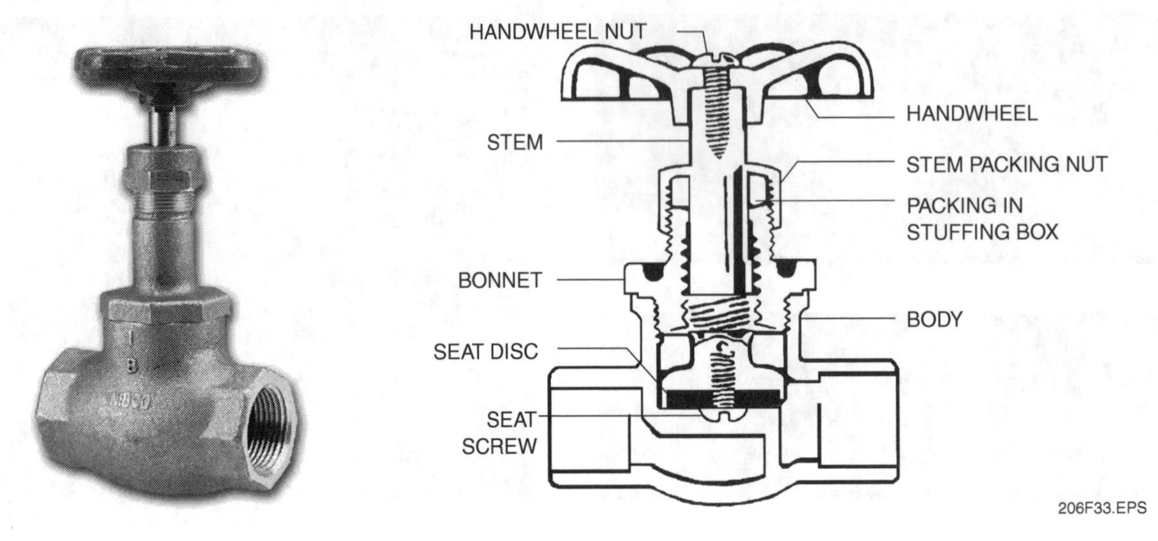

Figure 33 ◆ Basic parts of globe and angle valves and compression faucets.

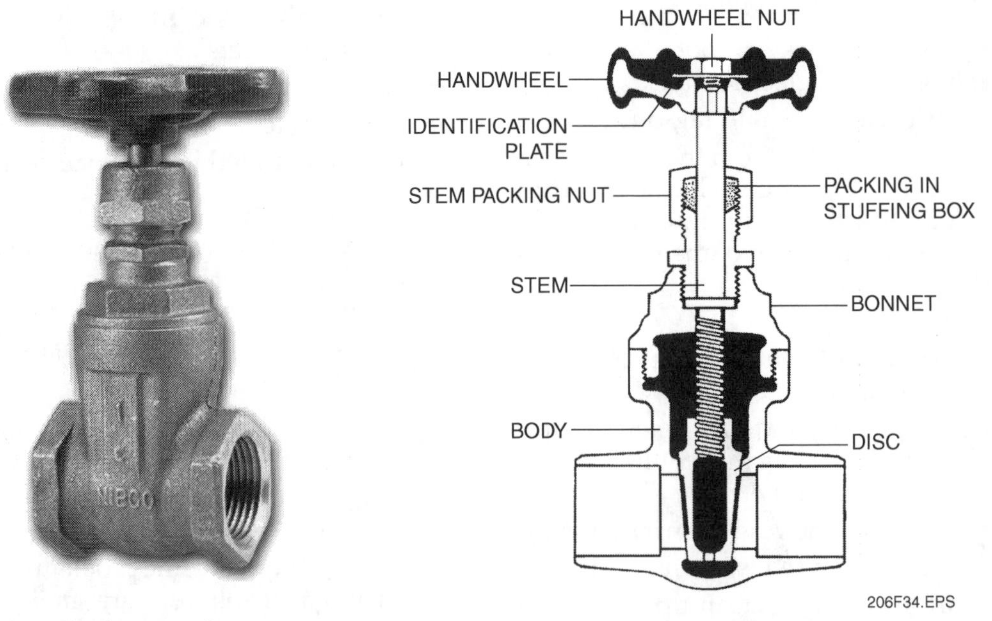

Figure 34 ◆ Basic parts of a gate valve.

Table 1 Troubleshooting Globe and Angle Valves and Compression Faucets

Problem	Possible Cause
Drip or stream of water flows when valve is closed.	Worn or damaged seat. Worn or damaged seat disc.
Leak around stem or from under knob.	Loose packing nut. Defective packing. Worn stem.
Rattle when valve is open and water is flowing.	Loose seat disc. Worn threads on stem.
Difficult or impossible to turn handwheel or knob.	Packing nut too tight. Damaged threads on stem.

valve showing signs of wire drawing can't be repaired; it must be replaced. If throttled flow is desired on the line, replace the gate valve with either a ball valve or a globe valve.

If a gate valve fails to stop the flow of water, the cause is one of three possible problems:

- The disc (gate) is worn.
- The valve seat is worn.
- Some foreign material is preventing the disc from seating properly.

To determine the source of the problems, remove the bonnet and inspect the gate and seat. Scrape out any mineral deposits that are built up inside the bonnet and body. Clean the mating surfaces. If you need to replace the disc, carefully follow the instructions in the manufacturer's specifications. If the seat area is worn, replace the valve. Note that some companies remanufacture larger gate valves. If the gate valve is 2 inches in diameter or larger, consider exchanging it for a remanufactured gate valve.

5.2.3 Repairing Flushometers

Four common problems are associated with flushometers:

- Leakage around the handle
- Failure of the vacuum breaker
- Control stop malfunction
- Leakage in the diaphragm that separates the upper and lower chambers

Several manufacturers produce flushometers and sell kits containing repair components. All flushometers contain the same basic components; however, specific parts may vary. Follow the manufacturer's specifications when installing replacement parts. The troubleshooting guides in *Tables 2* and *3* apply to manual and electric flushometers.

Table 2 Troubleshooting Guide for Manual Flushometers

Problem	Cause	Solution
Nonfunctioning valve	Control stop or main valve closed	Open control stop or main valve
Not enough water	Control stop not open enough Urinal valve parts installed in closet parts Inadequate volume or pressure	Adjust control stop to siphon fixture Replace with proper valve Increase pressure at supply
Valve closes off	Ruptured or damaged diaphragm	Replace parts immediately
Short flushing	Diaphragm assembly and guide not hand-tight	Tighten
Long flushing	Relief valve not seating	Disassemble parts and clean
Water splashes	Too much water is coming out of faucet	Throttle down control stop
Noisy flush	Control stop needs adjustment Valve may not contain quiet feature The water closet may be the problem	Adjust control stop Install parts from kit Place cardboard under toilet seat to separate bowl noise from valve noise—if noisy, replace water closet
Leaking at handle	Worn packing Handle gasket may be missing Dried-out seal	Replace assembly Replace Replace

DID YOU KNOW?

The early 1920s saw the introduction of the first quarter-turn plug valve. This valve operated with a simple 90-degree turn of the handle. Plug valves found wide application in the chemical and gas industries.

During World War II, a British army officer invented the diaphragm valve. This tight, corrosion-resistant valve featured a soft rubber disc bolted between the body and the bonnet.

World War II presented a special challenge to the valve industry and the U.S. Navy. Concussions from bombs dropped close to ships fractured many of the standard valves on board. The Navy replaced thousands of standard valves with impact-resistant valves designed to handle the shock waves.

Table 3 Troubleshooting Guide for Electric Flushometers

Problem	Cause	Solution
Valve does not function (red light does not flash when user steps in front of sensor).	Power is not being supplied to sensor.	Ensure that the main power is turned "ON." Check transformer, leads, and connections. Repair or replace as necessary.
	Sensor is not operating.	Replace sensor.
Valve does not function (INDICATION: red light flashes when user steps in front of sensor, stops flashing when user steps away, valve makes a "clicking" sound and does NOT flush).	Water is not being supplied to the valve.	Make certain that the water supply is turned "ON" and the control stop is open.
(INDICATION: red light flashes when user steps in front of sensor, stops flashing when user steps away, but does NOT make a "clicking" sound and does NOT flush.)	Solenoid shaft assembly is fouled or jammed.	Turn electric power to valve "OFF" (failure to do so could result in damage to solenoid coil). Remove nut from solenoid operator. Remove the coil from the solenoid operator. Use a spanner wrench or pliers to remove the solenoid shaft assembly from valve. Be sure to replace plunger when reassembling solenoid shaft assembly.
(INDICATION: red light flashes when user steps in front of sensor, then flashes in a three short/three long/three short cycle and continues to repeat this when user steps away.)	Sensor wiring connections are incorrect.	Rewire sensor to valve. One solenoid lead connects to the "TO VALVE" connection on sensor. One transformer lead connects to the "24 VAC IN" connection on sensor. Second solenoid lead and second transformer lead connect together.
Volume of water is insufficient to adequately siphon fixture.	Control stop is not open wide enough.	Adjust control stop for desired water delivery.
	Low consumption unit installed on water saver or conventional fixture.	Replace diaphragm component parts of valve with kit that corresponds to appropriate flush volume of fixture.
	Inadequate water volume or pressure available from supply.	Increase pressure or supply (flow rate) to the valve. Consult factory for assistance.
Length of flush is too long (long flushing) or valve fails to shut off.	Water saver valve is installed on low consumption fixture.	Replace diaphragm component parts of valve with kit that corresponds to appropriate flush volume of fixture.
	Relief valve in diaphragm is not seated properly or bypass hole in diaphragm is clogged.	Disassemble inside diaphragm component parts and wash parts thoroughly. Replace worn parts where necessary.
Water splashes from fixture.	Supply flow rate is more than necessary.	Adjust control stop to meet flow rate required for proper cleaning of the fixture.
	Closet valve is installed on urinal fixture.	Replace closet diaphragm component parts with proper urinal kit (inside diaphragm assembly or inside parts kit).

NOTE: Upon detection of the user, the red indicator light flashes slowly for a period of eight seconds. When the user leaves the detection range, the indicator light flashes rapidly and the sensor initiates the flush sequence. Then the indicator light stops flashing and the valve flushes. On water closet models, the valve will flush after a three-second delay.

5.2.4 Repairing Float-Controlled Valves/Ball Cocks

To repair a malfunctioning valve, use one of the repair kits available from the manufacturer. These kits contain replacement parts. If the stem, valve body, or float mechanism is damaged, replace the entire float valve.

5.2.5 Repairing Flush Valves

If you find severe corrosion of the valve or the lever that activates it, replace both parts. The more common problems, however, involve the component parts. Inspect the tank ball (or flapper tank ball), the chains (or wires), and the guide for deterioration. Depending on the condition of these components, make repairs or replace the entire assembly. Use a reseating tool to restore a corroded valve seat.

Review Questions

Sections 3.0.0–5.0.0

1. Rising stems are suitable for areas where space is limited.
   a. True
   b. False

2. Each of the following is a common type of packing *except* _____.
   a. braided
   b. extruded
   c. granulated fiber
   d. solid

Match the type of valve bonnet with its description.

3. _____ Screwed bonnet

4. _____ Union bonnet

5. _____ Bolted bonnet
   a. used when frequent disassembly of the valve is not required
   b. used on larger valves and in high-pressure applications
   c. used when frequent disassembly of the valve is required
   d. used only on valves with flanged ends

6. The most common end connection on valves that are used in plumbing systems is the _____ end.
   a. solder
   b. flanged
   c. internal thread
   d. external thread

7. A valve's capacity can always be determined by referring to its dimensions.
   a. True
   b. False

8. A valve that is marked 100 SWP with 200 WOG operates safely at _____.
   a. 100 psi of saturated steam or 200 psi of cold water, oil, or gas
   b. 100 psi of steam or 200 psi of any fluid except water
   c. 100 psi of steam made from salt water or 200 psi of water or gas
   d. 100 psi of saturated, steamed cold water or 200 psi of natural gas

Troubleshoot the causes of the following globe valve problems.

9. _____ Drip or stream of water flows when valve is closed

10. _____ Leak around stem or from under knob

11. _____ Rattle when valve is open and water is flowing

12. _____ Difficult or impossible to turn handwheel or knob
   a. Damaged threads on stem
   b. Ruptured or damaged diaphragm
   c. Worn or damaged seat
   d. Loose seat disc
   e. Loose packing nut

13. When troubleshooting a line, you discover a gate valve that vibrates and shows signs of wire drawing. The proper way to correct the situation is to _____.
   a. replace the disc in the current valve
   b. replace the stem in the current valve
   c. install a new ball valve or globe valve
   d. install a new gate valve with a larger disc

Review Questions

Troubleshoot the causes of the following manual flushometer problems.

14. _____ Not enough water

15. _____ Short flushing

16. _____ Noisy flush

17. _____ Leaking at handle

    a. Diaphragm assembly and guide not hand-tight
    b. Worn packing
    c. Control stop or main valve closed
    d. Urinal valve parts installed in closet parts
    e. Control stop needs adjustment

18. Each of the following is a common problem associated with flush valves *except* _____.

    a. valve lever corrosion
    b. chain deterioration
    c. flapper tank ball damage
    d. handle leakage

Valves regulate flow. They may provide on/off service, act as a throttling device, or prevent flow reversal through a line. The most common types of valves are gate, globe, angle, ball, butterfly, check, flushometer, plug, and T/P safety and relief valves. Materials used to make valves include iron, brass, carbon steel, bronze, aluminum, alloy steels, and polyvinyl chloride—practically every material that can be machined.

Valves consist of stems, packing, bonnets, and end connections. Packing material provides a tight seal around the stem. Plumbers join valves, like pipes, in several ways. The four basic types of end connections are internal threads, external threads, solder ends, and flanged ends. Always select a valve that is suitable for the application. Consider the suitability of the valve material. Ensure that the valve is sized appropriately for the system. Consult the rating on the valve body to ensure that it can handle the type and pressures of liquids or gases in the system.

Valves come in many styles, but for repair work, they can be divided into five categories: globe valves, gate valves, flushometers, flush valves, and float-controlled valves/ball cocks. The repair methods for globe and gate valves are similar. Manufacturers provide repair kits containing replacement parts for flushometers, flush valves, and float-controlled valves/ball cocks.

When repairing valves, remember that safety comes first. Take steps to protect yourself and others on the job site and to minimize property damage. Refer to your local code when selecting appropriate valves. Always follow the manufacturer's specifications when installing and repairing valves.

Trade Terms Introduced in This Module

Angle valve: A valve with flow characteristics similar to a globe valve, with a valve body that allows it to be used in a 90-degree elbow configuration.

ASME Boiler and Pressure Vessel Code: Publication that establishes criteria that govern the construction and maintenance of pressure vessels and related devices, such as steam safety valves.

Ball-check valve: A valve that allows one-way fluid flow in water supply or drainage lines by using an internal ball and seat.

Blowdown: The difference between the opening and closing pressures in a steam safety valve.

Bolted bonnet: A valve cover that guides and encloses the stem. It is used for larger valves or for high-pressure applications.

Bonnet: A valve cover that guides and encloses the stem.

Butterfly valve: An on/off or throttling valve that uses a disc and rotates 90 degrees from fully closed (perpendicular to flow) to fully opened (parallel to flow) to control fluid flow.

External thread: A valve end connection that has the threads on the outside.

Flanged end: A valve end connection that uses bolts and gaskets to connect the valve to a companion pipe flange.

Float-controlled valve (ball cock): A valve installed in water closet flush tanks to maintain a constant water level in the tank.

Flush valve: A device located at the bottom of a tank for flushing water closets and similar fixtures.

Huddling chamber: A chamber within a safety valve that harnesses the expansion forces of air or steam to quickly open (pop) or close the valve.

Internal thread: A valve end connection that has the threads inside.

Lever-and-weight swing-check valve: A type of swing-check valve that uses an external counterweight connected to an internal swing-check disc, frequently used on lines with pulsating flow.

Lift-check valve: A type of valve that prevents flow reversal in a piping system. It is used for gas, water, steam, or air, and for lines where frequent fluctuations in flow occur.

Nonrising stem: A type of stem in which neither the handwheel nor the stem rises when the valve is opened. It is suitable for areas where space is limited.

Outside screw and yoke stem: A type of valve stem suitable for use with corrosive fluids because it does not come in contact with the fluid line. Often abbreviated OS&Y.

Packing: A material that seals the valve stem and prevents fluids from leaking up through it.

Plug valve: A valve that provides both on/off and throttling control by raising and lowering a cylindrical or cone-shaped plug into the flow. Widely used in sanitary plumbing systems.

Pressure regulator valve: A valve that reduces water pressure in a building.

Rising stem: A type of valve in which both the handwheel and the stem rise.

Screwed bonnet: A two-piece unit that is used in low-pressure applications and where frequent disassembly of the valve is not required.

Slurries: Liquids that contain large amounts of suspended solids.

Solder end: A type of end connection for valves. The smooth ends are soldered to copper and brass pipes that have no internal or external threads.

Solid-wedge gate valve: A type of gate valve that is used for steam, hot water, and other services where shock is a factor.

Split-wedge gate valve: A type of gate valve used on lines that require a more positive closure—for example, low-pressure and cold water lines. It is also used on lines containing volatile fluids.

Supply stop valve: A valve used to disconnect the hot or cold water supply to individual fixtures such as water closets, lavatories, and sinks. Also called the supply stop.

Swing-check valve: A type of check valve that features a low-flow resistance, making it suited for lines containing liquids or gases with low to moderate pressure.

Trim: The parts of a valve that receive the most wear and tear: the stem, disc, seat ring, disc holder, wedge, and bushings.

Union bonnet: A type of valve construction in which a union nut connects the bonnet to the valve body; best suited for applications requiring frequent disassembly of the valve.

Wire drawing: A condition caused when gate valves open only part way, causing the disc, or gate, to vibrate, which erodes the disc edge.

Additional Resources

This module is intended to be a thorough resource for task training. The following reference works are suggested for further study. These are optional materials for continued education rather than for task training.

NIBCO Catalog C-VSG-0502, *Valve Selection & Specification Guide for Building Services*, 2002. Elkhart, IN: NIBCO.

Piping and Valves. 2001. Frank R. Spellman and Joanne Drinan. Lancaster, PA: Technomic.

References

Dictionary of Architecture and Construction, Third Edition, 2000. Cyril M. Harris, ed. New York: McGraw-Hill.

GlobalSpec.com, "About Plug Valves," http://valves.globalspec.com/LearnMore/Flow_Transfer_Control/Valves/Plug_Valves, viewed October 21, 2004.

NIBCO Catalog C-VSG-0502, *Valve Selection & Specification Guide for Building Services*, 2002. Elkhart, IN: NIBCO.

Pipefitters Handbook. Forrest R. Lindsay. New York: Industrial Press Inc.

2003 National Standard Plumbing Code. 2003. Falls Church, VA: Plumbing-Heating-Cooling Contractors—North America.

Figure Credits

The Wm. Powell Company	206F01
NIBCO International	206F02, 206F04 (photo), 206F05 (photo), 206F30, 206F31, 206F33 (photo), 206F34 (photo)
Dwyer Instruments, Inc.	206F03, 206F06
Sloan Valve Company	206F10, 206F11, 206T03
Fluidmaster, Inc.	206F13, 206F19
Flowserve Corporation	206F14
Kunkle Valve	206F15
Tyco Valves and Controls	206F16 (photos)
Marotta Controls, Inc.	206F16 (diagrams)
Watts Regulator Company	106F18 (photo)
Zurn Plumbing Products Group	206F20, 206F21, 206F22, 206F32

CONTREN® LEARNING SERIES — USER UPDATE

The NCCER makes every effort to keep these textbooks up-to-date and free of technical errors. We appreciate your help in this process. If you have an idea for improving this textbook, or if you find an error, a typographical mistake, or an inaccuracy in NCCER's Contren® textbooks, please write us, using this form or a photocopy. Be sure to include the exact module number, page number, a detailed description, and the correction, if applicable. Your input will be brought to the attention of the Technical Review Committee. Thank you for your assistance.

Instructors – If you found that additional materials were necessary in order to teach this module effectively, please let us know so that we may include them in the Equipment/Materials list in the Annotated Instructor's Guide.

Write: Product Development and Revision
National Center for Construction Education and Research
P.O. Box 141104, Gainesville, FL 32614-1104

Fax: 352-334-0932

E-mail: curriculum@nccer.org

Craft ______ Module Name ______

Copyright Date ______ Module Number ______ Page Number(s) ______

Description ______

______

______

______

(Optional) Correction ______

______

______

(Optional) Your Name and Address ______

______

______

Plumbing Level Two

02207-05

Installing and Testing Water Supply Piping

02207-05
Installing and Testing Water Supply Piping

Topics to be presented in this module include:

Overview

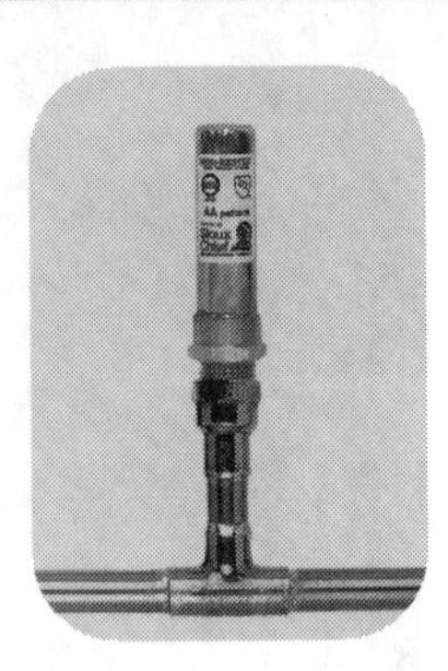

Installing a water supply system is similar to installing a DWV system. Plumbers develop a material takeoff, locate the fixtures, and determine the route of the piping. The design for the water supply system is based on the project plans and specifications. Plumbers consult local plumbing codes to select approved piping materials. To determine the size of the water supply piping, plumbers consider the demand for the proposed building.

Plumbers locate water supply piping in reference to the DWV piping. Although there is some flexibility when locating the water supply pipe, plumbers attempt to locate it as accurately as possible. The location of the water heater affects efficiency. After the water heater is located, the water softener and hose bibbs are installed, and the fixture riser and stubouts are in place, plumbers can install the main supply lines. Plumbers ensure that the main feeder line is sized to supply the required flow and pressure.

When installing water supply systems, plumbers also consider accessibility for maintenance and repair, frost and puncture protection, and backflow prevention. To ensure that a water supply system is ready for use, plumbers conduct an air test or a hydrostatic test. A plumbing inspector observes the test and verifies its success. The inspector should approve the test, which is then kept with the project records.

Focus Statement

The goal of the plumber is to protect the health, safety, and comfort of the nation job by job.

Code Note

Codes vary among jurisdictions. Because of the variations in code, consult the applicable code whenever regulations are in question. Referring to an incorrect set of codes can cause as much trouble as failing to reference codes altogether. Obtain, review, and familiarize yourself with your local adopted code.

Objectives

When you have completed this module, you will be able to do the following:

1. Develop a material takeoff from a given set of plans.
2. Use plans and fixture rough-in sheets to determine the location of fixtures and the route of the water supply piping.
3. Locate and size a water meter.
4. Locate a water heater, water softener, and hose bibbs.
5. Install a water distribution system using appropriate hangers.
6. Modify structural members, using the appropriate tools, without weakening the structure.
7. Correctly size and install a water service line, including backflow prevention.
8. Test a water supply system.

Trade Terms

Air test
Compressor
Curb box
Curb stop
Demand
Frost line
Hydrostatic test
Hydrostatic test pump
Pipe dope
Water supply fixture unit (WSFU)

Required Trainee Materials

1. Appropriate personal protective equipment
2. Pencil and paper
3. Copy of local applicable code

Prerequisites

Before you begin this module, it is recommended that you successfully complete *Core Curriculum; Plumbing Level One; Plumbing Level Two,* Modules 02201-05 through 02206-05.

This course map shows all of the modules in the second level of the Plumbing curriculum. The suggested training order begins at the bottom and proceeds up. Skill levels increase as you advance on the course map. The local Training Program Sponsor may adjust the training order.

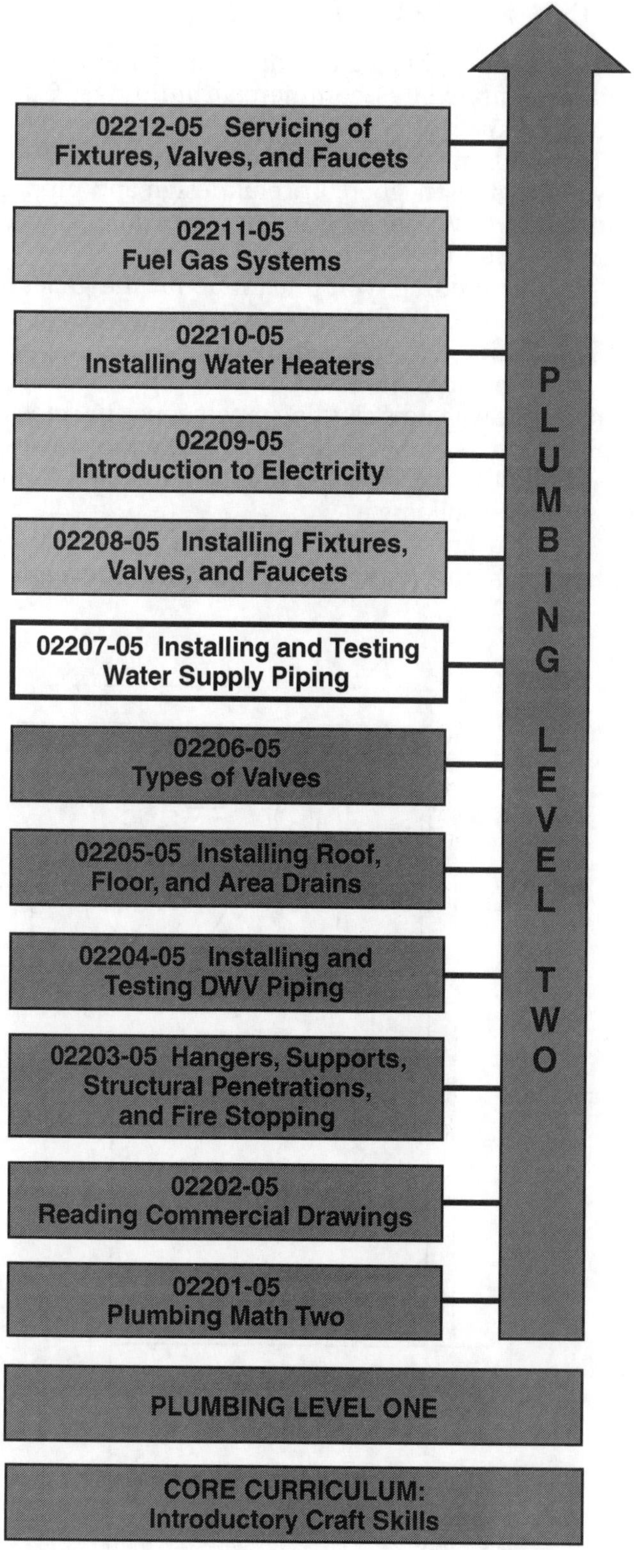

PLUMBING LEVEL TWO

02212-05 Servicing of Fixtures, Valves, and Faucets
02211-05 Fuel Gas Systems
02210-05 Installing Water Heaters
02209-05 Introduction to Electricity
02208-05 Installing Fixtures, Valves, and Faucets
02207-05 Installing and Testing Water Supply Piping
02206-05 Types of Valves
02205-05 Installing Roof, Floor, and Area Drains
02204-05 Installing and Testing DWV Piping
02203-05 Hangers, Supports, Structural Penetrations, and Fire Stopping
02202-05 Reading Commercial Drawings
02201-05 Plumbing Math Two
PLUMBING LEVEL ONE
CORE CURRICULUM: Introductory Craft Skills

207CMAP.EPS

1.0.0 ◆ INTRODUCTION

Water supply piping is an important concern in residential as well as commercial buildings. Customers expect—and should have—adequate water at fixtures, quiet pipes that don't freeze, and clean water that tastes good. Careful design and proper installation will ensure the best possible service from a building's water supply system.

In this module, you'll learn about the general design and installation concepts used for water supply systems. The potable (drinkable) water supply system includes the water service pipe, the water meter, the water distributing pipe, connecting pipes, pipe hangers and supports, fittings, control valves, and all other necessary components located in or next to the building.

As you know, plumbers refer to construction drawings and plans when installing plumbing systems. For most industrial, commercial, and multi-residential structures, plumbing plans are included with the architectural plans for the structure. These plans are designed with a certain amount of flexibility to allow the plumber to make on-the-job decisions when confronted by unusual circumstances. In many cases, particularly in light residential applications, plumbing plans are not included with the prints. In these cases, the plumber uses the floor plan to locate the runs of pipe and the various plumbing fixtures.

In this module, you will learn how to install and test a water supply system. You will work through a hypothetical water supply system installation for a single-story, two-bedroom cabin installed on a slab-on-grade foundation (see *Figure 1*). That way you will understand how each step and component fits into the overall installation process, just as if you were working on an actual installation in the field.

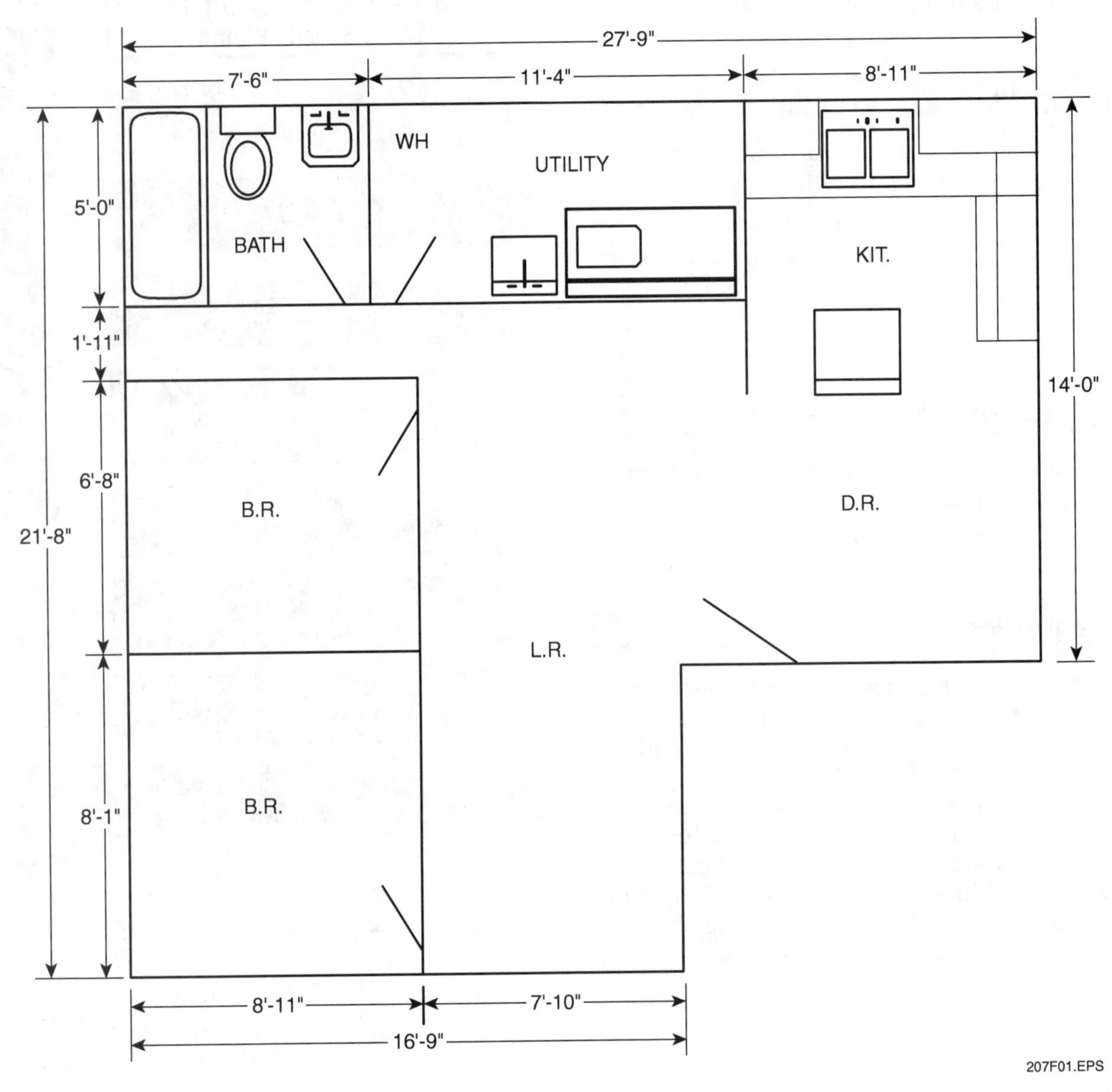

Figure 1 ◆ House floor plan.

Note that this building is the same one that appeared in the *Plumbing Level 2* module titled *Installing and Testing DWV Piping*. By using the same building plan, you will be able to better understand the relationship between water supply systems and drain, waste, and vent (DWV) systems.

As with all modules in the *Plumbing* curriculum that feature construction plans, takeoff drawings, and submittal data, the dimensions shown are for instructional purposes only.

2.0.0 ◆ MAIN TO METER WATER SERVICE

In some areas, the water service piping begins at the utility water main and either connects into the main itself or starts on the outlet side of a corporation stop provided by the water utility. In other areas, the water utility brings the water supply to the property line and terminates it with a **curb stop.** A **curb box** provides access to the supply (see *Figure 2*).

2.1.0 Piping Materials and Sizes

Consult the local plumbing code to learn which piping materials are approved. Copper tubing and certain plastics are the materials most commonly used.

To determine the size of the water supply piping, you must know the **demand** for the proposed building. Consider the following factors:

- The type of flush devices on the fixtures
- Water pressure in pounds per square inch (psi) at the source
- The length of pipe in the building
- The number and kinds of fixtures installed
- The number of fixtures expected to be used at any time

You can determine the correct pipe sizes for water supply fixtures by determining the system's demand. The demand is the water requirement for the entire water supply system—pipes, fittings, outlets, and fixtures. Plumbers calculate the rate of flow in gallons per minute (gpm) according to the number of **water supply fixture units** (WSFUs) that each fixture is designed to handle. WSFUs measure a fixture's load. WSFUs vary with the quantity and temperature of the water and with the type of fixture. Once you know the appropriate WSFU rating for a fixture or fixture group, you can select the correct size of pipe for the fixture branch.

Your local applicable code includes sizing tables, similar to that shown in *Table 1,* that indicate the WSFUs for residential fixtures. Pipe sizes

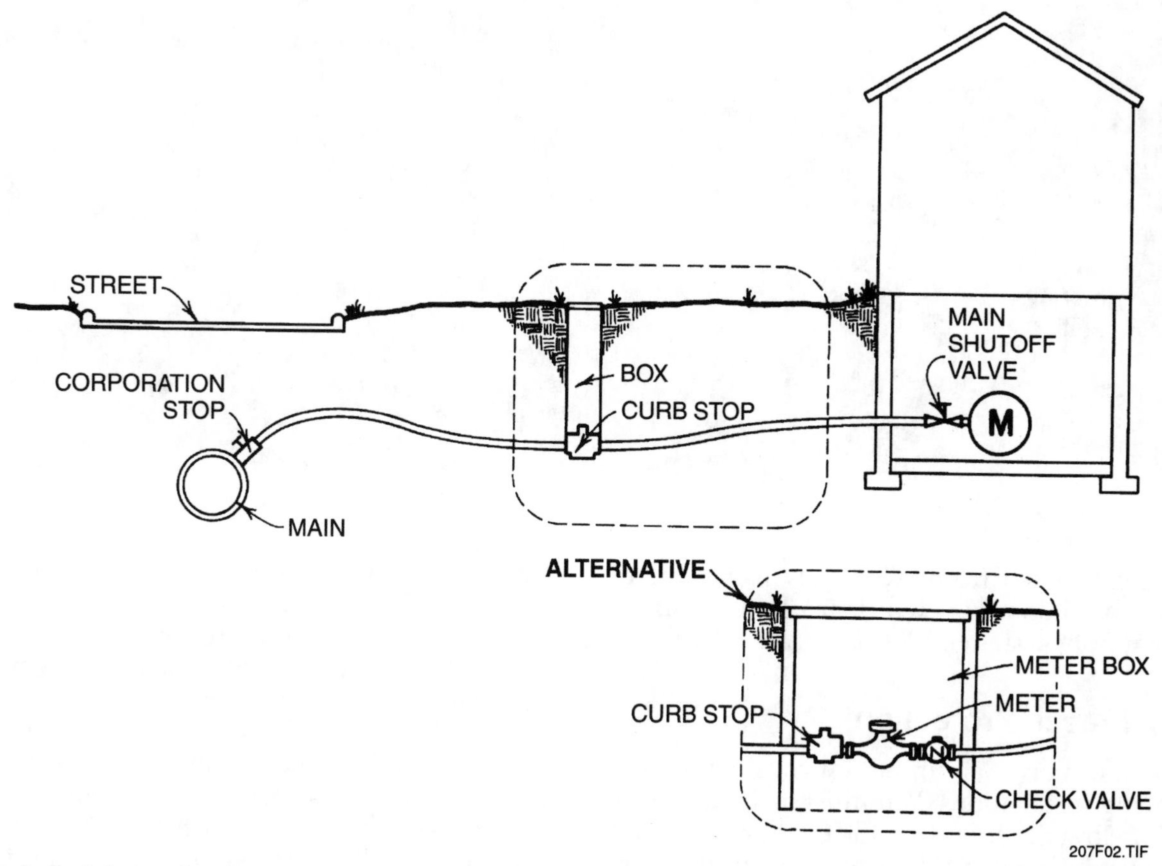

Figure 2 ◆ Curb box and curb stop.

Table 1 WSFUs for Selected Fixtures and Fixture Groups

Fixture	Occupancy	Type of Supply Control	Load Values in WSFUs		
			Cold	Hot	Total
Bathroom group	Private	Flush tank	2.7	1.5	3.6
Bathroom group	Private	Flush valve	6.0	3.0	8.0
Bathtub	Private	Faucet	1.0	1.0	1.4
Bathtub	Public	Faucet	3.0	3.0	4.0
Bidet	Private	Faucet	1.5	1.5	2.0
Combination fixture	Private	Faucet	2.25	2.25	3.0
Dishwashing machine	Private	Automatic		1.4	1.4
Drinking fountain	Offices, etc.	⅜ inch valve	0.25		0.25
Kitchen sink	Private	Faucet	1.0	1.0	1.4
Kitchen sink	Hotel, restaurant	Faucet	3.0	3.0	4.0
Laundry trays (1 to 3)	Private	Faucet	1.0	1.0	1.4
Lavatory	Private	Faucet	0.5	0.5	0.7
Lavatory	Public	Faucet	1.5	1.5	2.0
Service sink	Offices, etc.	Faucet	2.25	2.25	3.0
Shower head	Public	Mixing valve	3.0	3.0	4.0
Shower head	Private	Mixing valve	1.0	1.0	1.4
Urinal	Public	1 inch flush valve	10.0		10.0
Urinal	Public	¾ inch flush valve	5.0		5.0
Urinal	Public	Flush tank	3.0		3.0
Washing machine (8 lb.)	Private	Automatic	1.0	1.0	1.4
Washing machine (8 lb.)	Public	Automatic	2.25	2.25	3.0
Washing machine (15 lb.)	Public	Automatic	3.0	3.0	4.0
Water closet	Private	Flush valve	6.0		6.0
Water closet	Private	Flush tank	2.2		2.2
Water closet	Public	Flush valve	10.0		10.0
Water closet	Public	Flush tank	5.0		5.0
Water closet	Public or private	Flushometer tank	2.0		2.0

207T01.EPS

can be calculated from WSFUs by using one of several methods. You will learn how to size a water supply system in *Plumbing Level Three*.

2.2.0 Freeze Protection

Frost protection is an important consideration in plumbing installations. You must maintain a safe depth below the **frost line** for the buried service line. Depending on where in the country you work, the frost line depth will vary. Consult your local applicable code for the frost line depth in your area. The codes require accessibility to valves on the water service pipe and cleanouts on the sanitary sewer drainpipe.

In colder climates, locate the service entrance in areas other than under sidewalks, driveways, and patios. Snow acts like insulation and keeps frost from penetrating the soil too deeply. Therefore, the frost depth on snow-covered ground is signif-

Friction

The greater the number of fittings in the system (valves, tees, elbows, and so on), the greater the amount of friction, which causes a pressure drop in the system.

icantly less than it is in areas, such as driveways, that are routinely cleared of snow. In wet soils frost penetrates deeper than it does in dry soils. If possible, locate water pipes under lawns or other areas where snow will remain.

2.3.0 Pipe Protection

Place sleeves around pipes that pass through a building's structural concrete—for example, floors, footings, stem walls, and grouted masonry (see *Figure 3*). Sleeves protect the pipe from damage by contact with concrete. They are made from foam or plastic pipe that is one size larger than the pipe they protect. Secure sleeves in place before pouring the concrete. Sleeving methods vary according to the local applicable code.

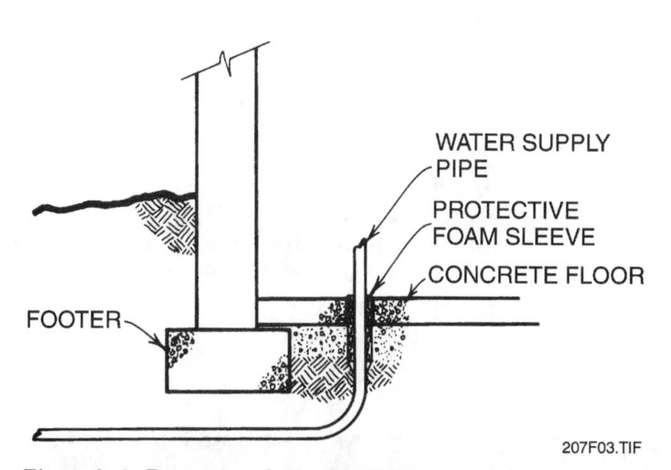

Figure 3 ◆ Protective sleeve for piping.

2.4.0 Water Meters

Water meters measure the amount of water passing through the water supply piping into a building (see *Figure 4*). Usually, they belong to a city or a local municipality, and you may not remove them without permission. Water department employees install and service the meters. The housing, or cover, of water meters may be made of bronze, cast iron, stainless steel, or plastic.

Figure 4 ◆ Water meters.

DID YOU KNOW?

The Country's First Waterworks

In 1652, Boston started the country's first waterworks, both for domestic use and for fighting fires. With wood-framed buildings everywhere, fire was a common hazard in the city. Chimney fires, too, posed a constant risk, so a quick supply of water was important. The water supply line—made from hollowed-out logs—ran from Jamaica Pond to the Faneuil Hall area. This is near where the Massachusetts rebels met to plan the Boston Tea Party in 1773.

The amount of water used varies for each customer. Meters are made with measurement capacities of 20, 30, or 50 gallons per minute (gpm). While the interior design varies from meter to meter, most meters today have a magnetic drive.

The plumber is responsible for plumbing the meter. Most utility companies will set meters up to 1½ inches in diameter. Plumbers usually plumb and set meters that require a larger pipe diameter. Leave the proper spacing between pipes to accommodate a water meter.

When a water meter is located inside a building, you can install an external meter register to make meter reading easier (see *Figure 5*). Install a main shutoff valve to control the water flow during and after construction. Local codes vary as to the type of valve required. You can locate the water meter directly after the main shutoff valve or before the main shutoff valve.

Figure 5 ◆ External meter register.

3.0.0 ◆ WATER HEATER, WATER SOFTENER, AND HOSE BIBBS

The working drawings will show the location of the water heater, the water softener, and hose bibbs. However, you have a certain amount of flexibility in locating these items. Each building presents its own challenges, and you may need to think creatively to resolve them or to make the system more efficient. Plumbing codes require a full way valve in the service line, regardless of the location.

3.1.0 Locating the Water Heater

Locate the water heater in a safe place that meets the requirements of your local applicable code. The placement of the water heater affects the efficiency of the hot water supply system. To reduce heat loss in the water heater, locate it away from excessively cold environments. To reduce heat loss along the supply line, locate the water heater as close to hot water outlets as possible. If it is impractical to place the water heater near hot water outlets—for example, in ranch-style houses, where bathrooms are often far from the laundry—you may need to locate two water heaters.

Customers may want the option of installing solar hot water heating in the future. To allow for this, install fittings that will make the transition easier. Some solar heating systems are designed to use the water heater as the storage tank, while others require a separate tank for efficient operation. To allow for a separate water storage tank, ensure that adequate space is provided in the water heater enclosure. This will eliminate the need to rebuild the enclosure later on. Depending on the job specifications, either the plumber or the general contractor may construct the enclosure.

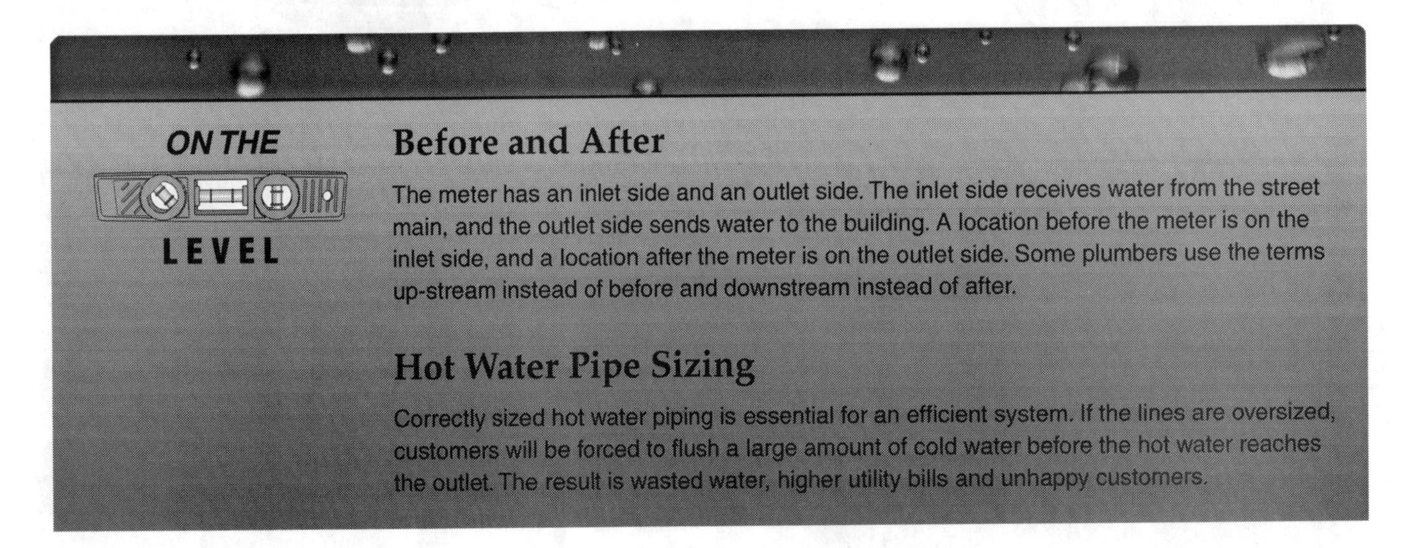

ON THE LEVEL

Before and After

The meter has an inlet side and an outlet side. The inlet side receives water from the street main, and the outlet side sends water to the building. A location before the meter is on the inlet side, and a location after the meter is on the outlet side. Some plumbers use the terms up-stream instead of before and downstream instead of after.

Hot Water Pipe Sizing

Correctly sized hot water piping is essential for an efficient system. If the lines are oversized, customers will be forced to flush a large amount of cold water before the hot water reaches the outlet. The result is wasted water, higher utility bills and unhappy customers.

3.2.0 Locating the Water Softener

Before installing the water supply piping, determine where in the building softened water is required. The plans may specify this, or you may need to consult with the customer. For example, residential hot water is sometimes softened. The water supplied to hose bibbs, on the other hand, may bypass the softener. Depending on the quality of the drinking water, a customer may prefer to have softened water at the kitchen sink's cold water tap. The rest of the cold water outlets in the home may or may not require softening.

If the softener drains to the floor drain, ensure that you maintain an appropriate air gap to prevent backflow due to accidental cross-connection.

3.3.0 Locating the Hose Bibbs

At least one hose bibb may be required by code in residential structures (see *Figure 6*). In many cases, the customer may want additional hose bibbs. These runs are usually long, and they normally bypass the water softener. Hose bibbs are usually ¾ inch, although you may use ½-inch piping to access the outlet.

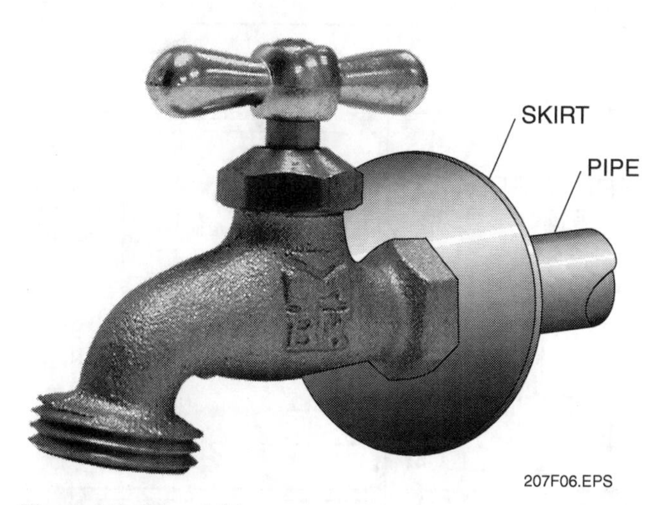

Figure 6 ◆ Hose bibb.

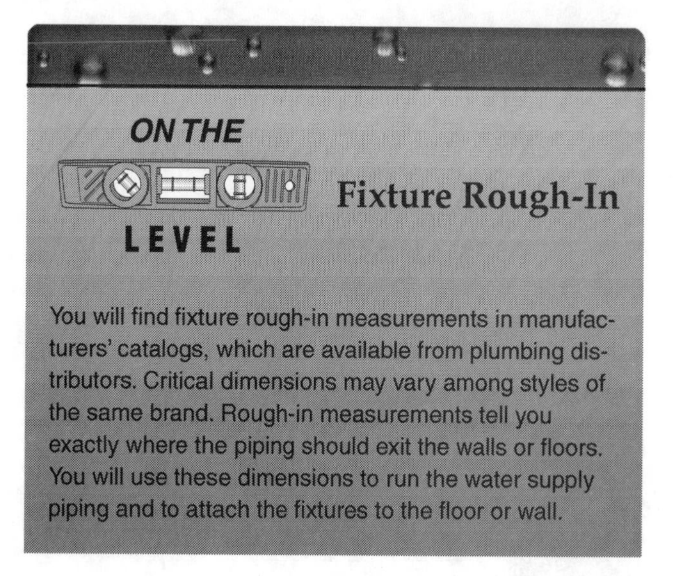

Fixture Rough-In

You will find fixture rough-in measurements in manufacturers' catalogs, which are available from plumbing distributors. Critical dimensions may vary among styles of the same brand. Rough-in measurements tell you exactly where the piping should exit the walls or floors. You will use these dimensions to run the water supply piping and to attach the fixtures to the floor or wall.

4.0.0 ◆ LOCATING THE FIXTURES

As you have already learned, the DWV system is installed prior to the water supply system. This is because DWV piping is larger, and therefore less flexible to install, than water supply piping. You must therefore locate water supply piping in reference to the DWV piping. Although you will have some flexibility with the location of the water supply pipe, make every effort to locate the water supply as accurately as possible.

Install the water supply piping in reference to the drains. The approved submittal data will contain reference dimensions that will help you locate the pipe properly (see *Figure 7*). The specifications apply to the style of fixture selected. For example, refer to illustrations A and B in *Figure 7*. Note that they are identical except for the measurement from the centerline to the drain. Note also that the water supply is referenced from the drain. If you locate the drain incorrectly, you may have connection problems. When installing water supply piping, always recheck the location of the drains to ensure correct positioning.

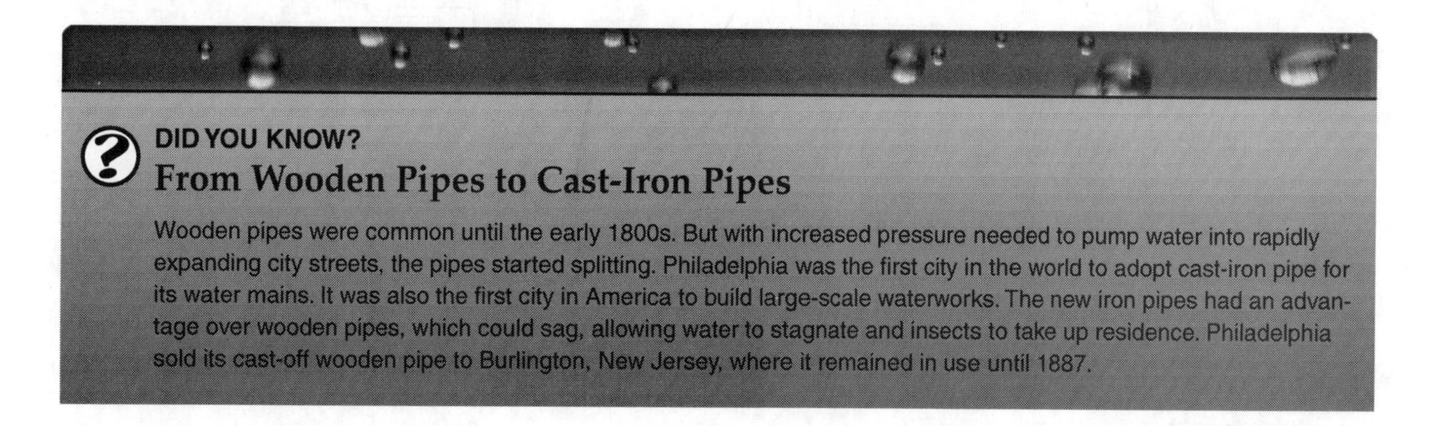

DID YOU KNOW?

From Wooden Pipes to Cast-Iron Pipes

Wooden pipes were common until the early 1800s. But with increased pressure needed to pump water into rapidly expanding city streets, the pipes started splitting. Philadelphia was the first city in the world to adopt cast-iron pipe for its water mains. It was also the first city in America to build large-scale waterworks. The new iron pipes had an advantage over wooden pipes, which could sag, allowing water to stagnate and insects to take up residence. Philadelphia sold its cast-off wooden pipe to Burlington, New Jersey, where it remained in use until 1887.

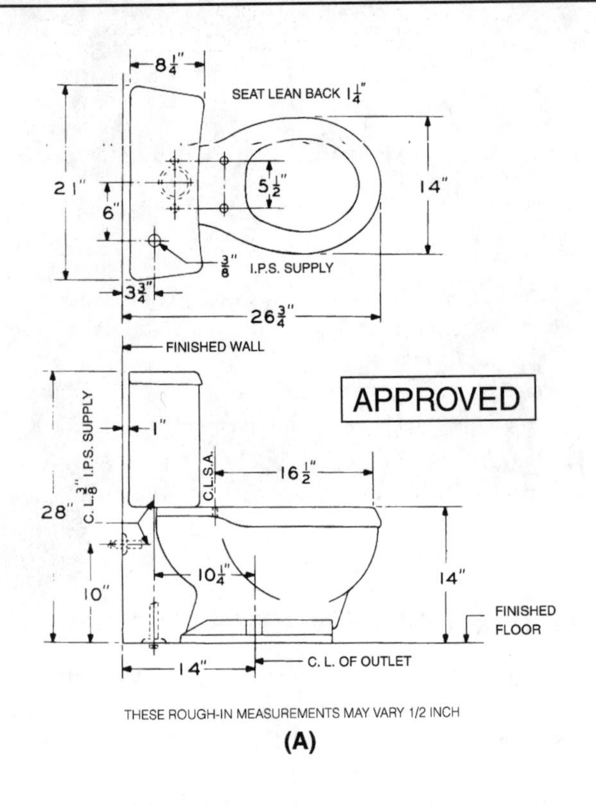

(A)

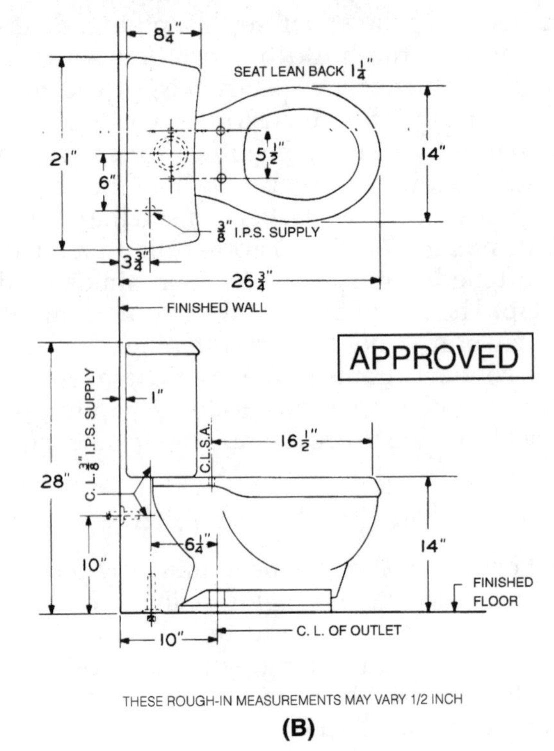

(B)

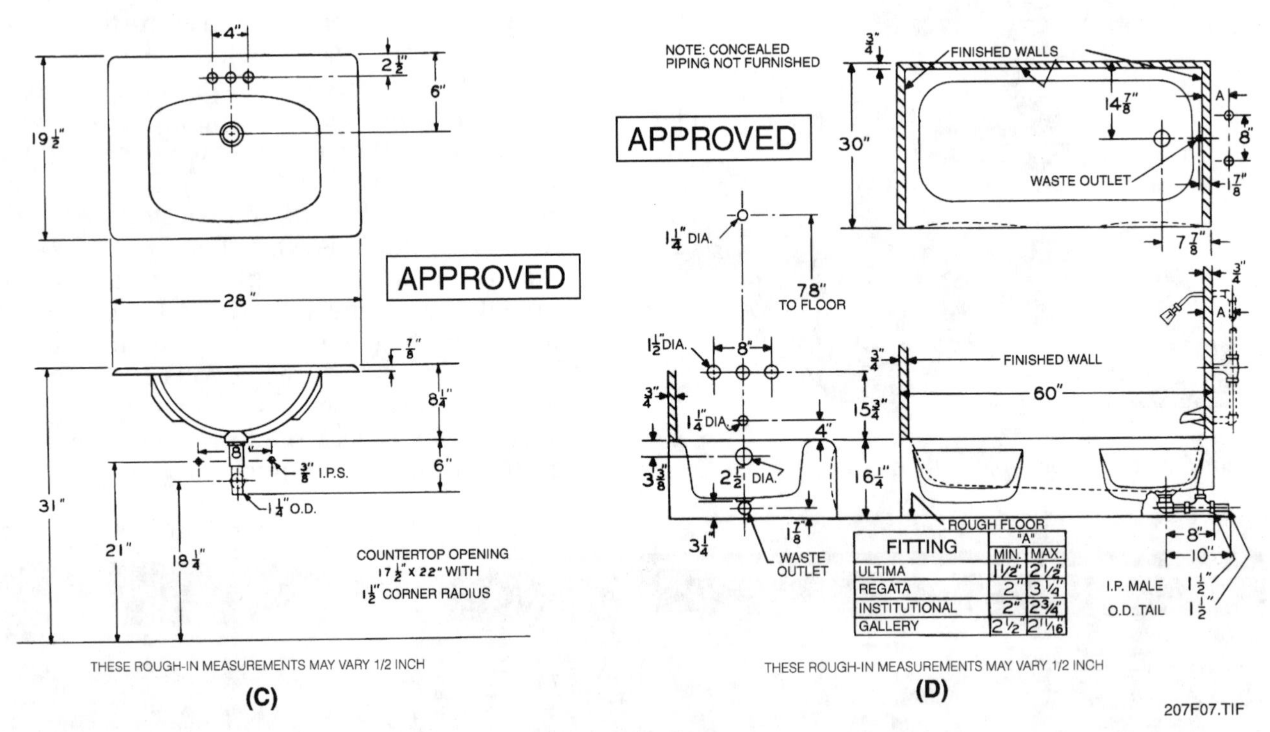

FITTING	"A" MIN.	"A" MAX.
ULTIMA	1 1/2"	2 1/2"
REGATA	2"	3 1/4"
INSTITUTIONAL	2"	2 3/4"
GALLERY	2 1/2"	2 11/16"

(C) **(D)**

207F07.TIF

Figure 7 ◆ Use the dimensions in approved submittal data to locate fixtures.

Preventing Corrosion

When different metals—like copper and iron—come into contact with one another in the presence of water, they can corrode. A dielectric fitting is a special type of adapter used to connect pipes of different metals. The adapter prevents corrosion from taking place. To prevent damage to the water supply system, use dielectric fittings on water heaters, water softeners, and solar collectors.

Dielectric fitting devices are not always included in codes or specifications. Knowing about them enables you to offer efficient and safe service to customers.

Backflow Preventers

Backflow occurs when liquids flow into the water supply's distributing pipes or into any fixture or appliance from a source opposite to the intended flow. Install backflow preventers in areas where hoses can siphon contaminated water into the potable water system. These areas include hose bibbs, dishwashers, and slop sinks. Refer to your local applicable code for approved backflow prevention devices and applications.

Remember that the dimensions in fixture rough-ins are taken from the finished wall surface. The thickness of interior wall coverings—drywall, plaster, wood, or ceramic tile—will vary. Locate the drain with the final wall thickness in mind.

4.1.0 Assembling and Installing the Stubouts

The stubout is that part of the pipe that extends beyond the finished wall. Once the wall is finished, install the angle stop and the flexible fixture riser. The riser extends from the angle stop and connects to the fixture faucet or common tank. Confirm the stubout dimensions against the approved submittal data before installing each fixture.

After you have assembled, capped, and soldered the stubouts, place the assembly through the access hole in the floor to the feeder lines below (see *Figure 8*).

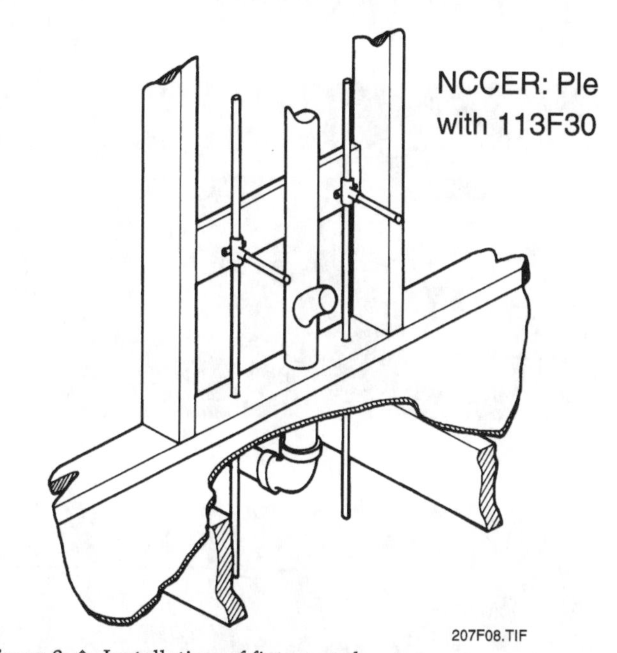

Figure 8 ◆ Installation of fixture stubouts.

DID YOU KNOW?

An Early Water Distribution System

In America, some of the earliest water distribution systems were developed not for domestic use but to fight fires. In 1795, the Jamaica Pond Aqueduct Corporation in Boston bored hemlock trees to produce about 15 miles worth of 3-inch and 5-inch wooden water pipe. Although crude, these pipelines provided a ready supply of water to quench fires. Firefighters punched a hole in the pipe and inserted a smaller pipe sized to fit it. Then, they'd attach the hose of their fire wagon, a two-man pumper, and fight the blaze. When the fire was out, they placed a cone-shaped stopper on the end of a long pole, inserted it into the hole, and banged it into place. The wooden pole, or fireplug, remained sticking

With the risers visible below the floor, you can easily locate the fixtures from below. To position the riser and stubout assemblies firmly in their permanent location, use either a drop ear ell or a high ear ell (see *Figure 9*) attached to a backing board (see *Figure 10*). Attaching the ells to the backing board anchors the pipe in place behind the finished wall. Refer to the job specifications to determine the appropriate fitting to use.

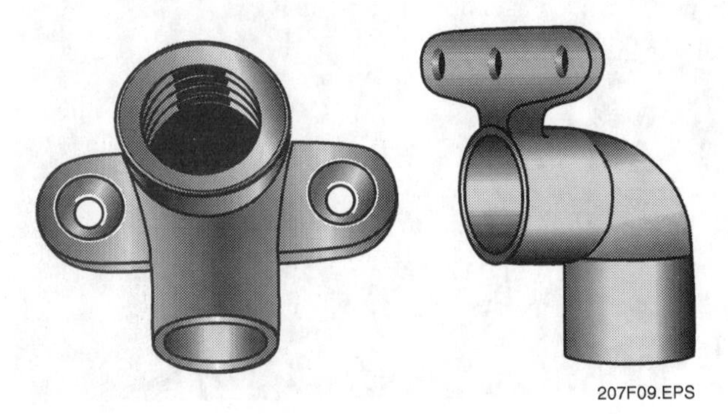

Figure 9 ◆ Drop ear ell and high ear ell.

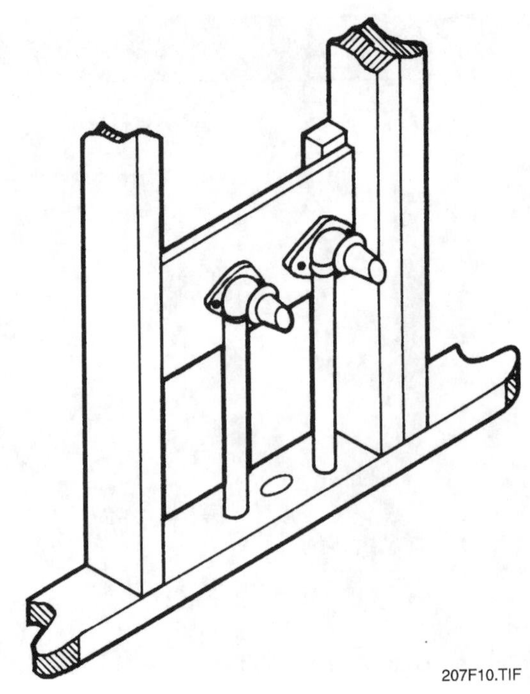

Figure 10 ◆ Backing board.

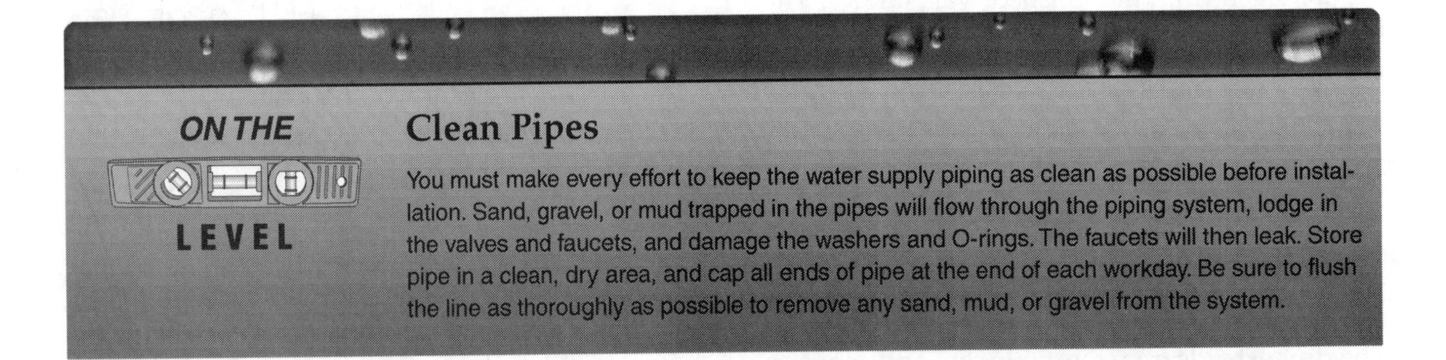

ON THE LEVEL

Clean Pipes

You must make every effort to keep the water supply piping as clean as possible before installation. Sand, gravel, or mud trapped in the pipes will flow through the piping system, lodge in the valves and faucets, and damage the washers and O-rings. The faucets will then leak. Store pipe in a clean, dry area, and cap all ends of pipe at the end of each workday. Be sure to flush the line as thoroughly as possible to remove any sand, mud, or gravel from the system.

DID YOU KNOW?

Herod's Human Water Supply Line

In 38 B.C.E., Herod ruled Judea. He left his mark on the land, most notably on Masada, a 1,300-foot-high rock fortress in the middle of a desert east of the Dead Sea. Its water system originated in two small wadis, or gullies. These wadis quickly flooded with water from sudden, unpredictable downpours. To hold the water until needed, the Romans constructed dams in two places. On demand, the dams allowed the water to flow by gravity through an aqueduct and channels directed to the site. In addition, a set of cisterns at the top of Masada connected to conduits to catch rainwater. But the water supply was not enough for the palaces located there. So Herod ordered a human conduit to bring up water from the cisterns far below. Some historians estimate that hundreds, maybe thousands, of slaves and beasts of burden carried jars of water up the cliff for the royal households for drinking, cooking, and luxurious steam baths. Today Masada is a popular tourist attraction and an Israeli national shrine.

Review Questions

Sections 1.0.0–4.0.0

1. The item shown in *Figure 11* is a _____.
   a. main shutoff valve
   b. curb stop
   c. corporation stop
   d. softener

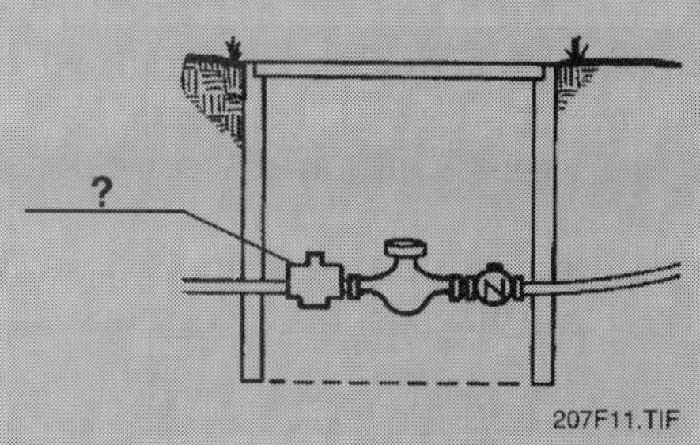

Figure 11 ◆ Review question illustration.

2. WSFUs vary with the quantity and temperature of the water and with the _____.
   a. drainage rate
   b. size of the fixture branch
   c. number of fixtures
   d. type of fixture

3. When calculating a building's demand, each of the following is a factor to consider *except* _____.
   a. water pressure in pounds per square inch (psi) at the source
   b. the length of pipe in the building
   c. the type of flush devices on the fixtures
   d. the type of piping to be used

4. The location of a water supply line and the depth below the frost line for a single-family residence usually are determined by the _____.
   a. water pressure
   b. local building codes
   c. distance from the curb stop
   d. number of fixtures in the building

5. Water meter housings are commonly made from the following materials *except* _____.
   a. brass
   b. plastic
   c. stainless steel
   d. cast iron

6. Most utility companies will install meters that are up to _____ inches in diameter; plumbers are responsible for installing meters larger that that.
   a. 1½
   b. 1¾
   c. 2
   d. 2½

7. To place a water heater in the most efficient location, you should _____.
   a. maximize the length of the piping runs
   b. select the southernmost corner of the building
   c. limit the number of hot water outlets near the heater
   d. minimize the length of the piping runs

8. Local codes require at least _____ hose bibb(s) in a residential building.
   a. one
   b. two
   c. three
   d. four

9. Hose bibbs are typically _____ inch.
   a. ¼
   b. ½
   c. ¾
   d. 1

10. To position a riser and stubout assembly firmly, use either a _____ or a _____ attached to a backing board.
    a. sanitary tee; sanitary ell
    b. curb box; curb stop
    c. drop ear ell; high ear ell
    d. riser clamp; pipe clamp

5.0.0 ◆ MAIN SUPPLY LINES

With the water heater located, the water softener and hose bibbs installed, and the fixture riser and stubouts in place, the main supply lines can now be installed. For residential installations, ensure that the main feeder line beyond the water heater is sized to supply the required flow and pressure.

Smaller pipe sizes may reduce energy use and save money. For example, a ¾-inch diameter pipe has a cross-sectional area of 0.4418 square inches, while a ½-inch diameter pipe has a cross-sectional area of 0.1963 square inches (see *Figure 12*). This means that a ¾-inch pipe will hold 2¼ times the amount of water that a ½-inch pipe will hold. With a larger pipe, a faucet will have to run longer to get hot water to a sink. In addition, the amount of hot water that will be left in the pipe to cool once the faucet is turned off is significantly greater with larger pipe.

Always consult the pipe sizing tables in your local applicable code to determine the water- and energy-saving measures possible for each project. On larger residential and commercial applications, plumbers install a recirculating pipe to return the water in the supply line to the water heater.

In multifamily units, control valves are required to isolate and repair each living unit. As the branch runs are installed, you should cut, prepare, and join several lengths of pipe and fittings before soldering. This saves time and provides for a smoother installation.

When installing the cold water lines, consider branching off the line to the hose bibbs to save pipe and time. If a softener is attached to the cold water lines, this option will not be possible. There is no need to put a softener on the cold water line that serves hose bibbs.

5.1.0 Shock Arresters

Install water hammer arresters on the supply line to prolong the service life of the components of the water supply system (see *Figure 13*). Water hammer is a high-intensity reverberating shock wave caused by the sudden closure of a valve (see *Figure 14*). The shock wave continues to pound between the point of impact and the point of relief until the destructive energy finally dissipates.

Air chambers are also used as a way to control shock in water supply lines. Air chambers consist of a capped piece of pipe that is the same diameter as the line it serves. Its length ranges from 12 inches to 24 inches. Air chambers may be constructed in several different shapes (see *Figure 15*). Air chambers require periodic draining. Open a low valve and the highest faucet in the system. This will let the water drain out and permit air to flow into the chamber in its place. This process, which is not preferred by many codes, may take a few minutes. Refer to your local applicable code for acceptable draining methods. Some codes do not permit the use of air chambers at all and instead require the use of water hammer arresters.

5.2.0 Other Water Supply Connections

In both residential and commercial installations, you may have to provide piping for miscellaneous appliances such as dishwashers, washing machines, and icemakers. Some codes require individual shutoff valves on each appliance. Dishwashers may require a tee or a three-way water stop off the hot water access line for the kitchen sink. Separate drains are not needed for dishwashers because their wastewater is pumped through the sink drain above the trap or into the garbage disposal unit.

Washing machines require both hot and cold water lines. They also require a hose bibb for each line so that the hoses from the washer can be connected to the lines. Locate these faucets so that the customer will have easy access after the washer is installed.

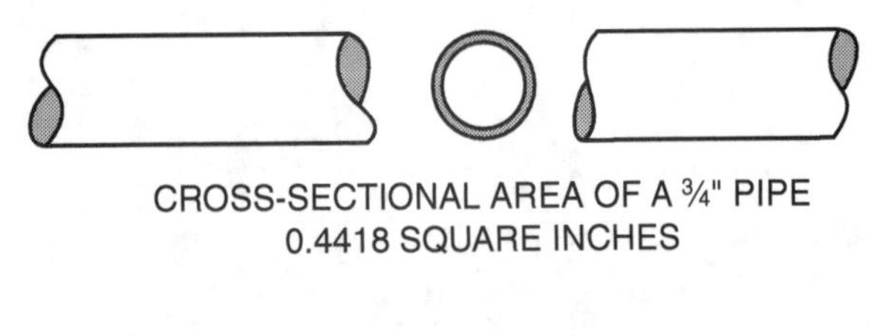

Figure 12 ◆ Cross-sectional areas of two pipe sizes.

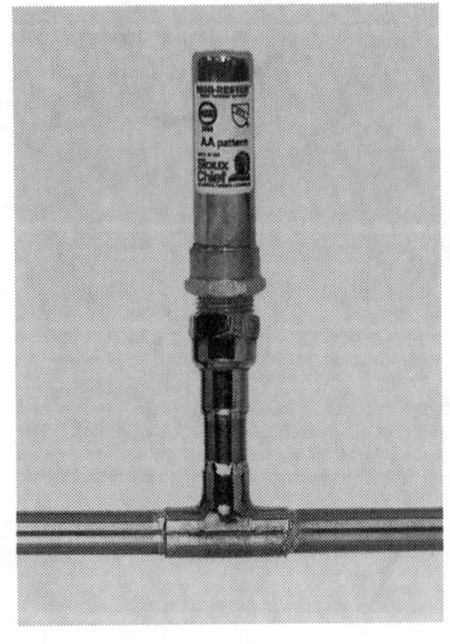

Figure 13 ◆ Water hammer arrester.

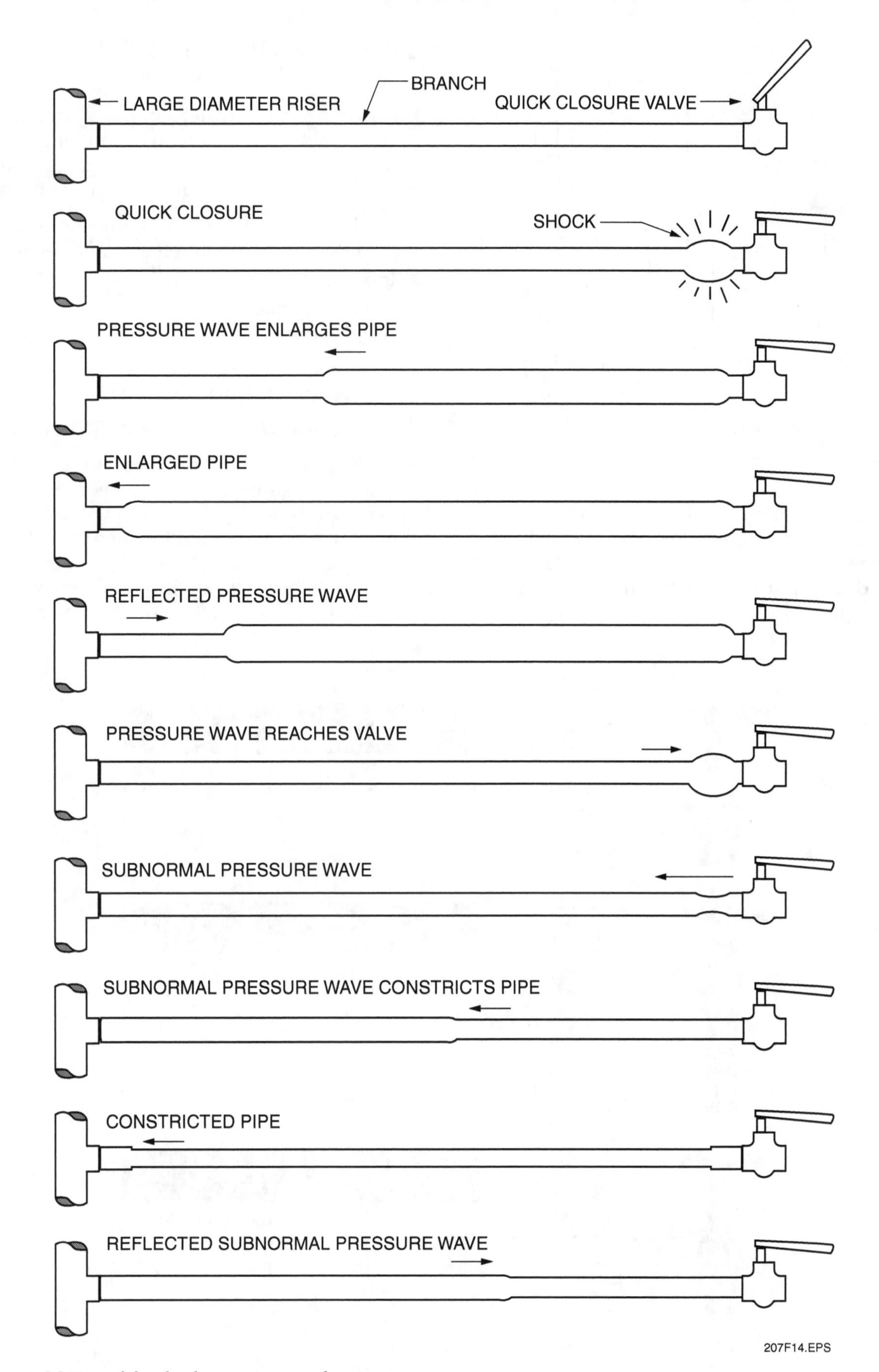

Figure 14 ◆ Motion of the shock wave in water hammer.

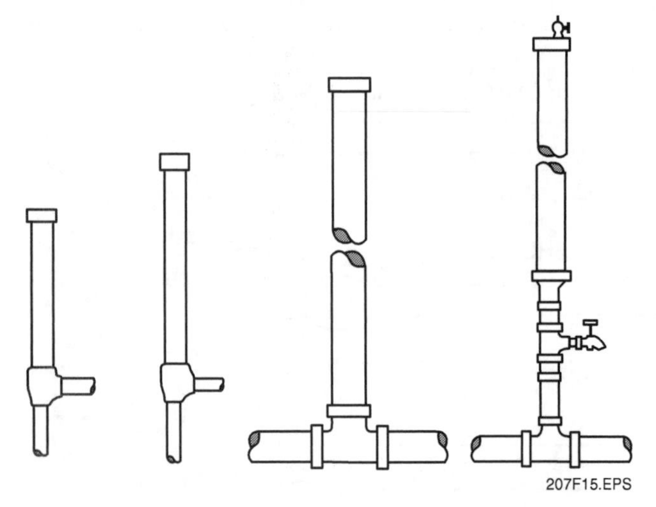

Figure 15 ◆ Air chambers.

Refrigerators with icemakers require both a line to bring in water and an accessible stop valve to turn the water supply on or off. Locate the valve beneath the sink or in a box behind the refrigerator (see *Figure 16*).

6.0.0 ◆ COMPLETING THE INSTALLATION

When installing a water supply system, a good plumber anticipates potential problems and provides ways to either prevent them or allow for quick repair. The following are some of the important factors to keep in mind:

- Accessibility for maintenance and repair
- Frost protection
- Puncture protection
- Backflow prevention

6.1.0 Accessibility for Maintenance and Repair

Shutoff valves are often located in areas that are not easily accessible once the project has been completed. For example, the shutoff valves to a tub or shower are located in the finished wall. Install an access panel that can be easily unscrewed and removed. Otherwise, a plumber will have to cut a hole in the wall to get to the valves to inspect,

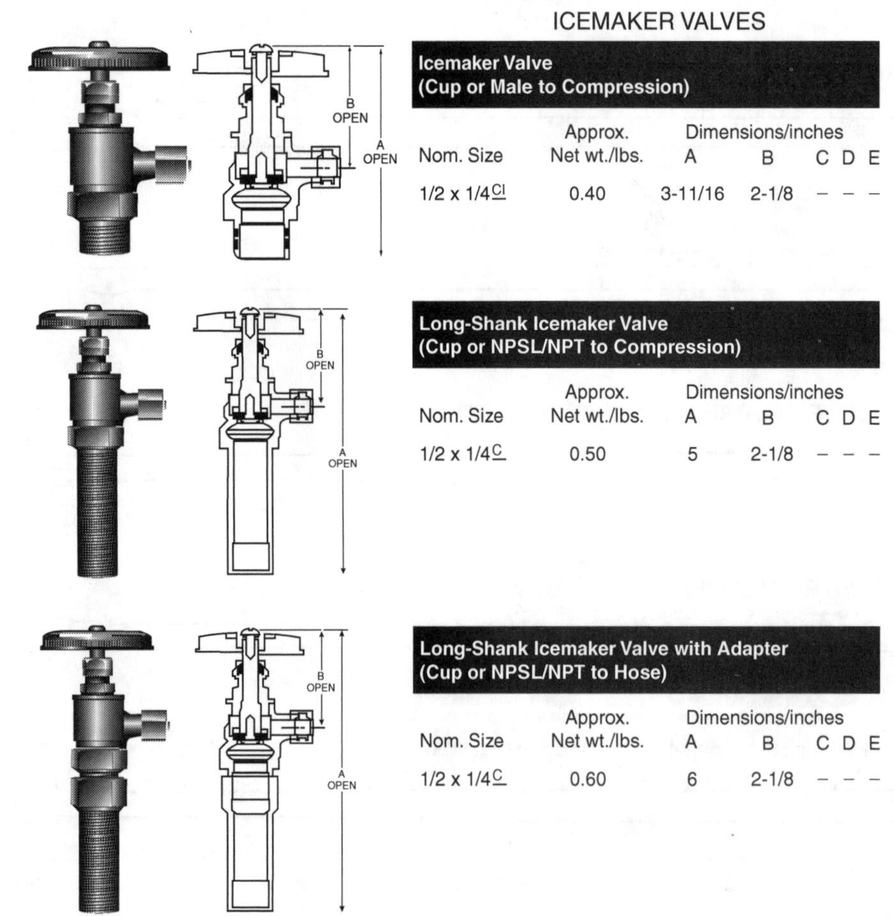

Icemaker Valve (Cup or Male to Compression)

Nom. Size	Approx. Net wt./lbs.	Dimensions/inches A	B	C	D	E
1/2 x 1/4CI	0.40	3-11/16	2-1/8	–	–	–

Long-Shank Icemaker Valve (Cup or NPSL/NPT to Compression)

Nom. Size	Approx. Net wt./lbs.	Dimensions/inches A	B	C	D	E
1/2 x 1/4C	0.50	5	2-1/8	–	–	–

Long-Shank Icemaker Valve with Adapter (Cup or NPSL/NPT to Hose)

Nom. Size	Approx. Net wt./lbs.	Dimensions/inches A	B	C	D	E
1/2 x 1/4C	0.60	6	2-1/8	–	–	–

Figure 16 ◆ Icemaker valves.

repair, or replace them. Refer to your local applicable code for guidelines and requirements.

Provide access to cleanouts, shutoff valves, and water meters located in basements and crawl spaces. If the structure is equipped with a gravity drain-down system, you must install strategically located valves at the low points of each line. These valves will also require easy access.

6.2.0 Freeze Protection

Earlier in this module, you learned about providing frost protection for the buried supply lines that run into the building. You must take precautions for frost protection inside the building as well. Do not run lines along exterior walls, especially in colder regions. When sinks or other fixtures must be located along exterior walls, run the supply piping up through the floor so that the pipes are inside the building. If permitted by your local applicable code, insulate water supply pipes that run inside walls.

Crawl spaces can also cause problems for water supply piping because they are often vented to the exterior. In many cases, crawl space vents are closed for the winter season, which should protect the plumbing. Avoid running water supply piping in front of or close to a vent when possible. Proximity to a vent may increase the potential for freezing.

6.3.0 Puncture Protection

During and after construction, pipes are vulnerable to punctures. To minimize the possibility of the pipes being punctured by a nail or screws, run pipes through the center of a 2 × 4 wall. Nail plates to both sides of the 2 × 4 studs to provide protection. Use only approved nailer plates as specified in your local applicable code. Be sure to follow the manufacturer's specifications when installing nailer plates.

6.4.0 Backflow Prevention

As you learned earlier in this module, backflow prevention is necessary on all installations where hoses are installed and on lines where backflow may occur through accidental cross-connection. Installations that typically require backflow prevention include dishwashers, hose bibbs, and slop sinks. Refer to your local applicable code for approved backflow prevention methods. Always install backflow preventers according to the manufacturer's instructions. If the project is a designed project and backflow prevention is required, submit an RFI for specific guidelines on the type of backflow preventers to be used.

CAUTION

Ensure that backflow preventers are appropriate for the type of plumbing installation. Refer to your local applicable code. If the project specifications call for a type of backflow preventer that is different from that recommended in your local applicable code, issue an RFI to the project engineer or architect.

7.0.0 ◆ TESTING

Test all water supply piping to ensure that it is free of leaks. Two types of tests may be conducted: the **air test** and the **hydrostatic** (water) **test.**

Test the water supply piping system before it is enclosed. On smaller systems, all of the water supply piping may be tested at one time. Larger projects—apartment buildings, for example—may require that you test the water supply piping in sections. Sometimes this is done floor by floor.

Typically, plumbers will conduct at least one test before scheduling the test with the inspector. This provides an opportunity to make any necessary repairs before the official test. The plumbing

DID YOU KNOW?

A Change in Plans

The White House did not have running water until 1831. That year, the Commissioner of Public Buildings bought a bubbling spring to pipe water to the White House in wood pipes made from drilled-out logs. The piping might have been installed earlier, in 1829 when Andrew Jackson took office, but the Committee on Public Buildings decided to improve the building's north entrance instead. The delay may have been beneficial. As the project got under way, the engineer exchanged the wooden pipes for pipes made of iron.

contractor is responsible for furnishing all materials, equipment, and labor to conduct the test. The plumbing inspector observes the test and verifies that it is successful. When the pipe is inspected and passes the pressure test, request the inspector to sign off on the successful test. Keep this sheet with the project records.

7.1.0 The Air Test

Air tests for water supply piping are similar to the air tests for DWV piping. It's a fairly simple procedure that requires a limited number of tools. Note that many codes prohibit air tests, or restrict the application of air tests to a few special circumstances. Refer to your local applicable code before testing a water supply system using air. Ensure that you follow all applicable safety and procedural guidelines.

A test gauge assembly enables compressed air to enter the piping system (see *Figure 17*). It also provides a gauge for measuring the air pressure within the piping system.

You can use a **compressor** to produce the required pressure (see *Figure 18*). A large air compressor may be on the construction site to operate a variety of air tools. If a large compressor is not available, use a small electric or gasoline-powered unit. In either case, the compressor should provide air at a pressure above 150 pounds per square inch (psi).

Plumbers generally don't use test plugs to close openings in the water supply piping system during testing. However, they are available and may be useful in selected applications. These caps are removed at the time of fixture installation and are not recommended for regular service use. The more common method of closing openings is to install regular service caps or plugs.

Figure 18 ◆ Air compressor.

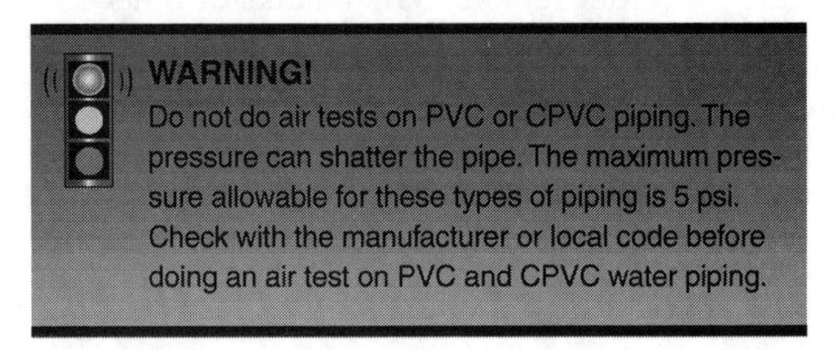

WARNING!
Do not do air tests on PVC or CPVC piping. The pressure can shatter the pipe. The maximum pressure allowable for these types of piping is 5 psi. Check with the manufacturer or local code before doing an air test on PVC and CPVC water piping.

7.2.0 Air Test Procedures

When conducting an air test, follow these steps:

Step 1 Inspect the system for defects and missing parts. During this inspection, you must close all openings with caps, plugs, test plugs, or valves.

Step 2 Attach the test gauge assembly. Make certain that it is securely joined to the piping system. Use pipe tape or **pipe dope** on threaded connections.

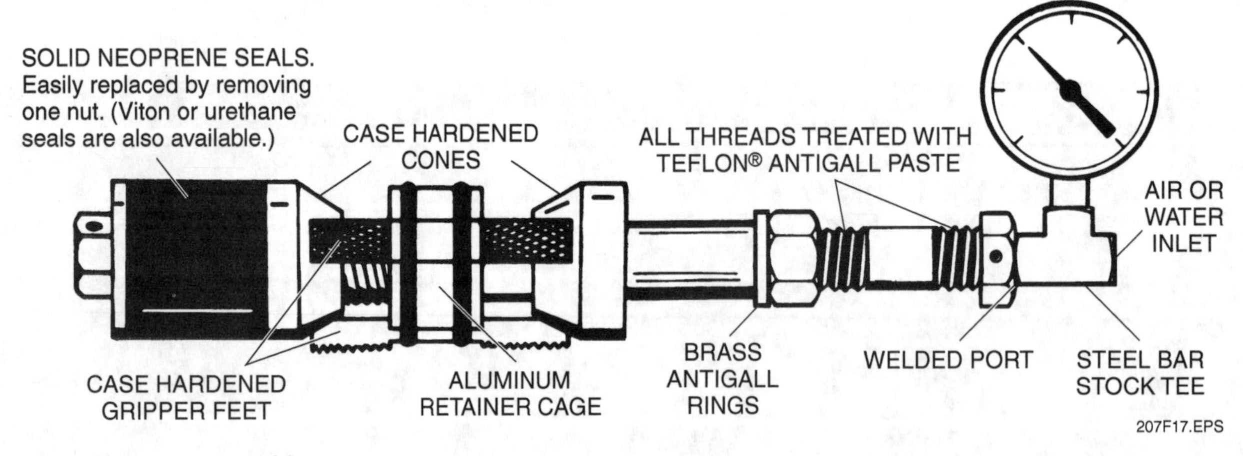

Figure 17 ◆ Test gauge assembly.

Step 3 Open the valve, and introduce air until the required pressure is recorded on the gauge. The exact pressure required will be dictated by local code; for example, 1½ times the normal water pressure. Make sure that the valve on the test gauge assembly is closed. This prevents leaks during testing. The test is considered successful if the water supply piping will hold the required pressure for a specified period, which may be up to 24 hours.

Step 4 If the piping fails the test, you must locate the defective or leaking joint. A soap solution may be helpful in identifying small leaks. Spray or spread the solution over the area, and watch for bubbles to form to locate the leak. Bubbles will not form on high-pressure leaks. Once you have repaired the leak, you must test the system again.

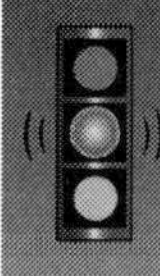

CAUTION

Test caps, or plugs, are not recommended for air tests in copper pipe. The air pressure might rupture the cap.

7.3.0 The Hydrostatic Test

Hydrostatic (water) tests are conducted by filling the pipe with water. Generally, this process is used for larger diameter piping such as water mains.

Use a **hydrostatic test pump** to conduct the test. The pump may be manually operated, electrically powered (*Figure 19*), or gasoline powered (*Figure 20*).

Powered pumps are rated by the number of gallons per minute (gpm) they put out and the maximum pressure (in psi) they produce. It is important to size the test pump to the requirements of the system being tested.

High-pressure hydrostatic testers also are available (see *Figure 21*). These units can produce up to 3,000 psi. They are used to test boilers, tanks, heat exchangers, and other systems that may require high pressures for testing.

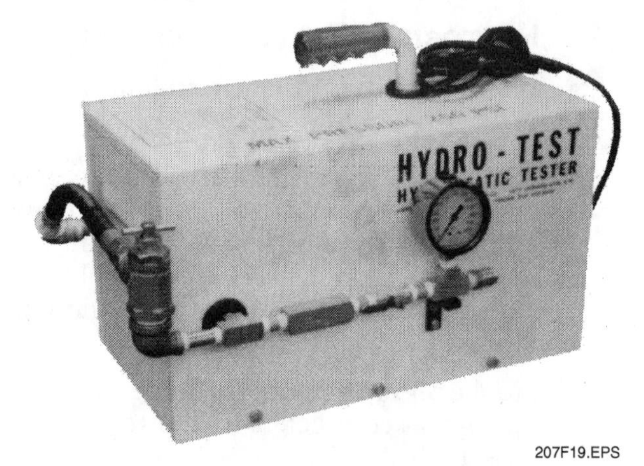

207F19.EPS

Figure 19 ◆ Electrically powered hydrostatic test pump.

207F20.EPS

Figure 20 ◆ Gasoline-powered hydrostatic test pump.

207F21.EPS

Figure 21 ◆ High-pressure hydrostatic tester.

7.4.0 Hydrostatic Test Procedures

When conducting a hydrostatic test, follow these steps:

Step 1 Fill the piping system with water. Because you'll need to replace all of the air in the piping with water, install valves at high points in the system, where air is likely to be trapped. Open these valves to allow air to escape.

Step 2 Make certain that the discharge end of the pump is connected to the test gauge assembly.

Step 3 Fit a hose onto the inlet (suction) end of the pump and connect it to a water supply or insert it into a bucket of water to provide extra water for compression. Water compresses very little; therefore, you need to force only a small amount of additional water into the piping to achieve the required pressure. If the pressure gauge fluctuates, or if you cannot obtain the required pressure, either a leak is present or air remains in the piping. You can easily detect a leak by observing where the water escapes. If trapped air is the problem, open the valves and allow the air to escape.

Step 4 Maintain the proper pressure and time for the test. The local code or job specifications will indicate the appropriate pressure to maintain and the length of the test.

Step 5 Once the test is concluded, slowly open a valve to relieve the pressure.

Step 6 Drain all water to prevent damage to the pipe from freezing.

CAUTION

Never leave a hydrostatic test pump unattended. The pressure can build quickly and may damage the system.

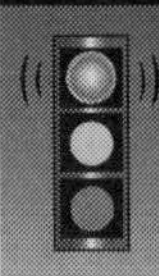

WARNING!

Never immerse an electrical test unit in water. You could be electrocuted.

7.5.0 Test Pump Operation

The following tips apply to using both electrically powered and gasoline-powered hydrostatic testing pumps:

- Always use clean, cold water.
- Use water pressure feed when operating electric test pumps. The inlet pressure may vary from 5 psi to 100 psi. Gasoline test pumps, on the other hand, are designed to work from gravity feed or siphon from a barrel or tank. If required, gasoline-powered units may also work from water pressure feed.
- Never leave a test pump unattended.
- Be sure the water flow is on before operating the unit. Never run a test pump without water for more than 30 to 45 seconds. Severe damage to the machine may result.
- Manufacturers of testing equipment suggest that you install a tee and a valve on the running side of the tee at the system being tested and use it for a bleed-down valve.
- Be sure you know what pressure the relief valve is set for before you begin the test. The setting of the relief valve determines the maximum pressure that the test pump can generate.
- Be sure to supply an electrical test unit with a grounded source of electricity. The power source must match the needs of the motor.
- Most test pumps are designed for use with water only. Avoid using other fluids.
- The time required to test a system depends on the amount of water needed to pressurize the system, not necessarily the size of the system. The incoming water displaces the air within the system and creates pressure. Therefore, the more air trapped within the system before testing, the longer it takes to pressurize the system and conduct the test.
- If it seems that it is taking longer than normal to test a system, look for a leak, an open valve, or some other way that the system might be losing water. If you suspect that the hydrostatic tester itself may not be working, test it by putting a valve on the outlet end of the outlet hose. Then, with the hydrostatic tester running, slowly turn the valve off and check the pressure gauge. If the gauge registers pressure, the unit is working.

Review Questions

Sections 5.0.0–7.0.0

1. A ¾-inch pipe will hold 2¼ times the amount of water that a ½-inch pipe will hold.
   a. True
   b. False

2. You are installing the water supply system for an apartment building. To allow individual apartments to be isolated for repair and servicing, install _____ on the supply lines to each apartment.
   a. control valves
   b. backflow preventers
   c. safety valves
   d. curb stops

3. Codes are increasingly phasing out air chambers in favor of shock arresters.
   a. True
   b. False

4. Because their wastewater is pumped through the sink drain above the trap or into the garbage disposal unit, _____ do *not* require separate drains.
   a. water coolers
   b. icemakers
   c. washing machines
   d. dishwashers

5. If a building is equipped with a gravity drain-down system, install valves _____.
   a. on the lowest line only
   b. at the point where the lines enter the building drain
   c. only if the building is more than one story
   d. at the low points of each line

6. You are installing a sink that must be located along an exterior wall. Run the supply piping for the sink _____.
   a. along the outer wall
   b. up through the floor
   c. along the ceiling
   d. parallel to the hot water line

7. To help reduce the possibility that water supply pipes will be accidentally punctured by nails or screws, _____.
   a. use thicker wall pipe on vertical runs behind walls
   b. run pipes through the center of a 2 × 4 wall
   c. run pipes through protective sleeves
   d. construct a chase and run pipes through it

8. Each of the following installations typically requires a backflow preventer *except* for a _____.
   a. hose bibb
   b. slop sink
   c. refrigerator
   d. dishwasher

9. In *Figure 22,* a test gauge assembly, the element being pointed to is a(n) _____.
   a. brass antigall ring
   b. case-hardened gripper foot
   c. air or water inlet
   d. welded port

10. During an air test, you should attach the test gauge assembly before closing all openings with caps, plugs, test plugs, or valves.
    a. True
    b. False

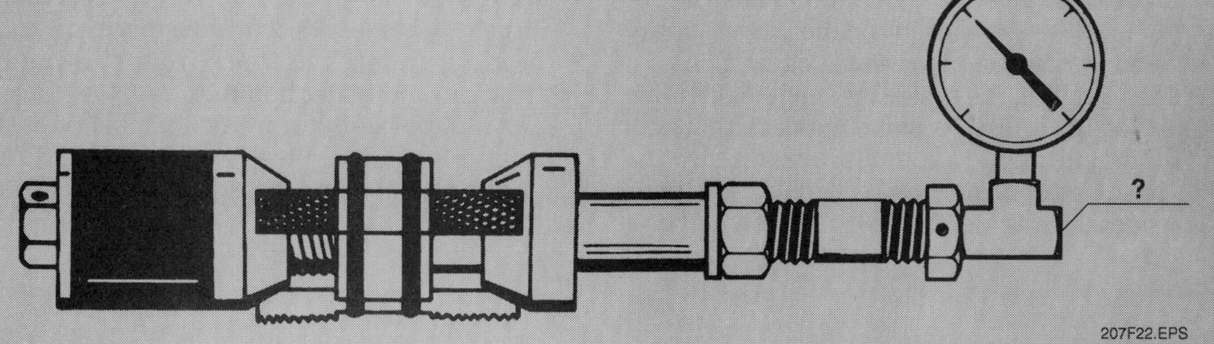

Figure 22 ◆ Review question test gauge assembly.

Review Questions

11. Hydrostatic tests are usually performed on _____ piping.
    a. thicker walled
    b. larger diameter
    c. thinner walled
    d. smaller diameter

12. If the pressure gauge fluctuates while performing a hydrostatic test, it is an indication of either a leak or _____.
    a. an open valve in the system
    b. insufficient test pressure
    c. the discharge end of the pump not being connected to the test gauge assembly
    d. air trapped in the piping

13. A hydrostatic test of a water supply system should occur with _____.
    a. air still in the system
    b. valves installed at low points to allow air to escape
    c. substantial pressure in the system
    d. valves at the top of the system open

14. Electric test pumps are designed to use _____ feed.
    a. water pressure
    b. siphon
    c. gravity
    d. water pump

15. The time required to hydrostatically test a system depends primarily on _____.
    a. the size of the system
    b. the pressure rating of the test gauge assembly
    c. the amount of water required
    d. the pump capacity

Summary

The process of installing a water supply system is similar to that for installing a DWV system, which you have learned how to do elsewhere in this curriculum. Begin by developing a material takeoff, locating the fixtures, and determining the route of the piping. Base the design for the water supply system on the project plans and specifications.

When installing the system, you will have some amount of flexibility to make changes that will save time and money. You will also modify the plans to solve problems on site. Some of the items to consider are the demand for water, whether a water softener is included, and the number of hose bibbs. In addition, the job may include fixtures like dishwashers, washing machines, and icemakers.

It is important to hang and support the pipes securely, making sure that you don't weaken any structural members that you modify. You'll also need to protect the pipes from nail punctures, freezing, and water hammer.

An important part of the installation is to conduct either an air test or a hydrostatic test to ensure that the system is leak-free and ready for the customer's use.

Trade Terms Introduced in This Module

Air test: A test to detect leaks and to determine a pipe's ability to withstand air pressure.

Compressor: A machine for compressing air or other gases.

Curb box: A vertical sleeve that provides access to a buried curb stop.

Curb stop: In a water service pipe, a control valve for a building's water supply, usually placed between the sidewalk and the curb.

Demand: The total water requirement for a water supply system, including the pipes, fittings, outlets, and fixtures used in the system.

Frost line: The depth of frost penetration into the soil. The depth varies in different parts of the country. Water supply and drainage pipes should be set below this line to prevent freezing.

Hydrostatic test: A test to determine a pipe's ability to withstand internal hydrostatic (water) pressure and to detect possible leaks in the system.

Hydrostatic test pump: A pump used to test hydrostatic pressure.

Pipe dope: A compound applied to pipe joints to allow them to be tightened to the point that they are leakproof.

Water supply fixture unit (WSFU): The measure of a fixture's load, varying according to the amount of water, the water temperature, and the fixture type.

Resources & Acknowledgments

Additional Resources

This module is intended to be a thorough resource for task training. The following reference works are suggested for further study. These are optional materials for continued education rather than for task training.

Practical Plumbing Engineering. 1991. Cyril M. Harris, ed. New York: McGraw-Hill.

Water Quality & Systems: A Guide for Facility Managers. 1996. Robert N. Reid. New York: UpWord Publishing.

Water Supply Systems Security. 2004. Larry W. Mays, ed. New York: McGraw-Hill.

References

2003 National Standard Plumbing Code. 2003. Falls Church, VA: Plumbing-Heating-Cooling Contractors—National Association.

Dictionary of Architecture and Construction, Third Edition. 2000. Cyril M. Harris, ed. New York: McGraw-Hill.

Figure Credits

Ivey Mechanical Company	Module divider
Jonathan Byrd	207F01
Badger Meter, Inc.	207F04, 207F05
Mueller/B&K Industries	207F06
Eljer Plumbingware, Inc. www.eljer.com	207F07
Sioux Chief Manufacturing Co., Inc.	207F13
Plumbing and Drainage Institute	207F14, 207F15
NIBCO International	207F16
Expansion Seal Technologies	207F17
DeWALT Inc.	207F18
General Pump & Equipment Co. Inc.	207F19–207F21
International Code Council, Inc. *2003 International Fuel Gas Code, Copyright 2003. Falls Church, Virginia: International Code Council, Inc. Reproduced with permission. All rightsReserved.*	Table 1

CONTREN® LEARNING SERIES — USER UPDATE

The NCCER makes every effort to keep these textbooks up-to-date and free of technical errors. We appreciate your help in this process. If you have an idea for improving this textbook, or if you find an error, a typographical mistake, or an inaccuracy in NCCER's Contren® textbooks, please write us, using this form or a photocopy. Be sure to include the exact module number, page number, a detailed description, and the correction, if applicable. Your input will be brought to the attention of the Technical Review Committee. Thank you for your assistance.

Instructors – If you found that additional materials were necessary in order to teach this module effectively, please let us know so that we may include them in the Equipment/Materials list in the Annotated Instructor's Guide.

Write: Product Development and Revision
National Center for Construction Education and Research
P.O. Box 141104, Gainesville, FL 32614-1104

Fax: 352-334-0932

E-mail: curriculum@nccer.org

Craft ______ Module Name ______

Copyright Date ______ Module Number ______ Page Number(s) ______

Description

______

______

______

(Optional) Correction

______

______

(Optional) Your Name and Address

______

______

02208-05

Installing Fixtures, Valves, and Faucets

02208-05

Installing Fixtures, Valves, and Faucets

Topics to be presented in this module include:

Overview

Plumbing fixtures are the connection between the water supply and the sewage system. Common residential and commercial fixtures include bathtubs, showers, lavatories, sinks, water closets, and urinals. Plumbers refer to approved submittal data and manufacturers' instructions when installing fixtures, which are available in a variety of models. Before installation, plumbers ensure that fixtures meet project specifications, examine their condition, and verify their dimensions.

Valves and faucets control the flow of fluids or gases through a pipe. They must be installed properly to ensure that they work correctly. Plumbers consider the application and materials when selecting one of the five types of valve connections: threaded, soldered, solvent welded, flanged, and compression. A faucet is a valve installed at the end of a water supply line. Faucets are available in floor-mounted and wall-hung varieties, and are classified as compression or noncompression faucets. To control flow, plumbers install float-controlled valves, flush valves, and flushometers.

There are basic procedures, tools, and techniques used to install tubs, showers, lavatories, sinks, water closets, and urinals. Installation requirements vary, however, so plumbers always read and follow the manufacturers' instructions. Plumbers also inspect and test fixtures to ensure that they work properly and always clean up the work area. Fixture work is finish work, so it should reflect the best skill and workmanship.

Focus Statement

The goal of the plumber is to protect the health, safety, and comfort of the nation job by job.

Code Note

Codes vary among jurisdictions. Because of the variations in code, consult the applicable code whenever regulations are in question. Referring to an incorrect set of codes can cause as much trouble as failing to reference codes altogether. Obtain, review, and familiarize yourself with your local adopted code.

Objectives

1. Describe the general procedures you should follow before installing any fixture.
2. Install bathtubs, shower stalls, valves, and faucets.
3. Install water closets and urinals.
4. Install lavatories, sinks, and pop-up drains.
5. Protect fixtures.

Trade Terms

Blocking
Callback
Closet bolt
Cross pattern
Cross threading
Escutcheon
Finish work
Lift rod assembly
Pop-up drain
Tub waste
Wax seal

Required Trainee Materials

1. Appropriate personal protective equipment
2. Pencil and paper
3. Copy of local applicable code

Prerequisites

Before you begin this module, it is recommended that you successfully complete *Core Curriculum; Plumbing Level One; Plumbing Level Two,* Modules 02201-05 through 02208-05.

This course map shows all of the modules in the second level of the Plumbing curriculum. The suggested training order begins at the bottom and proceeds up. Skill levels increase as you advance on the course map. The local Training Program Sponsor may adjust the training order.

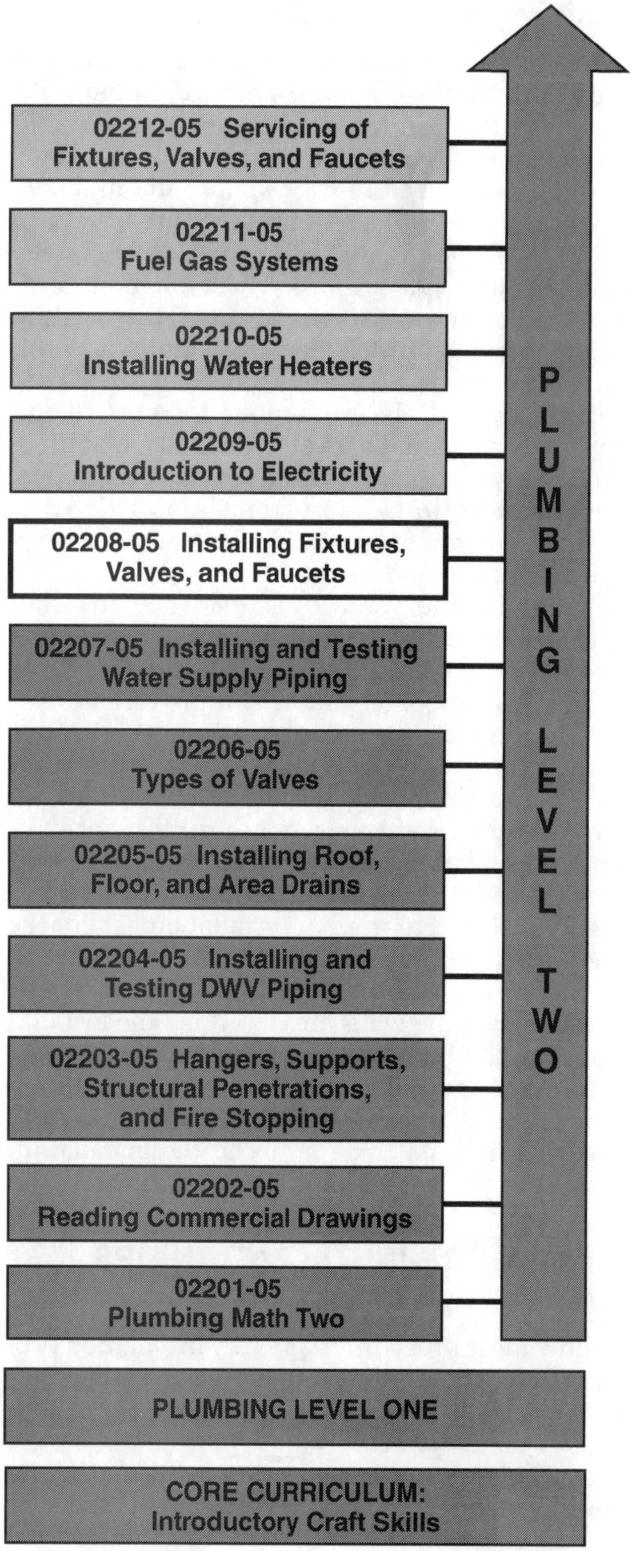

PLUMBING LEVEL TWO

02212-05 Servicing of Fixtures, Valves, and Faucets

02211-05 Fuel Gas Systems

02210-05 Installing Water Heaters

02209-05 Introduction to Electricity

02208-05 Installing Fixtures, Valves, and Faucets

02207-05 Installing and Testing Water Supply Piping

02206-05 Types of Valves

02205-05 Installing Roof, Floor, and Area Drains

02204-05 Installing and Testing DWV Piping

02203-05 Hangers, Supports, Structural Penetrations, and Fire Stopping

02202-05 Reading Commercial Drawings

02201-05 Plumbing Math Two

PLUMBING LEVEL ONE

CORE CURRICULUM: Introductory Craft Skills

208CMAP.EPS

1.0.0 ◆ INTRODUCTION

In this module, you will learn how to install basic plumbing fixtures. As you know, plumbing fixtures are the connection between the potable water supply and the sewage system. Waste flows into the fixture before it flows into the drainage system. Common residential fixtures include bathtubs, showers, lavatories, sinks, water closets, and, in some cases, urinals, which are more common in commercial installations. This module includes general instructions that apply to all fixture installations. Because fixtures come in a wide variety of models, remember to refer to the approved submittal data and follow the manufacturer's instructions when installing the fixtures. These sources provide important information that will prevent you from voiding the manufacturer's warranty during installation.

In this module, you will also learn how to install valves and faucets. A faucet is a valve installed at the end of a water supply line. Faucets are available in both floor-mounted and wall-hung varieties.

When installing fixtures, follow the proper installation procedures. These procedures are based on accepted engineering practices and code requirements. Remember that the little bit of extra time you take to follow correct procedures will prevent **callbacks**, or return trips to fix mistakes made during the original installation. Callbacks are time-consuming and costly.

Fixture installation is called **finish work** because once the walls are closed up, the installations are the one aspect of the plumbing project that everyone will see. Make sure your finish work reflects your best skill and workmanship. The care and skill you apply to the installation inside the walls must continue to the finish work.

2.0.0 ◆ PRE-INSTALLATION TIPS AND TECHNIQUES

Follow these procedures on every installation project. They will allow you to reduce or even eliminate mistakes, and will also help save time and money on the project.

A wholesaler may deliver the fixtures and trim directly to the project site, or you may pick them up from a supplier or from a plumbing contractor's warehouse. Preplan where the fixtures will be stored before they are delivered. Select a location that is both secure and close to the point of installation. That way, the fixtures will be handled as little as possible, reducing the potential for damage, vandalism, and loss. For efficiency, preplan the tags and labels to be used on pipe and fittings as well.

Check the approved submittal data to ensure that the fixtures meet the project specifications. Also see if fixture trim such as faucets, traps, and strainers are included. Ensure that each trim piece fits the fixture.

Examine the condition of the fixtures and trim. Open all boxes delivered to the project site while the delivery person is there. This is to ensure that the fixtures and trim have been delivered intact. If the fixtures and trim are being picked up directly from a supplier or contractor's warehouse, inspect them at the warehouse before they are delivered. Check the contents of previously opened cartons for missing items.

Verify that the rough-in dimensions are correct. This is a critical step. Errors could have occurred in the initial rough-in. Changes in the building, such as wall placement and wall covering, could affect your work. You might have to install a fixture that is different from the one originally specified. If this happens, issue an RFI to the project engineer or architect, and notify the owner.

Protect the fixtures before installation. Ideally, the fixtures will be scheduled for delivery right before they are to be installed. If that is not possible, protect the fixtures by storing them in a clean, dry space that can be locked. Ensure that the method and location of storage are convenient for productivity.

Save all of the packing materials. They can be used to protect the installed fixtures from weather and the work of other trades. Also, the package can be used to return an unsuitable fixture or piece of trim.

Save all of the manufacturer's installation and care instructions, warranty materials, and operation and maintenance manuals. Turn these over to the customer when the project is completed.

Protect the installed fixtures until the building is ready for occupancy. Cut and fit cartons over the installed fixtures to protect surfaces from scratches and chips.

Protect your customers. They expect to be the first to use their new facilities. Place signs marked DO NOT USE on the covered fixtures. On large projects, you may decide to install temporary fixtures in one restroom for workers to use.

3.0.0 ◆ INSTALLING BATHTUBS AND SHOWER STALLS

The following installation steps are provided as a general guide. Always read and follow the manufacturer's instructions to determine the specific instructions for bathtubs and shower stalls.

On the project site, you will probably work with bathtubs and shower stalls first, because they are too large to install later on. Follow these steps to install a typical bathtub:

Step 1 Check the rough-in dimensions.

Step 2 Install bracing for the tub (see *Figure 1*).

Step 3 Install the **tub waste** and overflow piping according to the manufacturer's directions. Note that moving the tub may require two people.

Step 4 Set the tub and shower valve, and brace properly.

Step 5 Install the shower riser, elbow, shower arm, and head. Brace the riser to the framing (see *Figure 2*).

Step 6 Install the filler drop, elbow, nipple, and spout.

Step 7 Move the tub into position. Make certain that it rests firmly on the bracing.

Step 8 Connect the waste tailpiece drain system.

Step 9 Fill the bathtub through the valve. Fill the tub to overflow.

Step 10 Check for leaks.

Step 11 Caulk the fixture, and cover the tub with packing material to protect it.

When installing a fiberglass tub and shower unit, always follow the manufacturer's instructions for positioning the required **blocking** (wood used to reinforce structural members in a wall),

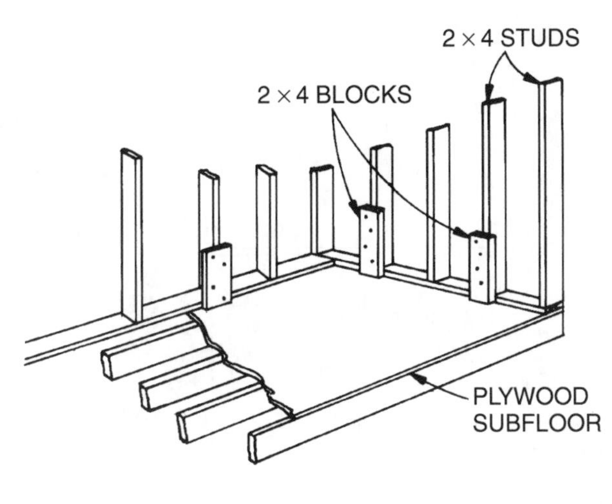

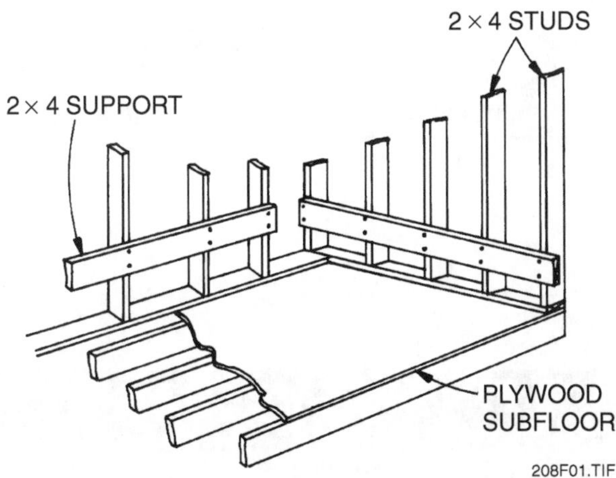

Figure 1 ◆ Bracing for the tub.

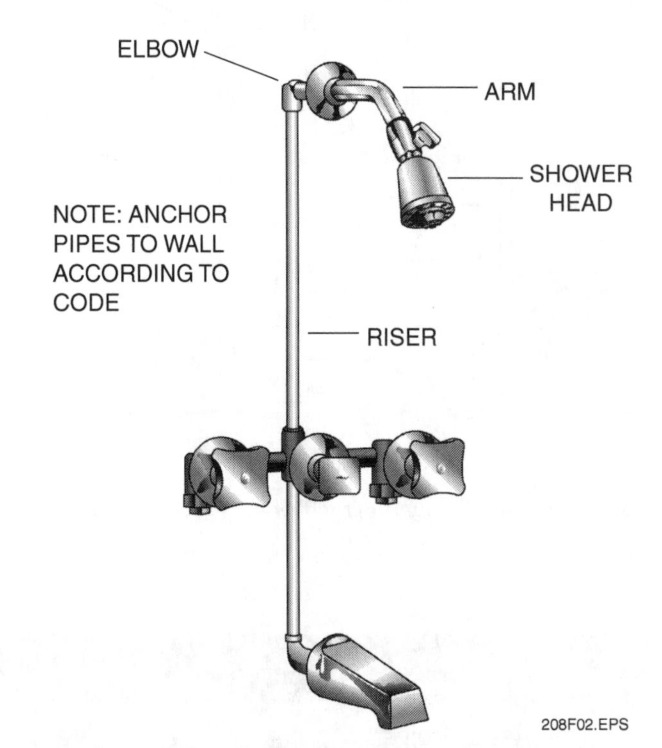

Figure 2 ◆ Shower riser, elbow, arm, and shower head.

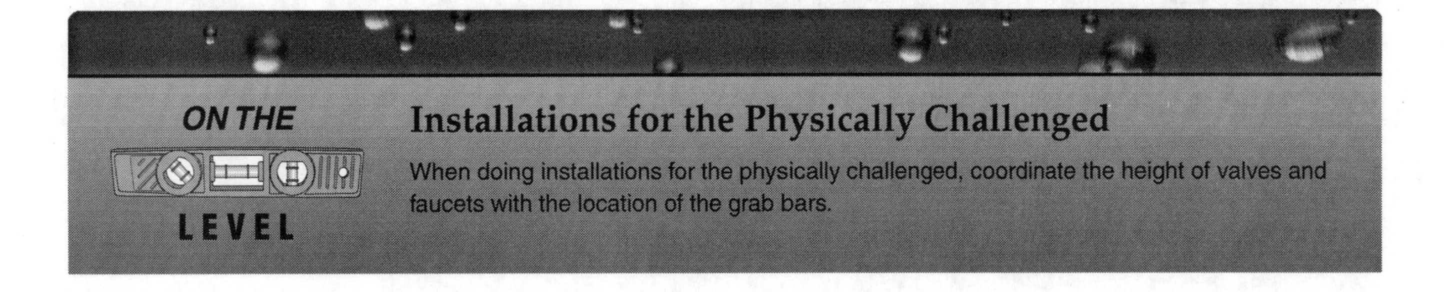

Installations for the Physically Challenged

When doing installations for the physically challenged, coordinate the height of valves and faucets with the location of the grab bars.

grout, or other supporting material (see *Figure 3*). Many fire codes require fire-rated blocking. Refer to your local applicable code. Trial fit the unit to ensure that the bearing points are properly supported. Fiberglass has a tendency to crack under stress, so the framing must fit correctly around units made of this material. Avoid bending the material during installation. Bending may result in cracks that will not become visible until later.

4.0.0 ◆ INSTALLING VALVES AND FAUCETS

Valves and faucets control the flow of fluids or gases through a pipe. They must be installed properly to ensure that they work correctly. There are five common types of valve connections, or joints:

- Threaded
- Soldered

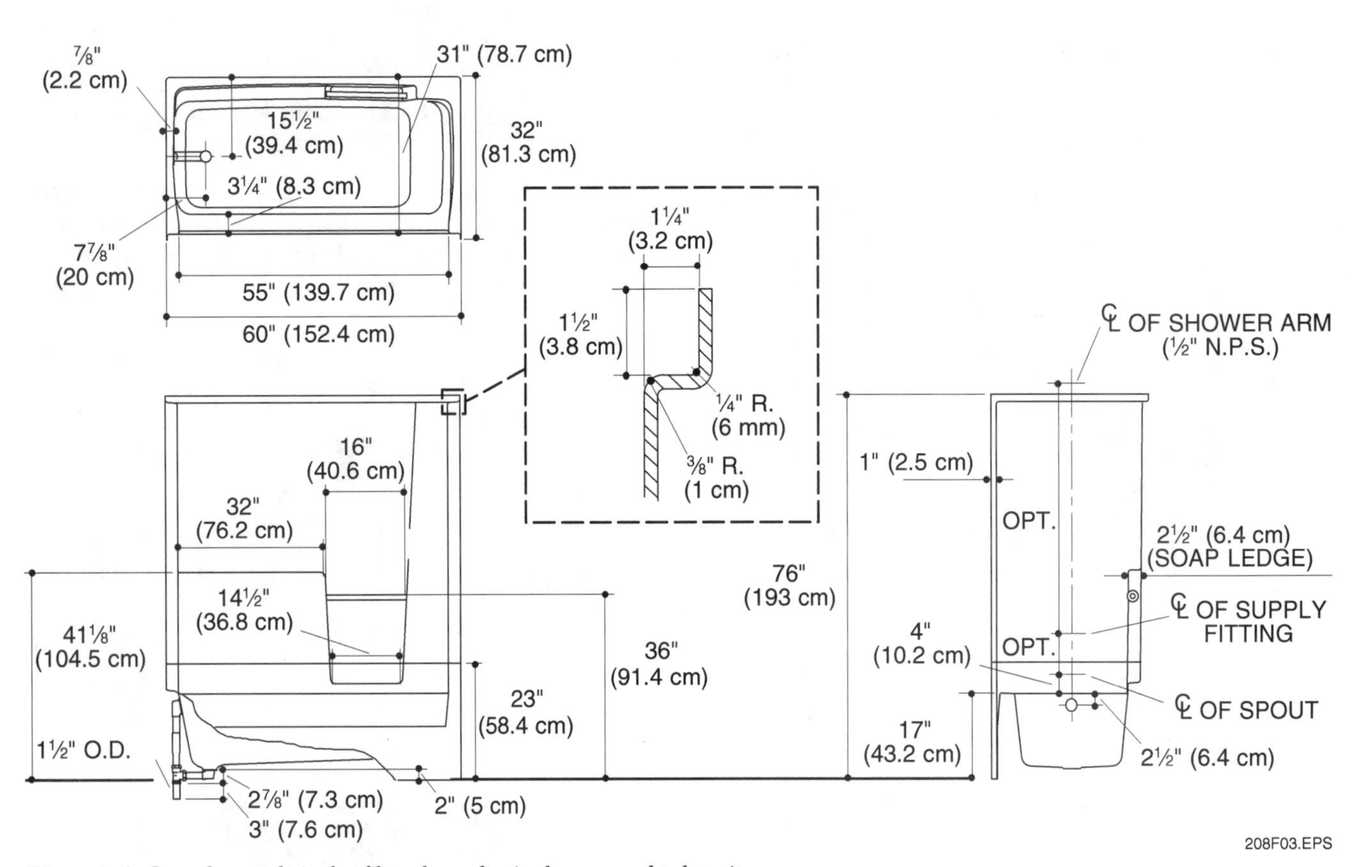

Figure 3 ◆ Sample rough-in for fiberglass plastic shower and tub unit.

DID YOU KNOW?

How Porcelain Enamel Is Made

Porcelain enamel is a form of glass bonded to metal. It is the highest quality, most durable, and most sanitary finish available for metal surfaces. To make it, manufacturers smelt a blend of minerals. They pour the red-hot mixture between water-cooled rollers where it quickly cools to a ribbon of glass. The glass is then shattered to form a material called frit. Because of that, the process is known as fritting.

Frit is applied to the metal and fired at temperatures ranging from 1,100°F to 1,600°F. The frit melts and bonds with the metal. This process does not simply coat the metal. Rather, it merges the chemical makeup of the frit and the metal and produces a new, chemically unique, finish.

- Solvent welded
- Flanged
- Compression

Determine the type of connection to use by considering the plumbing application, ease of installation, and the material that makes up the pipe and valves.

4.1.0 Threaded Valves

Threaded valves have internal threads. Check valves and temperature/pressure (T/P) relief valves are among the most common types of threaded valves. To install threaded valves, follow these steps:

Step 1 Apply Teflon® tape or pipe dope (if allowed) to the coupling threads. If you use pipe dope, do not overfill the threads. Compound buildup in the pipe can cause sanitary problems in the system. Too much compound can accumulate in the valve seat and cause a malfunction.

Step 2 Thread the valve by hand at first to prevent **cross threading**, which can damage the soft copper threads on the adapter. After starting the threads by hand, tighten the valve with a wrench. Firmly fix a second wrench on the adapter to prevent damage to the plumbing already in place.

Step 3 Rotate the valve to a final, vertical position.

When installing a threaded valve on a copper pipe, use a threaded copper adapter (see *Figure 4*). Solder the adapter to the end of the pipe first, then thread the valve onto the adapter. Some models of check valves are designed to be soldered and will not require a threaded copper adapter.

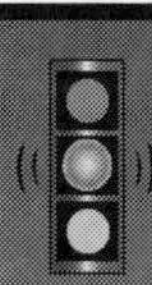

CAUTION

When installing fixtures, always read and follow the manufacturer's instructions and follow the installation guidelines in the instructions or the local applicable code, whichever is stricter. Bathtubs, shower stalls, water closets, urinals, sinks, and lavatories come in a wide variety of styles and materials. Each installation will require slightly different procedures. Manufacturers warrant their products against defects, but the warranty does not apply if you disregard their installation instructions and damage occurs as a result. If you damage a unit and void the warranty, you will have to replace the unit. Considering how much some of these units cost, it's a small investment of your time to follow the instructions and do the job right the first time.

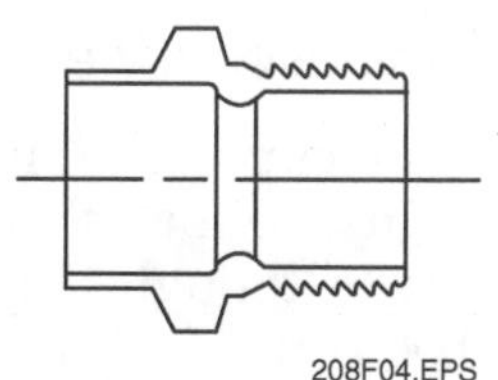

208F04.EPS

Figure 4 ◆ Threaded copper adapter.

4.1.1 Installing Temperature/Pressure Relief Valves

T/P relief valves are a special type of threaded valve (see *Figure 5*). They are safety devices that prevent a water heater explosion caused by a thermostat failure. The valve opens automatically when the temperature of the water exceeds a preset level. Before installing the valve, remove the protective plastic cap from the relief valve port on the water heater. This port may be on the top or side of the heater. Apply pipe joint compound (pipe dope) or Teflon® tape to the valve, and thread by hand. Position the valve so a drain line can easily be attached to the valve.

208F05.EPS

Figure 5 ◆ Relief valve.

ON THE LEVEL

Replace Leaking T/P Valves

Never reuse T/P valves. Always replace any leaking T/P valve. These valves are an essential part of the safety measures built into water heaters and boilers. Many states now require that the T/P valve(s) be placed on the water heater prior to installation.

Never modify a T/P valve installation in such a way as to prevent the immersion of the probe into the vessel.

4.1.2 Installing Check Valves

Check valves provide one-way passage of fluids or gases. Ensure that check valves are positioned properly. Although the proper check valve may have been specified for the project, if positioned improperly the valve will not function. The check valve in *Figure 6* is designed for installation in both vertical and horizontal lines with upward flow or in any intermediate position. Know the limitations of the valve being installed. Know the direction of flow, and make sure that it corresponds to the directional arrow shown on the check valve. Always match the direction of flow to the arrow. Never rotate the valve on its axis more than 15 degrees. To install check valves, follow these steps:

Step 1 Apply Teflon® tape or pipe dope (if permitted) to the male pipe threads. If using pipe dope, apply it to the first three to four threads. Do not overfill the threads. Compound buildup in the pipe can cause sanitary problems in the system. Too much compound can accumulate in the valve seat and cause a malfunction.

Step 2 Thread the valve by hand at first to prevent cross threading. Cross threading can damage the threads on the valve. After starting the threads by hand, tighten the valve with a wrench. Hold a second wrench firmly fixed on the pipe or adapter to prevent damage to the plumbing already in place.

Figure 6 ◆ Check valve.

4.2.0 Soldered Valves

When soldering valves in copper systems, remove the valve interior to prevent damage to the rubber washer by excessive heat buildup. Replace the interior as soon as the valve assembly has cooled.

To install a soldered shutoff valve, follow these steps:

Step 1 Ream the pipe ends and clean the metal with a copper-cleaning tool.

Step 2 Remove the valve interior.

Step 3 If you are installing the valve for a sink, lavatory, or water closet, place the **escutcheon**, which is a metal ring applied as finish trim, on the stubout.

Step 4 Align the valve properly and solder, taking care not to burn the wall.

Step 5 Replace the valve interior once the valve has cooled.

4.3.0 Solvent-Welded Valves

Use solvent welding when joining rigid plastic pipe. Plastic valves or special chrome-plated brass shutoff valves can be used to connect to a CPVC (chlorinated polyvinyl chloride) system. The special valves are made with an integral CPVC socket, which is a part of the valve assembly and can accept a solvent weld (see *Figure 7*).

To install a solvent-welded shutoff valve in a bath or kitchen, follow these steps:

Step 1 Cut the cap off the stub end squarely, and remove all burrs.

Step 2 Clean the surfaces to be joined with emery paper and a clean cloth.

Step 3 Apply CPVC primer and solvent to both the stubout and the valve.

Step 4 Push both together while rotating from right to left to ensure complete solvent contact (see *Figure 8*). Carefully align the valve properly now. Once the solvent dries, the valve cannot be repositioned.

Step 5 Allow adequate drying time as noted on the solvent can.

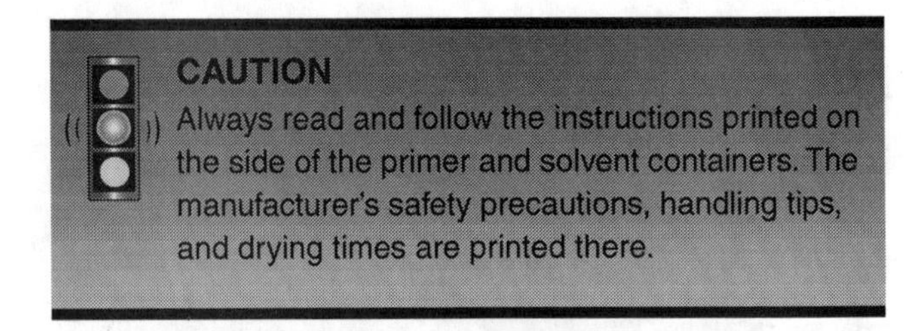

CAUTION

Always read and follow the instructions printed on the side of the primer and solvent containers. The manufacturer's safety precautions, handling tips, and drying times are printed there.

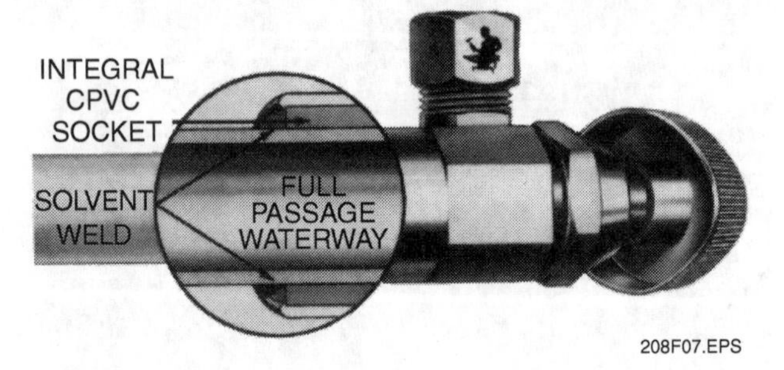

Figure 7 ◆ Integral CPVC socket.

Figure 8 ◆ Applying solvent to stubout and valve ends.

4.4.0 Flanged Valves

Flanged valves (see *Figure 9*) are used where frequent dismantling of piping is necessary. To install flanged valves, follow these steps:

Step 1 Set the valve between the pipe flanges.

Step 2 Put the gaskets between the valve flanges and pipe flanges.

Step 3 Line up the bolt holes.

Figure 9 ◆ Flanged valves.

Step 4 Install and tighten the bolts in a **cross pattern,** in which you tighten bolts a little at a time in an alternating sequence.

4.5.0 Compression Connection Valves

These valves are installed mainly on copper and plastic tubing. The valves consist of three parts: the valve body, the compression ring (also called a ferrule), and a compression nut that screws onto the valve body. To install a compression connection valve, follow these steps (refer to *Figure 10*):

Step 1 Ensure that the pipe is long enough. Consult the approved submittal data for the appropriate length.

Step 2 Place the compression nut and then the compression ring on the pipe.

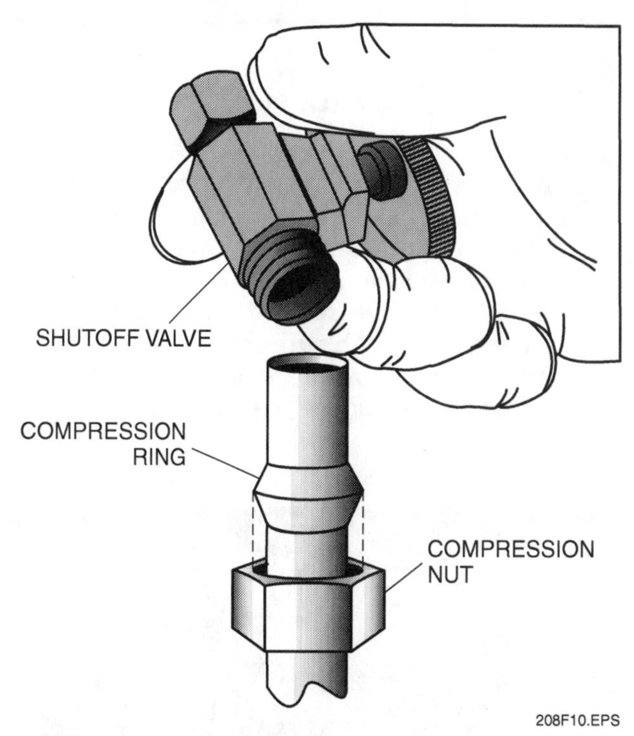

Figure 10 ◆ Compression connection valve.

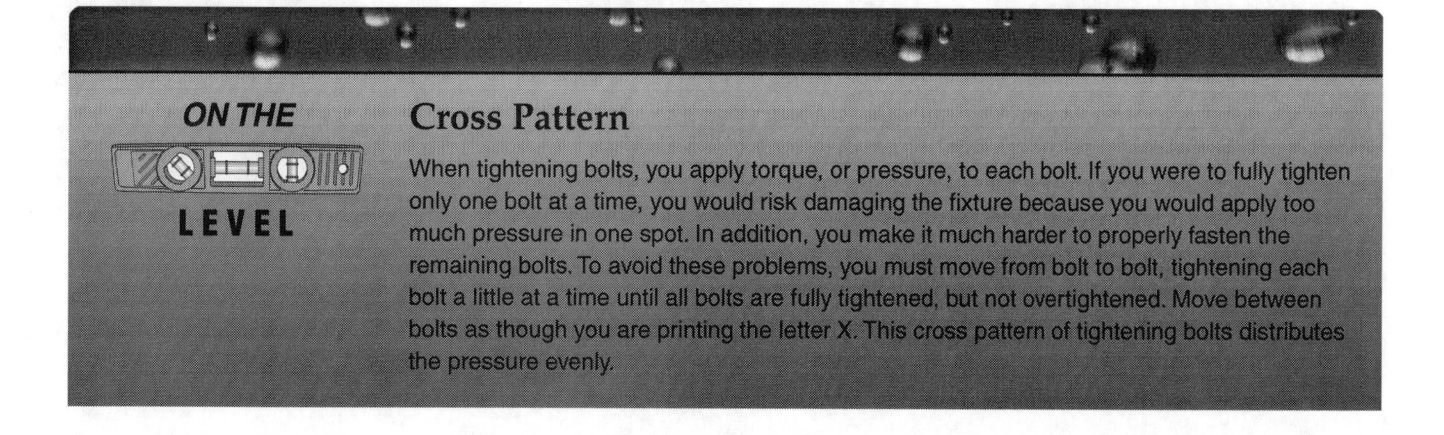

Cross Pattern

When tightening bolts, you apply torque, or pressure, to each bolt. If you were to fully tighten only one bolt at a time, you would risk damaging the fixture because you would apply too much pressure in one spot. In addition, you make it much harder to properly fasten the remaining bolts. To avoid these problems, you must move from bolt to bolt, tightening each bolt a little at a time until all bolts are fully tightened, but not overtightened. Move between bolts as though you are printing the letter X. This cross pattern of tightening bolts distributes the pressure evenly.

Step 3 Slide the valve onto the pipe until it hits the internal stop.

Step 4 Hold the valve and start threading the compression nut onto the valve.

Step 5 Line up the pipe to the connection to ensure that the nut is not cross threaded.

The compression ring is squeezed between the valve and the nut and compresses onto the pipe. When sufficient pressure is applied, the compression ring cannot be removed from the pipe. This type of joint is very quick and easy to make.

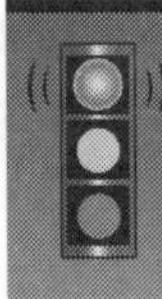

WARNING!
Do not install compression connection valves if the pipe is too short for the fitting. The fitting could blow off when under operating pressure and cause serious injury or even death.

4.6.0 Installing Faucets

As you learned in *Plumbing Level One,* faucets are classified as either compression faucets or noncompression faucets. Compression faucets are widely used in kitchen sinks and bathroom lavatories. Typically, compression faucets include two compression valves and a mixer that mixes the hot and cold water and delivers it through a mixing spout common to both valves. Noncompression faucets regulate water flow by obstructing the water with movable discs. When the faucet valve is open, openings in two parallel discs are matched. This allows the water to flow through the disc openings and through the faucet outlet. When the faucet valve is closed, the discs rotate so that their openings are no longer lined up. This cuts off the flow of water.

Regardless of the type of faucet being installed, always follow the manufacturer's specifications and instructions. Before installing a faucet, refer to the approved submittal data to ensure that the correct type and model of faucet is being used for the installation. The following are the three most common types of faucets that you will install:

- Cartridge faucets
- Rotating ball faucets
- Ceramic disc faucets

4.6.1 Cartridge Faucets

You can identify a cartridge faucet by the metal or plastic cartridge inside the faucet body. Many single-handle faucets and showers are cartridge designs. Installing a cartridge faucet is fairly easy (see *Figure 11*).

4.6.2 Rotating Ball Faucets

Like cartridge faucets, ball faucets have a single handle. However, instead of a cartridge inside the faucet body, they use a metal or plastic ball to regulate flow. The individual parts of a rotating ball faucet are illustrated in *Figure 12*.

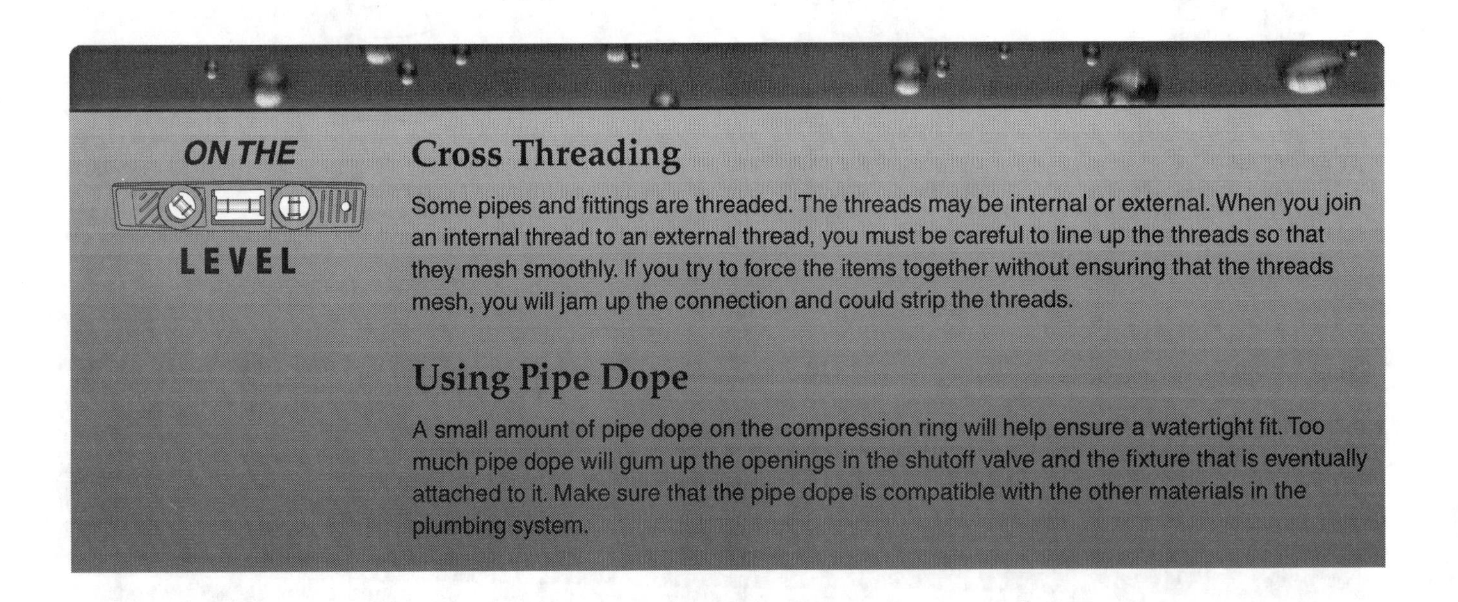

Cross Threading

Some pipes and fittings are threaded. The threads may be internal or external. When you join an internal thread to an external thread, you must be careful to line up the threads so that they mesh smoothly. If you try to force the items together without ensuring that the threads mesh, you will jam up the connection and could strip the threads.

Using Pipe Dope

A small amount of pipe dope on the compression ring will help ensure a watertight fit. Too much pipe dope will gum up the openings in the shutoff valve and the fixture that is eventually attached to it. Make sure that the pipe dope is compatible with the other materials in the plumbing system.

ON THE LEVEL

Faucet Orientation

When fixtures have separate side-by-side controls for hot and cold water, codes require that the valve and piping for the hot water be installed on the left side of the fixture.

Single-control valves, widely used in tubs and showers, will have the correct orientation marked on the fitting. Refer to the manufacturer's specifications to ensure the fitting is installed with the correct orientation.

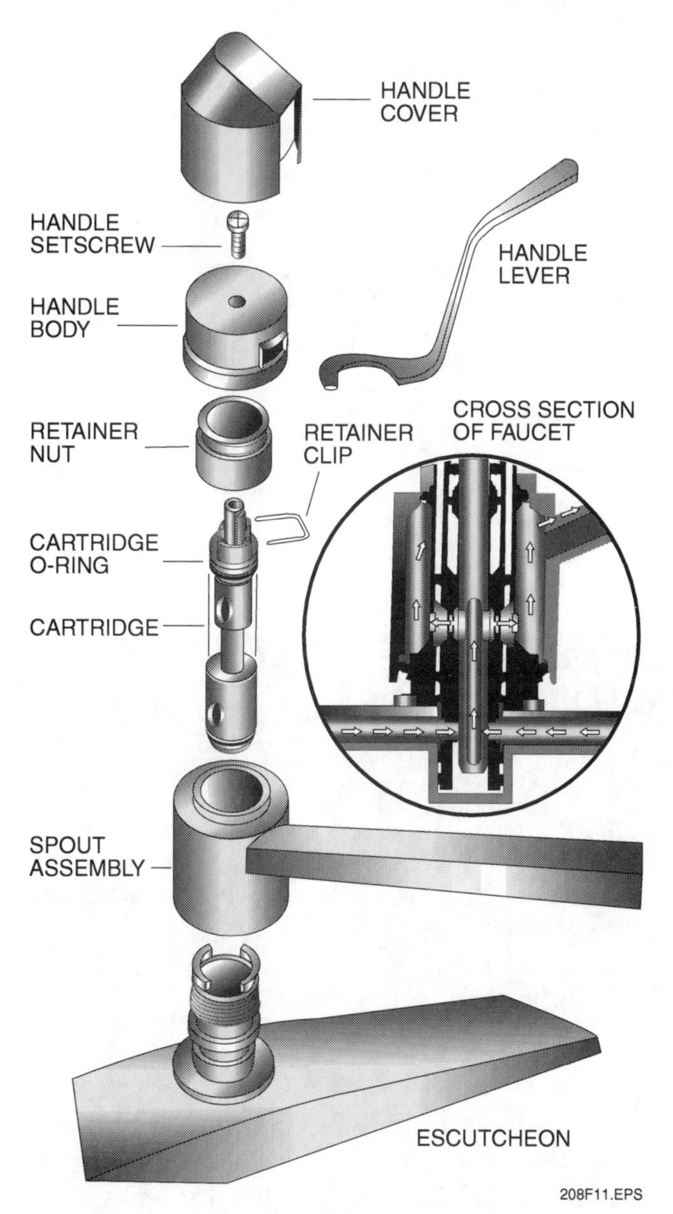

Figure 11 ◆ Cartridge faucet.

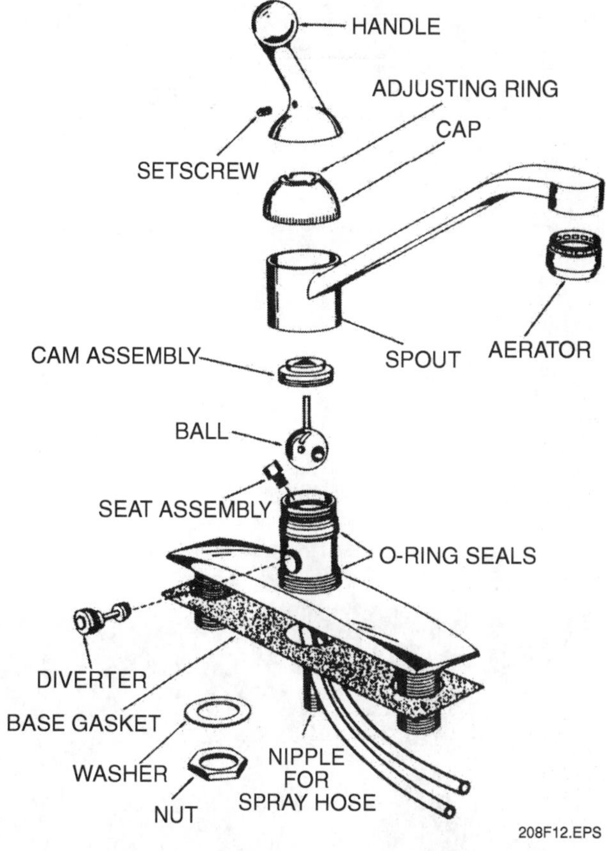

Figure 12 ◆ Rotating ball faucet.

4.6.3 Ceramic Disc Faucets

A ceramic disc faucet is a single-handle faucet that has a wide cylinder inside the faucet body. The cylinder contains a pair of closely fitting ceramic discs that control the flow of water. The top disc slides over the lower disk. The individual parts of the ceramic disc faucet are illustrated in *Figure 13*.

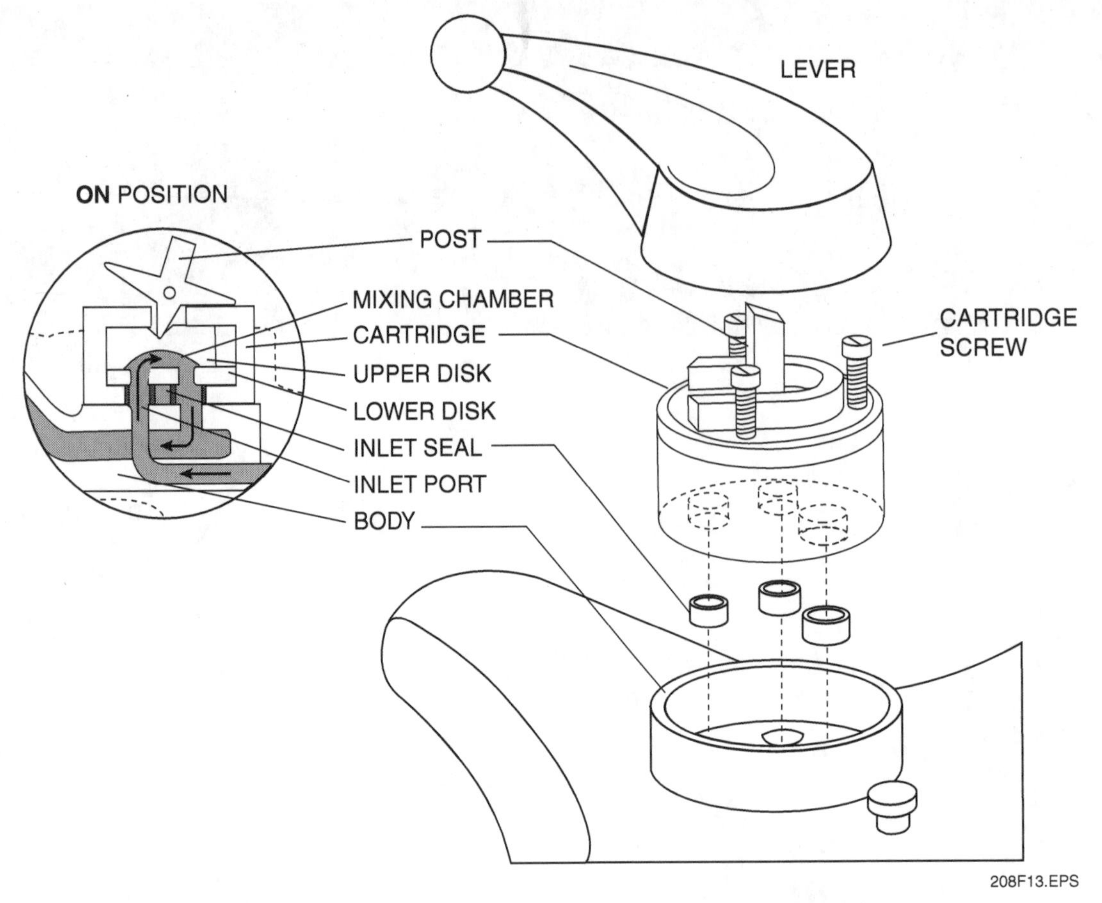

Figure 13 ◆ Ceramic disc faucet.

Review Questions

Sections 1.0.0–4.0.0

1. Faucets are available in both _____ and _____ varieties.
   a. residential; industrial
   b. flushometer; flush valve
   c. floor-mounted; wall-hung
   d. self-rimming; rim-mounted

2. The best way to prevent conflict about who is at fault if fixtures are damaged is to inspect them _____.
   a. just before they are installed
   b. after they are installed
   c. on the day before they are installed
   d. when they are delivered

3. When installing a bathtub/shower unit, fit the shower riser, elbow, shower arm, and head _____ the filler drop, elbow, nipple, and spout.
   a. after
   b. instead of
   c. above
   d. before

4. Bending a fiberglass shower unit during installation _____.
   a. is recommended to test for leaks after installation is done
   b. usually is required to support bearing points properly
   c. may result in cracks that become visible later
   d. will make it more secure in its bracing

Review Questions

Match each of the valve connections with the appropriate installation technique.

5. _____ Threaded
6. _____ Soldered
7. _____ Solvent welded
8. _____ Flanged
9. _____ Compression
   a. Lower the threaded base through the bottom hole, place the washer and nut on the base of the assembly, and tighten.
   b. After starting the threads by hand, tighten the valve with a wrench.
   c. Line up the bolt holes, then install and tighten the bolts in a cross pattern.
   d. Clean the surfaces to be joined with emery paper and a clean cloth. Then apply primer and solvent to the stubout and valve.
   e. Ream the pipe ends and clean the metal with a copper-cleaning tool, then remove the valve interior.
   f. Place the nut and then the ring on the pipe, then slide the valve on until it hits the internal stop.

10. When installing a threaded valve, a more watertight fit may result if you _____.
    a. wrap several inches of pipe with Teflon® tape
    b. put a small amount of pipe dope into the valve seat
    c. apply generous amounts of pipe dope to the threads
    d. use a small amount of pipe dope or Teflon® tape on the threads

11. _____ valves must be installed at the correct angle, or they will not operate correctly.
    a. Check
    b. Soldered
    c. Threaded
    d. Shutoff

12. A _____ valve is installed on a water heater to prevent an explosion due to overheating.
    a. check
    b. relief
    c. flanged
    d. control

13. Compression connection valves consist of three parts: the _____, the _____, and the _____.
    a. valve body; compression ring; thread adapter
    b. valve body; compression ring; compression nut
    c. compression ring; ferrule; washer
    d. compression ring; ferrule; bolt

14. Faucets are classified as either _____ faucets or _____ faucets.
    a. compression; noncompression
    b. cartridge; ceramic disk
    c. single-action; double-action
    d. single-handled; double-handled

5.0.0 ◆ INSTALLING VALVES FOR WATER CLOSETS AND URINALS

Three types of valves are used in water closets and urinals to control the flow of water into the fixture's bowl:

- Float-controlled valves (ball cocks)
- Flush valves
- Flushometers

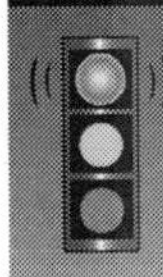

WARNING!
Codes prohibit the installation of urinals that do not have a visible trap seal. Such urinals present a threat to public health.

5.1.0 Installing Float-Controlled Valves (Ball Cocks)

Float-controlled valves, also called ball cocks, are installed in water closet flush tanks and maintain a constant water level in the tank to control flow (see *Figure 14*). When the water level in the tank drops, the valve lifts away from the seat as the float arm lowers. As the water level in the tank increases, the float arm rises and forces the valve down against the seat. This closes the valve and stops the flow of water.

To install the float valve assembly, follow these steps:

Step 1 Turn off the water supply.

Step 2 Lower the threaded base through the bottom hole in the closet tank with the gasket in place against the flange.

Step 3 Place the washer and nut on the base of the assembly on the outside of the tank and tighten. Do not overtighten the nut.

Step 4 Place the riser and coupling nut on the base of the float valve, and hand tighten.

Step 5 Align the riser with the shutoff valve. Remove the riser and cut it to fit.

Step 6 Reassemble the riser to the tank and valve, then turn on the water supply.

Step 7 Adjust the float in the tank to achieve the water level indicated on the inside of the tank.

Step 8 Adjust the float with the adjusting screw. Do not bend the float arm to adjust the water level.

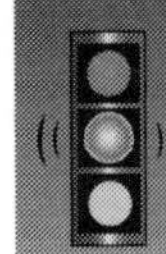

CAUTION
Never overtighten the nut on the base of the assembly on the outside of the tank. Overtightening puts stress on the fixture and can cause it to crack.

Figure 14 ◆ Float-controlled valve.

5.2.0 Installing Flush Valves

Flush valves are also installed in water closet flush tanks. They control the flow of water from the tank into the bowl (see *Figure 15*). Pushing the tank lever causes the valve to lift above the tank outlet and float. This action allows the water in the tank to flow rapidly into the bowl. When the water level in the tank drops to the point where the valve will no longer float, the valve reseats on the tank drain and the tank refills with water.

To install a flush valve, follow these steps:

Step 1 Turn off the water supply.

Step 2 Place the flush valve assembly into the center hole in the water closet tank. Secure the gasket and lock nut from the outside.

Step 3 Place the spud gasket into the hole on top of the bowl, and lower the tank into place.

Step 4 Place tank bolts with gaskets through holes in the tank's interior and bowl and install washers and nuts. Tighten firmly, but do not overtighten. Overtightening can damage the tank.

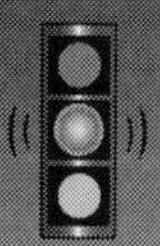

CAUTION

Never bend the float arm to adjust the water level. The rod may work around a half-turn (180 degrees) and cause the tank to overflow. Use the correct tool when adjusting the float arm to avoid marring the tank finish.

208F15.EPS

Figure 15 ◆ Manual flush valve.

5.3.0 Installing Flushometers

Plumbers install flushometers for water closets or urinals. They can be operated either manually (see *Figure 16*) or electronically (see *Figure 17*). Flushometers do not require storage tanks because the water flows, under pressure, directly into the fixture. The scouring action of the water created by the flushometer generally cleans more effectively than the gravity flow from a storage tank.

Flushometers are popular in commercial installations for two reasons:

- They can be connected directly to the water supply pipe.
- They can be flushed repeatedly without having to wait for a storage tank to refill between flushes.

Flushometers are available in either a diaphragm type or a piston type. The diaphragm type stops the

208F16.EPS

Figure 16 ◆ Manual flushometer.

DID YOU KNOW?

Valves Are Big Business

The U.S. industrial valve industry records annual sales of approximately $2.7 billion. It employs more than 40,000 people. The industry is made up of mostly small and medium-sized businesses.

Manufacturers invest a lot of money in research and development and in automation. They also invest in special high-tech equipment to test valves for their resistance to fire, noise, wear, corrosion, movement, and temperature. Many valve companies buy their castings, but some companies operate their own foundries. They design, develop, and produce raw castings that will be machined into valve bodies and parts.

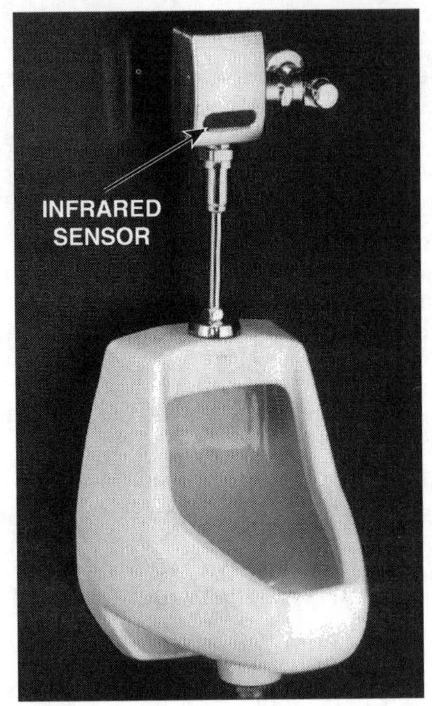

208F17.EPS

Figure 17 ◆ Electronic flushometer installed on a siphon-jet urinal.

flow of water when water pressure in the upper chamber forces the diaphragm against the valve seat. Motion of the handle in any direction pushes the plunger against the auxiliary valve. Even the slightest tilt of the handle will transfer water from the upper chamber around the diaphragm.

The diaphragm rises when pressure decreases in the upper chamber, allowing water to flow into the fixture. While the fixture is in the flushing mode, a small amount of water flows through the bypass. As the water fills the upper chamber, the diaphragm is forced to reseat against the valve seat. This action shuts off the flow of water to the fixture. The piston-type flushometer works in much the same way the diaphragm type works.

Install a control stop and, if required, a vacuum breaker on the flushometer. Control stops control the valve's flow rate and act as a shut-off. Vacuum breakers prevent back siphoning of polluted water into the water supply piping.

Install the control stop (see *Figure 18*) in the water supply line serving the flushometer. The control stop regulates the flow of water entering the flushometer. It also allows you to shut off the water supply to individual fixtures to make repairs. Install the vacuum breaker (see *Figure 19*) between the flushometer outlet and the flushometer tube according to the manufacturer's instructions.

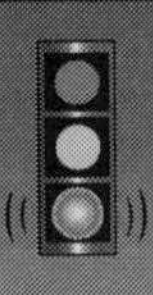

NOTE

Vacuum breakers may not be required on flushometers that use nonpotable water. This type of installation is rare in the U.S. Refer to your local applicable code.

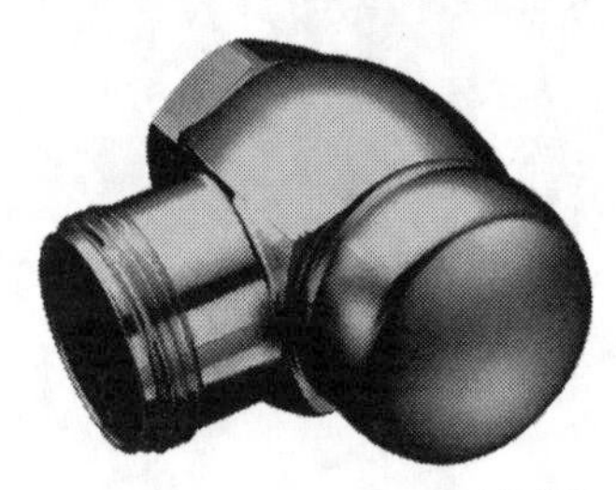

208F18.EPS

Figure 18 ◆ Flushometer control stop.

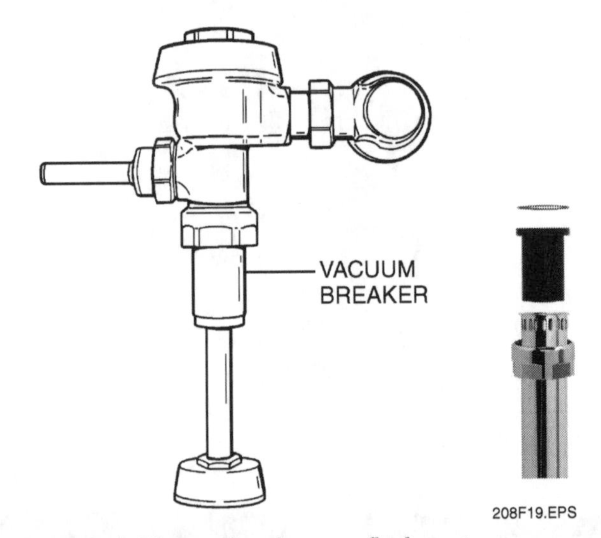

208F19.EPS

Figure 19 ◆ Vacuum breaker on a flushometer.

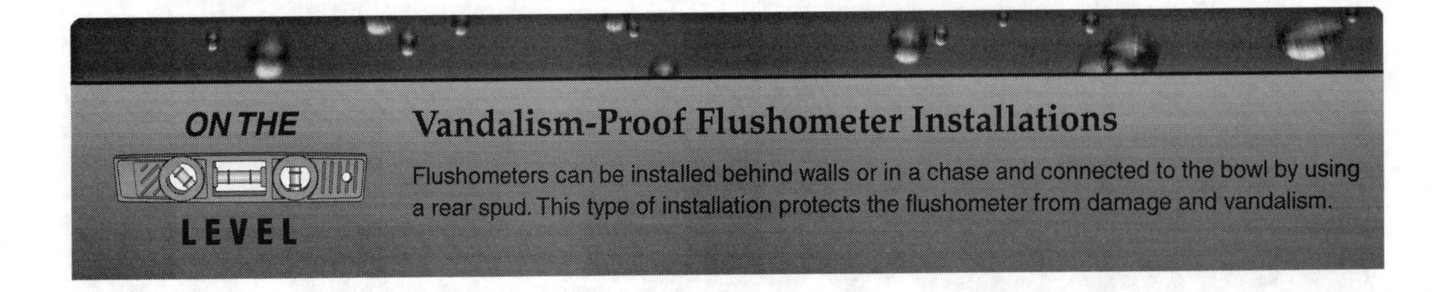

Vandalism-Proof Flushometer Installations

Flushometers can be installed behind walls or in a chase and connected to the bowl by using a rear spud. This type of installation protects the flushometer from damage and vandalism.

To install control stop connections, follow these steps:

Step 1 Cut the stubout to its proper length.

Step 2 Solder the male sweat adapter to the tube.

Step 3 Slip the covering tube over the male sweat adapter and stub, and insert the wall flange over the covering tube.

Step 4 Lock the setscrew over the covering tube to prevent it from sliding back into the wall cavity.

Step 5 Thread the angle stop to the stubout.

Step 6 Attach the flushometer (see *Figure 20*). Do this carefully to avoid damaging the chrome exterior on the coupler nut and on other parts of the valve.

Step 7 Install the vacuum breaker, which is usually a part of the fixture connector.

Step 8 Secure the connector to the valve and fixture.

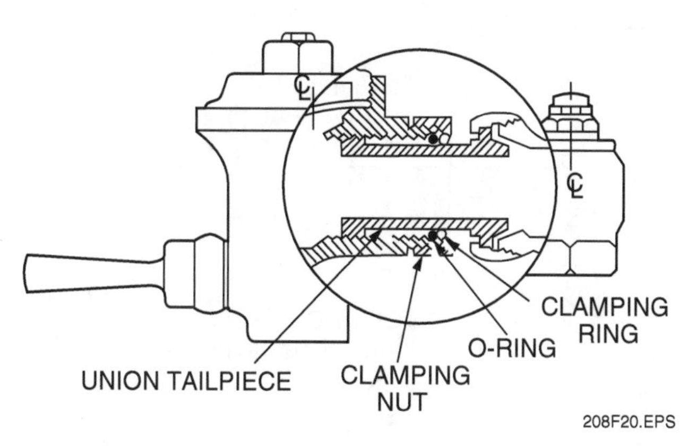

Figure 20 ◆ Flushometer installation.

Flushometers can be adjusted for both water consumption and noise. Check the manufacturer's instructions for more information. Pressure adjustments can also be made; you can find the ideal operating pressure for the valve by flushing and adjusting. As you gain experience, you will know how much to adjust the flushometer to get the ideal operating pressure. Install a pressure reducer if line pressure exceeds the recommended pressure. Low pressure is also a problem. Although many valves can function at a low pressure, they may not function well. Local plumbing codes include sizing criteria that specify the pressures allowed in your area.

6.0.0 ◆ INSTALLING LAVATORIES, SINKS, AND POP-UP DRAINS

Installation procedures for lavatories are similar to those for sinks. Fixtures may be either wall-hung or built-in. The method of attachment will affect how they are supported. Built-in fixtures are placed in a countertop, and you may have to cut the opening.

6.1.0 Wall-Hung Lavatories and Sinks

The following installation steps are provided as a general guide. Always read and follow the manufacturer's instructions to determine the specific instructions for each fixture.

The steps for installing wall-hung lavatories and sinks are as follows:

Step 1 Check the fixture for damage.

Step 2 Check the rough-in dimensions (see *Figure 21*).

Step 3 Secure the bracket to the framing.

Step 4 Place the lavatory or sink on the bracket temporarily and check it for level.

Step 5 Remove the fixture and adjust the bracket as necessary.

Step 6 Install the faucets, drain, and any other trim. Additional information about this part of the task is included in the manufacturer's directions supplied with the parts.

Step 7 Secure the fixture to the bracket and seal the joint between the wall and the fixture with the appropriate sealer.

Step 8 Attach the water supply.

Step 9 Install the waste piping.

Step 10 Turn on the water and test for leaks.

Step 11 Turn off the water and cover the fixture to protect it from damage.

6.2.0 Built-In Lavatories and Sinks

The following installation steps are provided as a general guide. Always read and follow the manufacturer's instructions to determine the specific instructions for each fixture.

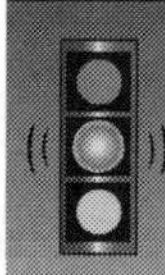

CAUTION

Protect the countertop from scratches. Make certain the base of your saw is free of burrs and sharp edges.

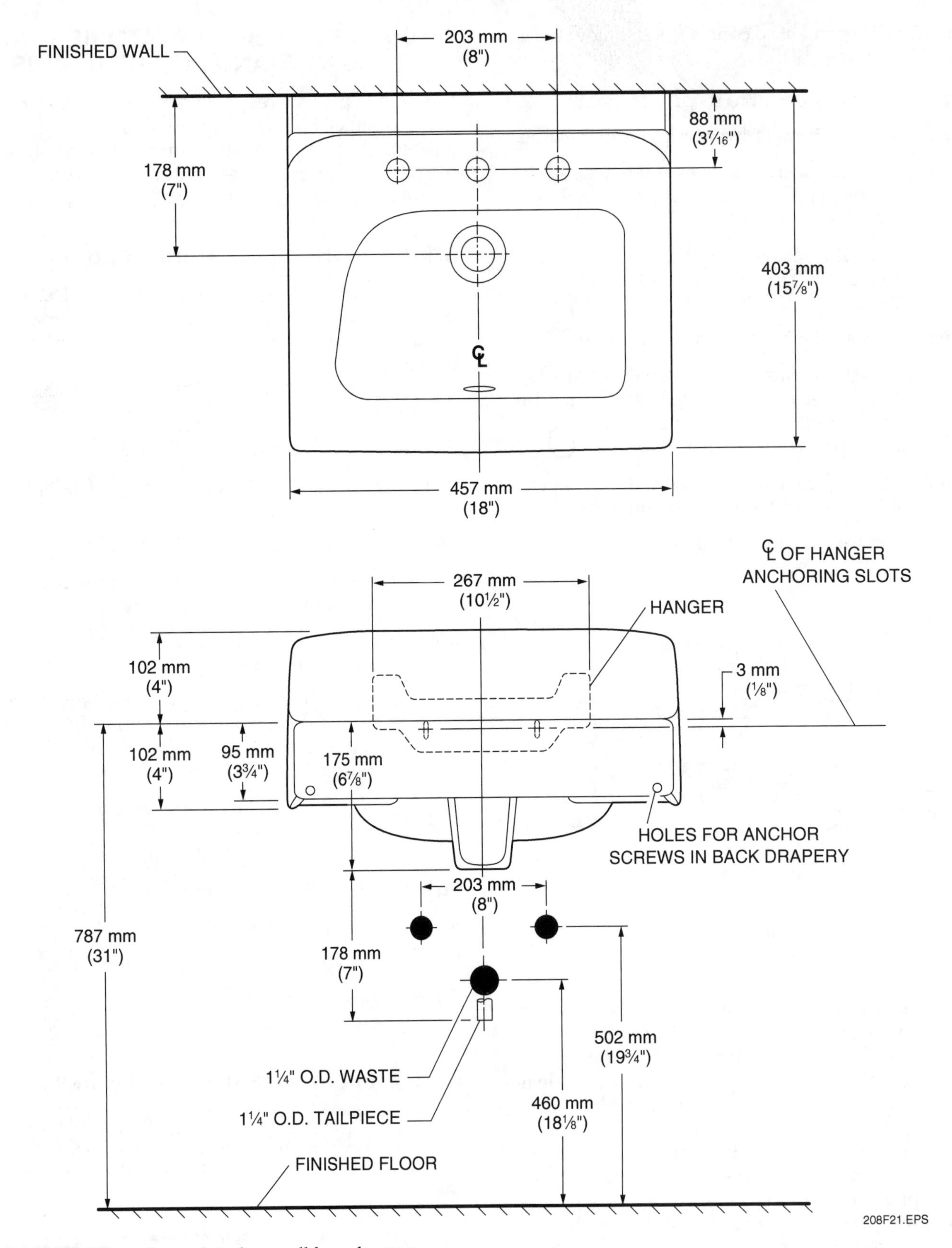

Figure 21 ◆ Sample rough-in for a wall-hung lavatory.

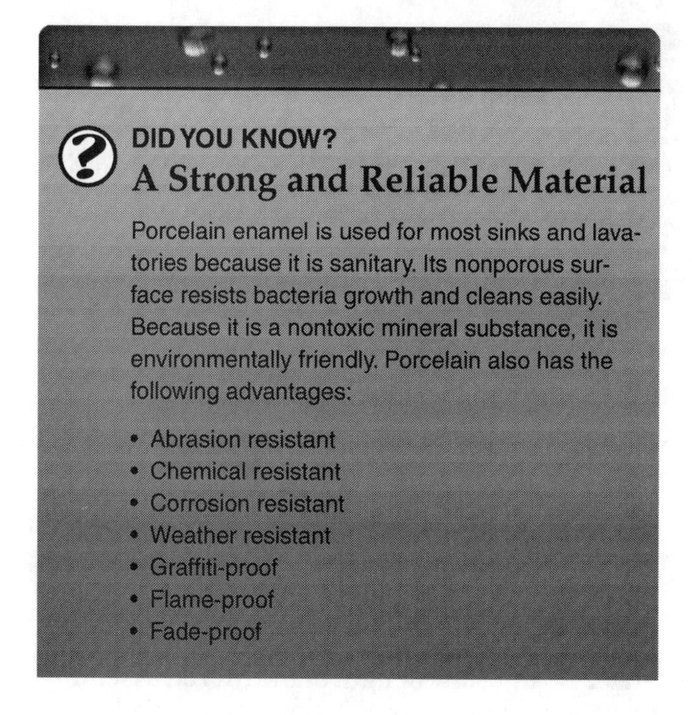

DID YOU KNOW?

A Strong and Reliable Material

Porcelain enamel is used for most sinks and lavatories because it is sanitary. Its nonporous surface resists bacteria growth and cleans easily. Because it is a nontoxic mineral substance, it is environmentally friendly. Porcelain also has the following advantages:

- Abrasion resistant
- Chemical resistant
- Corrosion resistant
- Weather resistant
- Graffiti-proof
- Flame-proof
- Fade-proof

Built-in lavatories and sinks are placed in an opening cut into a cabinet's countertop. Three styles are available (see *Figure 22*):

- Rim mounted
- Under the counter
- Self rimming

Although the cabinetmaker usually cuts the opening in the countertop, on some projects the plumber may have to. Before cutting out the opening, ensure that the cabinet's structural parts will not interfere with the installation. Provide enough space to allow access to the clips that fasten the fixture to the bottom of the countertop.

The steps for installing built-in lavatories and sinks follow:

Step 1 Measure the correct position and trace the opening on the countertop. Most manufacturers provide a pattern for this step. If a pattern is not provided, turn the fixture over, position it on the countertop, and carefully trace a line around the metal rim. If you use this process, make allowances for the rim. Otherwise, the cutout will be too big.

Step 2 Place the fixture to one side. Drill a starter hole at a point along the line where you would logically begin cutting. Insert the blade of the saber saw into the hole and make the cutout.

RIM MOUNTED

UNDER THE COUNTER

SELF RIMMING

208F22.EPS

Figure 22 ◆ Residential lavatories.

Step 3 Test fit the rim in the opening. If the fixture is self rimming, insert the fixture in the opening to test the fit.

Installation Tip

Install the faucets on the sink or lavatory before you install the sink or lavatory itself. This makes it easier for you to reach the nuts on the underside. Once the fixture is installed, there won't be much room between the bottom of the bowl and the rear of the cabinet to use tools.

Step 4 Make any adjustments necessary to the shape of the opening.

Step 5 Install the faucets and other trim.

Step 6 Apply fixture sealant to the rim. Ensure that the sealant is applied uniformly. Perform this step carefully and completely to ensure that water will not flow under the fixture.

Step 7 Place the fixture in the opening and adjust its position.

Step 8 Install and uniformly tighten the clamps that hold the fixture in place from underneath the countertop.

Step 9 Connect the water supply and the waste piping.

Step 10 Turn on the water and check for leaks.

6.3.0 Installing Pop-Up Drains

Most lavatories come with **pop-up drains.** These are more attractive than rubber stoppers and won't get lost. They come in finishes to match the faucets. To install this drain, follow these steps (see *Figure 23*):

Step 1 Attach the pivot rod assembly to the lift rod.

Step 2 Adjust the length of the **lift rod assembly** by loosening the clevis screw and moving the clevis so that the pop-up plug is closed when the lift knob is up and open when the lift knob is down.

Step 3 Check the retaining nut that secures the pivot rod for leaks.

Step 4 Tighten the retaining nut slightly to correct a leak, but do not overtighten the nut. If the nut is too tight, the pop-up plug may be difficult to operate.

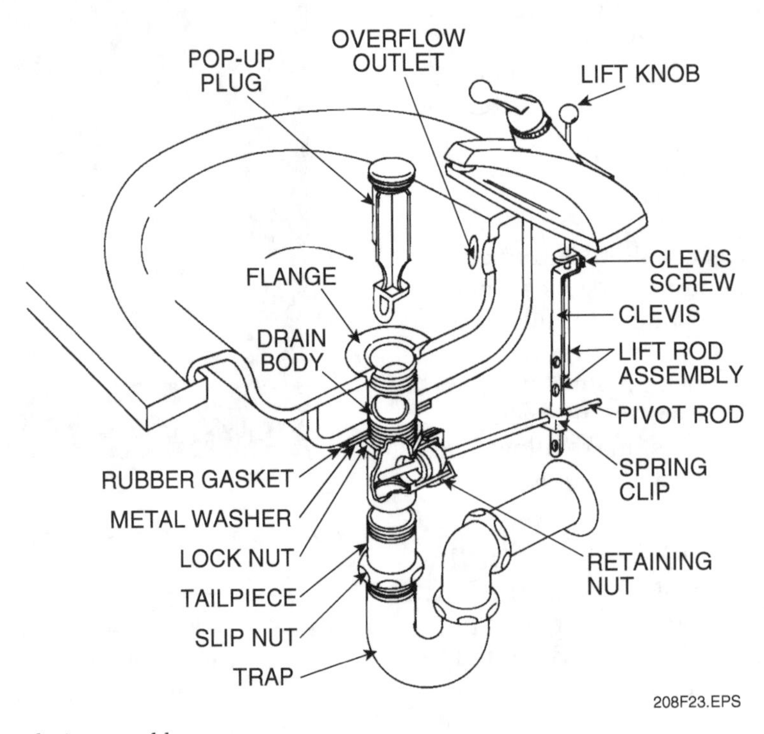

Figure 23 ◆ Typical pop-up drain assembly.

DID YOU KNOW?

American Inventors and the Water Closet

The first Americans awarded a patent for a water closet were James Henry and William Campbell. In 1857, they developed a plunger closet. This unit was not very sanitary, and some of the industry's earliest pioneers shunned it.

Patent grants for water closets increased as inventors realized the potential market for an improved model. John Mann got a patent for his three-pipe siphonic closet in 1870. In 1876, William Smith earned his for a jet siphon closet. Thomas Kennedy improved on Mann's design and patented a siphonic closet that needed only two delivery pipes. One flushed the rim and the other started the siphon. More improvement occurred in 1890 with William Howell's water closet. His model eliminated the lower trap.

The U.S. Patent Office received applications for 350 new water closet designs between 1900 and 1932. Two of the first were granted to Charles Neff and Robert Frame. These New Englanders produced a siphonic wash-down closet. But their unit's bowl sometimes overflowed. Around 1910, Fred Adee redesigned that bowl, eliminating the messy overflows, and gave birth to the siphonic closet in America.

In the early 1900s, the U.S. Patent Office granted patents for the flushometer valve, a backflow preventer, a wall-mounted closet with a blowout arrangement, a tank that rests on the bowl, and reverse trap toilets.

7.0.0 ◆ INSTALLING WATER CLOSETS

Water closets come in a variety of styles and materials. The following installation instructions are a general guide. Remember to always read and follow the manufacturer's instructions.

Before installing the water closet, check the condition of the floor. Don't set the water closet in place until a nonporous floor covering is installed. In older buildings, part of the floor may have to be replaced. The floor must carry the weight of the fixture and the person using it, so be sure that it is solidly constructed.

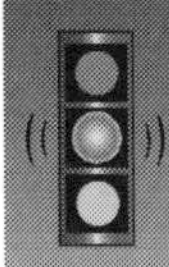

CAUTION

Never overtighten the bolts on the bowl or tank. Overtightening puts stress on the fixture and can cause it to crack.

To install a water closet, follow these steps:

Step 1 Check the rough-in opening and closet flange dimensions.

Step 2 Check the bowl or tank for defects and cracks.

Step 3 Set the **closet bolts** in the closet flange with the threads up. Many plumbers use push-and-twist closet bolts that eliminate wobbling, such as Hercules Chemical Company's johni quick bolts®.

Step 4 Turn the bowl bottom up and position the **wax seal** (see *Figure 24*) or rubber water closet gasket (see *Figure 25*).

Step 5 Apply a bead of fixture sealant along the outside rim of the bowl.

Step 6 Set the bowl over the closet flange. Make certain the closet bolts extend through the holes in the base of the bowl.

Step 7 Press the bowl firmly into position. Apply your full weight to ensure that the wax ring is completely sealed (see *Figure 26*).

Step 8 Check the bowl for level. Shim as required to level the bowl. If shims are used, grout the space between the bottom of the bowl and the floor to support the bowl evenly.

Step 9 Secure the bowl with the closet bolts. Stop tightening the bolts when you can no longer rock the fixture. Don't overtighten the bolts.

Step 10 Cut off the excess bolt length, and install the bolt caps.

Step 11 Check the tank to ensure that the ball cock and flush valve are completely and properly installed.

Step 12 Install the tank, and hand tighten the tank bolts. Follow the manufacturer's directions to make sure that washers and seals are installed correctly.

Step 13 Make sure the tank is level.

Step 14 Tighten the tank bolts until the tank is secure. Do not overtighten.

Step 15 Hook up the water supply and fill the tank.

Step 16 Flush the water closet several times. Adjust the float valve to control the water level and check the entire assembly for leaks.

Step 17 Install the closet seat securely.

Step 18 Turn off the water, cover the fixture, and post a sign marked DO NOT USE to protect the fixture until the project is finished.

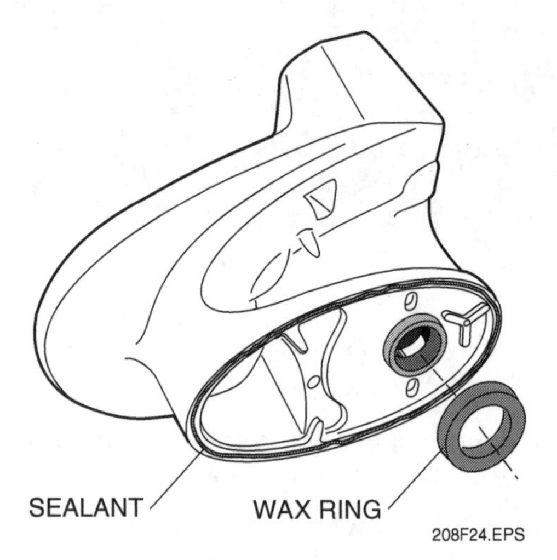

Figure 24 ◆ Location of the wax seal.

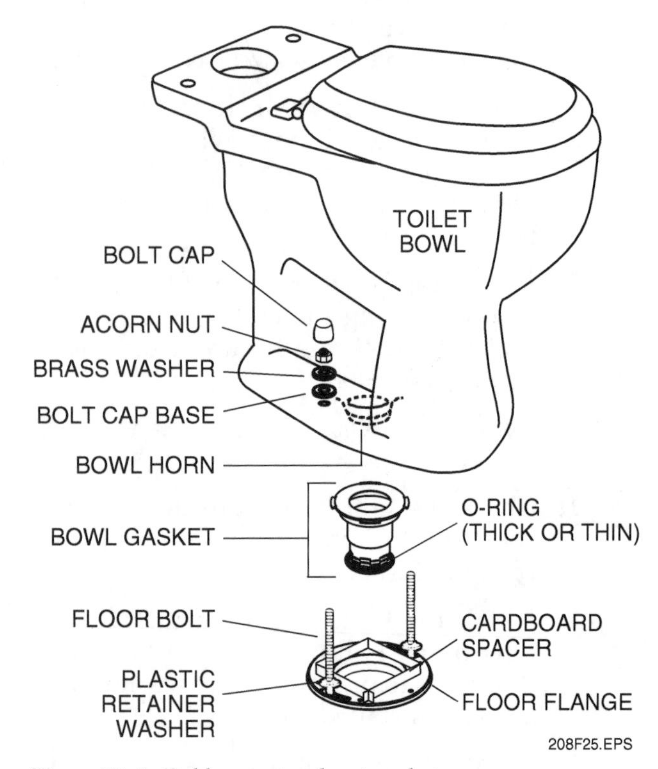

Figure 25 ◆ Rubber water closet gasket.

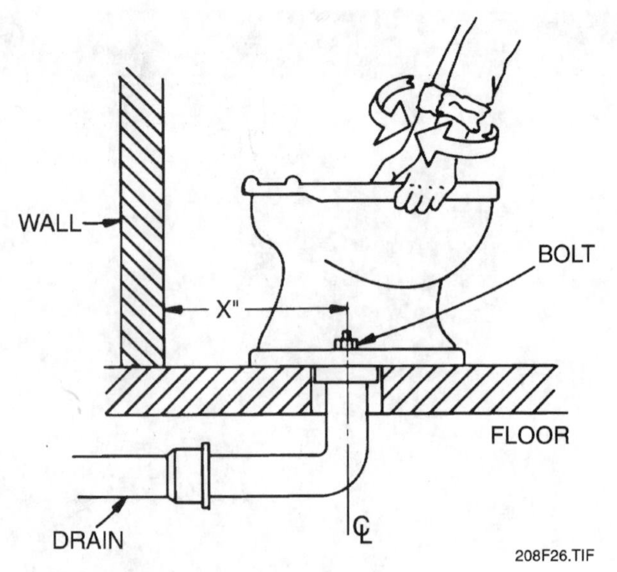

Figure 26 ◆ Seating the bowl.

8.0.0 ◆ INSTALLING URINALS

As with the other fixtures covered in this module, read and follow the manufacturer's instructions to ensure that the unit is installed correctly, thereby protecting the warranty. The installation steps included here are intended to serve as a general guide. Wall-hung urinal installations will also require a flushometer and a drain (see *Figure 27*).

To install a wall-hung urinal, follow these steps:

Step 1 Unpack and inspect the urinal for damage, defects, and cracks.

Step 2 Install the drain fitting.

Step 3 Install hangers at the correct height, if they are required. Otherwise, locate bolts to fasten the fixture to the wall framing.

Step 4 Bolt the urinal to the wall, making certain that it is level.

Step 5 Install the flush valve. Work carefully to avoid scratching or otherwise damaging the chrome-plated parts.

Step 6 Connect the waste piping.

Step 7 Turn the water on, and adjust the flushometer.

Step 8 Flush the unit several times to confirm that it works correctly.

Step 9 Turn off the water, cover the fixture, and post a sign marked DO NOT USE to protect the fixture until the project is finished.

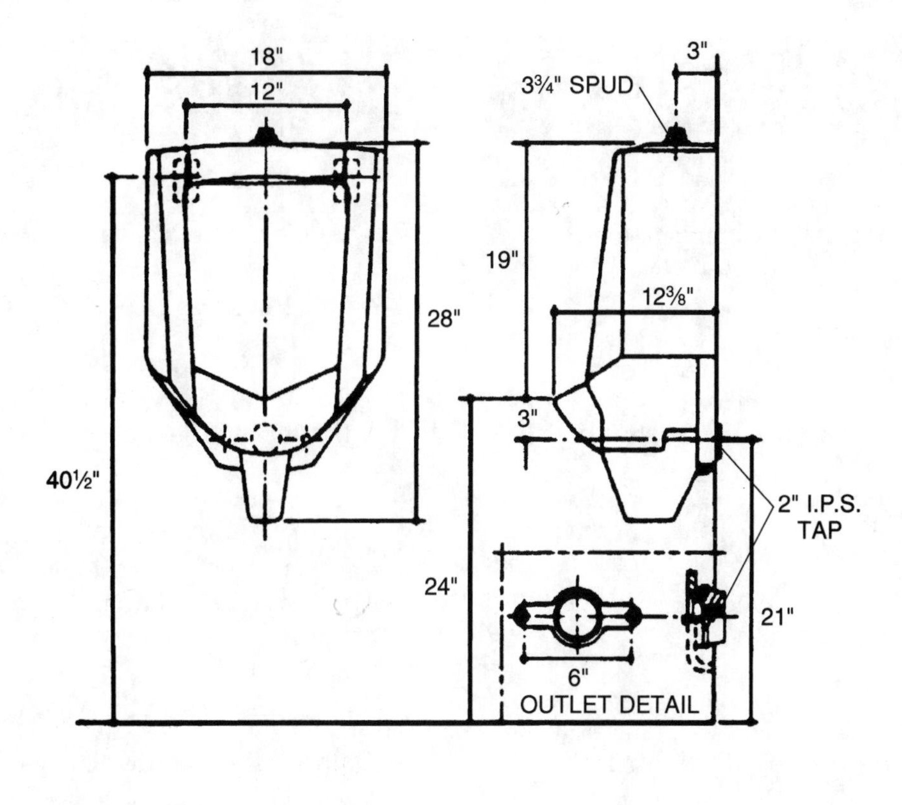

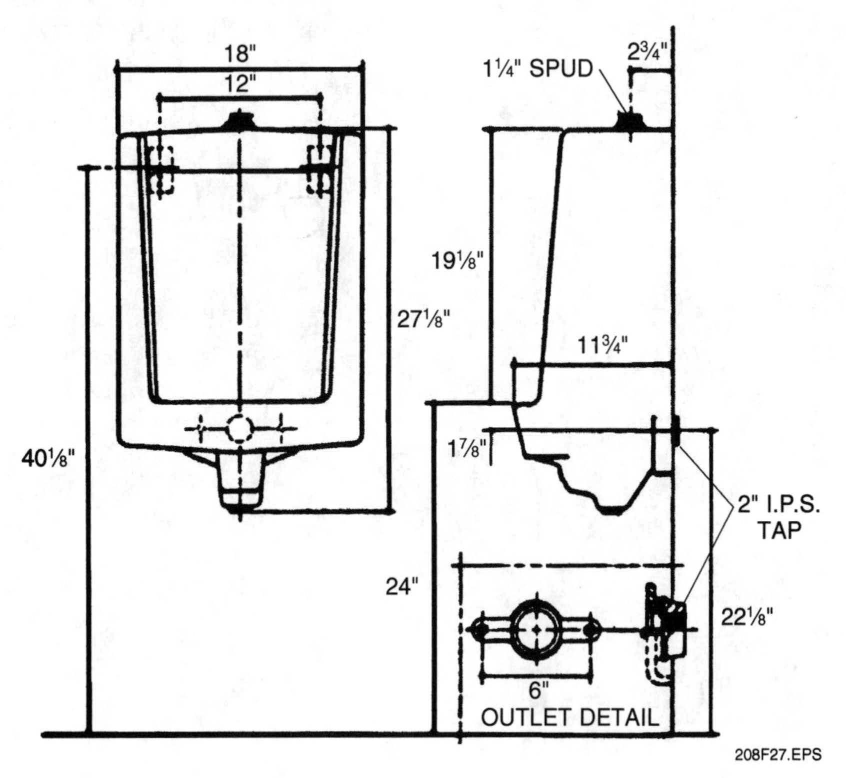

Figure 27 ◆ Sample rough-in of a wall-hung urinal.

Review Questions

Sections 5.0.0–8.0.0

1. Overtightening the nut on the base assembly of a water closet on the outside of the tank _____.
   a. is impossible
   b. is required for most water closets
   c. may cause the fixture to crack from stress
   d. may prevent leaks from the shutoff valve

2. When installing a float-controlled valve, the gasket should not be in place against the flange when you lower the threaded base through the bottom hole in the closet tank.
   a. True
   b. False

3. When installing flush valves, tank bolts are placed through holes in the tank's interior.
   a. True
   b. False

4. A _____ delivers water to a urinal or water closet without a storage tank.
   a. flushometer
   b. flush valve
   c. float valve
   d. pressure valve

5. The _____ regulates the flow of water entering the flushometer.
   a. piston
   b. diaphragm
   c. control stop
   d. vacuum breaker

6. Built-in lavatories are placed _____.
   a. on floor mountings
   b. back-to-back only
   c. in a countertop
   d. only where specified by code

7. A built-in sink or lavatory should be installed _____.
   a. at the same time as the shower or bathtub
   b. as soon as the cabinet is available
   c. before faucets are installed
   d. after faucets are installed

8. The pivot rod retaining nut on a pop-up drain should be installed _____.
   a. as tightly as possible
   b. just tightly enough to prevent leaking
   c. as loosely as possible without falling off
   d. so that the plug opens when the lift knob is up

9. When using shims to level a water closet, you should _____.
   a. fill the gap between the bottom of the bowl and the floor with wood or plastic
   b. tighten the closet bolts until the fixture does not rock
   c. ensure that the wax ring is completely sealed
   d. grout the space between the bottom of the bowl and the floor

10. To install a wall-hung urinal, you should install the drain fitting _____.
    a. before installing hangers or bolts
    b. before placing the washer and nut on the base of the assembly
    c. after connecting the waste piping
    d. after installing closet bolts

Summary

In this module, you learned the basic procedures, tools, and techniques necessary to install a wide range of fixtures, valves, and faucets used in residential, commercial, and institutional applications. While the particular installation requirements will vary depending on the fitting being installed, there are some common elements for each of them. Refer to the project plans, to the approved submittal data, and to your local applicable code to ensure that you are using the correct fitting. Always follow the manufacturer's instructions to ensure that the warranty is not voided.

Professional results require professional attention to detail. Protect all fixtures before, during, and after the installation procedure. Check the project specifications to ensure that the right fixtures, with all parts intact, arrive at the project site. Always read and follow the manufacturer's instructions to ensure proper installation and to protect the warranty. Inspect and test installed fixtures to ensure that they are intact, work properly, and don't leak. Finally, clean up your work area. Remove and dispose of unused materials and other waste. Your co-workers, the trades coming in after you, and especially your customers will appreciate this professional courtesy.

The installation of fixtures, valves, and faucets is your finish work. Once the walls are closed up, these fittings are the only part of your plumbing project that everyone sees every day, during the course of normal activities. Therefore, you should perform this work neatly, accurately, and completely the first time. This will prevent costly callbacks. Your work will be viewed as competent and professional, and it will reflect well on you and your company.

Notes

Trade Terms Introduced in This Module

Blocking: Pieces of wood used to secure, join, or reinforce structural members or to fill spaces between them.

Callback: term that refers to being called back to the worksite to repair or replace defective materials or to correct faulty workmanship.

Closet bolt: A bolt with a large-diameter, low circular head that is cupped on the underside so that it is sealed against the surface when the bolt is tightened. It is used to fasten a water closet bowl to the floor.

Cross pattern: A method for tightening bolts or screws in an even sequence, tightening each bolt or screw a little at a time to avoid putting too much pressure on any one bolt or screw.

Cross threading: A condition that occurs when the initial female thread on a pipe or fitting does not properly mesh with the initial male thread. This causes the connection to jam and can strip the threads.

Escutcheon: A flange on a pipe used to cover a hole in a floor or wall through which the pipe passes.

Finish work: The completion phase of any construction project. Generally the most visible part of the work, it includes paint, stain, trim, and fixtures.

Lift rod assembly: A mechanism consisting of the clevis and clevis screw that is used to adjust the operation of a pop-up plug.

Pop-up drain: A mechanism that allows a user to open or close a lavatory drain by pulling on a lift knob connected to a pivot rod.

Tub waste: The drain fixture and fittings that take wastewater from the tub.

Wax seal: A ring made of heavy-duty wax or rubber that fits between the bottom of the water closet bowl and the floor.

Additional Resources

This module is intended to be a thorough resource for task training. The following reference works are suggested for further study. These are optional materials for continued education rather than for task training.

Installing & Repairing Plumbing Fixtures. 1994. Peter Hemp. Newton, CT: Taunton Press.

Materials and Components of Interior Architecture, 6th Edition. 2003. J. Rosemary Riggs. Upper Saddle River, NJ: Prentice-Hall.

Plumbing Fixtures and Appliances. 1982. Patrick J. Higgins and Kevin T. O'Hearn. Westport, CT: Intext, Inc.

References

Dictionary of Architecture and Construction, Third Edition. 2000. Cyril M. Harris, ed. New York: McGraw-Hill.

Figure Credits

Kohler Co.	208F03, 208F12, 208F27
Mueller/B&K Industries	208F04
Watts Regulator Company	208F05
NIBCO International	208F06
BrassCraft Manufacturing Company	208F07, 208F08
Zurn Plumbing Products Group	208F09
Fluidmaster, Inc.	208F14, 208F15, 208F25
Sloan Valve Company	208F16–208F20
American Standard	208F21, 208F24
Eljer Plumbingware, Inc. www.eljer.com	208F22

CONTREN® LEARNING SERIES — USER UPDATE

The NCCER makes every effort to keep these textbooks up-to-date and free of technical errors. We appreciate your help in this process. If you have an idea for improving this textbook, or if you find an error, a typographical mistake, or an inaccuracy in NCCER's Contren® textbooks, please write us, using this form or a photocopy. Be sure to include the exact module number, page number, a detailed description, and the correction, if applicable. Your input will be brought to the attention of the Technical Review Committee. Thank you for your assistance.

Instructors – If you found that additional materials were necessary in order to teach this module effectively, please let us know so that we may include them in the Equipment/Materials list in the Annotated Instructor's Guide.

Write: Product Development and Revision
National Center for Construction Education and Research
P.O. Box 141104, Gainesville, FL 32614-1104

Fax: 352-334-0932

E-mail: curriculum@nccer.org

Craft ______________________ Module Name ______________________

Copyright Date __________ Module Number __________ Page Number(s) __________

Description

____

____

____

____

(Optional) Correction

____

____

____

(Optional) Your Name and Address

____

____

____

Plumbing Level Two

02209-05

Introduction to Electricity

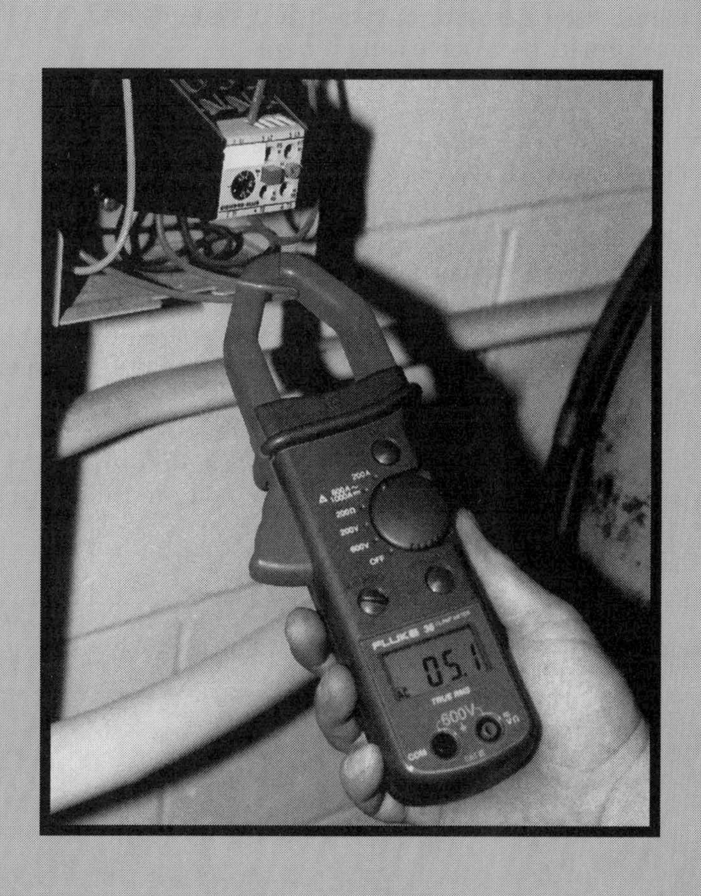

02209-05
Introduction to Electricity

Topics to be presented in this module include:

Overview

Plumbing systems use devices that contain electrical circuits. To install and service these systems, plumbers must understand how electrical components work, be able to read circuit diagrams, and know how to use electrical test equipment. Plumbers must follow electrical safety procedures, including wearing appropriate personal protective equipment and following manufacturers' instructions. Plumbers also follow general safety practices when working with or around any electrical systems and take the necessary precautions to protect themselves and others.

Because of the degree of danger when working with electricity, plumbers must be alert to the possibility of electric shock. The higher the voltage, the greater the current and the greater the chance for a fatal shock. Plumbers need a basic understanding of what current, voltage, and resistance are and what they do. Voltage, current, resistance, and power are closely related. By using any two of them, plumbers can apply simple mathematical equations (Ohm's law and the power formula) to determine the other two.

When troubleshooting an electrical circuit, plumbers measure voltage, current, and resistance using electrical test equipment, including analog meters, digital meters, ammeters, and multimeters. Plumbers encounter three categories of electrical circuits: series circuits, parallel circuits, and series-parallel circuits. Plumbers must recognize the different electrical symbols used to illustrate circuits, electrical components, and connections on circuit diagrams, construction drawings, and wiring schematics. Interpreting these symbols accurately is essential to electrical safety.

Focus Statement

The goal of the plumber is to protect the health, safety, and comfort of the nation job by job.

Code Note

Codes vary among jurisdictions. Because of the variations in code, consult the applicable code whenever regulations are in question. Referring to an incorrect set of codes can cause as much trouble as failing to reference codes altogether. Obtain, review, and familiarize yourself with your local adopted code.

Objectives

When you have completed this module, you will be able to do the following:

1. State and demonstrate the safety precautions that must be followed when working on electrical equipment.
2. State how electrical power is generated and distributed.
3. Describe how voltage, current, resistance, and power are related.
4. Use Ohm's law to calculate the current, voltage, and resistance in a circuit.
5. Use the power formula to calculate how much power is consumed by a circuit.
6. Describe the differences between series and parallel circuits.
7. Recognize and describe the purpose and operation of the various electrical components used in plumbing equipment.
8. Make voltage, current, and resistance measurements using electrical test equipment. Determine the positioning of leads. Test a fuse for continuity.
9. Explain and understand electrical symbols.

Required Trainee Materials

1. Appropriate personal protective equipment
2. Pencil and paper
3. Copy of local applicable code
4. Copy of *NFPA 70*, the *National Electrical Code®*

Prerequisites

Before you begin this module, it is recommended that you successfully complete *Core Curriculum; Plumbing Level One; Plumbing Level Two,* Modules 02201-05 through 02208-05.

This course map shows all of the modules in the second level of the Plumbing curriculum. The suggested training order begins at the bottom and proceeds up. Skill levels increase as you advance on the course map. The local Training Program Sponsor may adjust the training order.

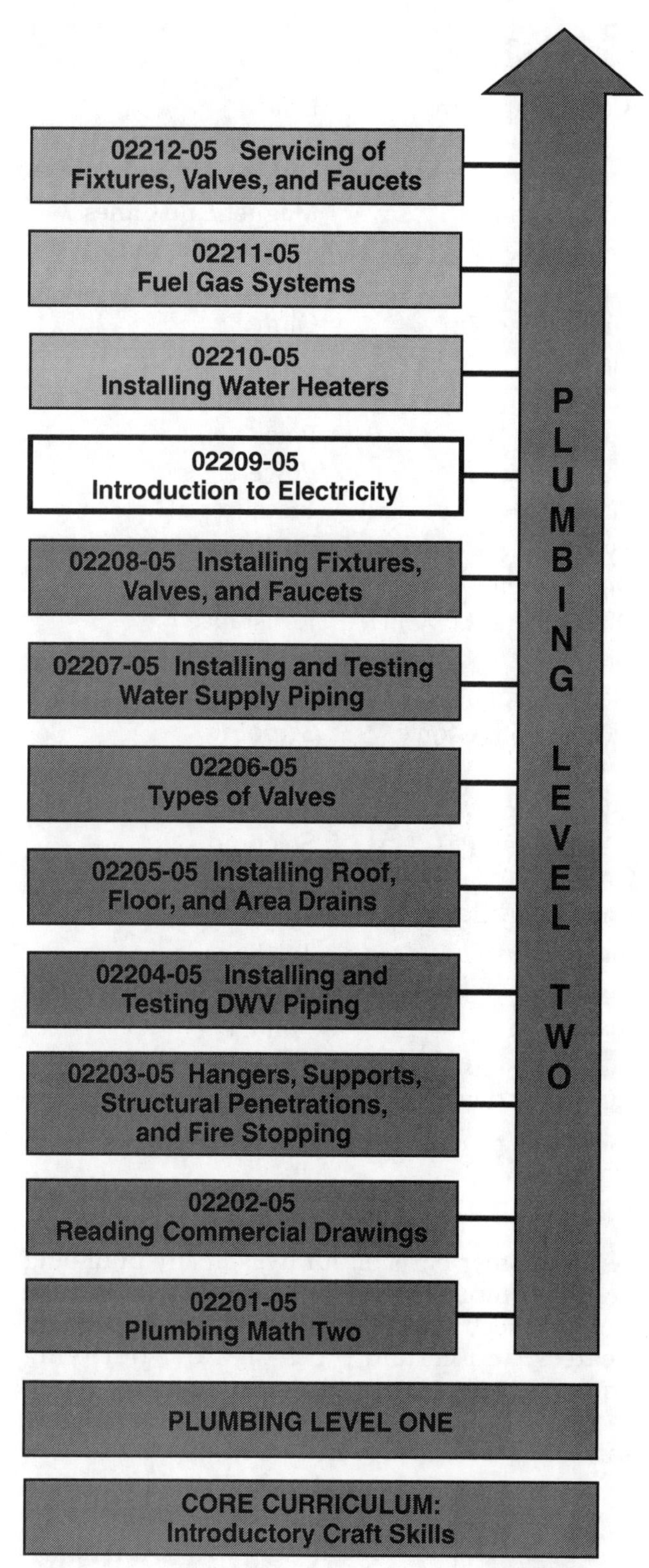

Trade Terms

- Alternating current (AC)
- Ammeter
- Ampere (amp)
- Analog meter
- Circuit
- Circuit diagram
- Clamp-on ammeter
- Conductor
- Contactor
- Continuity
- Current
- Cycle
- Digital meter
- Direct current (DC)
- Dry fire
- Electromagnet
- Electromotive force (EMF)
- Frequency
- Full-load amps (FLA)
- Ground fault
- Immersion element
- Induction
- In-line ammeter
- Insulator
- Ladder diagram
- Line-duty device
- Load
- Locked-rotor amps (LRA)
- Magnetic flux lines
- Magnetic starter
- Multimeter
- Ohm
- Parallel circuit
- Pilot-duty device
- Pole
- Power
- Pressurestat
- Rectifier
- Relay
- Resistance
- Series circuit
- Series-parallel circuit
- Setpoint
- Short circuit
- Slow-blow fuse
- Solenoid
- Throw
- Transformer
- Turbine
- Volt
- Voltage
- Voltage drop
- Watts

1.0.0 ◆ INTRODUCTION

As with other parts of our lives, many plumbing system components require electricity to operate. Pumps, electric water heaters, electric baseboard heaters, and certain types of water-purifying equipment all use electricity. So do many power tools and equipment such as drills, reciprocating saws, and electric pipe threaders.

> **NOTE**
> The *National Electrical Code®* (NEC®) is the standard code used for developing local applicable electrical codes. As with other model codes, local jurisdictions adopt the code and modify it to suit their needs. The NEC® itself can only be changed by the National Fire Protection Association (NFPA), the organization that writes and publishes it.

Many of the problems that you will encounter with these devices can be traced to their electrical **circuits**. To determine what is wrong, you must be able to read a **circuit diagram** and use electrical test equipment to make measurements at key points in the circuit. Your training in electricity will focus on learning to read electrical circuit diagrams and learning to use electrical test equipment.

2.0.0 ◆ ELECTRICAL SAFETY

Every year in the United States, approximately 700 deaths are caused by electrical accidents. Electrical accidents are the third leading cause of death in the workplace. Do not become a statistic. Work safely! Your reputation, your career, and your life depend on it.

When you work with electrical devices, always wear appropriate personal protective equipment. Follow the manufacturer's instructions when using tools around electrical equipment and when installing electrical devices. Working with electricity always involves some degree of danger. Always be alert to the possibility of electric shock. Always take necessary precautions to protect yourself and others.

2.1.0 The Effects of Current

The amount of **current** that passes through the human body determines the outcome of an electrical shock. The higher the **voltage**, the greater the current and the greater the chance for a fatal shock.

> **WARNING!**
> High voltage, defined as 600V or more, is almost 10 times more likely to kill than low voltage. On the job, you spend most of your time working on or near lower voltages. However, lower voltages can also kill. For example, portable, electrically operated hand tools can cause severe injuries or death if the frame or case of the tool becomes energized. Many electrical accidents can be prevented by following proper safety practices, including the use of ground fault circuit interrupters (GFCIs) and proper grounding.

Electrical current flows along the path of least **resistance** to return to its source. If you come in contact with a live **conductor**, you become a **load**. The amount of resistance that the human body presents under various circumstances varies (see *Figure 1*). Note that the potential for shock increases dramatically if the skin is damp. A cut will also reduce your resistance. Currents of less than 1A can severely injure and even kill a person (see *Table 1*).

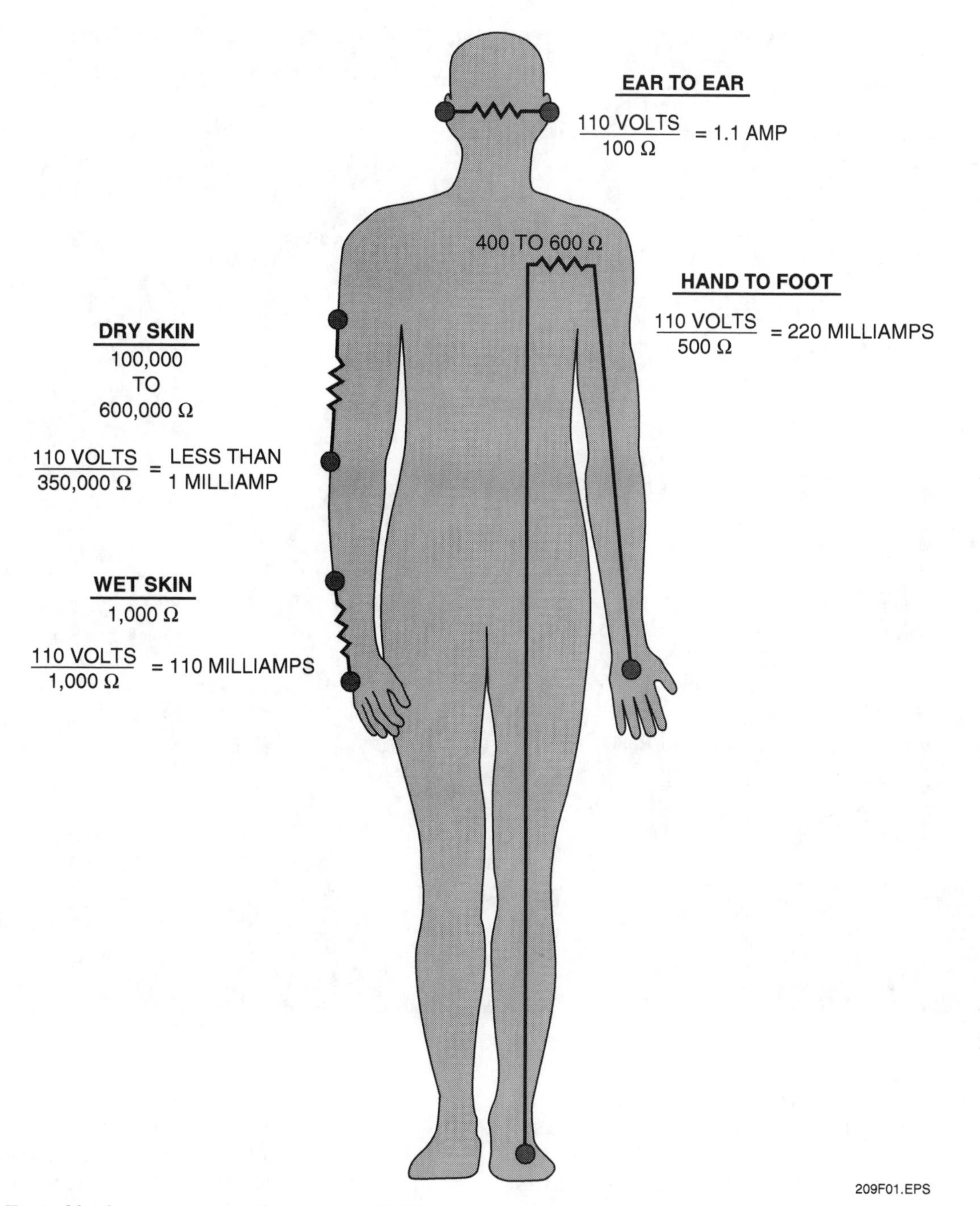

Figure 1 ◆ Typical body resistances and currents at 110V.

Table 1 Effects of Electrical Current on the Human Body

Current Value	Typical Effects
<1 mA	No sensation.
1–20 mA	Sensation of shock, possibly painful. May lose some muscular control between 10 mA and 20 mA.
20–50 mA	Painful shock, severe muscular contractions, breathing difficulties.
50–200 mA	Up to 100 mA, same symptoms as above, only more severe. Between 100–200 mA, ventricular fibrillation may occur. This typically results in almost immediate death unless special medical equipment and treatment are available.
>200 mA	Severe burns and muscular contractions. The chest muscles contract and stop the heart for the duration of the shock, followed by death unless special medical equipment and treatment are available.

Types of Transformers

There are many kinds of transformers. The pole-mounted transformer pictured elsewhere in this module is used to step down the high voltage on local utility lines to a level that can be used by lighting, equipment, and appliances in a building. It is large and very heavy. Other transformers, such as those found in power substations, can be nearly as big as a house. The transformers used in air conditioning equipment are usually small enough to hold in the palm of your hand.

Nearly all residential and light commercial electrical equipment contains at least one step-down transformer. To protect the transformer, many control circuits contain a fuse or circuit breaker in the transformer secondary circuit. If you are troubleshooting an electrical system and have power to the transformer primary, but control circuit voltage (usually 24V) is not present, check the fuse or circuit breaker before replacing the transformer.

Another common type of transformer is the step-up transformer. These transformers increase, or step up, the voltage to a higher level. Large step-up transformers are used for power distribution. Smaller ones are often used in oil burners to ignite the fuel.

You Are Responsible for Working Safely

OSHA regulations state that employees have a duty to follow the safety rules laid down by their employer. The amount an employee can collect from worker's compensation or disability insurance may be restricted if the employee was in violation of safety rules when injured. Your company also may terminate your employment or take other disciplinary action if you violate safety rules.

The following information on electrical safety has been included in this module to help you gain respect for the environment where you work and to stress the importance of safe working habits when working with electricity.

2.2.0 Safety Practices

Electrical technicians routinely work with potentially deadly levels of electricity. They can do so because good safety practices have become second nature to them. As a plumber, you will not be working on electrical systems, but it is important that you understand and follow proper safety practices. Follow these general safety practices whenever you are working with or around any electrical systems:

- Always shut off electricity at the source, unless you cannot work with the power off. Lock and tag the power switch in accordance with company or site procedures.
- Use a voltmeter to verify that the power to the unit is actually off. Remember that even though the power may be switched off, there is still potential at the input side of the shutoff switch.
- Use protective equipment such as rubber gloves.
- Use insulated tools.
- Short components to ground before touching de-energized wires.
- Do not kneel on the ground when making voltage measurements.
- Remove metal jewelry such as rings and watches.
- When testing a live circuit, keep one hand outside the unit when possible. This will reduce the risk of completing a circuit through your upper body, which might cause current to pass through your heart.

The NEC®, when used together with the electrical code for your local area, provides the minimum requirements for the installation of electrical systems. Always use the latest edition of the NEC® as your on-the-job reference. It specifies the minimum provisions necessary for protecting people and property from electrical hazards.

3.0.0 ◆ ELECTRICITY AND MAGNETISM

Plumbers need a basic understanding of what current, voltage, and resistance are and what they do. You have been introduced to many of these concepts as they apply specifically to components, tools, or equipment that you use on the job. In this section, you will learn about the principles behind them.

3.1.0 Current, Voltage, and Resistance

The movement of electrons causes an electrical current. Electrons are the negatively charged particles that exist in all matter. When a difference in the number of electrons exists between two points, electrons will flow from the negative point (the one with more electrons) to the positive point (the one with fewer electrons). The difference in electrical potential between the two points is called voltage. Voltage is also called **electromotive force (EMF)**.

Lightning is a good example of natural electron flow. A storm cloud has a negative charge with respect to the earth. Lightning occurs when the difference in potential between the cloud and the earth becomes so great that the air between them conducts a current.

In the common 12-**volt** (abbreviated 12V) car battery, a chemical reaction causes one of the **poles** to be negative with respect to the other. If you connect a light bulb or other electrical device between the negative (–) and positive (+) poles of the battery, electrons will flow from the negative pole to the positive pole (see *Figure 2*).

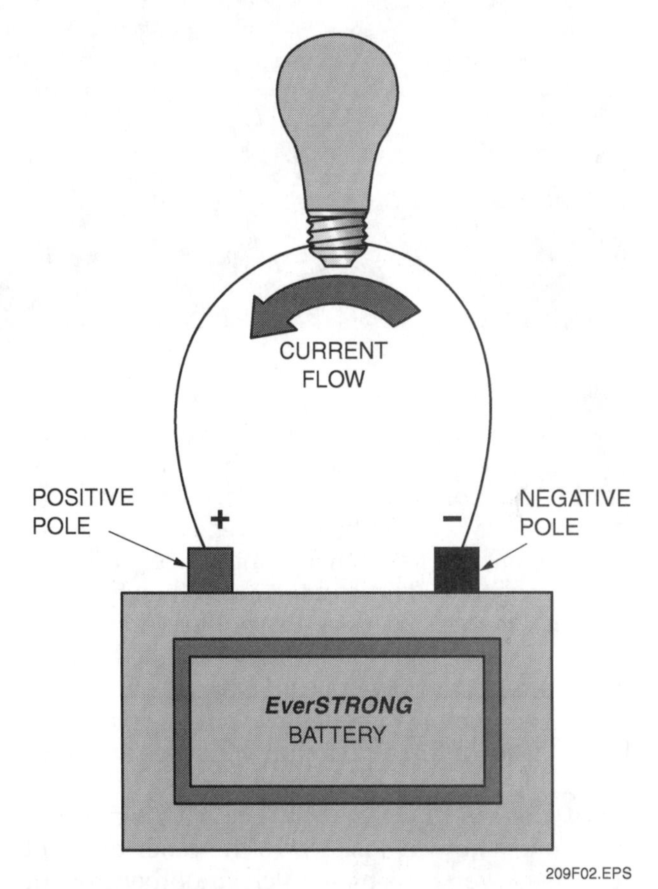

Figure 2 ◆ Current flow.

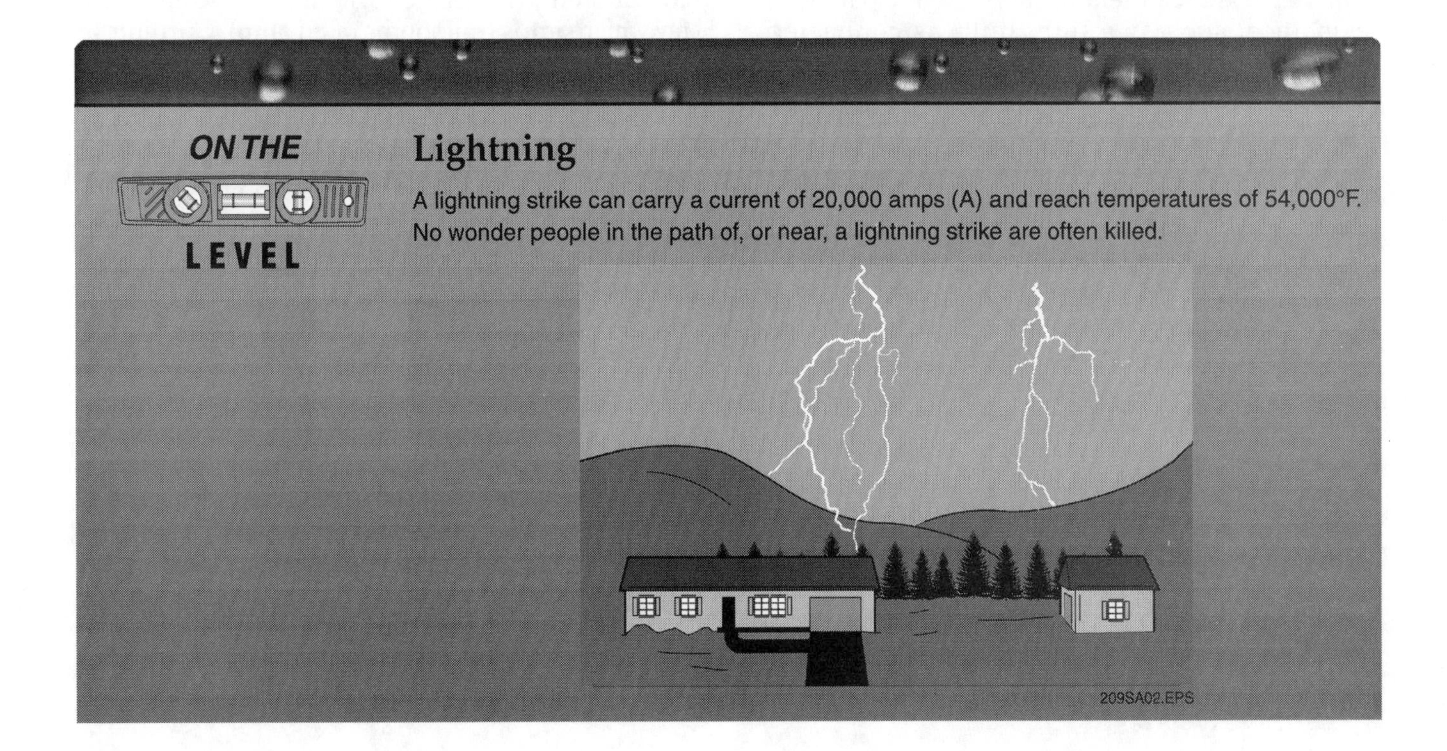

Voltage

A disagreement with a fellow scientist over the twitching of a frog's leg eventually led 18th-century physicist Alessandro Volta to theorize that when certain objects and chemicals come into contact with each other, they produce an electric current. Believing that electricity came from contact between metals only, Volta coined the term *metallic electricity*. To demonstrate his theory, Volta placed two discs, one of silver and the other of zinc, into a weak acidic solution. When he linked the discs together with wire, electricity flowed through the wire. Thus, Volta introduced the world to the battery, also known as the Voltaic pile. Now Volta needed a term to measure the strength of the electric push or the flowing charge; the volt is that measure. Voltage is known by other names as well, including electromotive force (EMF) and potential difference. Voltage can be viewed as the potential to perform work.

Voltage fluctuates rapidly between positive and negative. If you could see this fluctuation, it would resemble a wave pattern. The pattern is called a sine wave. Voltage alternates from zero to positive, back to zero, then to negative, and then back to positive 60 times a second. When the voltage goes to zero, so does the power. A series of waves is called a phase. One wave is called single-phase electrical power (see *Figure 3*). Electrical power is commonly available in single-phase and three-phase. In three-phase power, three waves of voltage are generated, separated from each other by 120 degrees (1/180th of a second). As a result, when one sine wave is at zero the other two still provide power, causing a much smoother flow of electricity (see *Figure 4*).

To harness the potential energy of the battery, a resistance, such as a light bulb, is connected between the two battery poles. Resistance is the property that causes a material to resist or impede current flow. When the switch closes, the moving electrons flow through the bulb, which converts the electrical energy into light. The bulb consumes the electrical energy in the process. The amount of energy consumed is known as power, and is expressed in **watts** (W). A 60W light bulb consumes 60 watts of power.

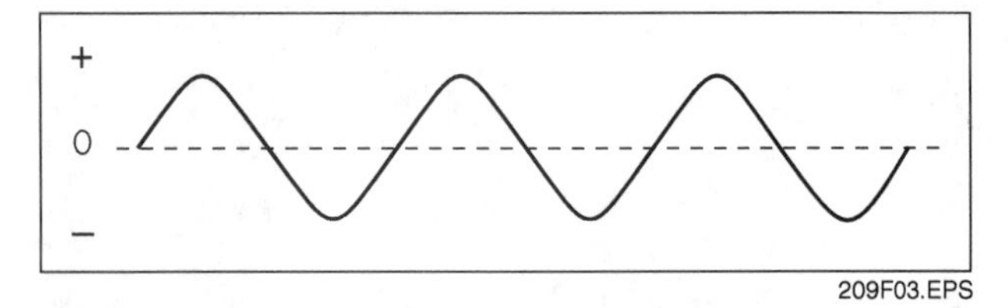

Figure 3 ◆ Sine wave for single-phase current.

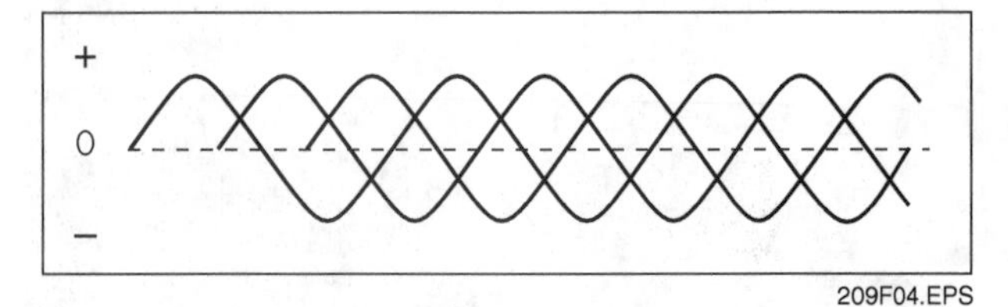

Figure 4 ◆ Sine wave for three-phase current.

Current is expressed in **amperes,** or **amps** (A); voltage is expressed in volts (V); and resistance is expressed in **ohms** (Ω). The AC voltages you will work with range from 24V to 600V. The occasional DC voltages you encounter will generally be low voltages in the 5V to 15V range. Occasionally, you may encounter a very high-voltage circuit, such as the ignition circuit for a gas furnace, that may produce a 10,000V spark.

Currents in common heating, ventilating, and air conditioning (HVAC) circuits may be as low as a few microamps (millionths of an amp). Currents in the milliamp or mA (thousandths of an amp) range are more common. Some circuits, especially motor circuits, may carry very high currents in the range of 100A or more. Be aware that, under the right conditions, even currents in the milliamp range can be dangerous.

Resistances are usually large values expressed in thousands of ohms (kilohms, or K) or millions of ohms (megohms, or M). In some cases, a resistance value may be only a few ohms. For example, the thermistors (heat-sensitive resistors) used as temperature-sensing devices and as an aid in motor startup fall into this category.

3.2.0 AC and DC Voltage

Voltage in an electric circuit can be either **direct current (DC)** or **alternating current (AC)**. DC electricity flows only one way through the circuit. AC electricity alternately flows one way and then the other through the circuit. Each repetition is called a **cycle** and the number of cycles per second is called the **frequency**.

Batteries produce DC voltage. Automobiles, portable stereos, calculators, and flashlights are good examples of devices that use DC voltage. Electricity supplied by power utilities is AC. Almost all electrical plumbing system components, as well as plug-in tools and machines, use AC.

Occasionally, you will find a unit that has a DC motor or an electronic circuit board that requires DC voltage. Rather than using batteries, which need to be replaced or recharged, such units contain special circuits called **rectifiers** that convert AC to DC. The device that allows you to plug a calculator or other portable device into a wall socket is a familiar example of a rectifier.

3.3.0 Magnets and Electromagnets

Many electrical components, including motors, **relays**, and **solenoids**, rely on the power of magnetism. Magnetized iron generates a field consisting of lines of force, also known as **magnetic flux lines** (see *Figure 5*). Magnetic objects within the field will be attracted or repelled by the magnetic field. The flux lines run between poles, termed the north and south poles. Opposing poles attract each other; like poles repel each other. The more powerful the magnet, the more powerful the magnetic field around it.

Electricity also produces magnetism. Current flowing through a conductor produces a small magnetic field around the conductor. If the conductor is coiled around an iron bar, the result is an **electromagnet** (see *Figure 6*). Electromagnets are the basis on which electric motors and other components operate.

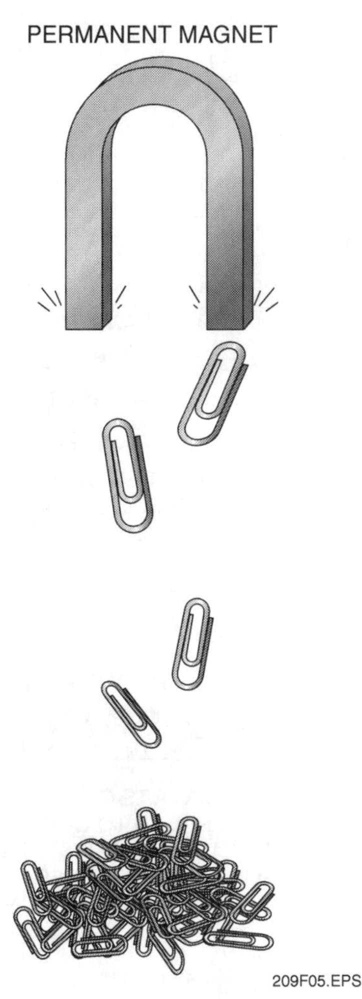

Figure 5 ◆ Magnetism.

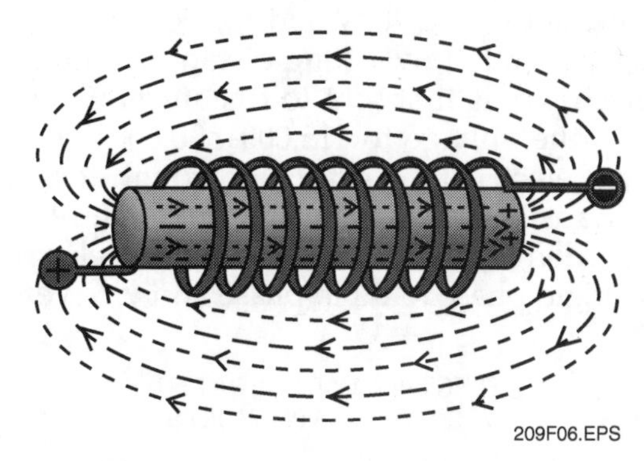

Figure 6 ◆ Electromagnet.

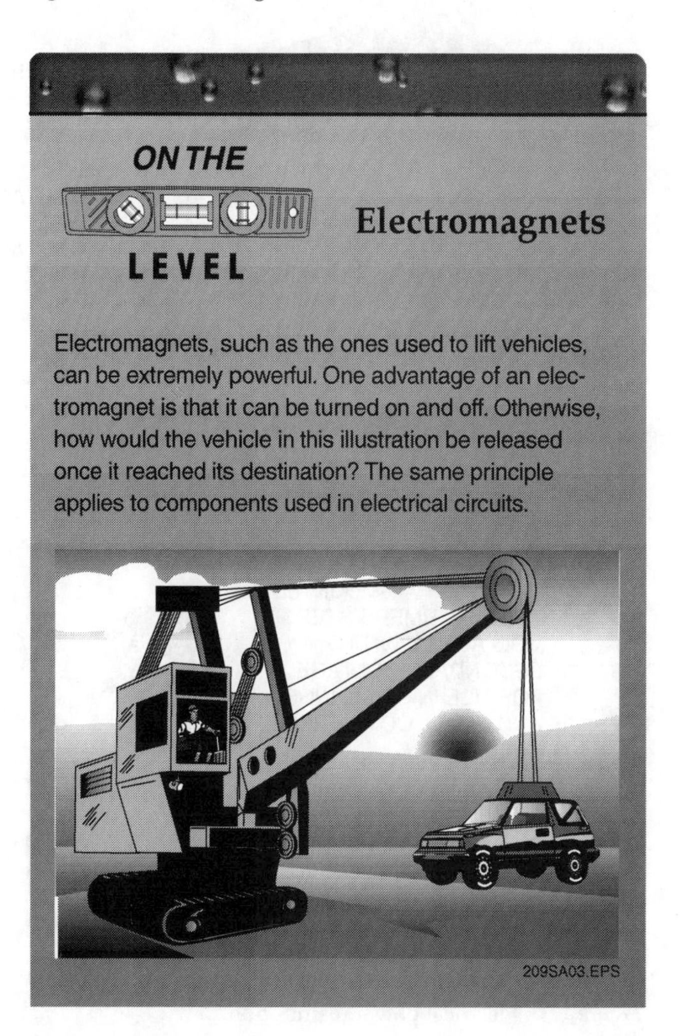

ON THE
LEVEL
Electromagnets
Electromagnets, such as the ones used to lift vehicles, can be extremely powerful. One advantage of an electromagnet is that it can be turned on and off. Otherwise, how would the vehicle in this illustration be released once it reached its destination? The same principle applies to components used in electrical circuits.
209SA03.EPS

3.4.0 Electrical Power Generation and Distribution

Many steps are required to create and distribute electricity (see *Figure 7*). A power-generating station, operated by an electrical utility company, creates electricity from huge generators called **turbines**. The turbine consists of a fan-like set of blades that is spun by the high-speed flow of steam or water. As the turbine spins, it rotates an electromagnet that is attached to it by a shaft. The electromagnet creates electrical current.

The voltage of the resulting electrical charge—typically between 2,400V and 13,800V—is too low to be transmitted over long distances. Therefore, the electricity is first sent to **transformers**, which step up the voltage to 115,000–500,000V. At that voltage, electricity can travel through long-distance transmission lines with little or no loss of voltage. Some transformers also convert electricity from single-phase to three-phase and vice versa.

At various points along the distribution line, electricity can be routed through electrical substations to distribution lines for use in homes, offices, and factories. At such points, the electricity is again passed through transformers, this time to step the voltage down to lower levels. *Figure 8* shows a pole-mounted transformer used to step down voltages for residential use. Voltage in a residential system is usually 208V, 220V, 230V, or 240V. Check with your local electrical utility for the residential voltages in your area.

At the wall outlet where small appliances such as televisions and toasters are connected, the voltage is about 120V. Large appliances such as electric stoves, clothes dryers, water heaters, and central air conditioning systems usually require the full 240V. Commercial buildings and factories may receive anywhere from 208V to 575V, depending on the amount of power their machines consume.

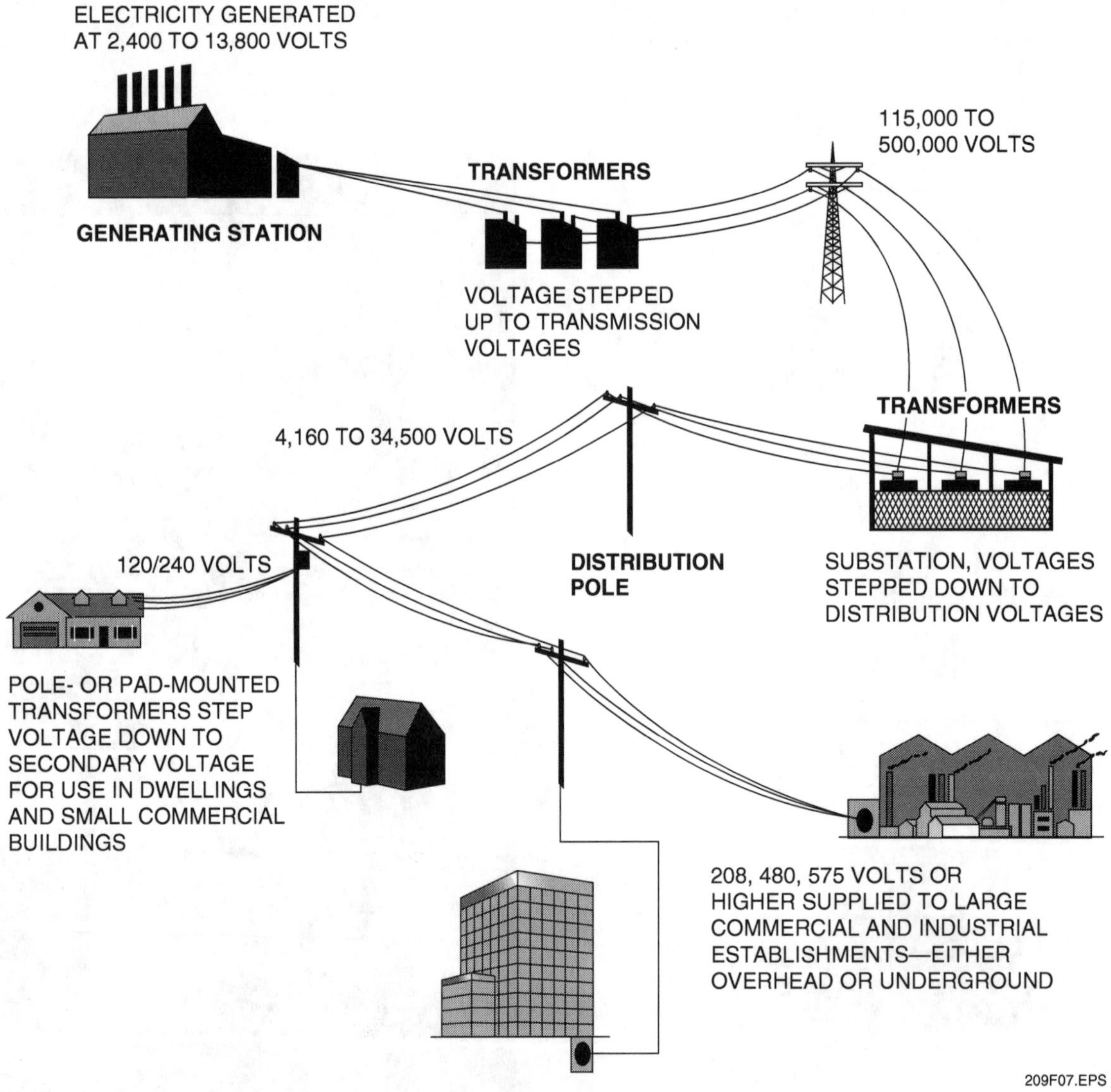

Figure 7 ◆ Electrical power distribution.

Figure 8 ◆ Residential power distribution.

3.5.0 Ohm's Law

Ohm's law is a formula for calculating voltage, current, or resistance. These values are expressed as E, I, and R, respectively. You can visualize the mathematical relationship among them in terms of the simple triangle illustrated in *Figure 9*. The triangle shows that if you know two of the values, the unknown value can be found.

The horizontal relationship in the triangle indicates multiplication, while the vertical relationships indicate division. For example, if the voltage and resistance are known, by covering the I you can see that you would divide E by R. However, to find E, you would multiply I and R.

For example, imagine that a 120V circuit contains a 60Ω load. What is the current flow? Use the formula for calculating current:

$$I = \frac{E}{R}$$

$$I = \frac{120V}{60\Omega}$$

$$I = 2A$$

Using Ohm's law, you find that the current flow for a 120V circuit with a 60Ω load is 2A.

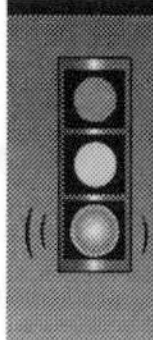

NOTE

Ohm's law applies only to resistive circuits. Because of the nature of alternating current, circuits containing motors, relay coils, and other inductive devices do not act the same as pure resistances.

	LETTER SYMBOL	UNIT OF MEASUREMENT
CURRENT	I	AMPERES (A)
RESISTANCE	R	OHMS (Ω)
VOLTAGE	E	VOLTS (V)

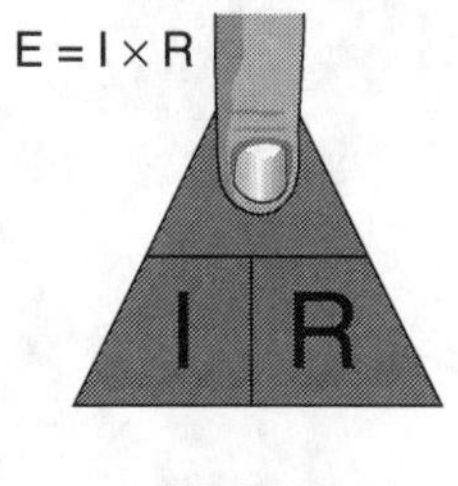

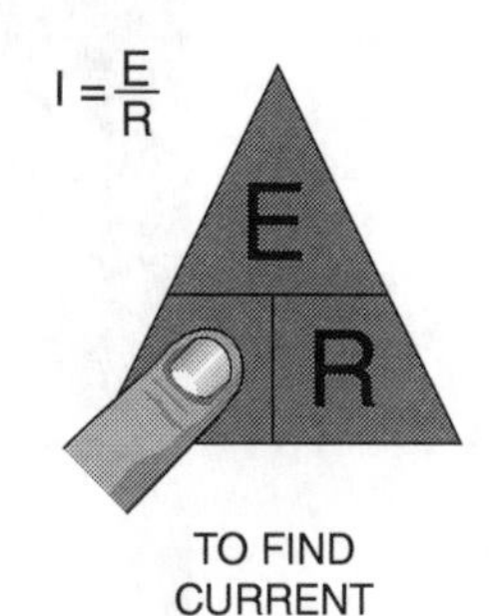

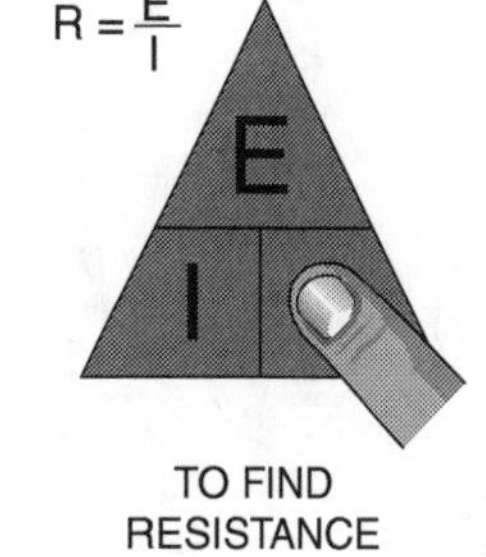

Figure 9 ◆ Graphic representation of Ohm's law.

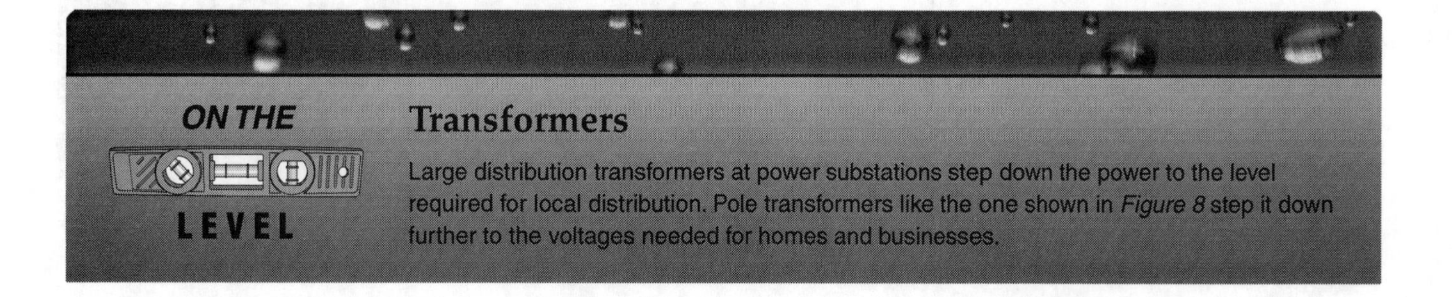

Transformers

Large distribution transformers at power substations step down the power to the level required for local distribution. Pole transformers like the one shown in *Figure 8* step it down further to the voltages needed for homes and businesses.

4.0.0 ◆ ELECTRICAL MEASURING INSTRUMENTS

When troubleshooting an electrical circuit, it is usually necessary to measure voltage, current, and resistance using electrical meters. **Analog meters**, which require the reader to interpret a scale, are still around, but have been largely replaced by direct-reading **digital meters** (see *Figure 10*). **Ammeters** are used to measure alternating current; **multimeters** are commonly used to measure AC and DC voltage, resistance, and direct current. Some can read AC current in the milliamp range.

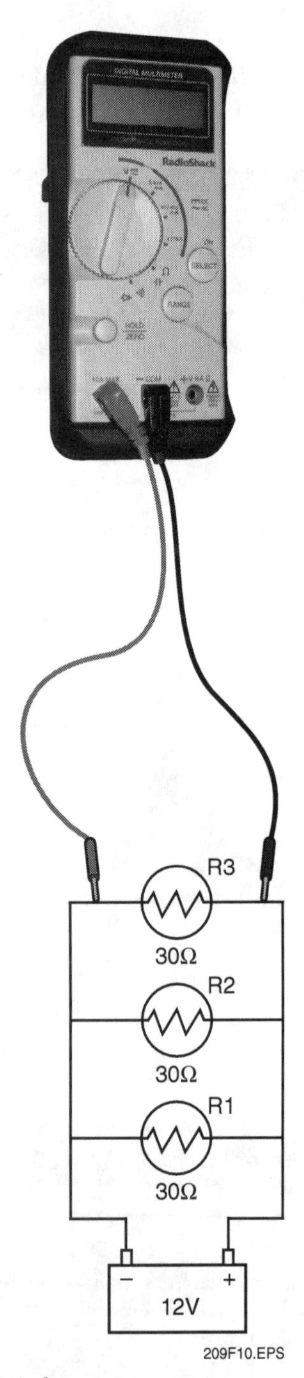

209F10.EPS

Figure 10 ◆ Digital meter.

4.1.0 Ammeter

Ammeters are often used to check motor circuits. Use a **clamp-on ammeter** to measure AC current (see *Figure 11*). Place the jaws of the ammeter around the wire conductor. Current flowing through the wire creates a magnetic field, which induces a proportional current in the ammeter jaws. This current is read by the meter movement and appears as a direct readout or, on an analog meter, as a deflection of the meter needle.

In-line ammeters are less common than clamp-on models (see *Figure 12*). Connect these meters in series with the circuit. This requires the circuit to be opened.

Aside from following good safety practices, there are a few things to remember when measuring an electrical current:

- Ensure that the ammeter jaws are clean and aligned correctly. Dirty or misaligned jaws will cause an inaccurate reading.
- When using an analog meter, always start at a high range and work down to avoid damaging the meter.
- Do not clamp the meter jaws around two different conductors at the same time. Doing so will cause an inaccurate reading.

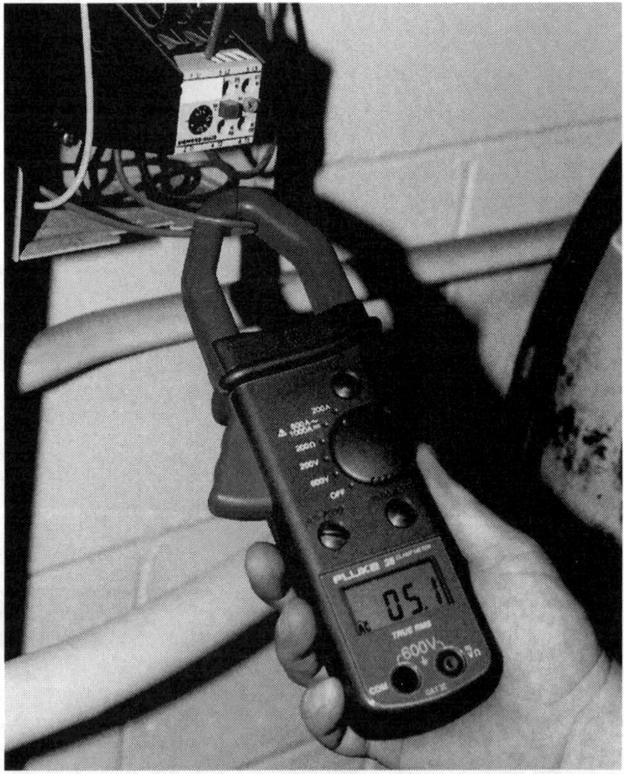

209F11.EPS

Figure 11 ◆ Clamp-on ammeter.

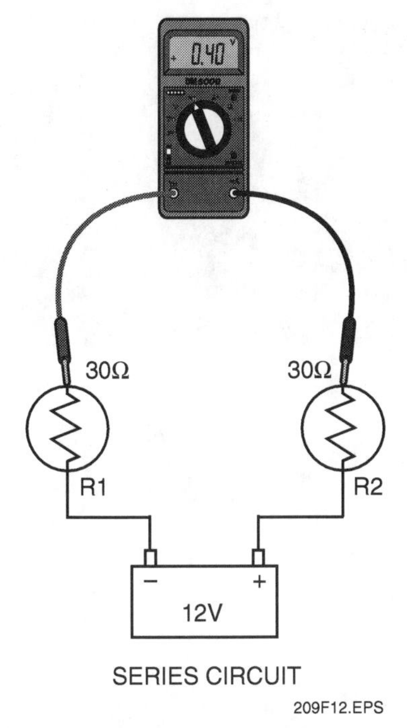

Figure 12 ◆ In-line ammeter test setup.

4.2.0 Multimeters

Multimeters have a selector on the meter that allows the user to select AC or DC voltage (voltmeter), resistance (ohmmeter), or current (ammeter) (see *Figure 13*). On an analog meter, the range of values to be read must also be selected.

4.2.1 Voltage Measurements

A voltmeter must be connected in parallel with (that is, across) the component or circuit to be tested (see *Figure 14*). If a circuit function is not operating, the voltmeter can be used to determine if the correct voltage is available to the circuit. Voltage must be checked with power applied.

CAUTION

When testing a live circuit, use an insulated alligator clip on the common meter lead. That way, only one hand, the one holding the other probe, is in the unit with power applied. Turn the power off before connecting the alligator clip.

4.2.2 Resistance Measurements

Ohmmeters contain an internal battery that acts as a voltage source. Therefore, resistance measurements are always made with the system power shut off. Sometimes an ohmmeter is used to measure resistance in a load; motor windings are a good example. More often, an ohmmeter is used to check **continuity** in a circuit. This means to check whether the circuit is continuous and unbroken. A wire or closed switch offers negligible resistance.

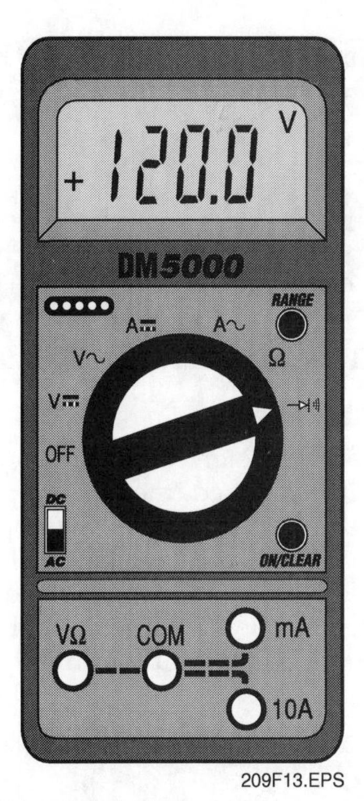

Figure 13 ◆ Digital multimeter.

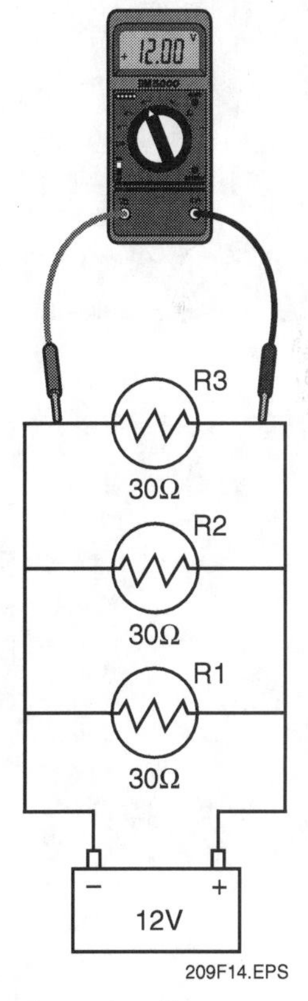

Figure 14 ◆ Voltmeter connection.

DID YOU KNOW?

Low Current Measurements

When low currents are being measured with a clamp-on ammeter, it may be necessary to wrap the conductor around the meter jaws. The reading must then be divided by the number of times the conductor passes through the jaws. For example, if the wire passes through the meter jaws 10 times and the reading is 5A, the actual current is 0.5A (5 ÷ 10).

With the ohmmeter connected as in *Figure 15* and all switches closed, the current produced by the ohmmeter battery will flow unopposed, and the meter will show zero resistance. The circuit has continuity. If a switch is open, however, there is no path for current and the meter will see infinite resistance; that is, a lack of continuity.

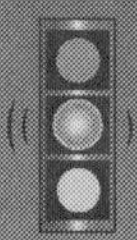

CAUTION

Never measure resistance in a live (energized) circuit. The voltage present in a live circuit may permanently damage the meter. Higher quality meters are equipped with a fuse or circuit breaker to protect the meter if this occurs. Also, a circuit in parallel with the circuit you are testing can provide a current path that will make it appear as though the circuit you are testing is good, even when it is not. Avoid this problem by disconnecting the device you are testing from the rest of the circuit.

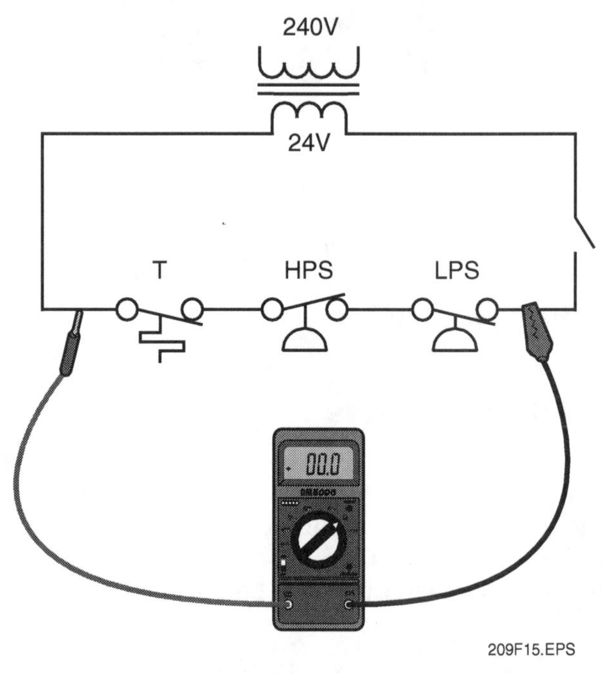

Figure 15 ◆ Ohmmeter connection for continuity testing.

Using Ohm's Law

Suppose the relative load in a circuit is expected to be 50Ω. You want to know if the resistance value of the load is correct, but because the battery in your multimeter is dead, you can't measure resistance. You measure the applied voltage and find that it is 230V. Using an ammeter, you find that the load is drawing 4.5A. What is the resistance of the load?

The allowable tolerance of the resistance value for this load is ±5%. Is the resistance of the load within this range?

Review Questions

Sections 1.0.0–4.0.0

1. A current of _____mA will cause no sensation when in contact with skin.
   a. less than 1
   b. 1–20
   c. 20–50
   d. 50–200

2. The difference in potential between the two poles of a car battery is caused by _____.
   a. the current flowing between the two poles
   b. the circuit resistance
   c. the amount of power consumed by the load
   d. a chemical reaction

3. Current is measured in _____.
   a. milliwatts per square meter
   b. volts
   c. amps
   d. miles per hour

4. A(n) _____ converts AC voltage to DC voltage.
   a. voltmeter
   b. inverter
   c. rectifier
   d. transformer

5. In a magnet, opposing poles repel each other and like poles attract each other.
   a. True
   b. False

6. A transformer is used to _____.
   a. transform electrical energy into another form of energy
   b. raise or lower a voltage
   c. consume power
   d. increase the resistance of a circuit

7. Solve the problem using the applicable formula: If a circuit with a 12Ω resistance draws 10A, the source voltage is _____V.

8. Solve the problem using the applicable formula: If a 120V circuit draws 3A, the resistance of the circuit is _____Ω.

9. When using an analog ammeter, always start at a high range and work down to avoid damaging the meter.
   a. True
   b. False

Match the electrical term with its definition.

10. _____ Current
11. _____ Voltage
12. _____ Resistance

a. The property that causes a material to impede the flow of electricity. Expressed as ohms (Ω).
b. The result when electrical potential is multiplied by its flow. Expressed as power (P).
c. Caused by the movement of electrons. Expressed as amperes or amps (A).
d. The difference in electrical potential between the positively and negatively charged points. Expressed as volts (V).

5.0.0 ◆ ELECTRICAL CIRCUITS

An electrical circuit is a closed loop that contains a voltage source, a load, and conductors to carry current. It usually contains a switching device to turn the current on and off. *Figure 16* illustrates a basic electrical circuit. The electrical circuits you will encounter on the job can be divided into three categories: **series circuits**, **parallel circuits**, and **series-parallel circuits**.

A conductor is any material that readily carries an electric current. Most metals are good conductors. Copper and aluminum are the most common conductors. Gold and silver are better conductors, but as precious metals they are too expensive for use in all but the smallest electronic circuits. Water is an excellent conductor because of its mineral content. That is why it is dangerous to work with electricity or use electrical appliances in a wet or damp environment.

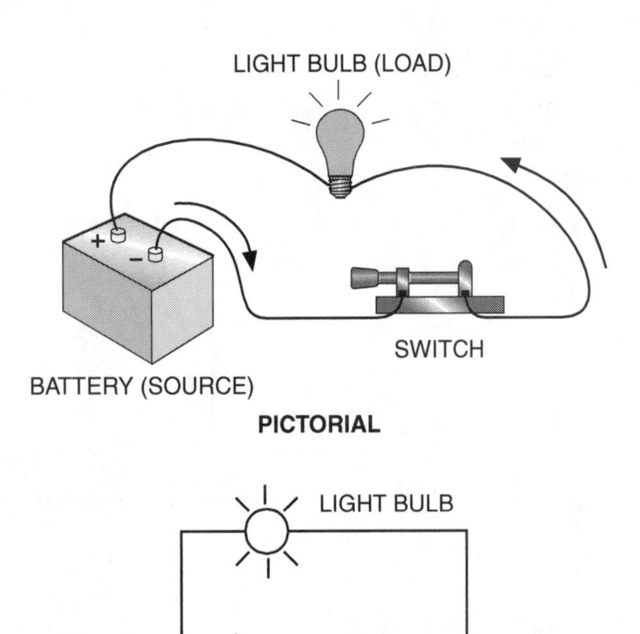

Figure 16 ◆ Basic electrical circuit.

An **insulator** is the opposite of a conductor. It inhibits the flow of electricity. Rubber is a good insulator. Tools used in electrical trades are often insulated with rubber to prevent electrical shock.

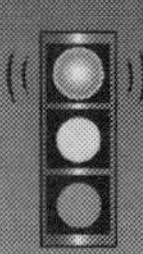

WARNING!

Electrical shock can injure and kill. Electricity can disrupt your nervous system, can cause burns, and can disrupt or even stop your heart. To eliminate the danger posed by electrical shock, always wear appropriate personal protective equipment and use properly insulated tools when working on electrical equipment.

5.1.0 Circuit Diagrams

Circuits are illustrated using a special type of drawing called a circuit diagram. A circuit diagram is like a road map that shows the routes that electricity travels. By reading a circuit diagram correctly, you can determine how the electrical circuits are supposed to act when the unit is running properly. You can then diagnose the operation of the circuit to find the portion that is not working correctly. Circuit diagrams supplied by manufacturers come in a variety of formats and typically include a wiring diagram and a simplified schematic diagram (see *Figure 17*).

5.1.1 Wiring Diagram

As its name suggests, the wiring diagram shows how the wiring is physically connected. It also identifies the color of each wire, often using standardized abbreviations (for example, on the wiring diagram in *Figure 17* "BLK" means "black"). Wiring diagrams are helpful if you have to rewire an electrical device. They can also be used to identify the correct place from which to take a measurement when troubleshooting the device.

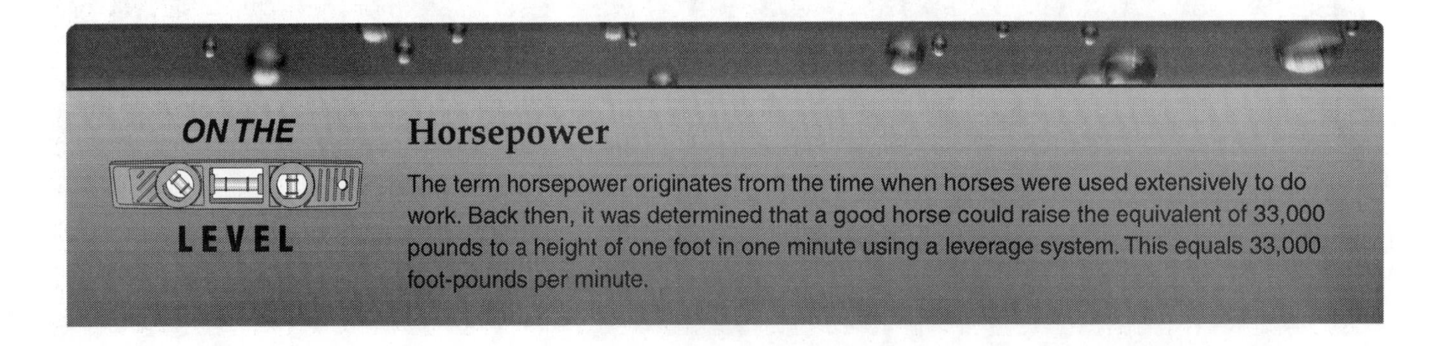

Horsepower

The term horsepower originates from the time when horses were used extensively to do work. Back then, it was determined that a good horse could raise the equivalent of 33,000 pounds to a height of one foot in one minute using a leverage system. This equals 33,000 foot-pounds per minute.

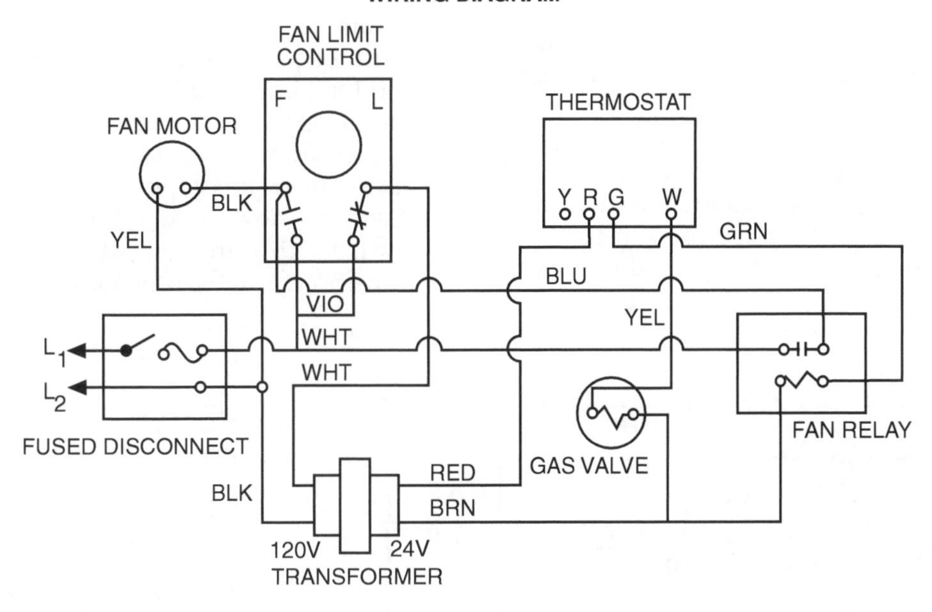

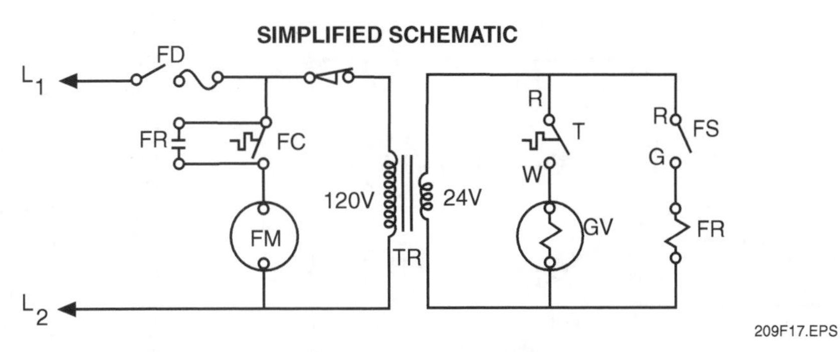

Figure 17 ◆ Wiring diagram and simplified schematic diagram of a gas furnace.

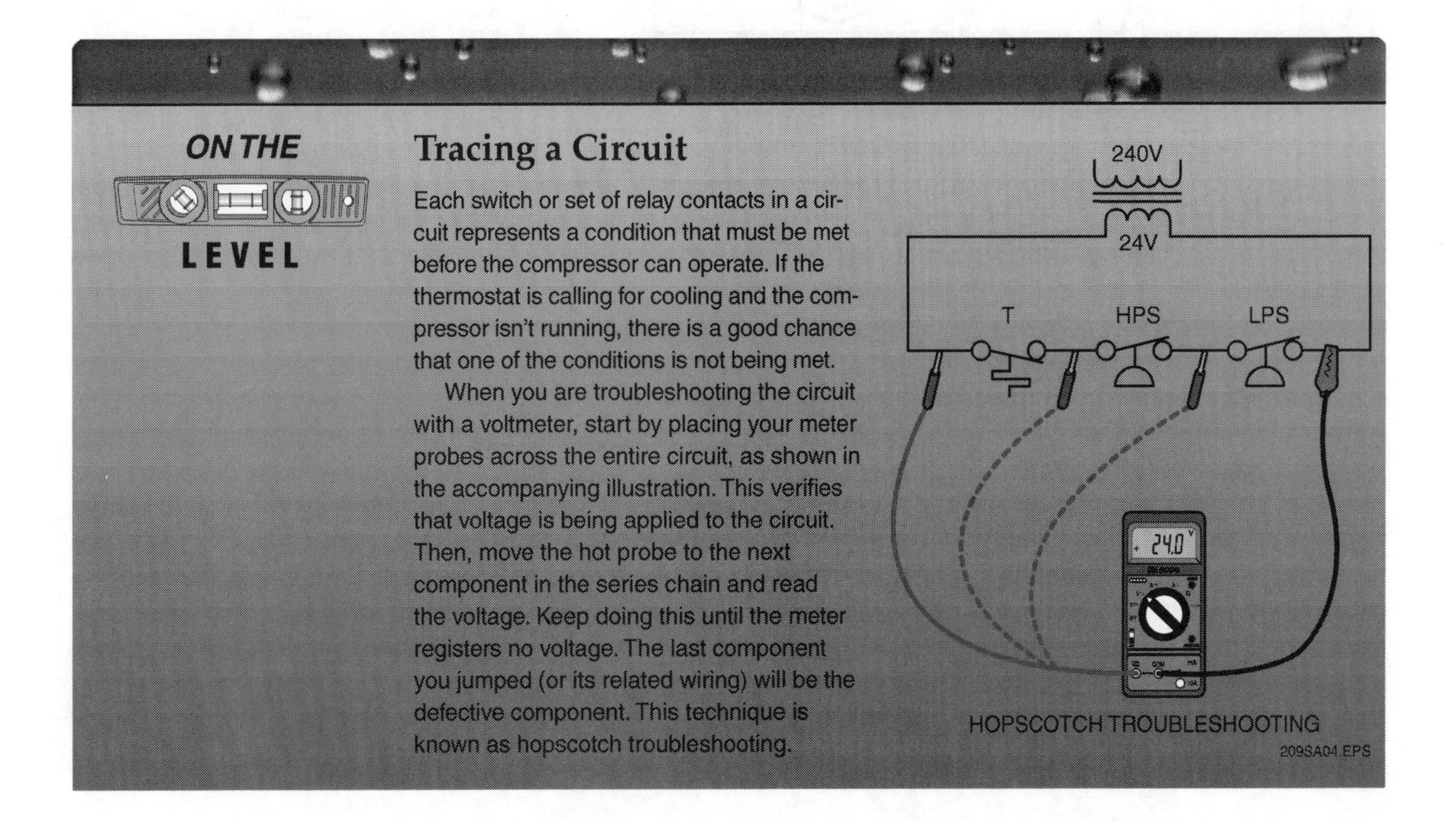

Tracing a Circuit

Each switch or set of relay contacts in a circuit represents a condition that must be met before the compressor can operate. If the thermostat is calling for cooling and the compressor isn't running, there is a good chance that one of the conditions is not being met.

When you are troubleshooting the circuit with a voltmeter, start by placing your meter probes across the entire circuit, as shown in the accompanying illustration. This verifies that voltage is being applied to the circuit. Then, move the hot probe to the next component in the series chain and read the voltage. Keep doing this until the meter registers no voltage. The last component you jumped (or its related wiring) will be the defective component. This technique is known as hopscotch troubleshooting.

5.1.2 Simplified Schematic Diagram

The simplified schematic diagram uses standard electrical diagram symbols to illustrate the operation of the electrical device. It does not include information about the wire color or the physical connections, or pictorial views of the wiring itself. Because simplified schematics are easier to read than wiring diagrams, they are used to trace circuits. Manufacturers may also provide specially laid-out schematic diagrams called **ladder diagrams** (see *Figure 18*). In a ladder diagram, the lines from the power source are shown as vertical lines, and load lines and their related control devices are shown as horizontal lines.

The diagram provided by the manufacturer may also contain a component location diagram. This diagram shows where the electrical components are located in the unit. It will help you make the transition from the schematic diagram to the actual

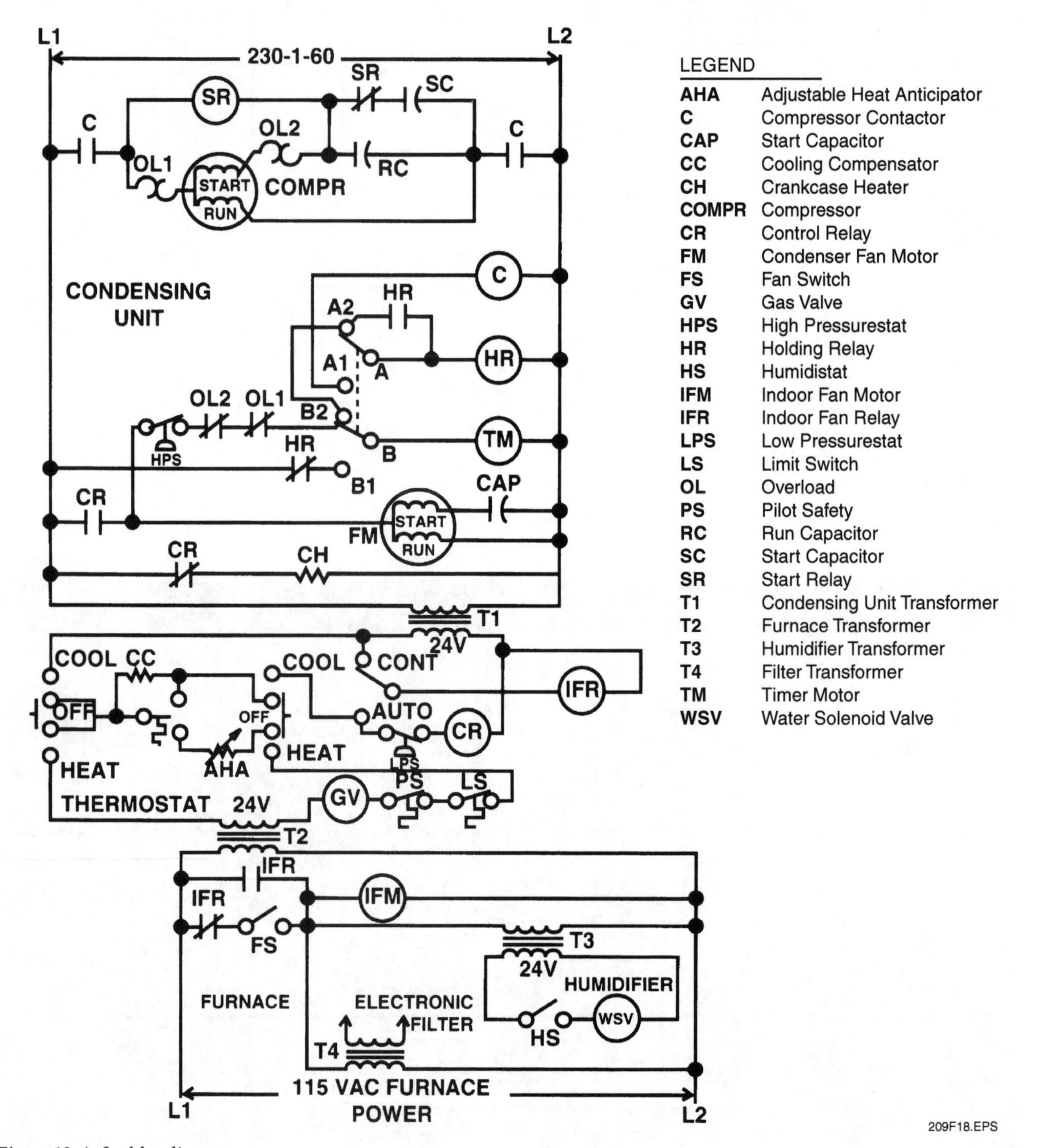

Figure 18 ◆ Ladder diagram.

device, because the schematic diagram shows the components in symbol form but does not indicate their physical location within the device.

5.2.0 Series Circuits

A series circuit provides only one path for current flow. If a series circuit is open at any point, no current will flow. For example, if five light bulbs are connected in series and one of them fails, the remaining four lights will turn off from lack of electrical power. Circuits containing loads in series are uncommon in plumbing work, but you need be to able to identify this type of circuit if you encounter it on the job.

The total resistance of a series circuit is equal to the sum of the individual resistances. The 12V series circuit illustrated in *Figure 19* has two 30Ω loads. The total resistance, therefore, is 60Ω. Calculate the amount of current flowing through the circuit using Ohm's law:

$$I = \frac{E}{R}$$

$$I = \frac{12V}{60\Omega}$$

$$I = 0.2A$$

The amount of current flowing through the circuit in *Figure 19* has a current of 0.2A. If the circuit has five 30Ω loads, the total resistance would be 150Ω. The current flow is the same through all the loads. The voltage measured across any load, called the **voltage drop**, depends on the resistance of that load. The sum of the voltage drops equals the total voltage applied to the circuit.

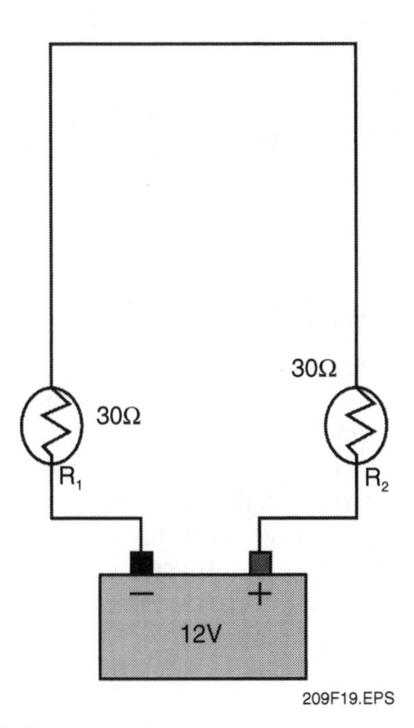

Figure 19 ◆ Series circuit.

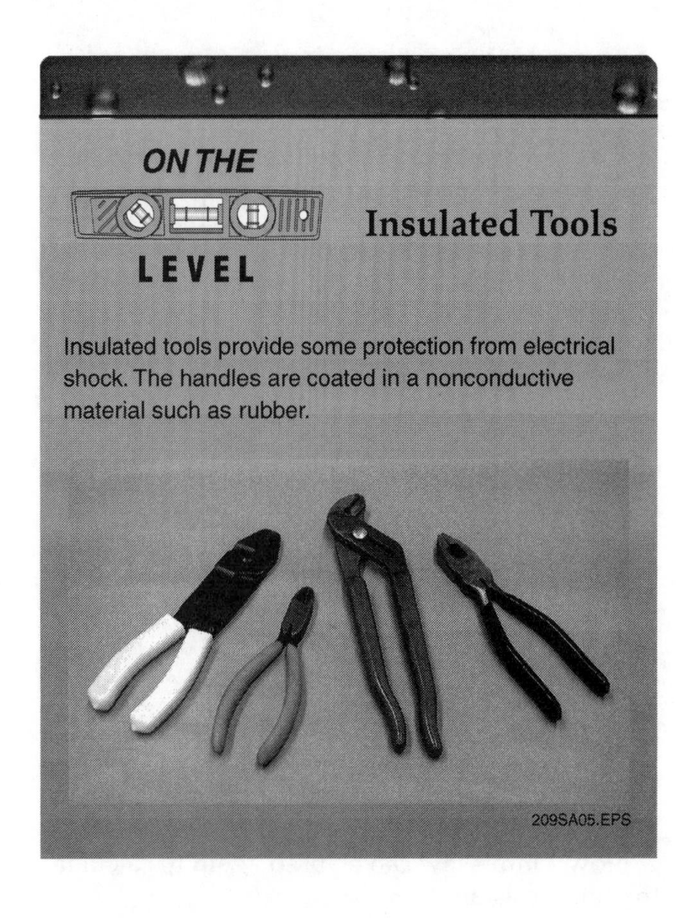

Insulated Tools

Insulated tools provide some protection from electrical shock. The handles are coated in a nonconductive material such as rubber.

5.3.0 Parallel Circuits

In a parallel circuit, each load is connected directly to the voltage source; therefore, the voltage drop is the same through all loads. The power source treats the circuit as two or more individual circuits containing one load each. If five light bulbs were connected on a parallel circuit, the failure of one light bulb would not affect the others.

The parallel circuit illustrated in *Figure 20* has three circuits, each containing a 30Ω load. The resistance of any load determines the current flow through that load. Thus, the total current drawn by the circuit is the sum of the individual currents. The total resistance of a parallel circuit is calculated differently from that of a series circuit. In a parallel circuit, the total resistance is less than the smallest of the individual resistances.

For example, each of the 30Ω loads draws 0.4A at 12V; therefore, the total current is 1.2A:

$$I = \frac{E}{R}$$

$$I = \frac{12V}{30\Omega}$$

$$I = 0.4A \text{ per circuit}$$

$$0.4A \times 3 \text{ circuits} = 1.2A$$

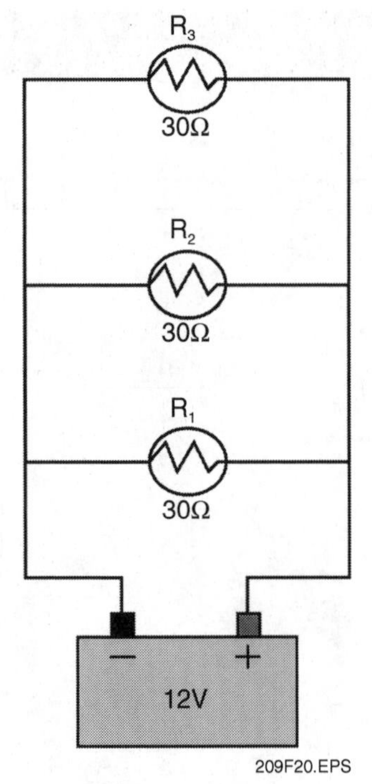

Figure 20 ◆ Parallel circuit.

Now, Ohm's law can be used again to calculate the total resistance:

$$R = \frac{E}{I}$$

$$R = \frac{12V}{1.2A}$$

$$R = 10\Omega$$

This example is simple because all the resistances were the same value. In cases where the resistances are different, the process is the same but separate current calculations are required for each load. The individual currents are added to get the total current.

Unlike series circuits, parallel circuits continue working even if one circuit opens. Household circuits are wired in parallel. Almost all the load circuits you encounter will be parallel circuits. You can use one of two formulas to convert parallel resistances to a single resistance value. Use the following formula for systems that have two resistances in parallel:

$$\text{Total resistance} = \frac{R_1 \times R_2}{R_1 + R_2}$$

Use this formula for systems that have three or more resistances in parallel:

$$\text{Total resistance} = \frac{1}{\frac{1}{R_1} + \frac{1}{R_2} + \frac{1}{R_3}}$$

Practice Exercises

1. Calculate the total resistance of the parallel circuit in *Figure 21*.

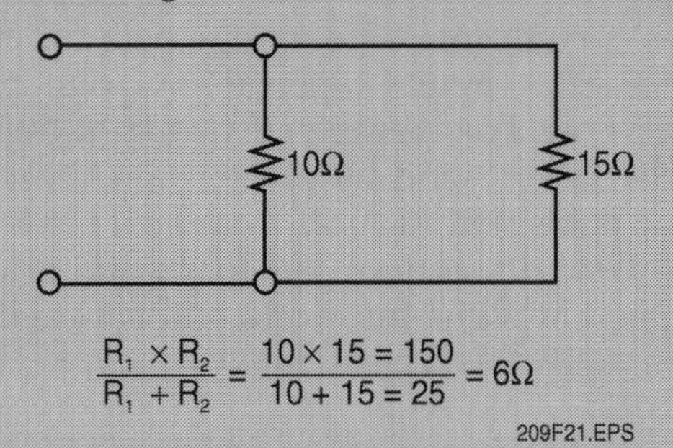

$$\frac{R_1 \times R_2}{R_1 + R_2} = \frac{10 \times 15 = 150}{10 + 15 = 25} = 6\Omega$$

209F21.EPS

Figure 21 ◆ Calculating two resistances in parallel.

The total resistance of the parallel circuit is 6Ω.

2. Calculate the total resistance of the parallel circuit in *Figure 22*.

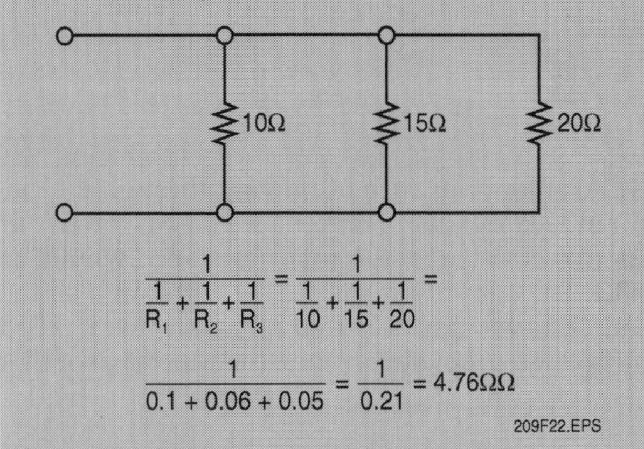

$$\frac{1}{\frac{1}{R_1} + \frac{1}{R_2} + \frac{1}{R_3}} = \frac{1}{\frac{1}{10} + \frac{1}{15} + \frac{1}{20}} =$$

$$\frac{1}{0.1 + 0.06 + 0.05} = \frac{1}{0.21} = 4.76\Omega\Omega$$

209F22.EPS

Figure 22 ◆ Calculating three resistances in parallel.

The total resistance of the parallel circuit is 4.76Ω.

5.4.0 Series-Parallel Circuits

Electronic circuits often contain a hybrid arrangement known as a series-parallel circuit (see *Figure 23*). To calculate the total circuit resistance, convert the parallel loads to their equivalent series resistance. Then add the load resistances together. You will rarely have to determine the electrical characteristics of one of these circuits as part of an installation or maintenance job.

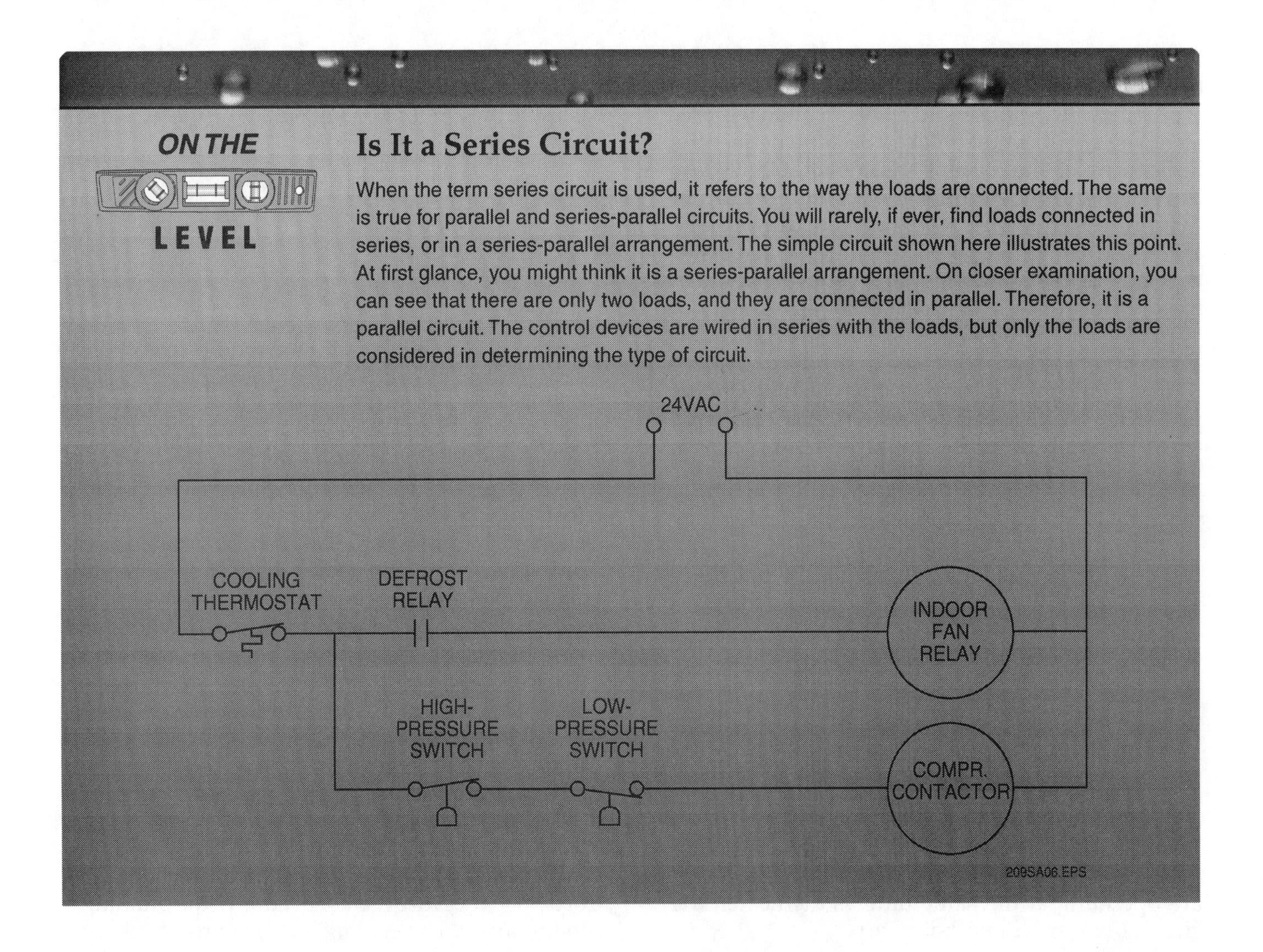

ON THE
LEVEL
Is It a Series Circuit?
When the term series circuit is used, it refers to the way the loads are connected. The same is true for parallel and series-parallel circuits. You will rarely, if ever, find loads connected in series, or in a series-parallel arrangement. The simple circuit shown here illustrates this point. At first glance, you might think it is a series-parallel arrangement. On closer examination, you can see that there are only two loads, and they are connected in parallel. Therefore, it is a parallel circuit. The control devices are wired in series with the loads, but only the loads are considered in determining the type of circuit.
24VAC
COOLING
THERMOSTAT
DEFROST
RELAY
INDOOR
FAN
RELAY
HIGH-
PRESSURE
SWITCH
LOW-
PRESSURE
SWITCH
COMPR.
CONTACTOR
209SA06.EPS

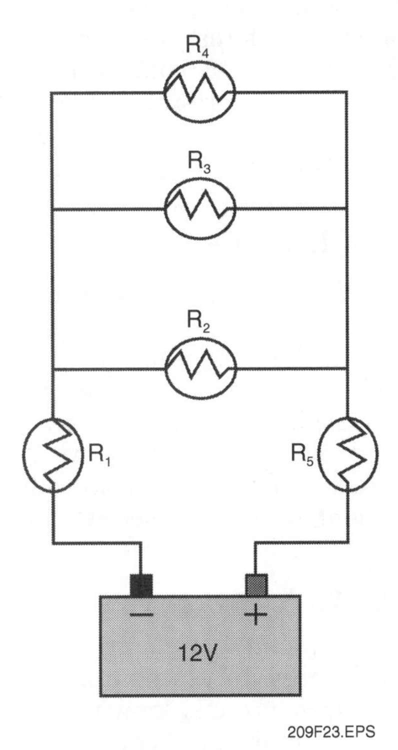

Figure 23 ◆ Series-parallel circuit.

6.0.0 ◆ ELECTRICAL CIRCUIT CHARACTERISTICS

Voltage, current, resistance, and **power** are closely related. If you know any two of them, you can determine the other two. For example, if you know how much voltage is available and how much power the load consumes, you can figure out how much current the circuit will draw. Suppose a 1,600W electric heater will be plugged into a household 120V circuit that is protected by a 15A circuit breaker. How do you determine if you can safely add the heater without overloading the circuit and tripping the circuit breaker?

The wrong way would be to plug in the heater and see if the circuit breaker trips. A better, and less dangerous, approach is to calculate how much current is used (drawn) by each appliance in the circuit, including the heater. This can be done using two simple mathematical equations: Ohm's law and the power formula. You learned about Ohm's Law earlier in this module.

Most load devices are rated by the electrical power they consume, rather than by the resistance they offer. Consider the example mentioned at the beginning of this section. A 1,600W electric heater is plugged into a household 120V circuit that is protected by a 15A circuit breaker. Because neither the resistance nor the current are known, Ohm's law can't be used to determine the amount of current that the heater will draw. The current draw can be calculated using the power formula, which states that power (P) equals voltage times current. In mathematical terms, this can be expressed as:

$$P = E \times I$$

The relationship among current, resistance, power, and voltage is illustrated in *Figure 24*. In the above example, because the current is unknown, the following formula would be used:

$$I = \frac{P}{E}$$

Now calculate the current:

$$I = \frac{P}{E}$$

$$I = \frac{1{,}600W}{120V}$$

$$I = 13.3A$$

Using the power formula, you find that on a 120V circuit, a 1,600W heater would draw 13.3A. The 15A circuit breaker would barely accommodate the heater. If other appliances in the circuit are operating at the same time as the heater, the circuit breaker will probably trip.

Electric motors may be rated in equivalent horsepower, rather than watts. A large swimming pool pump, for example, might contain a 2 horsepower (hp) electric motor. To convert horsepower to watts, use this simple formula:

$$1hp = 746W$$

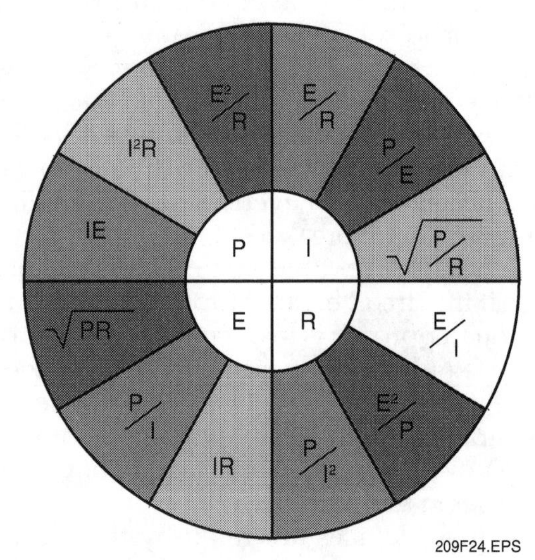

Figure 24 ◆ Power formula.

Using the formula, you can see that a 2hp motor would consume 1,492W. Use the power formula to calculate the amount of amps that the motor would draw:

$$I = \frac{P}{E}$$

$$I = \frac{1.492W}{120V}$$

$$I = 13.26A$$

In a 120V circuit, the motor would draw more than 13A.

7.0.0 ◆ ELECTRICAL COMPONENTS

Circuit diagrams use symbols to represent the electrical components and connections. Many construction drawings will include a key to electrical symbols on the electrical drawings (see *Figure 25*). You can expect to see different symbols depending on the applicable codes and standards where you work. In this section, you will begin to learn about electrical components and how they are shown on wiring schematics. Electrical components generally fall into two categories: load devices and control devices. Some devices, such as relays, contain both load and control elements. For example, a device may have both a motor and switch contacts. These are treated as control devices.

You need to know about electrical load and control devices because their heat adds to the building's cooling loads. The cooling load in turn affects systems that use plumbing: for example, hydronic or radiant heating systems, water heaters, electric baseboard heaters, and air conditioners. You do not need to know how to perform cooling load calculations, but you should be aware of the effect these components have on cooling loads.

7.1.0 Load Devices

Any device that consumes electrical energy and does work in the process is a load device. Loads convert electrical energy into some other form of energy, such as light, heat, or mechanical energy. They have resistance and consume power. An electric motor is a load; so is the burner on an electric stove. As electrical current flows through the burner element, it converts electricity into heat.

7.1.1 Motors

Electric motors convert electrical energy into mechanical energy. In plumbing, they are used primarily to drive pumps, compressors, and fans. Electric motors are the most common loads found in plumbing systems. *Figure 26* illustrates some of

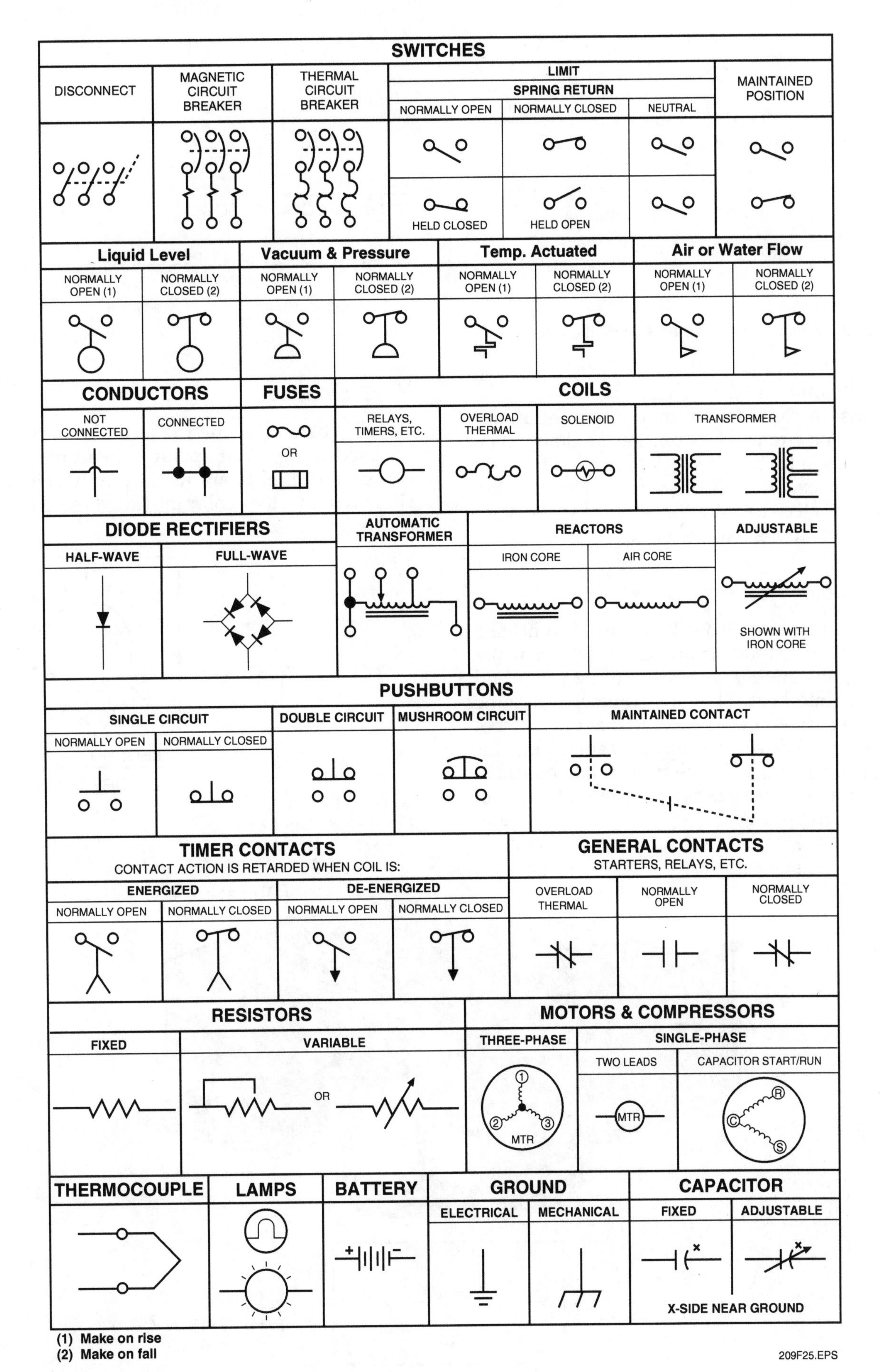

Figure 25 ◆ Electrical symbol key from a construction drawing.

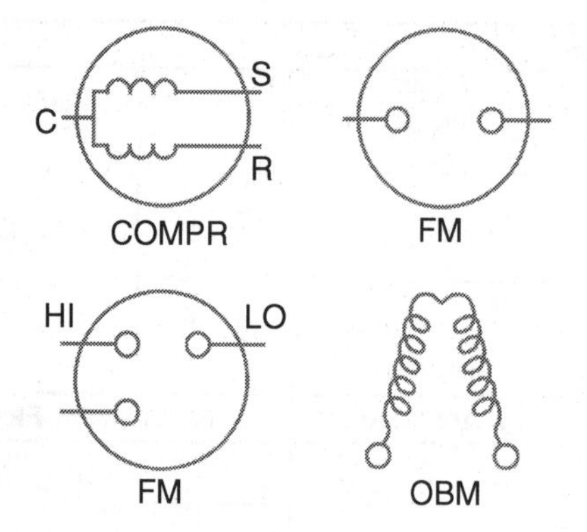

LEGEND
COMPR = COMPRESSOR
FM = FAN MOTOR
FM = FAN MOTOR, MULTIPLE SPEED
OBM = OIL BURNER MOTOR

209F26.EPS

Figure 26 ◆ Schematic symbols for electric motors.

the ways motors are depicted on schematic wiring diagrams. In cases where more than one motor appears on a circuit diagram, standard abbreviations are used to identify each motor by its application. For example, IFM stands for indoor fan motor. The electrical drawings will list the standard abbreviations that are used.

7.1.2 Electric Heaters

Electric heaters are also called resistance heaters because they are made from high-resistance materials that consume a large amount of electricity and convert it into heat. They are somewhat like the burners on an electric stove. The diagram symbol for an electric heater is usually the same as that for a resistor (see *Figure 27*). Electric baseboard heaters are a common type of electric heater that you will encounter on the job.

7.1.3 Lights

Light bulbs are classified as secondary loads. They are used to signal an operator about the status of equipment. Lights are usually represented as a circle with the lens color indicated (see *Figure 28*).

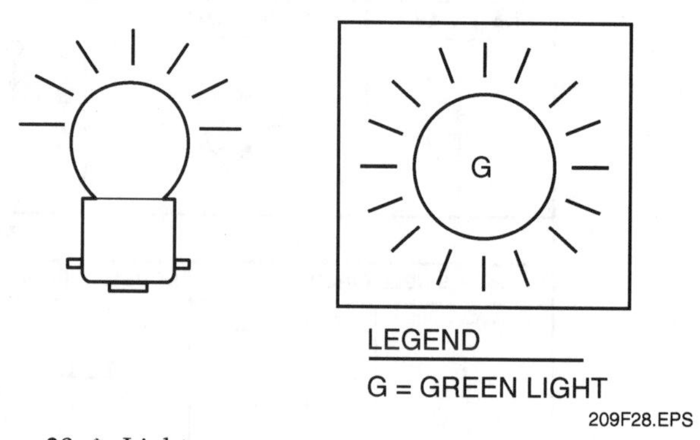

Figure 28 ◆ Light.

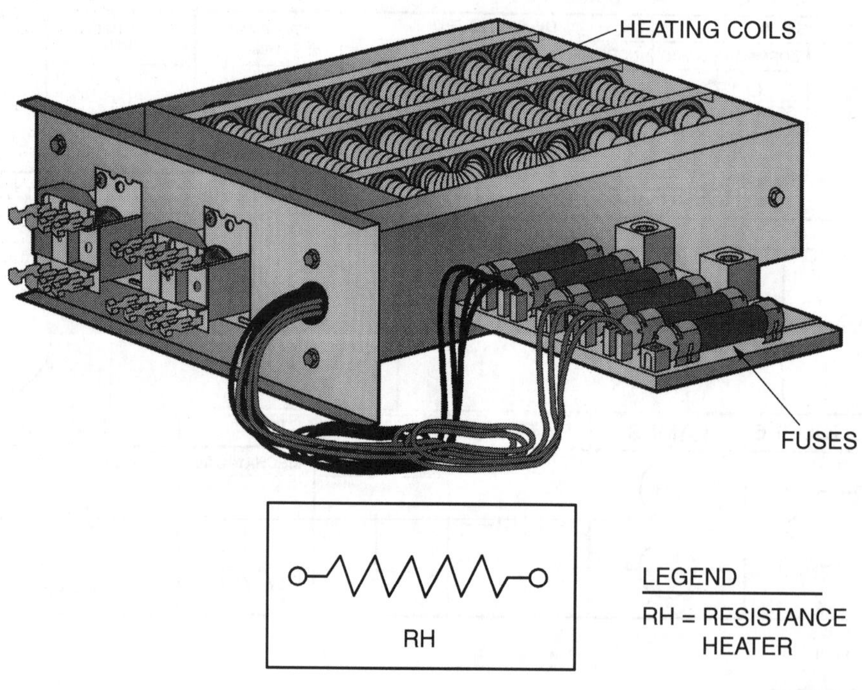

209F27.EPS

Figure 27 ◆ Resistance heater.

Electric Heaters in Air Conditioning Systems

Air conditioning systems and heat pumps often contain electric resistance heaters. Local codes may even require that a heat pump have enough electric heat capacity to support up to 100 percent of the heating demand in the event of a compressor failure. An air conditioning system with 15,000W (15 kilowatts, or kW) of supplementary electric heaters would draw about 65A in a 230V system. Many homes do not have that much extra capacity in their electrical service. In older homes, for example, the entire electrical service may be only 100A.

It is not uncommon when replacing a furnace with a heat pump to find it necessary to expand the electrical service. This can represent a significant cost to the building owner. The sales engineer who specifies the system must take such factors into account when pricing the job.

7.2.0 Control Devices

Control devices turn loads on and off. They are classified by the way they operate. You are likely to encounter many different types of control devices as part of the electrical components of plumbing systems. Always ensure that control devices are installed correctly so that their associated electrical devices function correctly.

7.2.1 Switches

Switches stop and start the flow of current to other control devices or loads. Electrical switches are classified according to the force used to operate them (manual, magnetic, temperature, light, moisture, pressure).

The simplest type of switch is one that makes (closes) or breaks (opens) a single electrical circuit. Switches that are more complex are used to control several circuits. The switching action is described by the number of poles, or electrical circuits per switch, and the number of **throws**, or switch positions. The most common switch arrangements include the following (see *Figure 29*):

- Single-pole, single-throw (SPST)
- Double-pole, single-throw (DPST)
- Double-pole, double-throw (DPDT)

Main power disconnect switches are common manual switches (see *Figure 30*). They are usually mounted directly on an electrical device. The switch allows the operator to cut power from the system

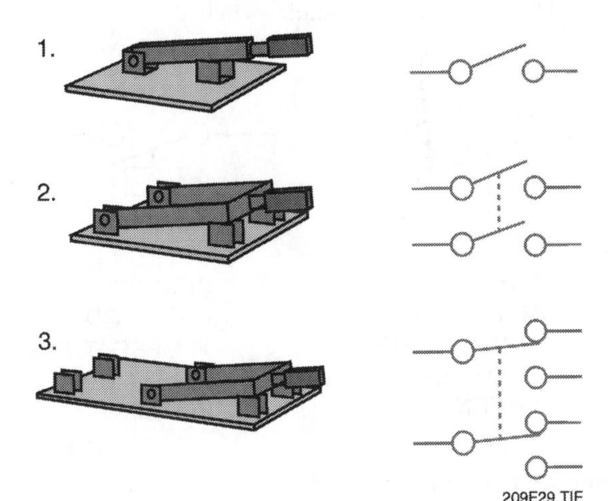

Figure 29 ◆ Switches.

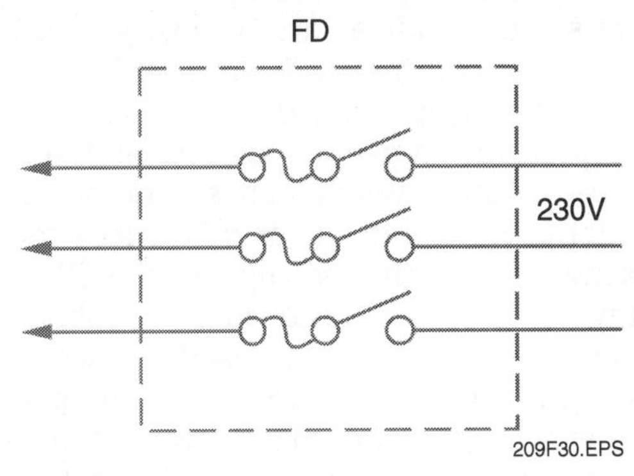

Figure 30 ◆ Fused power disconnect.

for safety reasons. Some disconnects are connected to fuses to protect the electrical device in case of an electrical overload. The *National Electrical Code®* (NEC®) and many local building codes require the installation of a power disconnect device within sight of an air conditioner installation. Some disconnects have a handle on the side to turn the power on and off (refer to *Figure 30*). Others have a plug that is removed to disconnect the power.

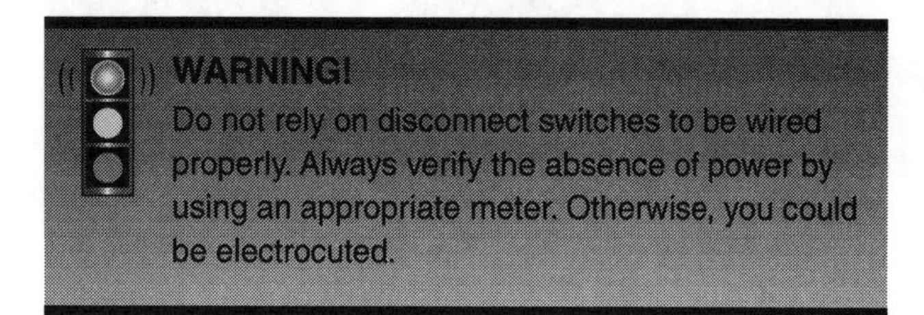

WARNING!
Do not rely on disconnect switches to be wired properly. Always verify the absence of power by using an appropriate meter. Otherwise, you could be electrocuted.

Thermostatic switches, also commonly called thermostats, open and close in response to changes in temperature. They are widely used to control heating and cooling in occupied spaces. Though thermostats have traditionally been available in analog models (see *Figure 31*), digital models are increasingly popular (see *Figure 32*).

Thermostats are designed to operate on either a temperature drop or a temperature rise (see *Figure 33*). The thermostat on the left in *Figure 33* closes on a temperature drop. It would thus be used to control a heating system. A cooling thermostat, as illustrated on the right in *Figure 33*, closes on a temperature rise.

As you have already learned, thermostatic switches are used in storage water heaters. Storage water heaters use upper and lower thermostats to control heating. Instantaneous water heaters are fitted with electric flow switches or electrically actuated pressure differential switches. Other types of temperature-controlled switches provide protection from current or heat overloads. They are commonly used in compressor circuits. Heat-sensing switches are also used as limit switches to shut off furnaces in case of a malfunction.

Pressure-sensing switches, also known as **pressurestats**, are used in a variety of ways (see *Figure 34*). They are commonly used to shut off an air conditioning compressor if a problem causes the system pressure to rise too high or fall too low.

Light-operated switches are sometimes used to sense flame. If the flame in an oil furnace goes out, for example, the light-sensing switch will shut off the flow of oil to the combustion chamber. Moisture-operated switches are used in humidifiers and dehumidifiers. A strand of hair or nylon, which expands and contracts with changes in moisture, can be used to activate the unit.

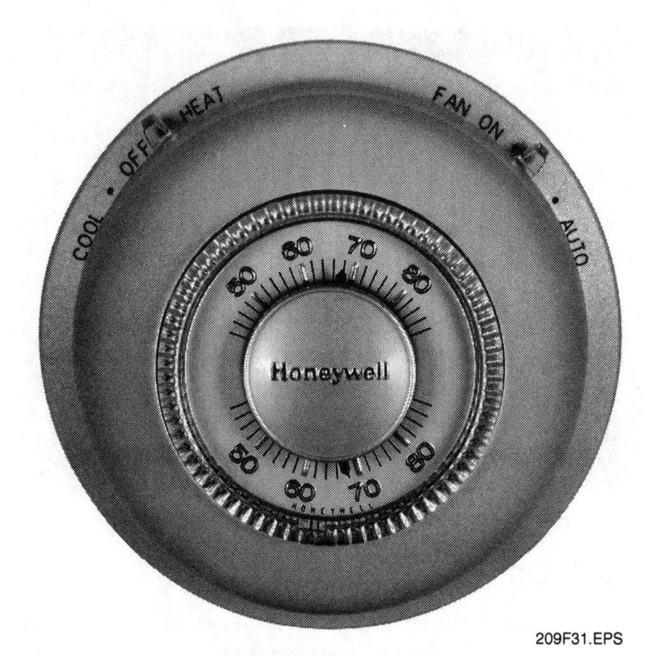

209F31.EPS

Figure 31 ◆ Analog thermostatic switch.

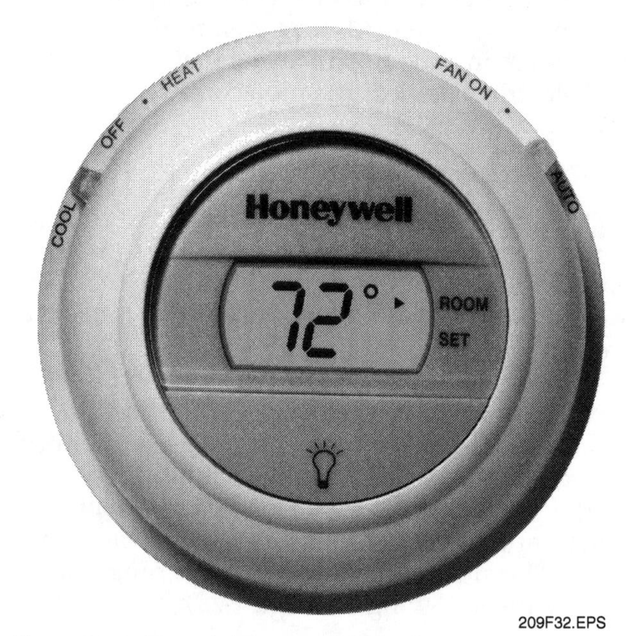

209F32.EPS

Figure 32 ◆ Digital thermostatic switch.

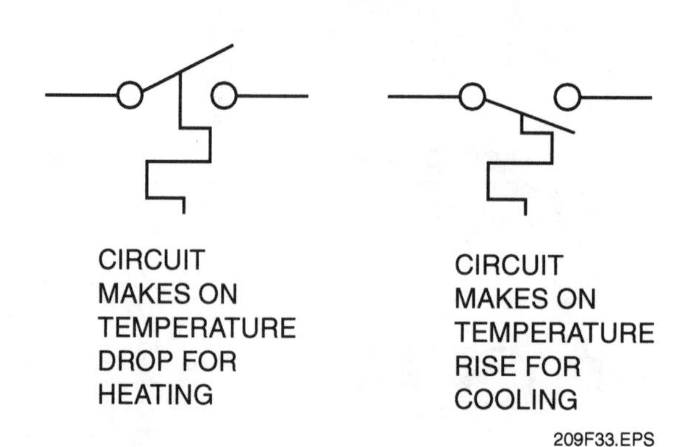

209F33.EPS

Figure 33 ◆ Thermostatic switches for temperature drops or temperature rises.

7.2.2 Fuses and Circuit Breakers

Electricians install fuses and circuit breakers to protect components and wiring from damage caused by **short circuits** and current surges (see *Figure 35*). Short circuits occur when current flow bypasses the load, such as when two conductors touch after their insulation is rubbed off by friction. Because there is no load to absorb the current, the current flow is uninhibited. This can burn up the wiring and cause a fire. Current surges are large spikes in the current that overload and damage electrical components.

A fuse contains a metal strip that melts when the current exceeds the rated capacity of the fuse (see *Figure 36*). Because the metal strip is in series with the circuit, the current flow is cut off when the strip opens. The fuse will blow more quickly in the presence of a short circuit than it will in the presence of a current overload. Once a fuse has blown, it must be replaced.

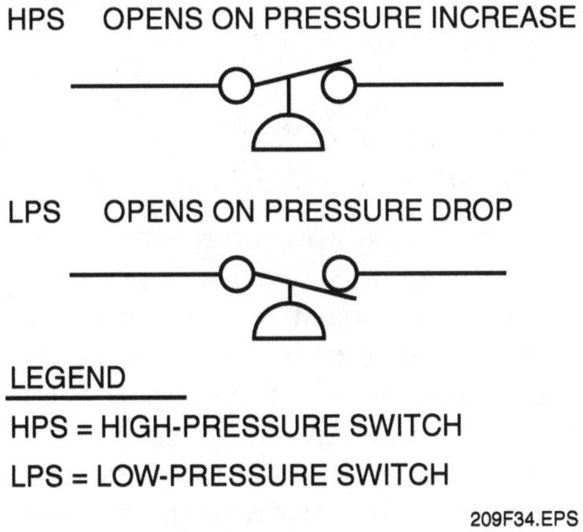

HPS OPENS ON PRESSURE INCREASE

LPS OPENS ON PRESSURE DROP

LEGEND

HPS = HIGH-PRESSURE SWITCH

LPS = LOW-PRESSURE SWITCH

209F34.EPS

Figure 34 ◆ Pressure-sensing switches.

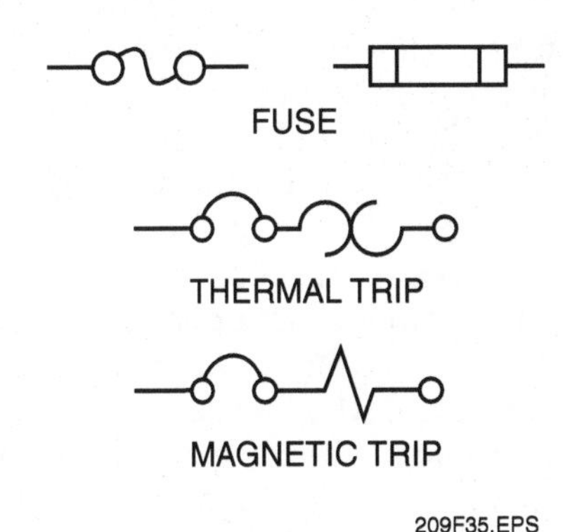

Figure 35 ◆ Fuses and circuit breakers.

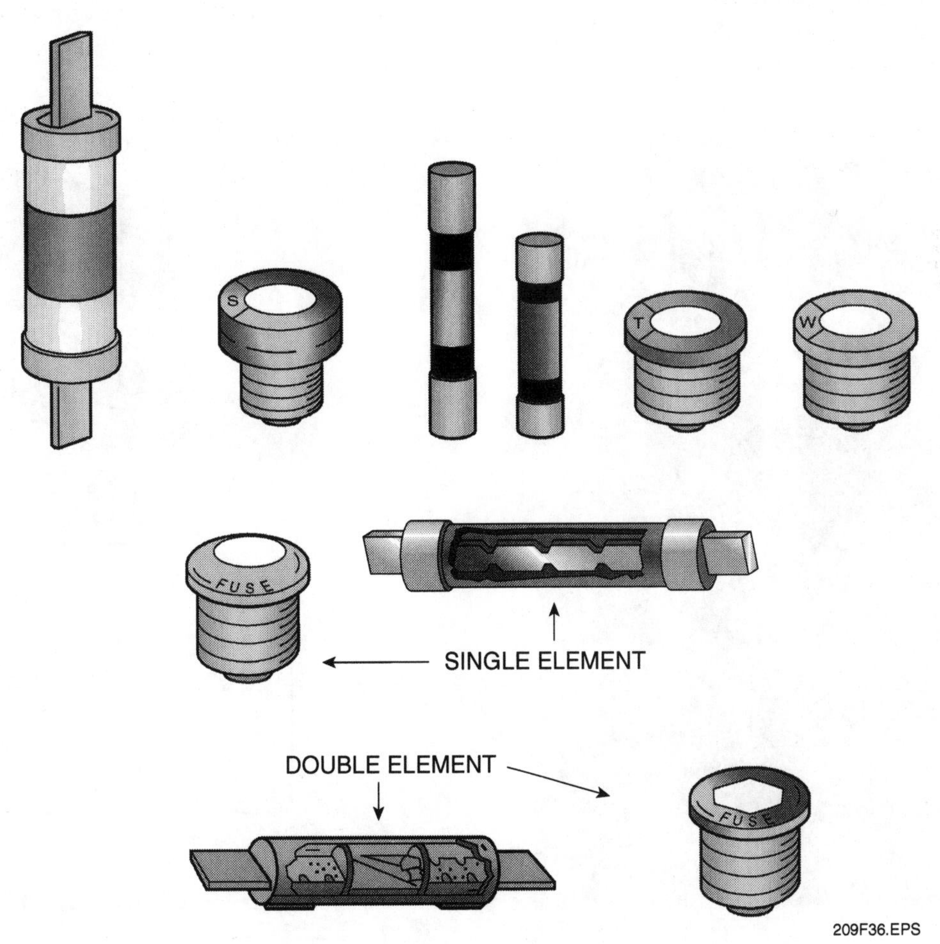

Figure 36 ◆ Examples of fuses.

Compressor motors require a large start-up current. The initial current surge, known as **locked-rotor amps (LRA)**, may be six times that of its normal running current, known as **full-load amps (FLA)**. The LRA on many compressors is more than 100A. Special delayed-opening fuses, also known as **slow-blow fuses**, are used in air conditioning equipment. These fuses will not blow unless the current surge keeps up.

Circuit breakers serve the same purpose as fuses (see *Figure 37*). Most modern construction uses circuit breakers. The big advantage of circuit breakers is that they operate like switches and can be reset when they trip. If a circuit breaker trips more than once, however, it is a sign of a circuit problem. Special circuit breakers used for air conditioning equipment offer the same protection as slow-blow fuses.

Thermal trip circuit breakers contain a spring-loaded metal element that opens when a high current causes it to overheat. Magnetic trip breakers contain a small coil of wire that magnetically opens the contacts when the current is excessive.

When a live conductor touches another conducting substance, such as the metal frame of an electrical device, it causes a special type of short circuit called a **ground fault**. Install ground fault circuit interrupters (GFCI) to prevent ground faults.

7.2.3 Solenoids

Solenoids are electromagnets that are used to operate valves and to switch devices on and off (see *Figure 38*). One of the most common applications of a solenoid is to reposition a valve to either start or shut off the flow of a gas or liquid. When a current flows through the solenoid, the magnetism produced by the current is used to attract or repel a circuit contact. A solenoid labeled GV on the electrical plan indicates that the solenoid is used to control a gas valve, which feeds natural or LP gas to a gas-fired furnace.

7.2.4 Relays, Contactors, and Magnetic Starters

Relays, **contactors**, and single- and three-phase **magnetic starters** are electrically operated switching devices that consist of a solenoid and one or more sets of movable contacts. The contacts open and close in response to the magnetic field generated by the current flow in the coil.

Figure 39 illustrates common types of relays. When a current flows through the relay coil, a magnetic field is created. This field causes the contacts to change position. Some of the contacts close, providing a path for current; these are called

Figure 37 ◆ Circuit breaker panel.

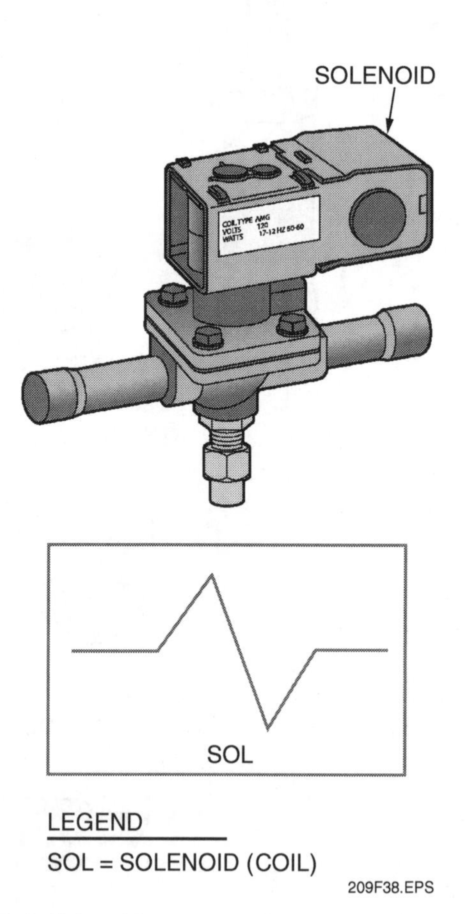

Figure 38 ◆ Solenoid.

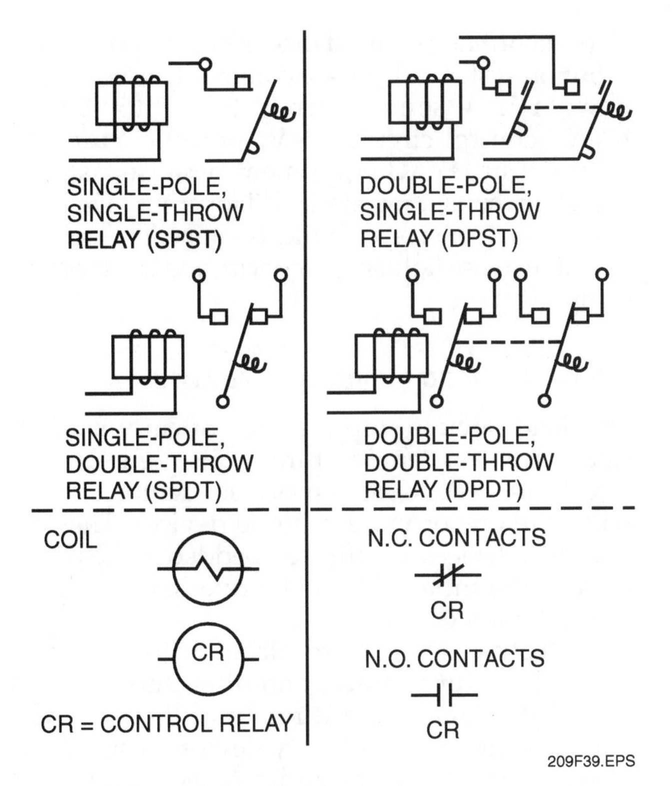

Figure 39 ◆ Types of relays.

normally open (N.O.) contacts because they are open when the relay is de-energized. Other contacts open; these are called normally closed (N.C.) contacts because they are closed when the relay is de-energized.

On circuit diagrams, N.C. contacts have a slash through them. The coil and contacts of a relay will be shown at different locations on a diagram because each set of contacts controls a different circuit. You can find them because the coil and contacts will have the same name; in this case, CR for control relay. The same pole-throw descriptions used for switches also apply to relays and contactors.

Figure 40 illustrates a hypothetical electrical control circuit with a relay. The thermostatic switch on the left closes when the temperature in the space exceeds a selected value, which is called the **setpoint**. The closed thermostat completes a current path to the relay coil. The coil produces a magnetic field that changes the position of the relay contacts. The N.O. contacts close, completing a path to the fan motor. The N.C. contacts open, causing the FAN OFF light to go out. When the thermostat opens, the current stops and the contacts return to their original positions.

Contactors are heavy-duty relays designed to carry large currents. They are often used to start and stop compressors. Magnetic starters are used to start large motors. They usually contain overload

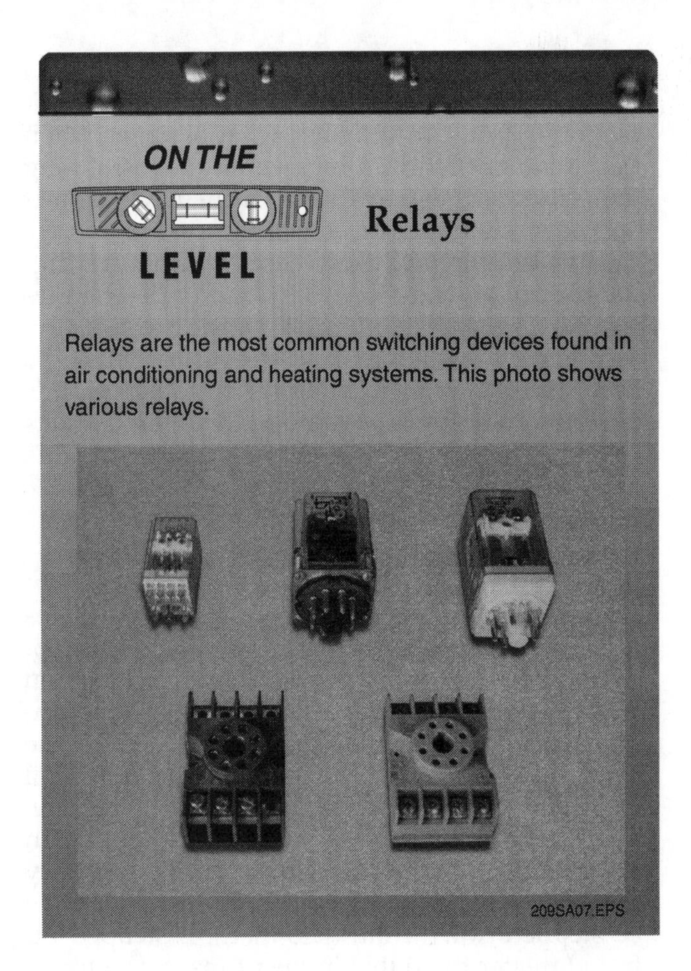

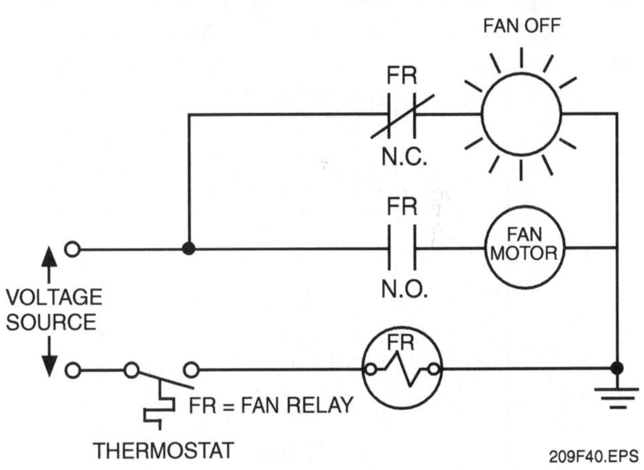

Figure 40 ◆ Hypothetical relay circuit.

protection devices and may include an on-off switch. The NEC® will specify whether a contactor or magnetic starter is required for a motor.

A typical magnetic starter consists of a magnetic coil, bimetallic relays, screw terminals, a reset switch, heater packs, and single-phase protection. Magnetic starters may be mounted in either horizontal or vertical positions. Refer to your local applicable code for appropriate applications.

Note that the contacts of a relay, contactor, or switch will appear in a circuit diagram in their normal state; that is, their state when no voltage is being applied to the circuit. When reading circuit diagrams, you can visualize what happens when the circuit is energized by mentally repositioning the contacts. Imagine a slash in the N.O. contact and no slash in the N.C. contact.

Time-delay relays use a thermal element or digital control to delay the energizing or de-energizing of a relay. Delays may range from a few seconds to a few hours. They are typically used to keep a furnace fan running for 30 to 60 seconds after the burners have shut off. The fan extracts residual heat from the heat exchangers. This improves heating efficiency.

7.2.5 Transformers

Transformers are used to raise or lower voltages (see *Figure 41*). A transformer usually consists of two or more coils of wire wound around a common iron core. When a current passes through the primary coil (primary winding) of the transformer, the resulting magnetic field cuts through the other coil (secondary winding), creating a current in that coil. This process is known as **induction**. Depending on the number of turns of wire in each secondary winding, the voltage induced in the secondary will be stepped down from (less than) or stepped up from (greater than) that in the primary. The transformer in *Figure 41* is a step-down transformer with two secondary windings. Each secondary winding acts as the power source for a circuit.

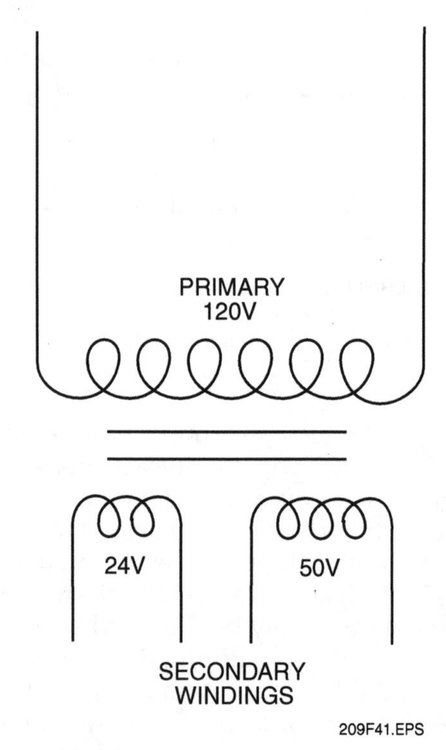

Figure 41 ◆ Transformer.

Transformers are used extensively in power distribution systems. They are commonly used to step down a 120V source current to 24V to operate HVAC control circuits. 24V control circuits are common in HVAC equipment because, as low-voltage circuits, they are less dangerous. They are also less expensive to build because low-voltage circuits can use lighter-gauge wire and lighter-duty components.

7.2.6 Overload Protection Devices

Overload devices stop the flow of current when safe current or temperature limits are exceeded (see *Figure 42*). Most compressor motor circuits will have one or more of these devices. Thermal overload devices are often embedded in the windings of the motor. Magnetic overloads operate much the same as relays.

Protective devices are placed into two categories: **line-duty devices** and **pilot-duty devices.** Line-duty devices are directly in line with the voltage source. In pilot-duty devices, a small coil of wire senses current and acts as a magnet to open contacts in the motor control circuit when the current is too high.

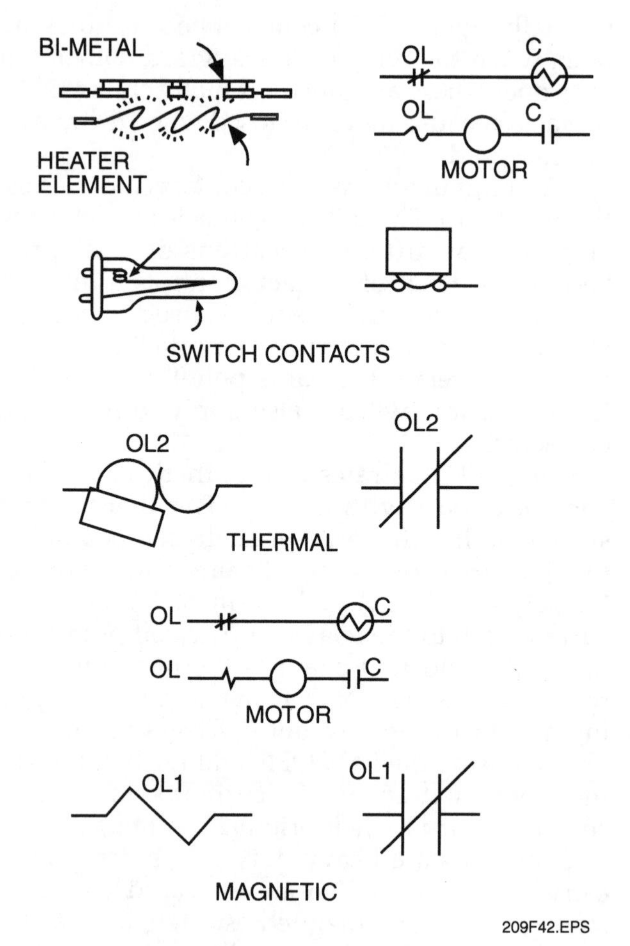

Figure 42 ◆ Overload protection devices.

Some overload devices reset automatically when the cause of the overload has been removed. Others must be manually reset, usually by pressing a button on the device. Overload devices embedded in motor windings are typically automatic reset types because they are not easily accessible.

8.0.0 ◆ ELECTRONIC CONTROLS

Electronic circuits use solid-state timing, switching, and sensing devices to control loads and protective circuits. Electronic circuits operate at much lower voltage and current levels than electromechanical devices such as relays and solenoids. In addition, they are more reliable because they have no moving parts. Most electronic circuits consist of microminiature components mounted on printed circuit boards (see *Figure 43*). The components are mounted to the top of the board. The copper runs found on the underside of the board serve the same purpose as wiring. Circuits that can be used in a variety of products are often packaged in sealed modules.

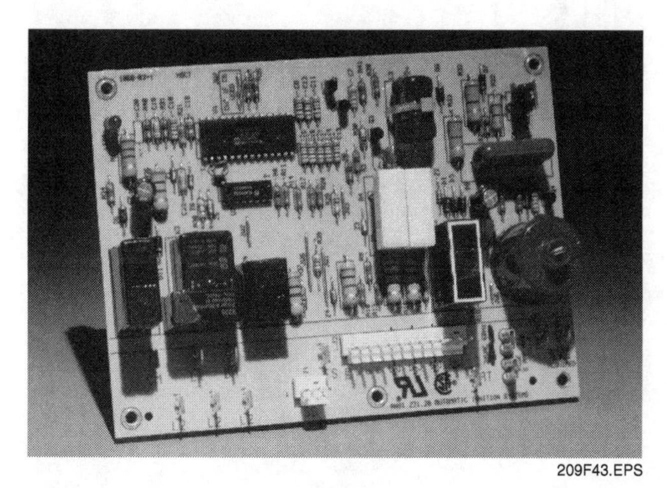

209F43.EPS

Figure 43 ◆ Printed circuit board.

Microcomputers allow building occupants or managers to custom tailor the operation of electrical systems to their needs (see *Figure 44*). Programmable thermostats, for example, allow the occupant to program a different operating temperature for different times of day. Microcomputers are able to receive a large amount of information and make decisions based on the information and the instructions in their programs.

Microprocessor-controlled circuits offer more capabilities and more precise control than conventional circuits. They are often self-diagnosing. When there is a problem, the microprocessor evaluates information from around the system to determine the location of the problem. A digital readout or other display indicates to the technician the component or electronic circuit that needs to be checked.

An electronic circuit is usually treated as a "black box." In other words, if the technician finds that the control circuit has failed, the entire circuit board or module is replaced. Unlike conventional circuits, it is not necessary to analyze the circuit to figure out which component has failed.

9.0.0 ◆ WATER HEATER ELECTRICAL SYSTEMS

Water heaters heat, store, and supply hot water to all or part of a building (see *Figure 45*). They are the simplest electrical devices that plumbers deal with

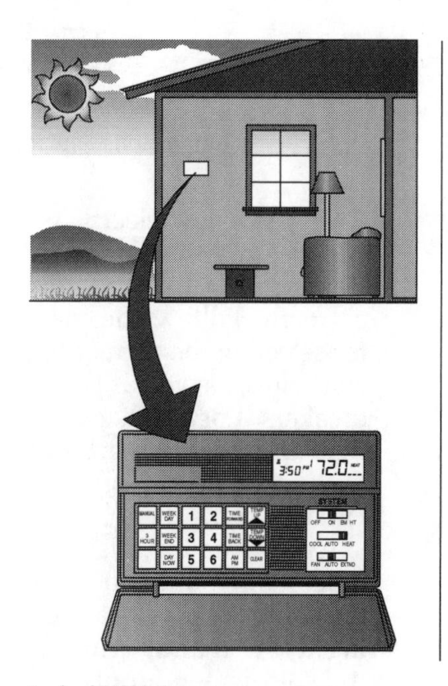

209F44.EPS

Figure 44 ◆ Electronic control of HVAC systems.

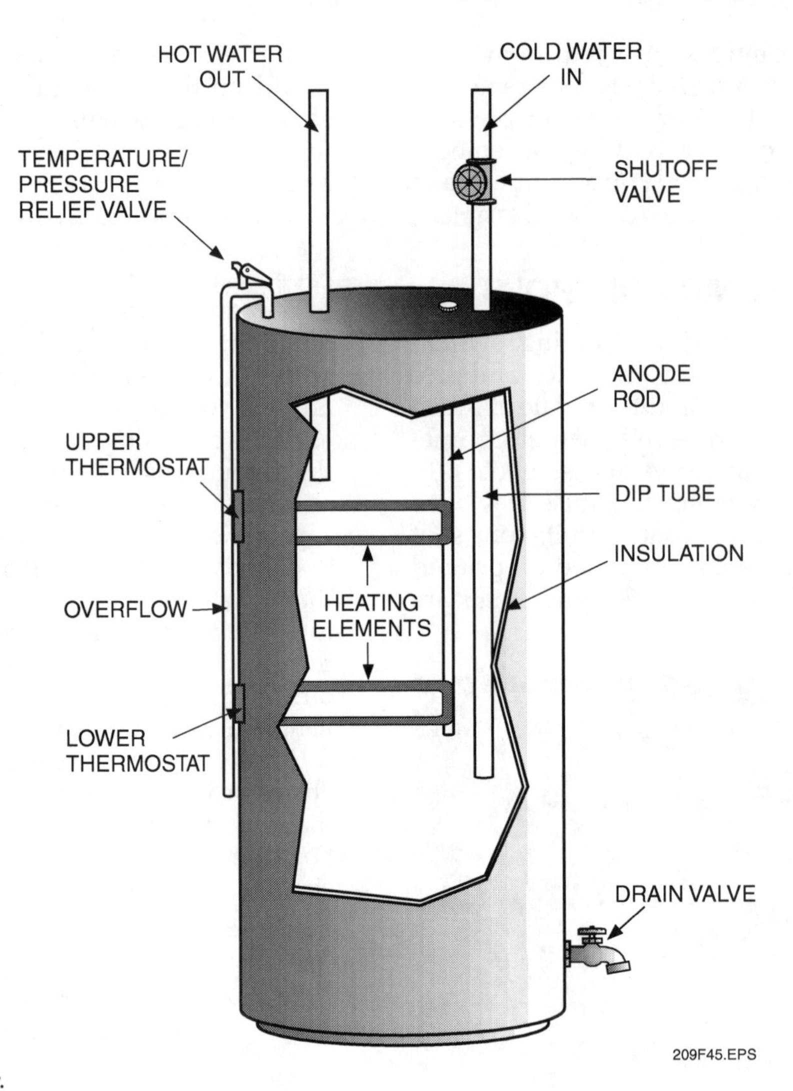

Figure 45 ◆ Water heater.

on the job. Water heaters are classified according to their heat source. Common types include gas-fired, electric, and solar water heaters. Regardless of the heat source, however, all water heaters use electricity in one way or another. You will learn more about water heaters in the module *Installing Water Heaters* elsewhere in *Plumbing Level Two*. In this section, you will learn how typical water heater electrical systems work and how to troubleshoot problems with system components.

Electrical water heaters do not burn fuel to generate heat. Instead, the water is heated by two **immersion elements** in the water storage tank (see *Figure 46*). An automatic thermostat controls the immersion elements (see *Figure 47*). The thermostat senses the tank's outside surface temperature and adjusts the flow of electricity to the elements. In normal operation, the upper element of a typical water heater will operate first to heat the water in the upper section of the tank. When that water reaches a preset temperature, the upper element is cycled off and the lower element is cycled on to heat the remaining water (see *Figure 48*).

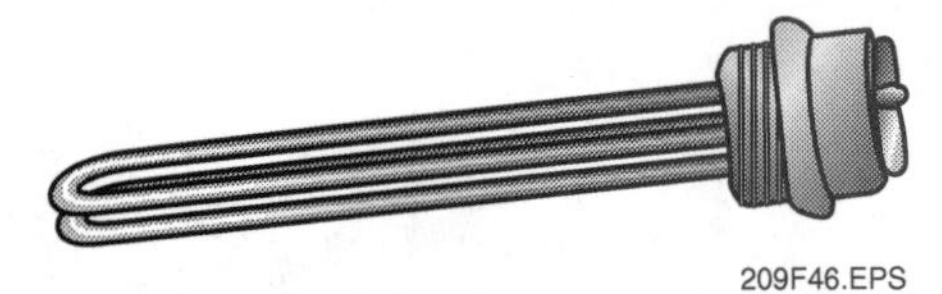

Figure 46 ◆ Immersion element.

To test a new electric water heater after installation, begin by turning on the water and checking for leaks. Ensure that the power supply voltage is set correctly. Follow the manufacturer's instructions for selecting the correct size wire for the branch circuit wires, as well as properly rated fuses or circuit breakers. Ensure that all electrical connections are tight and the unit is grounded properly. Use a meter to ensure that the heater is set at the proper voltage. Finally, set the thermostat to the desired setting.

Once installed, water heaters will require periodic maintenance and repair. Remove and inspect the anode yearly. Replace corroded anodes according to the manufacturer's specifications. Turn off the cold water supply to the water heater before removing the anode for inspection. Never **dry fire**

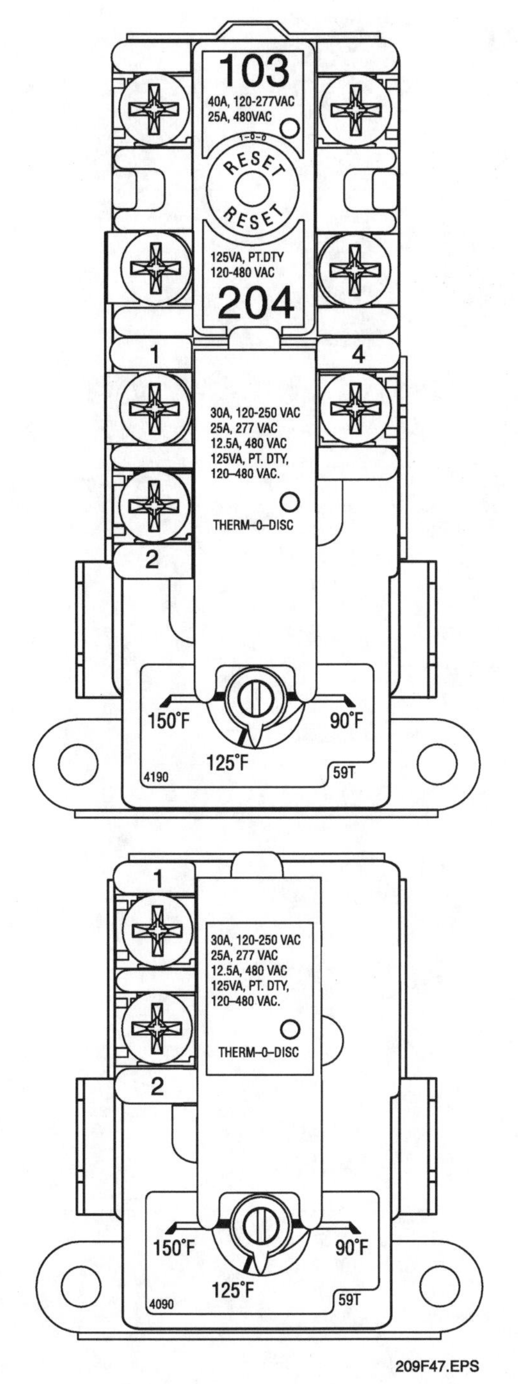

Figure 47 ◆ Automatic thermostat on an electric water heater.

an electric water heater. This means to operate the water heater without water in the storage tank. Dry firing a water heater can cause the immersion elements to burn out.

CAUTION

Never operate a water heater with the anode removed. Not only will this violate the water heater's warranty, but it will also shorten the life of the water heater's glass tank.

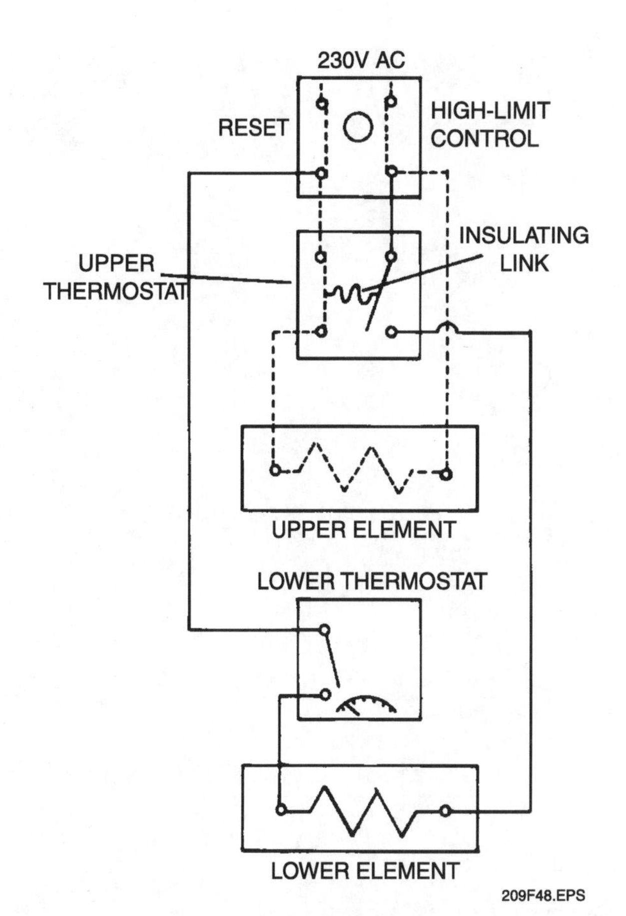

Figure 48 ◆ Wiring diagram of immersion elements in an electric water heater.

When the upper and lower heating elements and thermostats require servicing, be sure to take all appropriate safety precautions when working with electrical components. Turn off the electrical power to the water heater. Apply lockout and/or tagout devices to prevent someone from accidentally turning the power back on before your work is complete. Remove the protective cover from the thermostat and carefully remove and save any insulation or lining inside. Follow the manufacturer's instructions for removing the thermostat from its mounting bracket. Use a meter to test the thermostat by applying the meter probes to the thermostat terminals. Always replace thermostats with authorized replacement parts.

The heating element is sealed by a gasket. When inserting a heating element into the water heater after testing or replacement, install a new gasket to prevent leakage. Use a meter to test the heating element by touching one probe to a ground and the other to each terminal. Replace an element that registers resistance to the current. Be sure to replace all components correctly according to the manufacturer's instructions. Finally, restore electrical power and remove the lockout and tagout devices. You will learn how to troubleshoot water heater problems in more detail in *Installing Water Heaters*.

Review Questions

Sections 5.0.0–9.0.0

1. The current drawn by a 120V series circuit containing a 100Ω load and a 50Ω load is _____.
   a. 0.8W
   b. 0.8A
   c. 8A
   d. 1.25A

2. In a 120V circuit containing two parallel resistances of equal value, the voltage drop across each resistance is ____V.
   a. 30
   b. 60
   c. 120
   d. 240

3. A 120V circuit containing a 1,500W load draws a current of _____.
   a. 125mA
   b. 0.08A
   c. 180,000A
   d. 12.5A

4. A 120V parallel circuit that contains a 100Ω load and an 80Ω load and draws 2.7A consumes _____W.
   a. 324
   b. 44
   c. 80
   d. 120

5. A resistance heater produces 5kW of heat in a 240V circuit. The circuit draws _____A and the resistance of the heating elements is _____Ω.

6. The total resistance of the circuit in *Figure 49* is _____Ω.

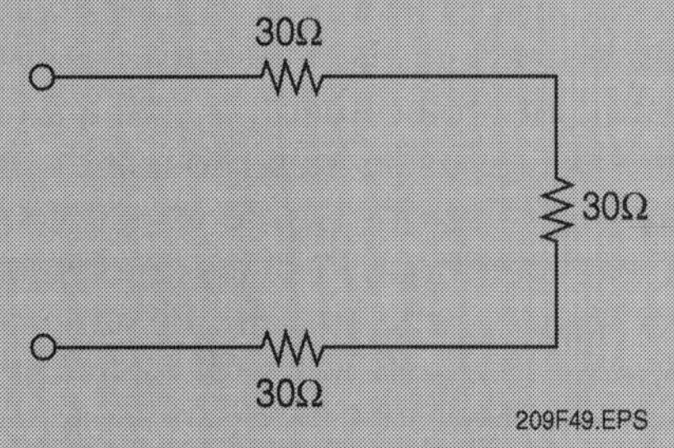

Figure 49 ◆ Review question circuit.

7. In *Figure 49*, if the circuit has a 120V supply, the current draw will be _____A.

8. The current through each resistor in the circuit in *Figure 49* is _____A.

9. The total resistance of the parallel circuit in *Figure 50* is ____Ω.

10. In *Figure 50*, if the circuit draws 3A, the source voltage must be ____V.

11. *Figure 51* illustrates common schematic symbols for _____.
    a. motors
    b. electric heaters
    c. lights
    d. switches

12. The diagram symbol for an electric heater is usually the same as that for a _____.
    a. fuse
    b. switch
    c. resistor
    d. thermostat

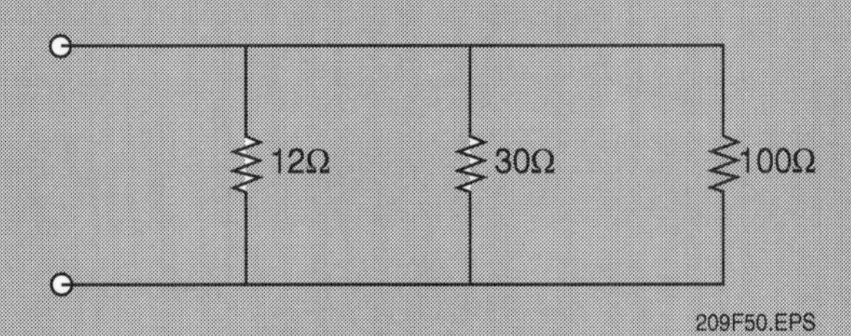

Figure 50 ◆ Review question parallel circuit.

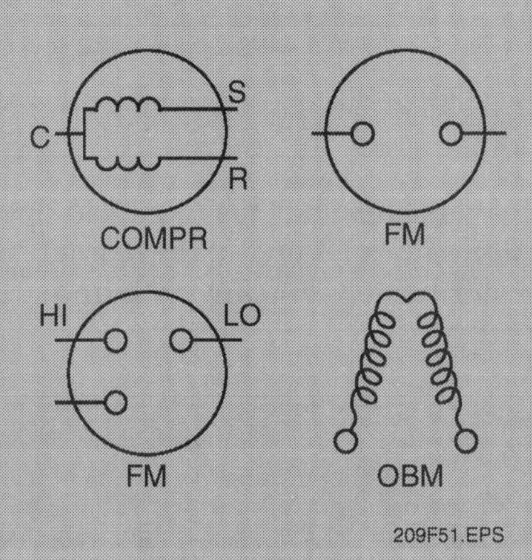

Figure 51 ◆ Review question symbols.

Review Questions

13. Control devices are classified by _____.
    a. capacity
    b. their output
    c. their rate of conversion
    d. the way they operate

14. A DPDT switch can control _____ circuit(s).
    a. 4
    b. 1
    c. 3
    d. 2

Match each type of switch with its correct schematic drawing in *Figure 52*.

15. _____ Single-pole, single-throw (SPST)

16. _____ Double-pole, single-throw (DPST)

17. _____ Double-pole, double-throw (DPDT)

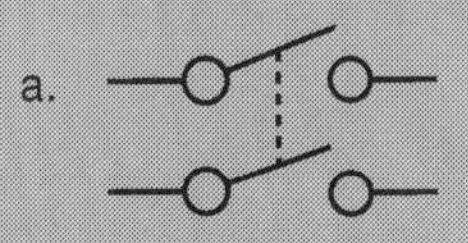

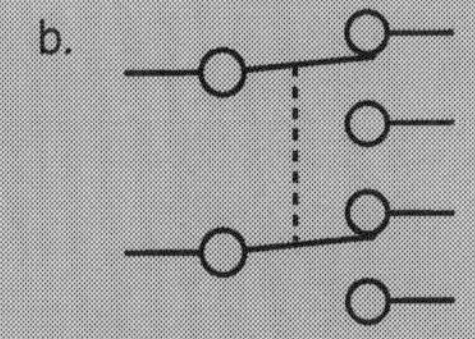

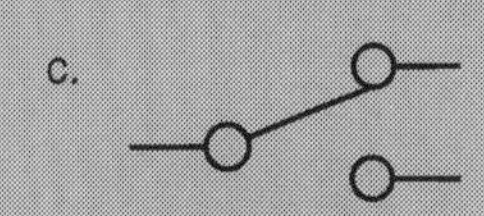

d.

209F52.TIF

Figure 52 ◆ Review question switches.

18. On a compressor motor, the initial current surge is known as _____ amps.
    a. locked rotor
    b. free rotor
    c. full load
    d. partial load

19. Fuses are rated on the basis of _____.
    a. how fast they blow
    b. how much current they can withstand before they blow
    c. full-load amps (FLA)
    d. locked-rotor amps (LRA)

20. _____ circuit breakers contain a spring-loaded metal element that opens when a high current causes it to overheat.
    a. Delayed opening
    b. Thermal trip
    c. Ground fault interrupt
    d. Magnetic trip

21. Control devices that use electromagnets to operate valves are called _____.
    a. relays
    b. contacts
    c. solenoids
    d. starters

22. On circuit diagrams, N.C. relay contacts are shown with _____.
    a. a circle around them
    b. the abbreviation "N.C."
    c. a slash through them
    d. an X through them

23. Transformers are used to step down 120V source current to operate _____V control circuits, which are common in HVAC equipment.
    a. 12
    b. 24
    c. 36
    d. 48

24. Overload protection devices are categorized as _____ devices and _____ devices.
    a. short; surge
    b. N.O.; N.C.
    c. AC; DC
    d. line-duty; pilot-duty

25. During periods of normal demand, _____ immersion element(s) in an electric water heater is/are used.
    a. the upper
    b. the lower
    c. both
    d. neither

Many plumbing systems use devices that contain electrical circuits. To install and service these systems, plumbers must know how electrical components work, how to read circuit diagrams, and how to use electrical test equipment. When working with electricity, be sure that you understand and follow proper safety practices. Failure to develop good safety habits and follow safety rules can result in injury or death to yourself or your coworkers.

Notes

Trade Terms Introduced in This Module

Trade Terms Introduced in This Module

Alternating current (AC): An electrical current that changes direction in cycles.

Ammeter: A test instrument used to measure current flow.

Ampere (amp): The unit of measurement for current flow. The magnitude is determined by the number of electrons passing a point at a given time.

Analog meter: A meter that uses a needle to indicate a value on a scale.

Circuit: A loop of electrical current that contains a voltage source, a load, and conductors.

Circuit diagram: A schematic drawing that indicates the path that electricity flows in a circuit.

Clamp-on ammeter: A current meter in which jaws placed around a conductor measure the amount of current flow.

Conductor: A material that readily conducts electricity; also, the wire that connects components in an electrical circuit.

Contactor: A control device consisting of a coil and one or more sets of contacts used as a switching device in high-voltage circuits.

Continuity: A continuous current path. Absence of continuity indicates an open circuit.

Current: The rate at which electrons flow in a circuit, measured in amperes.

Cycle: In AC electricity, a single repetition of alternating current flow.

Digital meter: A meter that provides a direct numerical reading of the value measured.

Direct current (DC): An electrical current that flows in one direction.

Dry fire: A term that refers to turning on an electric water heater with no water in the storage tank, causing the elements to burn out.

Electromagnet: A coil of wire wrapped around a soft iron core. When a current flows through the coil, magnetic flux lines are created.

Electromotive force (EMF): Another term for voltage.

Frequency: In AC electricity, a measure of the number of cycles per second.

Full-load amps (FLA): The normal running current of a compressor motor.

Ground fault: A type of short circuit caused by a live conductor touching another conducting substance.

Immersion element: An electrically heated element that, when exposed to water, quickly and efficiently transfers heat.

Induction: The generation of an electrical current by a conductor placed in a magnetic field. The motion of either the conductor or the magnetic field causes the current flow.

In-line ammeter: A current reading meter that is connected in series with the circuit under test.

Insulator: A device that inhibits the flow of electrical current. An insulator is the opposite of a conductor.

Ladder diagram: A simplified schematic diagram in which the load lines are arranged like the rungs of a ladder between vertical lines representing the voltage source.

Line-duty device: A protective device that is placed directly in line with a voltage source.

Load: A device that converts electrical energy into another form of energy, such as heat, mechanical motion, or light.

Locked-rotor amps (LRA): The initial current surge in a compressor motor.

Magnetic flux lines: Lines of force in magnetized iron.

Magnetic starter: An electrically operated switching device used to start large motors.

Multimeter: A test instrument capable of reading voltage, current, and resistance.

Ohm: The unit of measurement for electrical resistance.

Parallel circuit: A circuit in which each load is connected directly to the voltage source.

Pilot-duty device: A protective device in which a wire coil opens an electric motor's contacts when the current rises above a pre-set level.

Pole: In a magnet, the ends of the magnetic flux lines, identified as either north or south. In a battery, the direction of current flow toward a positive (+) or away from a negative (–) charge.

Power: The amount of energy, measured in watts, consumed by an electrical load. Power equals voltage multiplied by current.

Pressurestat: A pressure-sensitive switch used to protect compressors.

Rectifier: A device that converts AC voltage to DC voltage.

Relay: A magnetically operated device consisting of a coil and one or more sets of contacts.

Resistance: The opposition to current flow, such as in a load.

Series circuit: A circuit that provides a single path for current flow.

Series-parallel circuit: A circuit that consists of both a series and a parallel design.

Setpoint: A pre-selected temperature in a thermostatic switch that causes the switch to close when exceeded.

Short circuit: A high current flow caused by a conductor bypassing the load.

Slow-blow fuse: A fuse with a built-in time delay.

Solenoid: An electromagnetic coil used to control a mechanical device.

Throw: A single operation of a switch, from one position to another.

Transformer: An electromagnet that is used to raise or lower voltage.

Turbine: A set of rotary blades spun by the high-speed flow of steam, water, or gas.

Volt: The unit of measurement for voltage.

Voltage: A measure of the electrical potential for current flow. Also known as electromotive force (EMF).

Voltage drop: Voltage measured across any load.

Watts: The unit of measure for power consumed by a load.

Additional Resources

This module is intended to be a thorough resource for task training. The following reference works are suggested for further study. These are optional materials for continued education rather than for task training.

Modern Refrigeration and Air Conditioning, Latest Edition. Tinley Park, IL: The Goodheart-Willcox Company, Inc.

Vest Pocket Guide to the National Electrical Code, Latest Edition. Quincy, MA: National Fire Protection Association.

References

Dictionary of Architecture and Construction, Third Edition, 2000. Cyril M. Harris, ed. New York: McGraw-Hill.

Electrical Residential Water Heaters, Catalog AP9029-8, 2003. Montgomery, AL: Rheem Water Heater Company.

Figure Credits

John Traister	209F07
Topaz Publications, Inc.	209F08, 209F10, 209F11, 209F30 (photo), 209F37, 209SA01, 209SA05, 209SA07
Courtesy of Honeywell International Inc.	209F31, 209F32
Carrier Corporation	209F43
Rheem Manufacturing Company	209F47

CONTREN® LEARNING SERIES — USER UPDATE

The NCCER makes every effort to keep these textbooks up-to-date and free of technical errors. We appreciate your help in this process. If you have an idea for improving this textbook, or if you find an error, a typographical mistake, or an inaccuracy in NCCER's Contren® textbooks, please write us, using this form or a photocopy. Be sure to include the exact module number, page number, a detailed description, and the correction, if applicable. Your input will be brought to the attention of the Technical Review Committee. Thank you for your assistance.

Instructors – If you found that additional materials were necessary in order to teach this module effectively, please let us know so that we may include them in the Equipment/Materials list in the Annotated Instructor's Guide.

Write: Product Development and Revision
National Center for Construction Education and Research
P.O. Box 141104, Gainesville, FL 32614-1104

Fax: 352-334-0932

E-mail: curriculum@nccer.org

Craft ______ Module Name ______

Copyright Date ______ Module Number ______ Page Number(s) ______

Description ______

______

______

______

(Optional) Correction ______

______

______

(Optional) Your Name and Address ______

______

______

02210-05

Installing Water Heaters

02210-05
Installing Water Heaters

Topics to be presented in this module include:

Overview

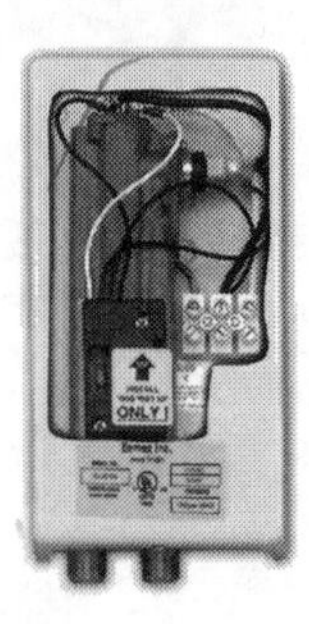

Hot water is not supplied by the public utility. Instead, water is heated in each building and distributed to the fixtures that use it. Plumbers install and service centrally located water heaters. Water heaters heat, store, and supply hot water to all or part of a building. Water heaters commonly consist of a heat source, a storage tank, temperature controls, and safety devices. Plumbers understand the basic concepts and the general variations of water heaters. The most common types of water heaters include gas-fired, electric, or solar water heaters. Water heaters are further divided into the following categories: self-contained storage type, remote storage type, instantaneous, and semi-instantaneous.

Plumbers select water heaters that meet the needs and requirements of each installation. Plumbers consider their cost, capacity, durability, and ability to hold heat, as well as the availability of fuel. Before installation, plumbers inspect the water heater for damage or defects and measure the heater to ensure there is ample space. Plumbers install water heaters by completing two basic phases: rough-in plumbing and installation. Rough-in drawings enable plumbers to identify the physical dimensions, capacity specifications, input Btu per hour, and recovery rate. Local applicable codes specify safety requirements. After installation, plumbers test the heaters to ensure that they work correctly. Proper testing eliminates callbacks and ensures that the water heater is operating efficiently and safely.

Focus Statement

The goal of the plumber is to protect the health, safety, and comfort of the nation job by job.

Code Note

Codes vary among jurisdictions. Because of the variations in code, consult the applicable code whenever regulations are in question. Referring to an incorrect set of codes can cause as much trouble as failing to reference codes altogether. Obtain, review, and familiarize yourself with your local adopted code.

Objectives

When you have completed this module, you will be able to do the following:

1. Describe the basic operation of water heaters.
2. Identify and explain the functions of the basic components of water heaters.
3. Install an electric water heater.
4. Install a gas water heater.
5. Describe the safety hazards associated with water heaters.

Trade Terms

Anti-siphon hole
Aquastat
Baffle
British thermal unit
Collector
Condensing water heater
Convection
Dip tube
Dry fire
Flammable Vapor Ignition Resistant (FVIR) water heater
Flue
Heat exchanger
High-limit control
Immersion element
Indirect water heater
Instantaneous water heater
Jacket
Main burner
Pilot burner
Probe
Recovery rate
Relief valve opening
Safety pilot
Scale
Solenoid valve
Storage tank
Thermocouple
Thermostatic control

Required Trainee Materials

1. Appropriate personal protective equipment
2. Pencil and paper
3. Copy of local applicable code

Prerequisites

Before you begin this module, it is recommended that you successfully complete *Core Curriculum; Plumbing Level One; Plumbing Level Two,* Modules 02201-05 to 02209-05.

This course map shows all of the modules in the second level of the Plumbing curriculum. The suggested training order begins at the bottom and proceeds up. Skill levels increase as you advance on the course map. The local Training Program Sponsor may adjust the training order.

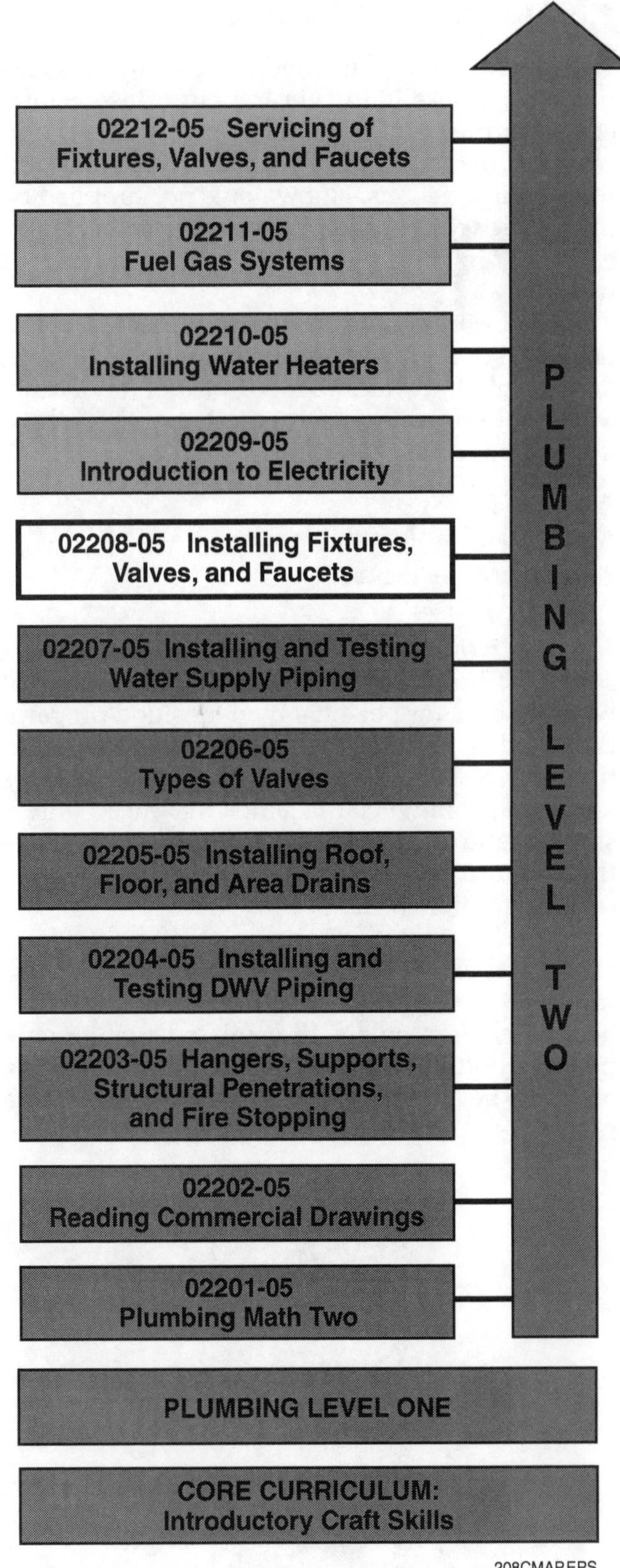

208CMAP.EPS

1.0.0 ◆ INTRODUCTION

Take a moment to think about the many different fixtures that use hot water in your house. Showers, lavatories, sinks, bathtubs, dishwashers, and washing machines all require hot water to perform their functions. Hot water is not supplied by the public water utility. Water must be heated in each building and distributed to the fixtures that use it.

Most buildings have centrally located water heaters (see *Figure 1*). Water heaters heat, store, and supply hot water to all or part of a building. Most modern water heaters consist of the following components:

- A heat source
- A **storage tank** and its shell
- Temperature controls
- Safety devices

Codes require all inhabited structures to have a source of hot water for sanitary purposes. Hot water is used for personal hygiene, food preparation, clothes washing, general cleaning, and multiple industrial and commercial uses. Residences and small commercial facilities may need only a single water heater. Larger facilities, or facilities that require large quantities of hot water, may require multiple heaters.

The basic operating principles of a gas water heater are easy to understand (see *Figure 2*). A piping system supplies cold water to the water heater, where it enters an insulated storage tank near the bottom of the tank. At the bottom of the tank is a heat source, which heats the water to a preset temperature. As the water warms, it expands and rises to the top of the tank. As the water near the top cools over time, it sinks toward the bottom of the tank, where it is again heated by the heat source. The cycle continues, providing a constant reservoir of hot water.

When a hot water fixture is activated, another set of pipes located near the top of the tank draws off hot water and delivers it to the fixture. Fresh cold water is introduced at the bottom of the tank to replace the water that was drawn off.

As a plumber, you will locate and install water heaters and their associated piping systems. You will also be called on to perform minor maintenance. This module will give you a basic understanding of how water heaters operate and their special plumbing needs. This information will help you to install and maintain water heaters according to the safety and efficiency standards in your area.

As in all plumbing work, public safety should be your prime consideration. Perform installations in strict compliance with all safety and code requirements. Personally ensure that the work is done according to professional plumbing standards.

This module does not cover the sizing of hot and cold water lines attached to water heaters. As a journey plumber, you must have an experienced and qualified plumber size the lines according to the local applicable code. When installing water heaters in larger, more complex plumbing systems, consult with the project engineer or architect for advice on designing a suitable system.

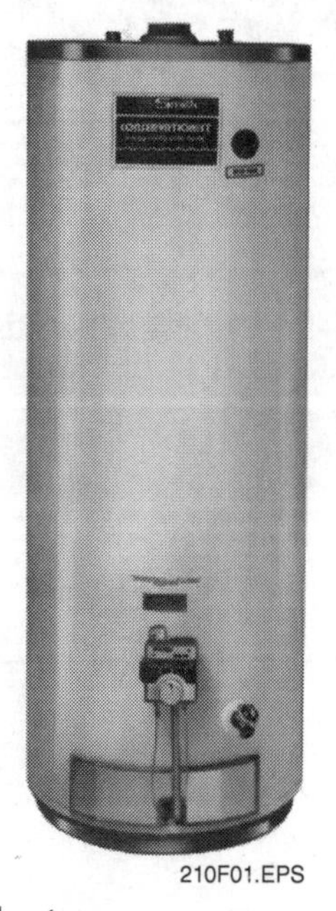

Figure 1 ◆ Water heater.

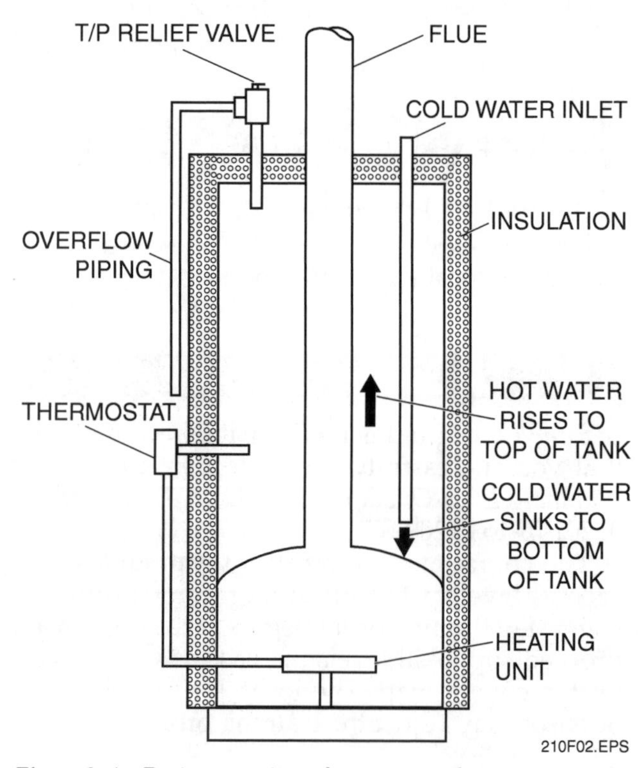

Figure 2 ◆ Basic operation of a gas water heater.

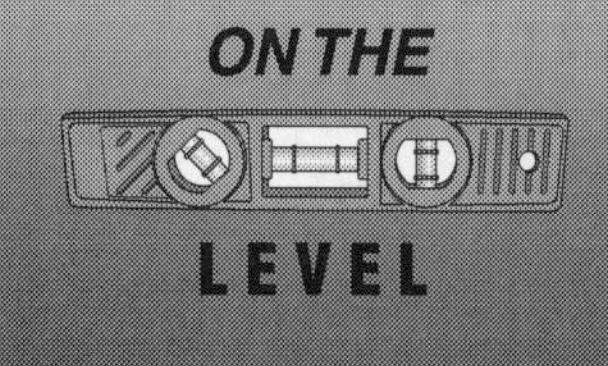

Obey the Law

Always install water heaters according to the law as specified in your local applicable code and the manufacturer's specifications. Inspectors will not grant a permit to a water heater that has not been installed in accordance with the law. Always consult your local applicable code and the manufacturer's specifications before installing water heaters to ensure that you follow all requirements closely.

Note that the requirements for sizing and installing hot tubs and whirlpools are different from those used for water heaters. Refer to your local applicable code for this information. Hot tubs and whirlpools are covered in detail elsewhere in the Plumbing curriculum.

2.0.0 ◆ STORAGE WATER HEATERS

The concept of a storage water heater is simple:

- An energy source transfers heat to the water.
- An insulated vessel stores the heated water.
- Control devices adjust the energy input to provide a fixed water supply temperature.
- Safety devices prevent overpressurization and overheating.
- A venting system allows combustion byproducts to escape if the unit is gas- or oil-fired.
- Pipes distribute water to and from the heater.

Despite the simplicity of the basic system, innumerable methods and systems for accomplishing the heating of water are available. As a plumbing professional, you should understand the basic concepts, the general variations of water heaters, and the proper methods of installation. Always refer to the manufacturer's instructions when installing a water heater. Your local applicable code will specify the safety requirements that you must follow.

2.1.0 Heat Sources

Water heaters are classified by the heat source they use. The following are the most common types of water heaters you will work with:

- Gas-fired water heaters, which are fueled by natural gas or propane (see *Figure 3*)
- Electric water heaters, which use electricity to heat water (see *Figure 4*)
- Solar water heaters, which heat water using the energy of the sun (see *Figure 5*)

Within these classifications, water heaters are divided into the following additional categories:

- Self-contained storage type
- Remote storage type
- Instantaneous
- Semi-instantaneous

Storage-type water heaters use storage vessels to provide draw-down capacity for peak usage and to limit the size of the heat input. They are the most widely available type of gas and electric water heater. These water heaters heat and store water at a thermostatically controlled temperature for delivery on demand. Instantaneous-type water heaters require an energy input that provides peak hot water requirements on demand. Instantaneous water heaters have no storage capacity; semi-instantaneous water heaters offer limited water storage.

DID YOU KNOW?

In 1964, the Powers Regulator Co., now called Powers Process Controls, introduced the Hydroguard™ 410 Pressure Balancing Valve. This valve protected bathers from hot blasts or icy bursts. It did this with a pressure chamber that sensed and equalized any change in water pressure for the tub or shower.

The idea of controlling water temperature through pressure dates back to 1887. That's when company founder William Penn Powers equipped his new heating and plumbing business with a central water heating plant—the first in Wisconsin. To eliminate boil-over, he developed a process based on the relative boiling points of water under different pressures. He achieved his goal with a water-filled pipe in the boiler. One end of the pipe was closed; the other end was attached to a diaphragm that controlled a damper.

210F03.EPS

Figure 3 ◆ Gas water heater.

210F04.EPS

Figure 4 ◆ Electric water heater.

210F05.EPS

Figure 5 ◆ Solar water heater.

2.1.1 Gas Automatic Storage Tank Water Heaters

Most gas-fired storage tank water heaters use either natural gas or liquefied petroleum gas (LPG) as fuel to heat the water in the storage tank. Oil-fired water heaters are less common, though you may encounter one from time to time. *Figure 6* illustrates the components of a typical gas water heater. Gas is efficient and clean burning. Although natural gas and LPG units are similar in design, they cannot be used interchangeably. Never connect a fuel supply to a water heater that is not designed to use that fuel.

The fuel is gravity fed into the **main burner**, located under the tank, inside the outer metal shell, or **jacket** (see *Figure 7*). In the burner, the fuel gas is mixed with air and then ignited by the **pilot burner** (see *Figure 8*) or by an electronic spark ignition. Pilot burners are also called **safety pilots** because the gas supply to the water heater will automatically shut off if the pilot goes out. The combustion of the gas/air mixture in the main burner chamber heats the water in the tank to a preset temperature. Burners in residential and light commercial gas water heaters may be gravity fed or of a natural-draft fire type.

DID YOU KNOW?

Solar Thermal Systems

Solar thermal systems use the sun's energy to heat water or produce power. Small-scale water heating systems use flat plate collectors to capture the sun's energy. Power plants must concentrate the sunlight for the high temperatures they need to produce power.

A home hot water system can be retrofitted with a solar thermal system. Most of these systems have a backup heating element to fill in on cloudy days or to cover higher than normal hot water use. A solar thermal system can take care of 60 to 80 percent of residential hot water needs.

About 250,000 commercial and industrial buildings in the United States use these systems to provide hot water or space heating. Industries in which these systems are most common include laundry, food service, food processing, metal plating, and textiles..

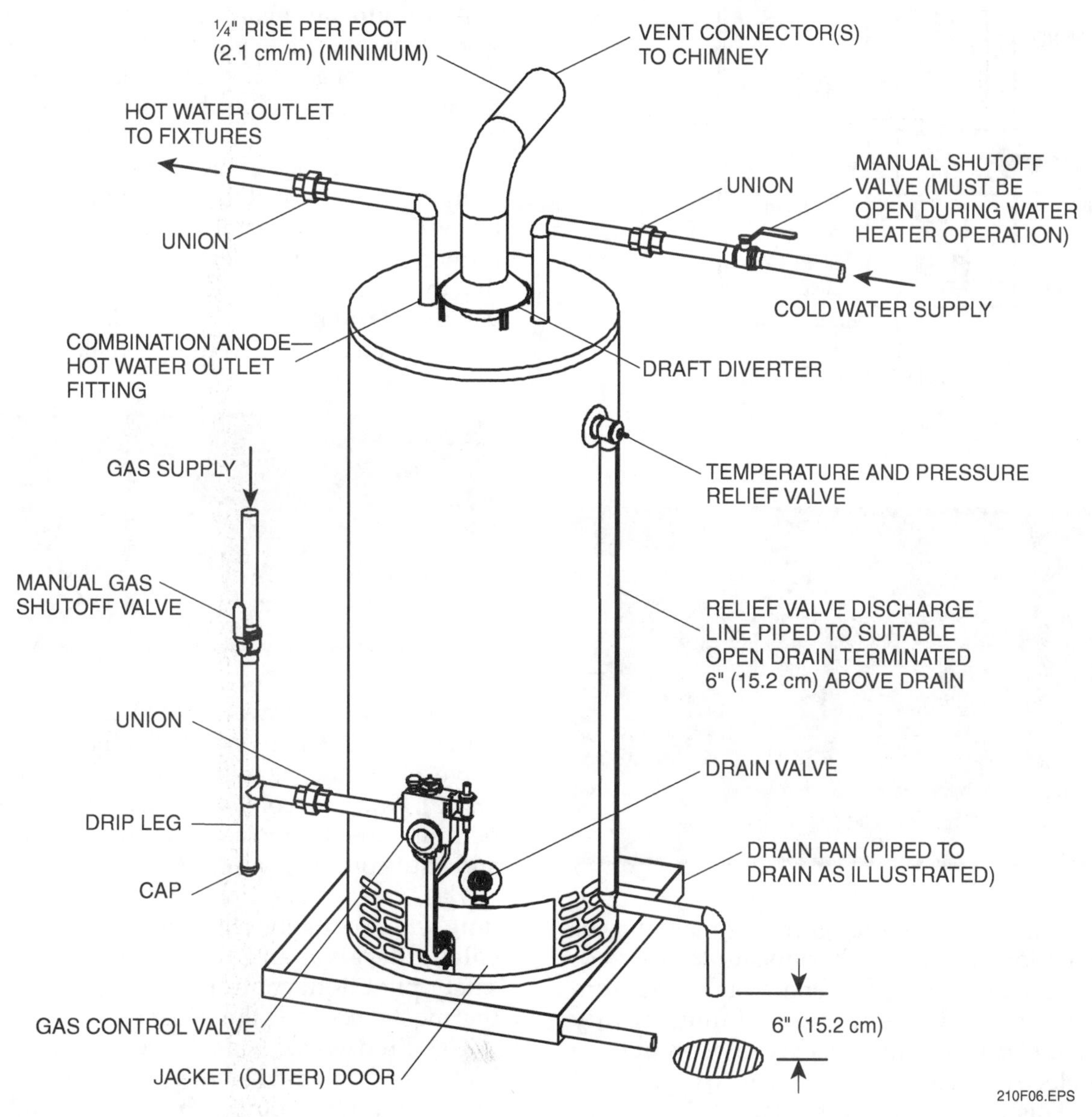

Figure 6 ◆ Components of a gas-fired water heater.

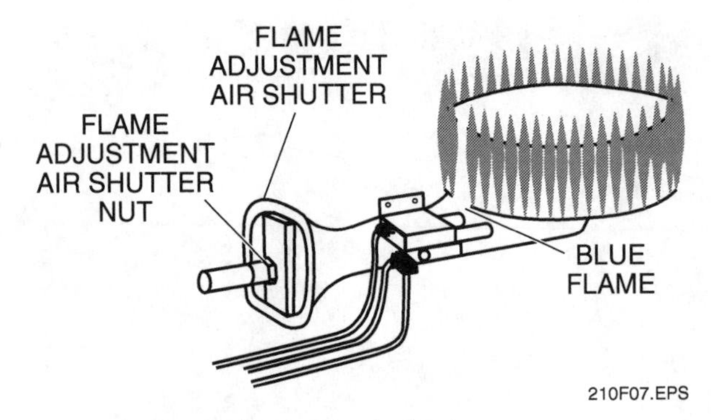

Figure 7 ◆ Burner in a gas water heater.

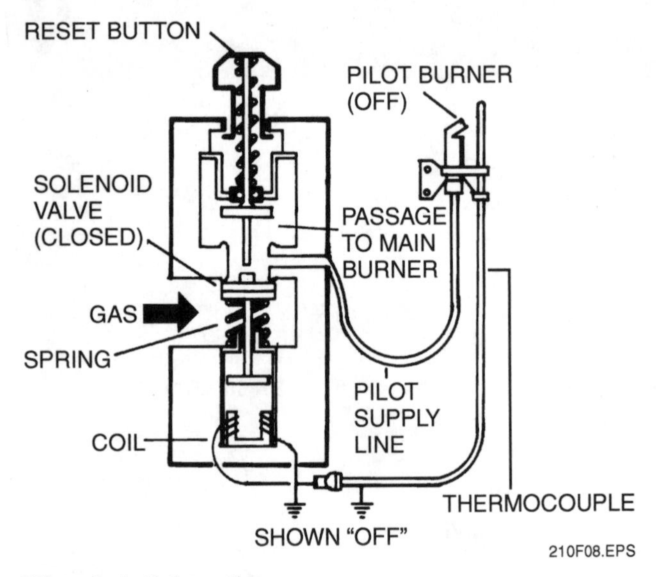

Figure 8 ◆ Safety pilot.

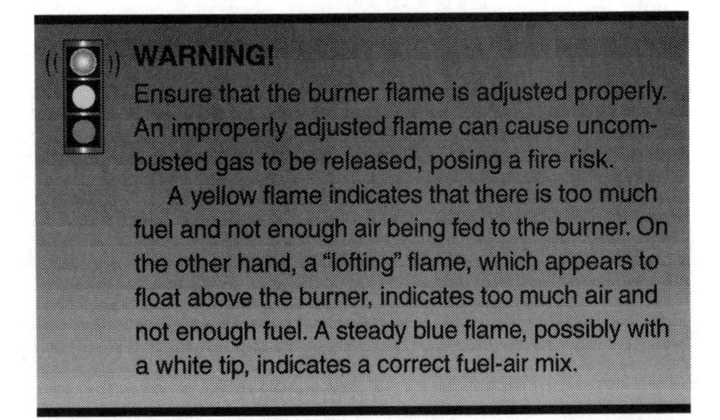

WARNING!

Ensure that the burner flame is adjusted properly. An improperly adjusted flame can cause uncombusted gas to be released, posing a fire risk.

A yellow flame indicates that there is too much fuel and not enough air being fed to the burner. On the other hand, a "lofting" flame, which appears to float above the burner, indicates too much air and not enough fuel. A steady blue flame, possibly with a white tip, indicates a correct fuel-air mix.

Standing pilot ignitions ignite the main burner using a pilot valve and a **thermocouple**. Thermocouples are small electric generators made from two pieces of different metals joined firmly together. The heat from the safety pilot produces a small electric current between the two pieces of metal. This current holds the safety shutoff gas valve in an open position. If the pilot flame burns too low or goes out, the current breaks and the valve spring releases, shutting off the gas flow to the main burner.

The combustion of the fuel produces byproducts that travel directly up a chimney, or from an independent **flue** to a chimney. The flue is a tube-like passage that runs up the center of the heating tank. **Baffles** inside the flue collect heat from the gases moving through the flue and transfer it to the water in the tank. Gas heaters require a piping system that delivers fuel to the burner unit and a venting system that carries away smoke and other combustion products.

Although liquefied petroleum is stored in large tanks as a liquid, it vaporizes before it is burned. LPG is a more concentrated fuel than natural gas. Water heaters that use LPG, therefore, have smaller fuel and heating orifices, or openings, than natural gas water heaters.

A **thermostatic control** is used to control the water temperature in a gas water heater (see *Figure 9*). The flow of gas into the burner is determined by the temperature within the tank, which is in turn determined by a sensing element, or **probe**, attached to the back of the thermostatic control and immersed in the water. If the temperature drops below the indicated setting, the valve opens and the burner fires. When the temperature sensor indicates that the desired temperature has been reached, the valve shuts off the gas flow. The desired water temperature can be selected using a dial.

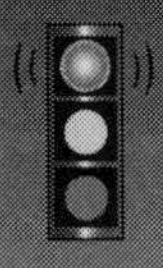

WARNING!

Before firing a gas water heater, always make sure that the gas supply type matches the type indicated on the control valve. Never connect propane to a natural gas valve or vice versa. Such a connection could cause an explosion.

When a user draws hot water from the tank, cold water flows in. A sensor in the probe opens a valve, allowing gas to flow to the burner (see *Figure 10*). Once the cold water heats to a preset temperature, the thermostatic control closes the valve and shuts off the gas supply.

Gas water heaters also feature a valve control unit, which combines the automatic pilot valve, or **solenoid valve**, and the thermostat valve. The safety pilot light controls the solenoid valve. The thermostat controls the thermostat valve.

Gas-fired water heaters are susceptible to the buildup of mineral sediment in the bottom of the heater. A large amount of heat transfer takes place on the bottom of the tank. Water mineral deposits act as an insulator, causing the tank bottom to over-

Figure 9 ◆ Typical thermostatic control in a gas water heater.

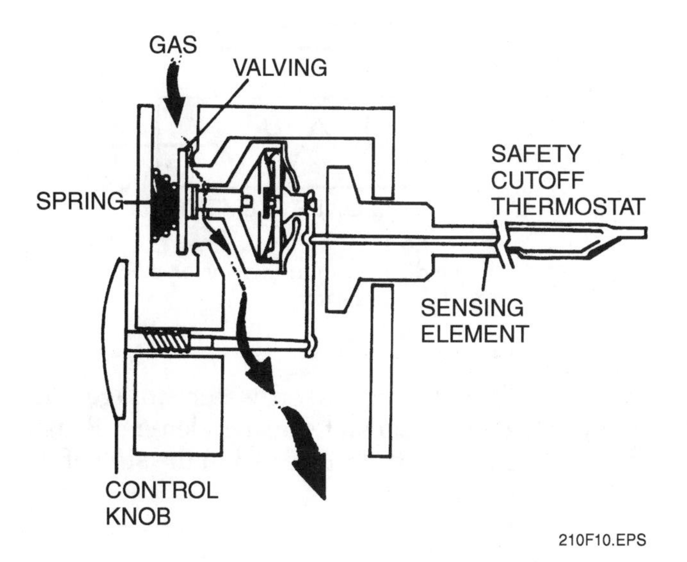

Figure 10 ◆ Operation of the thermostatic probe.

heat. Premature failure of the water heater may occur if these deposits are not regularly drained off. Follow the manufacturer's operating instructions.

New residential and small commercial gas water heaters often feature electric spark ignition, induced draft fans, and larger heat transfer surfaces. These water heaters are often called **condensing water heaters**. The combustion chamber in a condensing water heater is usually sealed, drawing combustion air from the outside through a separate intake pipe. These water heaters, also called pulse-type water heaters, are generally more fuel efficient than water heaters that use gravity to feed the fuel to the burner. The burner and control mechanisms in a condensing water heater are more complex than those in gravity-fed gas water heaters. They should be serviced by a plumber who has been trained by the water heater manufacturer.

Large commercial and industrial gas-fired water heaters are available in several different configurations. They may use vertical or horizontal water storage tanks with fan-forced burners. These water heaters usually incorporate pre- and post-purge burner cycles and use a complex ignition/flame safety control. Burners in these water heaters require a flue system that is suitable for fan-forced combustion.

2.1.2 Electric Automatic Storage Tank Water Heaters

Because they do not burn fuel to generate heat, electric water heaters do not require flues, baffles, main burners, safety pilots, thermocouples, or valve control units. Instead, two **immersion elements** in the storage tank heat the water. Automatic thermostats control the immersion elements (see *Figure 11*). The thermostats sense the tank's outside surface temperature and adjust the flow of electricity to each element. In normal operation, the upper element of a typical water heater will operate first to heat the water in the upper section of the tank. When that water reaches the a preset temperature, the upper element is cycled off and the lower element is cycled on to heat the remaining water (see *Figure 12*).

A **high-limit control** connected to the immersion element circuit protects against extreme water temperatures caused by defective thermostats or grounded water heater elements. The high-limit control breaks the flow of electricity to the heating element circuit when the temperature of the tank surface reaches a preset limit (refer to *Figure 12*).

Some electric water heater models use a thermostat exposed to the water, with a remote sensing bulb that is located in an immersion well. Refer to the water heater's specifications to identify the type of thermostat used. Install the thermostat according to the manufacturer's instructions.

Make sure that the storage tank is full of water before you turn on an electric water heater. Never **dry fire** an electric water heater, which means operating the water heater without water in the storage tank. Dry firing will cause the immersion elements to burn out and damage or even destroy the water heater.

Most electric water heaters require a 240-volt electrical service to operate the heating element efficiently. Although some water heaters are designed to operate on a 120-volt electrical service, they are usually less efficient than the 240-volt models. For buildings that are equipped for 440-volt service, 440-volt models are also available.

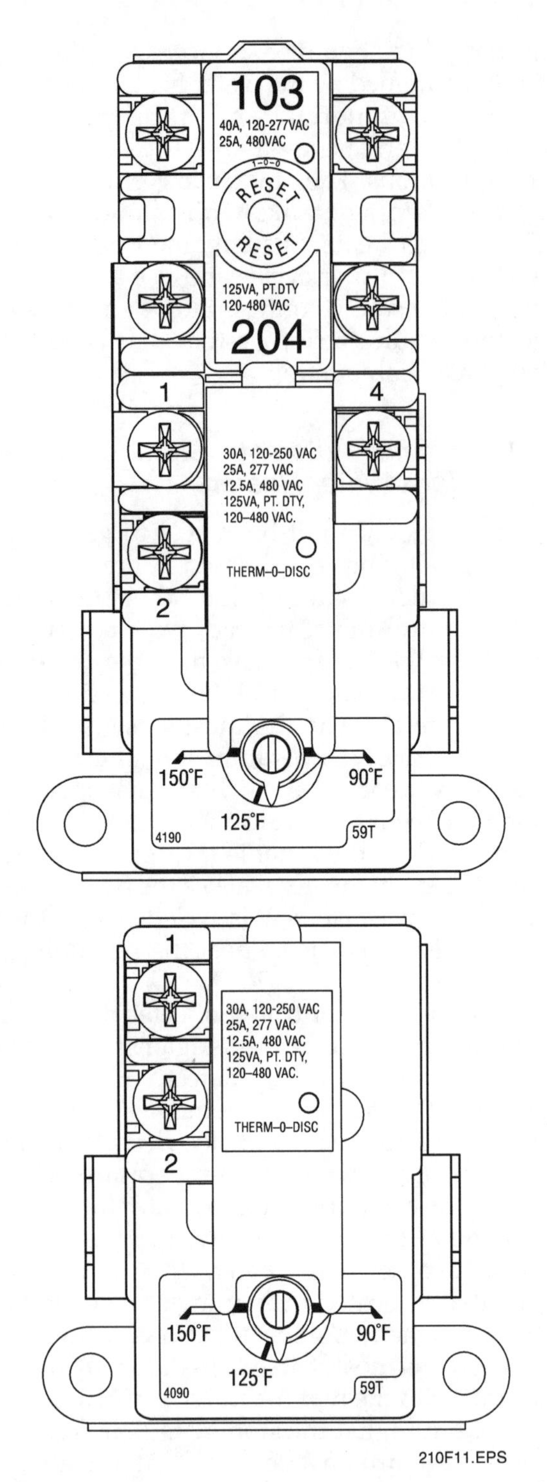

Figure 11 ◆ Automatic thermostat on an electric water heater.

Larger capacity electric water heaters may utilize large bundles of 480-volt heater elements. The electrical wiring and control system will obviously be much more complex than those used on simple residential heaters. They may use large open-type contactors that are pulled in by the control thermostat. The thermostat is usually a multistep unit that allows sequential operation of the

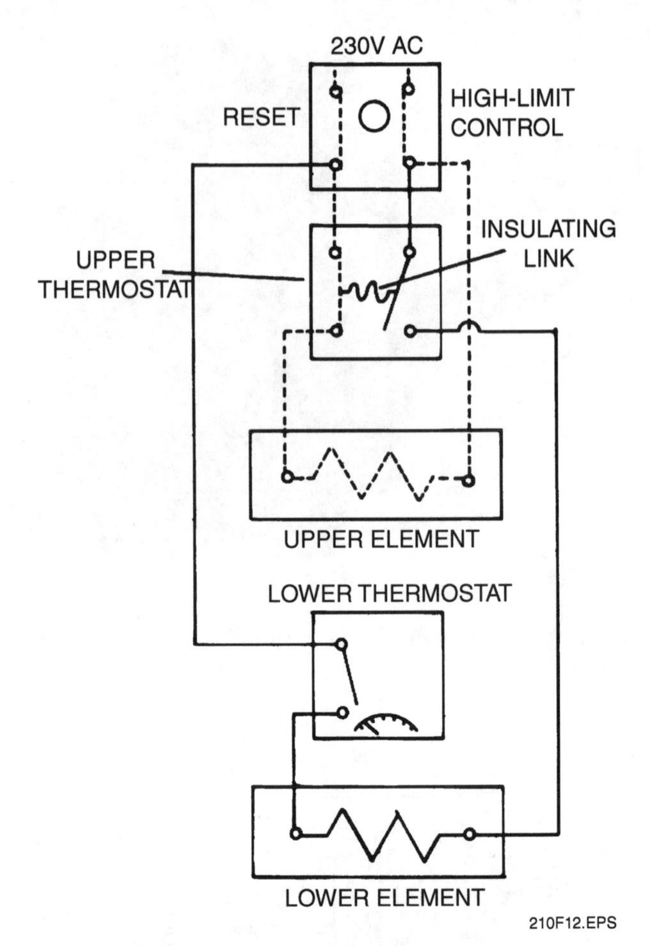

Figure 12 ◆ Wiring diagram of immersion elements in an electric water heater.

elements to maintain a fixed water storage temperature. The thermostat sensing element is usually an immersion type, inserted in the side of the storage tank.

2.1.3 Solar Water Heaters

For buildings that receive direct sunlight year-round, solar power is a cost-effective option for heating water. In a typical solar heating system, water is pumped through a **collector**, where it is heated by sunlight. A typical collector consists of a narrow, rectangular box with a metal plate, a water pipe, and a transparent cover (see *Figure 13*). Sunlight enters the collector through the transparent cover and heats the metal plate. The heat is transferred to the water as it circulates through a pipe that runs through the collector. The transparent cover acts as a reflector to keep the heat from radiating back out of the collector.

From the collector, the hot water circulates by **convection** (motion caused by the different densities of hot and cold water) or by pump through the hot water supply system, or it may be stored

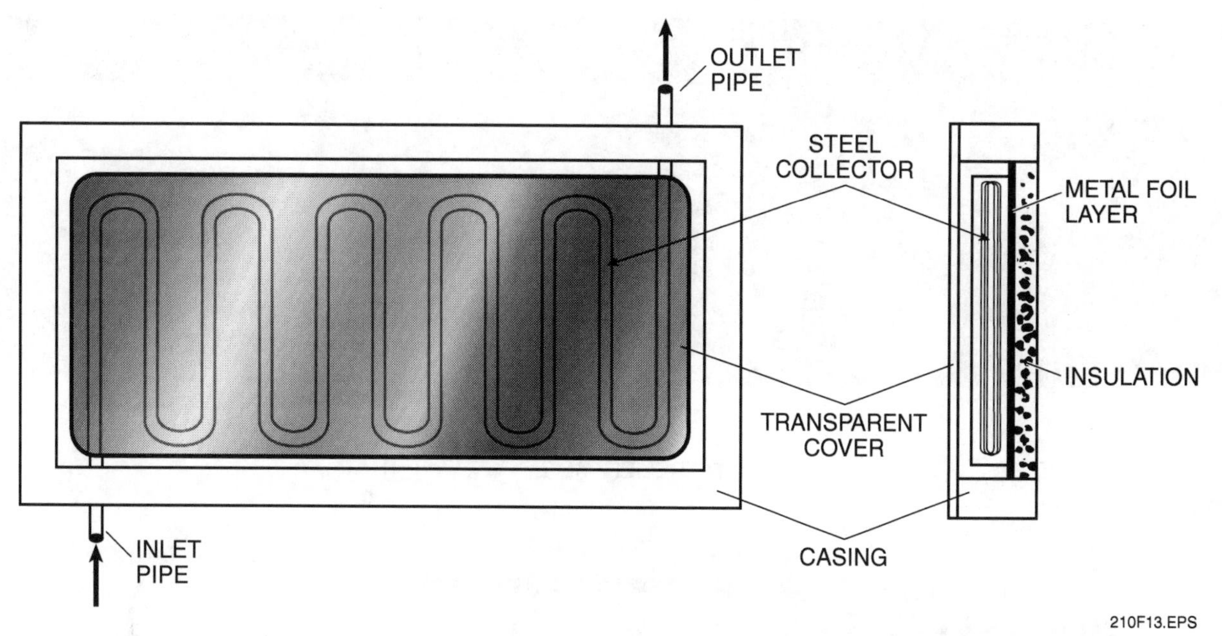

Figure 13 ◆ Typical solar collector.

in a tank. Some systems use a liquid other than water to absorb the sun's heat, and then transfer the heat to the water using a **heat exchanger** (see *Figure 14*).

Solar heating equipment is more expensive than that used in other types of water heating. However, the fuel—sunlight—is free. This helps offset the higher cost of the solar heating equipment.

Several types of solar heaters are available. The most practical ones include a backup system that uses conventional fuels for overcast days or when hot water demand is higher than normal (see *Figure 15*). You will learn more about the different types of solar heating systems in *Plumbing Level Four*.

2.2.0 Shell and Storage Tank

A jacket and top cover form the outer shell that protects a hot water storage tank. Jackets and top covers are made of steel coated in baked enamel. Insulation is installed between the jacket and tank to reduce heat loss from the water.

The storage tank itself is a capped metal cylinder lined with a glass coating fused to the inner surface. This coating protects the tank from the

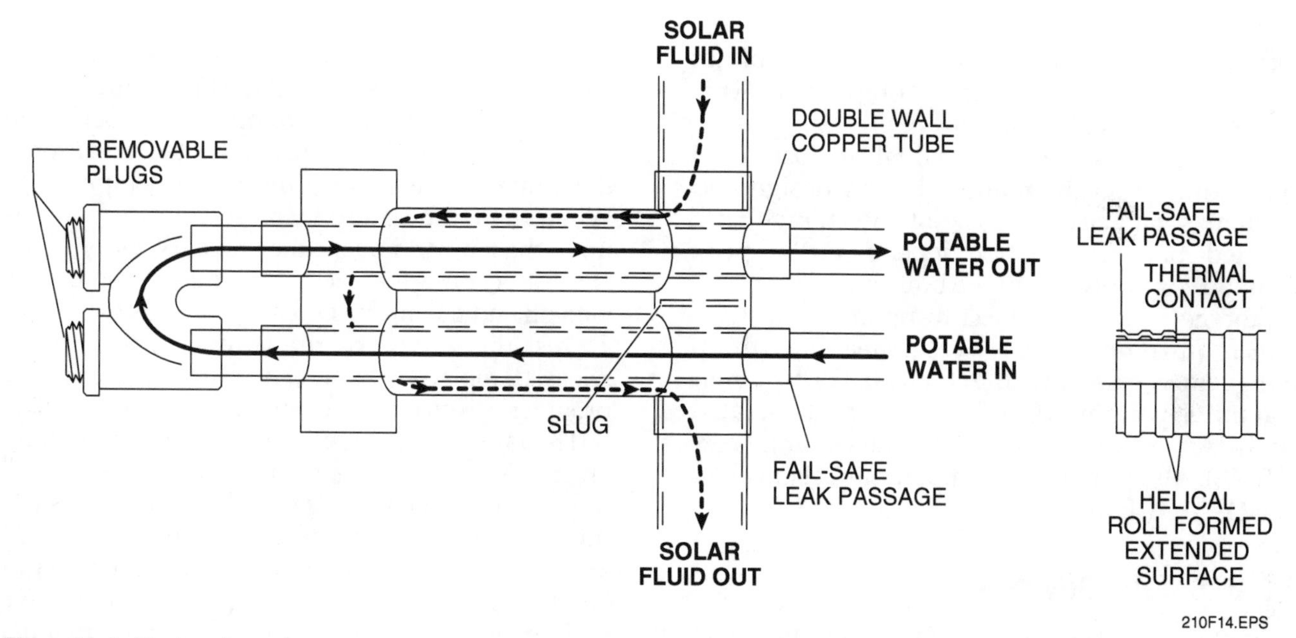

Figure 14 ◆ Heat exchanger.

ON THE LEVEL

Keep Relief Valves Clear

Ensure that there is sufficient space around a relief valve. Don't let your next customer have an experience like this:

A plumber installed a temperature and pressure (T/P) relief valve on a water heater in a school building. However, the valve was plugged off. When the internal temperature of the water heater triggered the valve, the valve could not release the pressure. The water heater exploded, blowing out a brick wall. Fortunately, no one was injured.

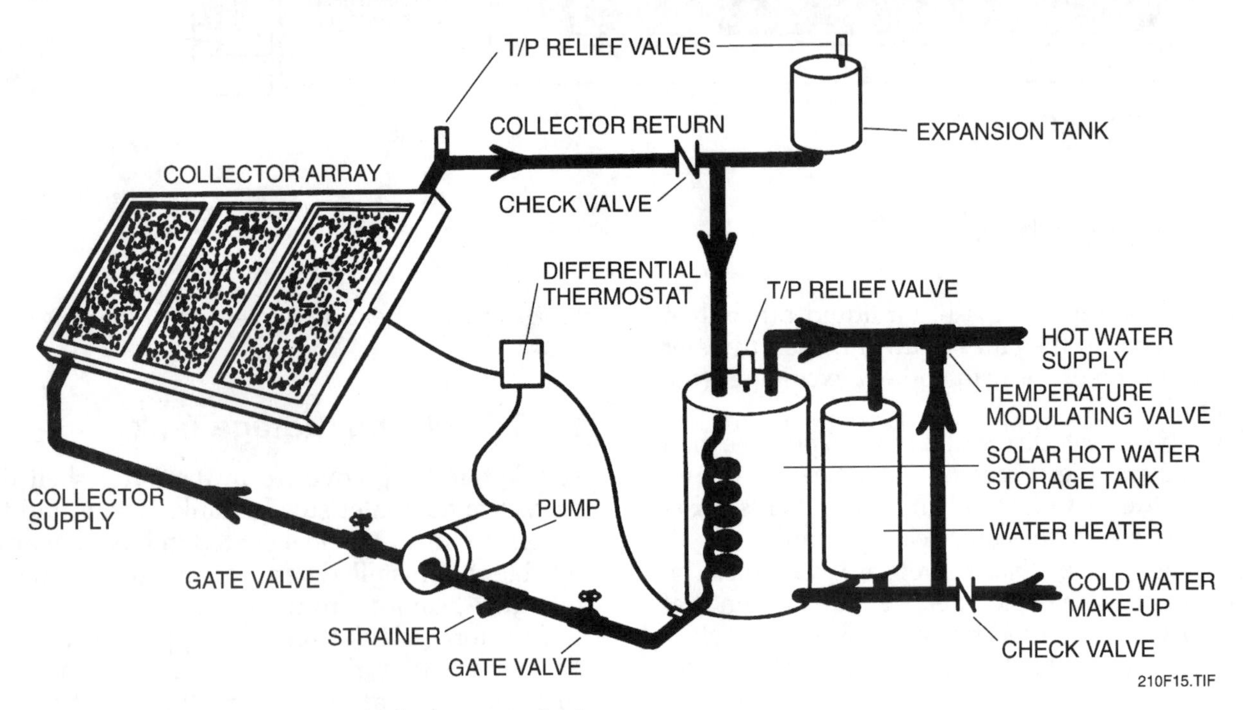

Figure 15 ◆ Solar hot water system with backup water heater.

corrosive effects of minerals that leach out of hot water and build up on metal components. This mineral crust is called **scale.**

Storage tanks may be vertical or horizontal depending upon the manufacturer's design and the tank's capacity. The size of the storage tank selected for an installation depends on the peak demand for hot water in the building.

Storage tanks are subject to rigorous construction requirements. Any water heater with an energy input greater than 200,000 Btu (**British thermal units**) per hour (Btu/h) must use a tank rated by the American Society of Mechanical Engineers (ASME). All pressure and temperature relief valves must be ASME rated.

2.3.0 Safety Devices

As you have learned, water heaters are fitted with T/P relief valves, which are installed in the storage tank or as close to the tank as possible (see *Figure 16*). T/P relief valves prevent the tank from exploding if the thermostat fails to operate properly and allows the water to become overheated. A sensor in the valve detects an extreme rise in temperature. When the temperature increases, pressure builds up in the tank and opens the valve. Water or steam is released until the excess pressure in the tank is bled off and the internal pressure returns to normal. To test the valve, lift the lever. If water is not released, replace the valve. A **relief valve opening** provides access for installation of a T/P relief valve.

The tank drain valve is attached to an opening on the side of the storage tank, located as close to the bottom of the tank as possible. Customers use this valve to drain the tank and clean out sediment and scale. Many manufacturers recommend regular drainage as part of a program of preventive maintenance. Consult the manufacturer's manual for the recommended schedule.

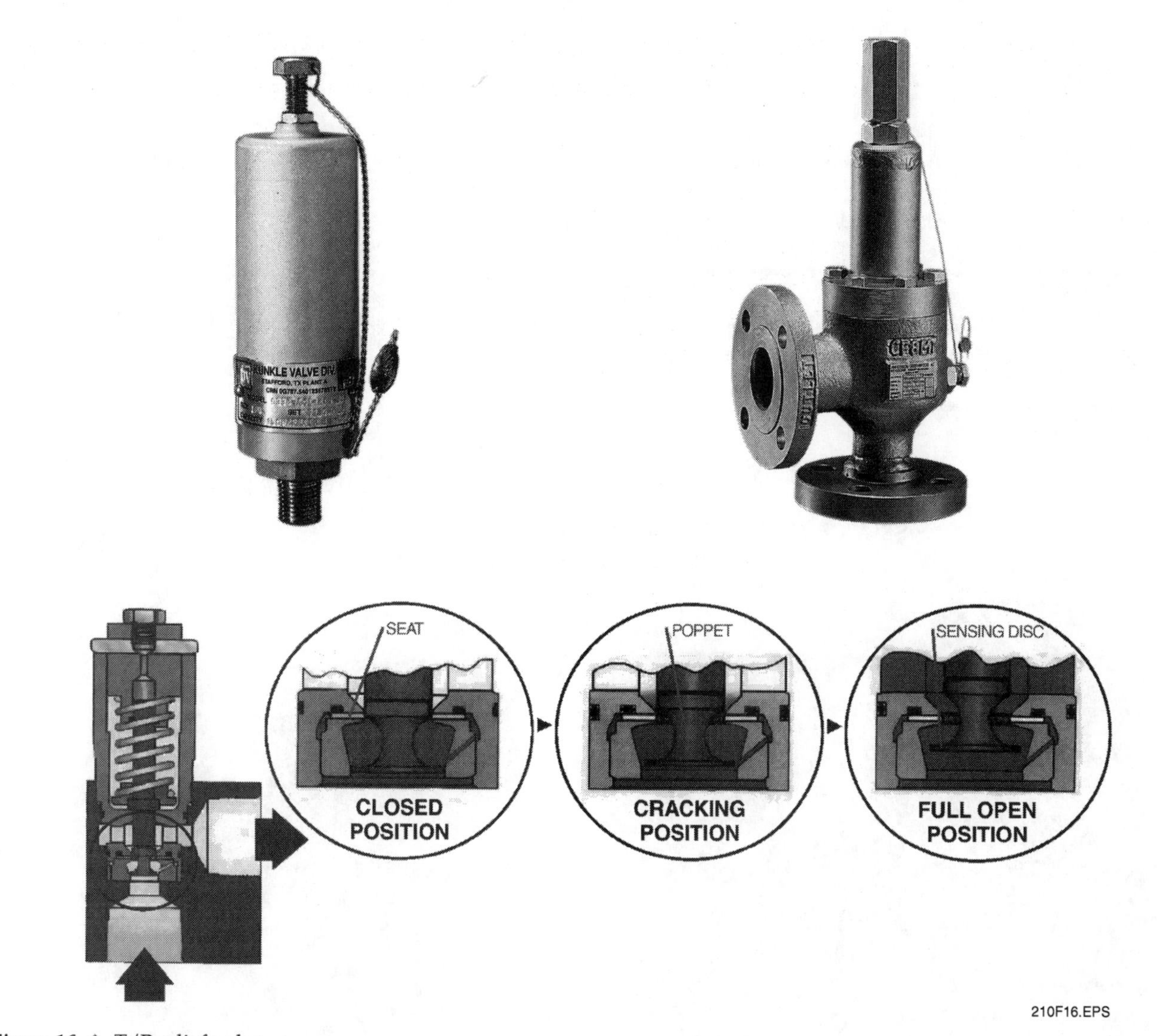

Figure 16 ◆ T/P relief valve.

The **dip tube** prevents cold water from mixing with hot water. The dip tube delivers incoming cold water through the stored hot water to the bottom of the tank. The cold water is then rapidly heated and allowed to mix with the hot water. Near the top end of the dip tube, a small opening called an **anti-siphon hole** prevents hot water from siphoning out if the cold water supply is interrupted. Dip tubes are typically installed on vertical storage tanks, but not on horizontal ones.

3.0.0 ◆ INDIRECT WATER HEATERS

Indirect water heaters are an efficient method for providing hot water to fixtures. An indirect water heater system uses a boiler and heat exchanger, instead of a gas burner or electric heating element, to heat the water. As with the water heaters discussed earlier, hot water is stored in an insulated water tank. Indirect heaters are usually found in commercial applications.

Figure 17 illustrates the operating principles of an indirect water heating system. Cold supply water enters the boiler, where it passes over a heat exchanger. The heat exchanger warms the water, which then flows into the storage tank. Recirculating hot water flows through a copper coil in the boiler, and its heat is absorbed by the heat exchanger. The heat exchanger then transfers the heat to cold supply water, thus continuing the cycle.

The high-limit **aquastat** is a special type of thermostat that regulates the temperature of the hot water in the boiler. Aquastats act as safety devices, turning the boiler off when the temperature of the water in the boiler reaches a preset maximum temperature.

Indirect water heater systems are available with and without tankless heaters. The components of this system include a heat exchanger and a circulator/adapter. In this system, the boiler heats the water passing through the heat exchanger, and the water heater serves only as a storage tank.

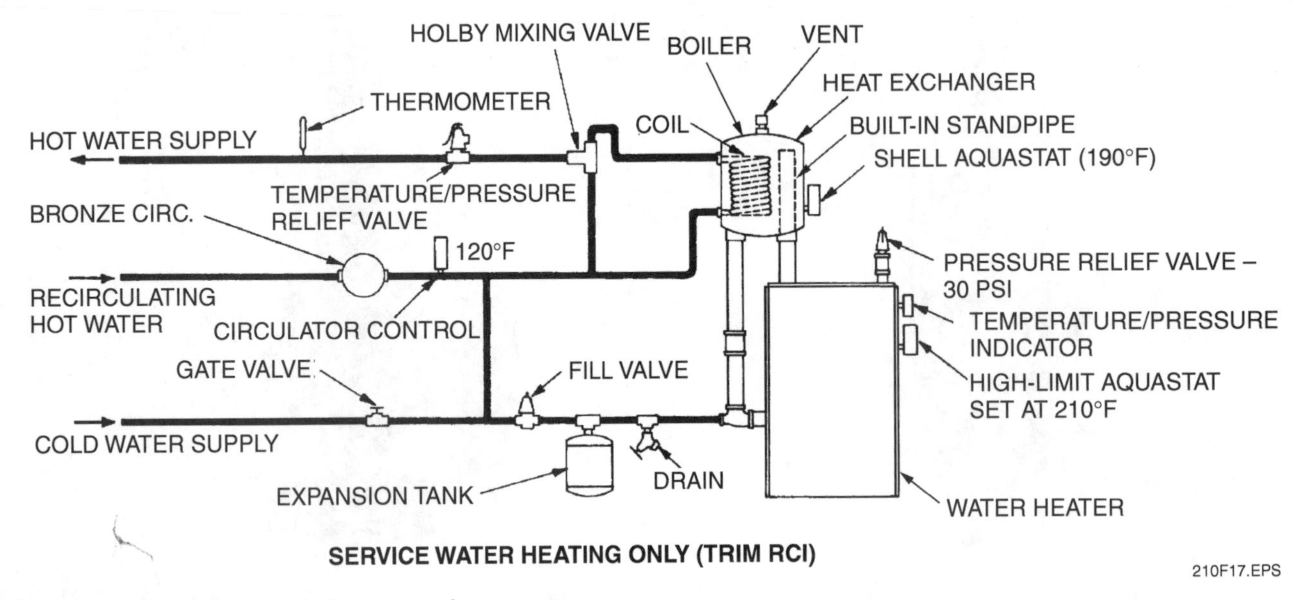

Figure 17 ◆ Schematic of an indirect water heater system.

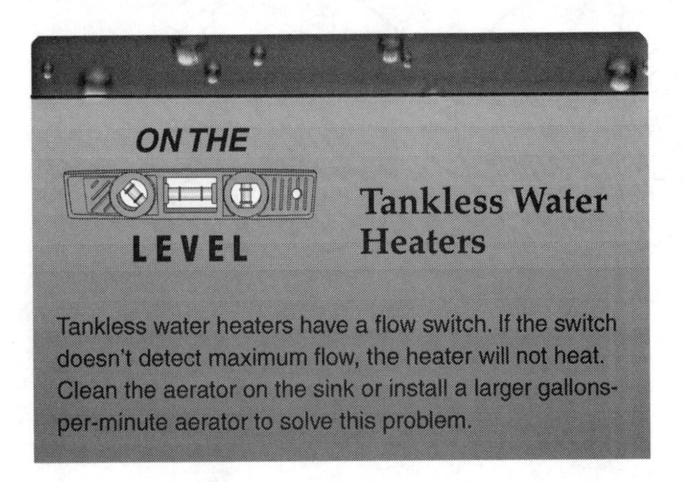

ON THE
LEVEL
Tankless Water Heaters
Tankless water heaters have a flow switch. If the switch doesn't detect maximum flow, the heater will not heat. Clean the aerator on the sink or install a larger gallons-per-minute aerator to solve this problem.

When the system includes a tankless heater, the heat exchanger is not required because the tankless heater functions as the heat exchanger.

4.0.0 ◆ INSTANTANEOUS WATER HEATERS

Storage water heaters and indirect water heaters continue to use energy even when all the water in the tank has been heated. This is because the water gradually loses heat through the tank walls and piping, and must be reheated periodically. Up to 20 percent of the energy used by a storage tank water heater is used for reheating. In some installations, that could mean $100 or more per year.

Instantaneous water heaters, also called demand water heaters and tankless water heaters, heat water when it is used (see *Figure 18*). They do not consume energy reheating stored water. Instantaneous water heaters are equipped with heaters that activate when a fixture's hot water valve is operated. Instantaneous water heaters are widely used in Japan and Europe, and are becoming more popular in the United States. They are available in gas-fired and electric models and are sized for individual fixtures or entire buildings. Typical instantaneous water heaters are designed to last 20 years.

Figure 18 ◆ Instantaneous water heater.

The hot water flow in an instantaneous water heater is limited by the heater's output capacity. Most models cannot provide adequate hot water for multiple high-demand fixtures and appliances operating simultaneously, for example a dishwasher and a washing machine running at the same time. Ensure that the unit you select is suitable for the application. Point-of-use units for single fixtures can be installed in closets or under the sink. Larger units should be installed centrally as with other types of water heaters. Refer to your local applicable code for sizing and locating requirements.

Review Questions

Sections 1.0.0–4.0.0

1. In a residential water heater, the hottest water can typically be found _____.
   a. in a special chamber of the tank
   b. mixed throughout the tank
   c. at the bottom of the tank
   d. at the top of the tank

2. In *Figure 19*, the component indicated is called a(n) _____.
   a. flue
   b. anode
   c. overflow piping
   d. drain pipe

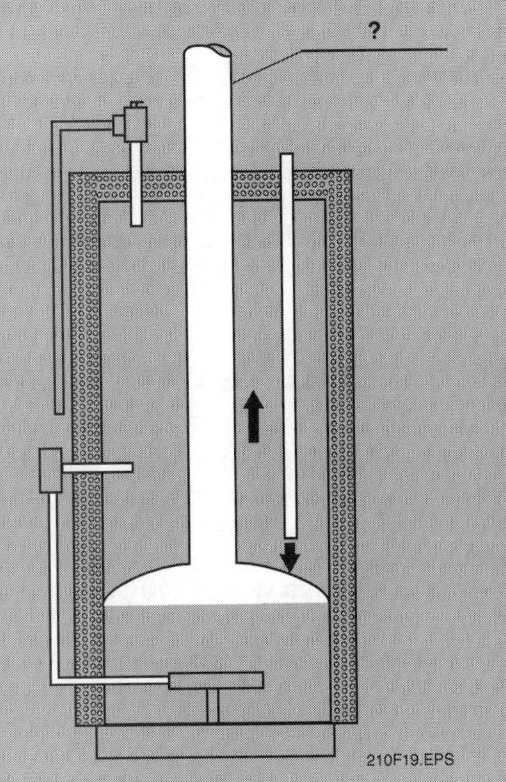

Figure 19 ◆ Review question gas heater operation.

3. Water heaters are classified by their_____.
   a. Btu rating
   b. heat source
   c. water capacity
   d. output rate

4. The _____ of a gas water heater transfer heat from burning gases to water in the tank.
   a. baffles
   b. thermocouples
   c. dip tubes
   d. thermostatic controls

5. The top immersion element in most electric water heaters operates _____.
   a. whenever water is in the tank
   b. before the lower element
   c. only when hot water demand is low
   d. only when the lower element is not working

6. The high-limit control of a water heater protects against extreme water temperatures by _____.
   a. shutting off the flow of water to the tank
   b. shutting off the flow of gas to the burner
   c. shutting off electricity to the heating elements
   d. releasing steam or water to reduce pressure in the tank

7. In a typical solar heating system, water is heated by sunlight as it is pumped through a _____.
   a. filter
   b. heat exchanger
   c. collector
   d. thermosyphon

Review Questions

8. Motion caused by the different densities of hot and cold water is called _____.
   a. conduction
   b. radiation
   c. absorption
   d. convection

9. The T/P relief valve of a gas water heater should be replaced if during testing it does not produce _____.
   a. a high limit reading
   b. a low limit reading
   c. water or steam
   d. compressed air

10. Most instantaneous water heaters supply enough hot water to meet simultaneous demand from heavy-use fixtures in a single household.
    a. True
    b. False

5.0.0 ◆ SELECTING WATER HEATERS

Select a water heater that best fits the needs and requirements of the installation. Remember that no single model or type of water heater is right for every application. Always select water heaters that meet current industry standards. The water heater should bear the seal of an industry-recognized testing organization. Gas water heaters are tested and approved by the American Gas Association (AGA). Electric water heaters are tested and approved by Underwriters Laboratories (UL).

Consider the following factors when selecting a water heater:

- *Cost and availability of fuel* – Gas lines may not have been extended to the building, or local codes may not allow gas tanks to be installed. In some cases, electric heat may be too expensive.
- *Capacity* – The tank must be large enough to keep up with demand. Storage tank size and **recovery rate** determine water heater capacity. The recovery rate is the amount of time it takes a given amount of cold water to be heated. Refer to your local applicable code for the recommended storage tank sizes and Btu ratings for water heaters. If more hot water is required—for example, to supply an automatic washer—use the next larger size storage tank or select a heater with a high recovery or Btu rating.
- *Durability* – A heater's ability to stand up to daily use depends on the material used to make it. Most units are made from galvanized steel, copper, or glass-lined steel. Most installers prefer glass-lined steel for its economy and durability. You can learn a lot about the tank's quality by reviewing the manufacturer's guarantee. A high-quality tank, properly cared for, should last 15 to 25 years or more. However, water conditions in different areas may shorten the life of the water tank.
- *Ability to hold heat* – Select fully insulated units to reduce heat loss and fuel consumption. The insulation is sandwiched between the tank and the outer covering. Insulation is made from a variety of materials. The manufacturer's product sheet often includes a description of the type of insulation and an indication of its heat-retaining capability.

6.0.0 ◆ INSTALLING WATER HEATERS

As with other plumbing systems, the proper installation procedure for a water heater can be broken down into two phases: rough-in and installation. Before installing the water heater, inspect it for damage or defects. Measure the heater to ensure that there is ample space to set it properly. Most manufacturers provide a rough-in drawing showing the physical dimensions of the various parts of the heater as well as the specifications for capacity, input Btu per hour, and recovery rate (see *Figure 20*). Your local applicable code will include information on spacing and combustion air requirements for gas-fired water heaters.

Residential Energy Saver Gas-Upright

The Bradford White Defender Safety System models meet ANSI Standard Z21.10.1a.

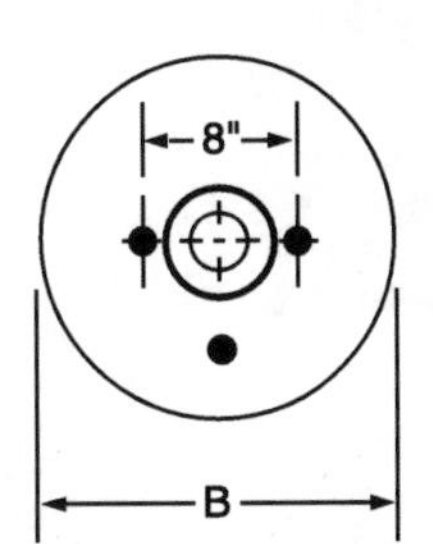

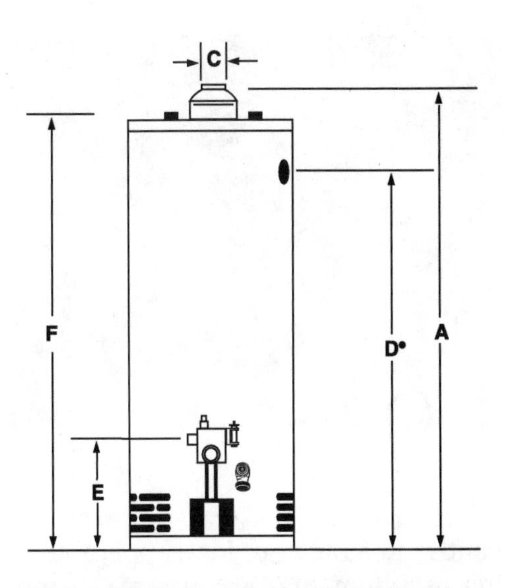

Energy Saver Models

1" Non-CFC Foam Insulation (except where noted) C.E.C. Listed
76% Recovery Efficiency • (M-I-100 has an 80% Recovery Efficiency)

NATURAL GAS

Model Number	Imp. Gal. Cap.	U.S. Gal. Cap.	Liters	Nat. & LP BTU Input	KW Input	Recovery 90 F/50 C Rise: Imp. GPH	U.S. GPH	Liters/ Hour	A Floor to Flue Conn.: in.	mm.	B Jacket Dia.: in.	mm.	C Vent Size: in.	mm.	D Floor to T&P Conn.: in.	mm.	E Floor to Gas Conn.: in.	mm.	F Floor to Water Conn.: in.	mm.	Approx. Shipping Weight: Lbs.	Kg.
M-I-30T6FBN(LP)•	25	30	114	32,000	9.4	27	33	125	59⅛	1502	16	406	3	76	49¾/56⅜	1264/1432	13	330	57½	1461	104	47
M-I-30S6FBN	25	30	114	30,000	8.8	26	31	117	48⅜	1229	18	457	3	76	38¾/48⅝	984/1159	13	330	46¾	1187	100	45
M-I-303T6FBN(LP)	24	29	110	40,000	11.7	34	41	155	58	1473	16	406	3	76	49¾/55¼	1264/1403	13	330	56½	1435	109	49
M-I-40T6FBN(LP)•	33	40	151	40,000	11.7	34	41	155	59⅜	1508	18	457	3	76	50/56⅝	1270/1438	13	330	57¾	1467	120	54
M-I-403S6FBN(LP)•	33	40	151	40,000	11.7	34	41	155	50	1270	20	508	3	76	41/47¼	1041/1200	13	330	48½	1232	128	58
M-I-404T6FBN(LP)•	33	40	151	48,000	14.1	41	51	185	60¾	1543	18	457	4	102	51½/58	1308/1473	13	330	58½	1486	127	58
M-I-5036FBN(LP)•	42	50	189	40,000	11.7	34	41	155	59⅝	1514	20	508	3	76	50/56⅞	1270/1445	13	330	58	1473	145	66
M-I-50L6FBN(LP)	40	48	182	40,000	11.7	34	41	155	49¾	1264	22	559	3	76	40½/47	1029/1194	13	330	48¼	1226	153	69
M-I-504S6FBN(LP)•	42	50	189	48,000	14.1	41	51	185	58½	1486	20	508	4	102	50/55¾	1270/1416	13	330	57	1448	150	68
M-I-75S6BN	62	75	284	76,000	22.0	63	78	288	63¾	1619	24½	622	4	102	53	1346	16	406	61¼	1518	264	120
M-I-100T6BN	83	100	379	85,000	24.9	79	92	360	68¹¹⁄₁₆	1745	28¼	718	4	102	59⁵⁄₁₆	1507	15⅝	397	65¹¹⁄₁₆	1651	420	194

Floor to space heating return for **M-I-75S6BN** 17¾"/451mm • **M-I-100T6BN** 17¾"/450mm

Floor to space heating outlet for **M-I-75S6BN** 53"/1346mm • **M-I-100T6BN** 50⅞"/1295mm

All LP Gas models are equipped with a cast iron burner. To order an LP Gas model change suffix "BN" to "CX"

M-I-100T models feature hand hole cleanout.

• Models feature optional top T&P location and must be specified when ordering.
"D" dimension listed as side/top.

Suitable for Water (Potable) Heating and Space Heating

Toxic chemicals, such as those used for boiler treatment, shall NEVER be introduced into this system. This unit may NEVER be connected to any existing heating system or component(s) previously used with a non-potable water heating appliance.

Meets NAECA requirements. • All Natural Gas models meet SCAQMD requirements.

General

All gas water heaters are certified at 300 PSI test pressure (2068 kPa) and 150 PSI working pressure (1034 kPa). All water connections are ¾" (19mm) NPT on 8" (203mm) centers. The 75 gallon model (284 liters) has 1" (25mm) NPT on 11" (279mm) centers and 100 gallon model (379 liters) has 1¼" (32mm) water connections on 16" (406mm) centers. All gas connections are ½" (13mm). All models design certified by CSA International (formerly AGA/CGA) and peak performance rated. For "10" year models, change suffix number "6" to "10".

Dimensions and specifications subject to change without notice in accordance with our policy of continuous product improvement.

(LP) LP Gas Inputs

Model	BTU/H	KW	Recovery at 50°C Rise and 90°F Rise: Imp. GPH	U.S. GPH	Liters/ Hr.
M-I-30T6FCX	31,000 BTU/H	9.1 KW	27	32	121
M-I-303T6FCX	38,000 BTU/H	11.1 KW	32	39	148
M-I-40T6FCX	36,000 BTU/H	10.6 KW	32	37	140
M-I-403S6FCX	38,000 BTU/H	11.1 KW	32	39	148
M-I-404T6FCX	46,000 BTU/H	13.5 KW	39	47	178
M-I-5036FCX	38,000 BTU/H	11.1 KW	32	39	148
M-I-50L6FCX	38,000 BTU/H	11.1 KW	32	39	148
M-I-504S6FCX	46,000 BTU/H	13.5 KW	39	47	178

210F20.EPS

Figure 20 ◆ Rough-in sheet for a water heater.

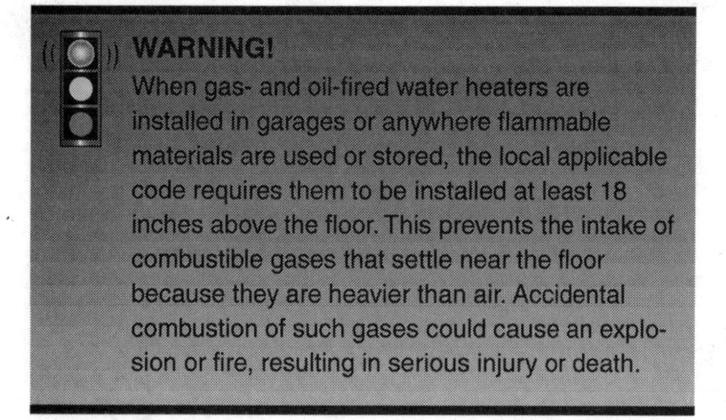

Ensure that gas water heaters are properly vented. Refer to your local applicable code for venting requirements in your area. If the outside air exchange is rapid, venting may not be required. For example, a large commercial building with many loading dock doors that remain open for long periods may not require venting for its water heater. Ensure that gas- or oil-fired heaters are placed as close as possible to chimneys or flues. This will ensure a proper draft for safe operation. Refer to your local applicable code for spacing requirements.

When installing connecting pipes from the water heater flue to the chimney, avoid using too many bends. Bends tend to reduce the capacity of the chimney. Every bend decreases the allowable length of run or footage. Ensure that the flue connector is at least as large as the heater flue outlet.

Existing furnace chimneys may be used for the water heater, as long as the water heater flue is connected to the chimney above the furnace flue. Refer to your local applicable code. The horizontal pipe run must maintain an upward slope of ¼ inch per foot. Use a wye connector to connect directly into a furnace flue. Do not use a tee connector for this connection. Heat does not flow through a tee connector as efficiently as it does through a wye. An angled heat flow is more efficient.

Use galvanized or black iron pipe and fittings to connect the gas or oil supply piping to the water heater. Attach a shutoff valve to the piping between the water heater and the fuel supply. This valve can be used to shut off the fuel supply when servicing or replacing the water heater. Install the water heater so that the burner is easily accessible for maintenance or repair.

Use corrugated brass connectors to simplify hookup of the hot and cold water lines, but check to make sure their use is allowed by local codes. If the codes require regular pipe connectors, you must install unions (see *Figure 21*). Install a shutoff valve on the cold water inlet. This allows the cold water supply to be shut off when servicing or repairing the water heater.

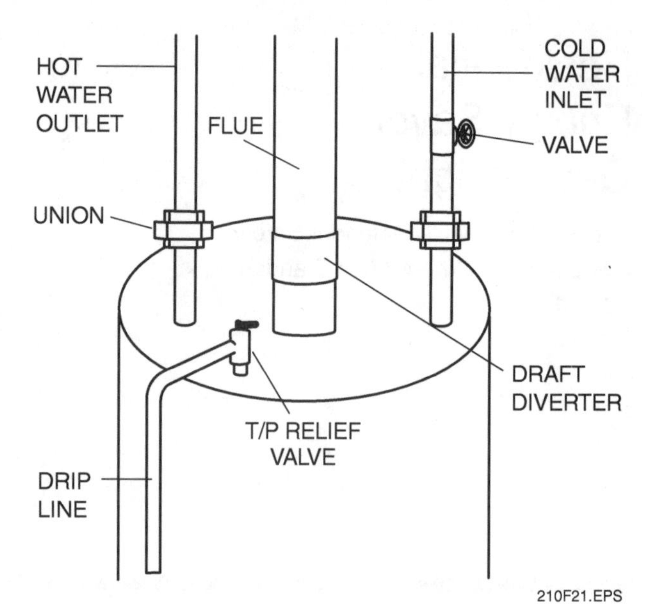

Figure 21 ◆ Use of unions with a water heater.

Install T/P relief valves in the specially designed opening in the top or side of the storage tank. Follow the manufacturer's installation instructions closely. Never install plugs or valves between a T/P relief valve and its outlet discharge. Do not downsize the pipe between the T/P and the outlet discharge.

Attach a drip line to the outlet of the T/P relief valve. This will allow any hot water escaping from the valve to be directed safely into the sewer. Run the drip line so that it extends to within 6 to 12 inches of the floor drain, empties directly into a safety pan drain, or runs to the exterior of the building just above grade level. Refer to your local applicable code for guidelines on installing a drip line. To prevent back siphoning, ensure that an air gap is maintained between the end of the drip line and the top of the floor drain or standpipe.

Remember that electric water heaters are manufactured in a range of voltage ratings. Make sure that the power source is the same voltage as the water heater. When disconnecting an old electric water heater and installing a new one, never assume that the breakers or fuses are labeled correctly. Always check the voltage with a voltage tester.

6.1.0 Parallel Connection

In some commercial installations with a high-volume demand for hot water, two hot water tanks can be connected in parallel (see *Figure 22*). The pipes leading from both feed lines to the cold water inlet and the hot water outlet must be equal in length. If they are not, the water will take the shortest path, and most or all of the hot water will be drawn from one tank. The result is that only one of the tanks will work, reducing the supply of

Choosing Fixture Size

Be sure to match the fixture to the appliance. For example, a typical 40-gallon water heater is not sufficient for large whirlpool tubs. This will leave the owner with a lukewarm bath. Your local applicable code will specify the minimum requirements.

Checking Water Heaters

Dirt and grime, small cracks, or damaged connectors can cause major problems once the water heater is installed and operating. Be sure to check each water heater for damage before installing it.

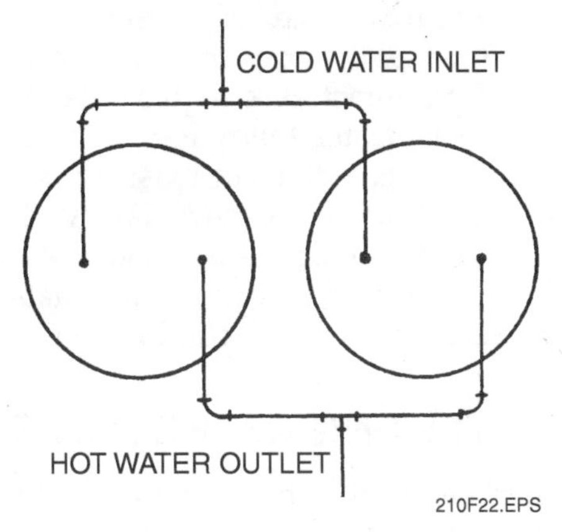

Figure 22 ◆ Parallel connection.

unused hot water and increasing the chance that one of the heaters will fail early.

In a parallel connection, one heater can supply a building's hot water needs while the other is turned off for a brief time for maintenance, repair, or replacement. So long as the downtime for one heater occurs during a low-demand period, customers will not experience any decrease in water temperature.

6.2.0 Series Connection

Water heaters can also be connected in series. In a series connection, water is preheated in one tank to supply a second tank, which brings the water up to final temperature. The preheated water can also be supplied to the second tank by a coil output from an air conditioner or by a solar source (see *Figure 23*).

6.3.0 Non-Basement Installation

When installing a water heater in a location other than a basement, place a safety pan beneath the heater to collect any leakage (see *Figure 24*). Run a ¾- or 1-inch pipe from the pan to the floor drain or

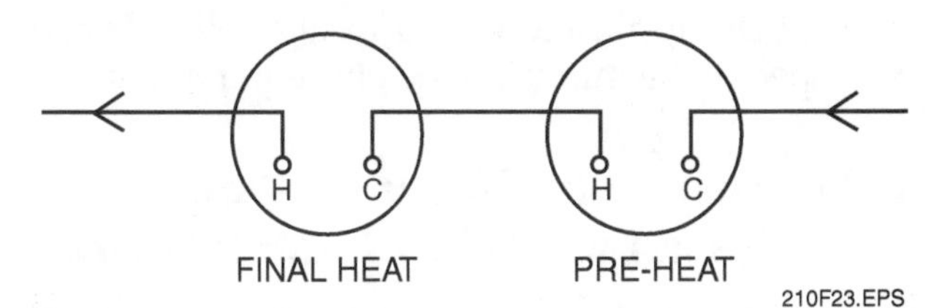

Figure 23 ◆ Series connection.

DID YOU KNOW?

Solar Energy

The concept of solar energy really got cooking in 1767 when a Swiss scientist invented the world's first solar collector, or hot box. Sir John Herschel, a famous astronomer, cooked his food in a solar hot box during his expedition to Southern Africa in the 1830s.

In the U.S., solar water heating began when Clarence Kemp patented the first commercial water heater in 1891. The idea of using the sun to heat water was popular in areas that had to import fuel. By 1897, nearly 30 percent of the houses in Pasadena, California, had solar water heaters. In the 1970s, fuel shortages made solar water heating even more popular. The first commercial use of this technology appeared in the early 1980s. Today, utilities provide hundreds of megawatts of solar thermal generating capacity.

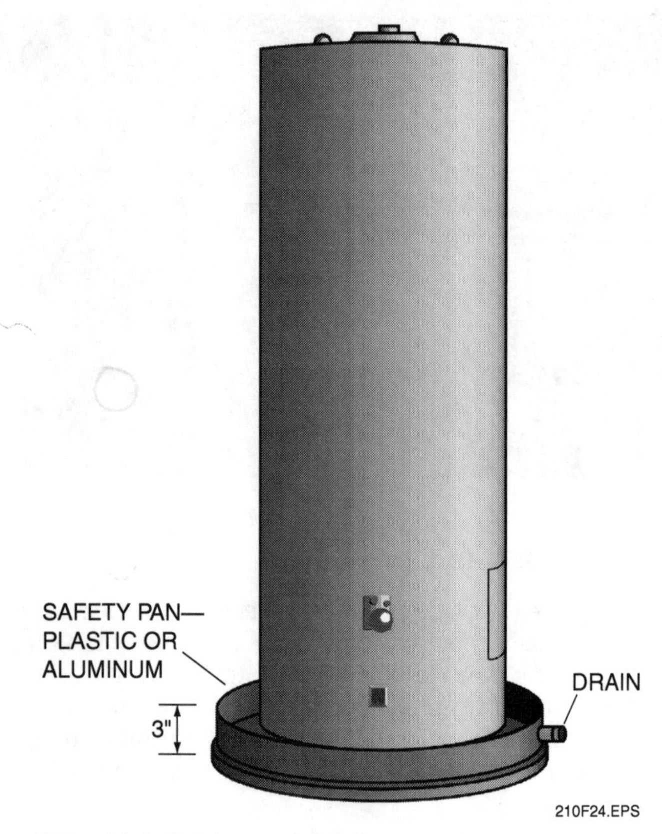

Figure 24 ◆ Safety pan and drain.

waste opening. Pans come in different materials, shapes, and sizes. Be sure to refer to the manufacturer's specifications and your local applicable code for guidelines for selecting the proper type of safety pan for the application. Your local applicable code will specify the minimum height requirements.

6.4.0 Flammable Vapor Ignition Resistant (FVIR) Water Heaters

Properly installed natural draft gas fired water heaters are safe and reliable. However, owners often disregard safety recommendations and store or use flammable liquids near gas water heaters. By doing so, they greatly increase the risk of fire. Every year, the accidental ignition of flammable vapors by natural draft gas fired water heaters kills and injures dozens of people and causes millions of dollars in property damage.

To combat this, in 1994 the Consumer Product Safety Commission (CPSC) and the gas water heater industry collaborated to develop guidelines for natural gas and propane fired gas water heaters. In July 2003 the American National Standards Institute (ANSI) standard for gas water heaters, *ANSI Z21.10.1—2001*, began requiring all new 30-, 40-, and 50-gallon natural draft residential gas water heaters to be certified as an **FVIR (Flammable Vapor Ignition Resistant) water heater**. The following year, the standard also took effect for new residential sealed combustion chamber (power vent) water heaters.

FVIR certified water heaters are fitted with metal flame arrestor plates around the air intakes. The arrestor plates have holes or slits that allow the flammable air-vapor mix to enter the combustion chamber, but which prevent the flame from passing back through the arrestor plate into the room. Water heater manufacturers use arrestor plates of varying configurations and materials. Rheem's FVIR water heaters are also equipped with a door that closes off the combustion air chamber when the fusible link that holds the door open is exposed to flame.

Refer to the manufacturer's installation instructions to ensure that the arrestor plate is installed correctly. Some manufacturers require the installation of a new water heater following exposure to fire, while others specify that an existing unit can be repaired and returned to service. Follow the manufacturer's recommendations when you are called on to repair or replace a gas water heater that has been exposed to accidental vapor combustion.

7.0.0 ◆ TESTING WATER HEATERS

Test water heaters after they are installed to ensure that they work correctly. Testing should be your company's policy because it eliminates callbacks, which waste both time and money. Testing also ensures that the water heater is operating safely, and that any risk of explosion or fire due to malfunction or improper installation has been eliminated.

Manufacturers often include troubleshooting charts to help you spot and correct problems with their products (see *Appendix A*). Use these tables and other information provided by the manufacturer when testing a water heater.

One common problem with water heaters is a constantly weeping relief valve. If the relief valve drips steadily, check the system for proper expansion capability. Water expands as it is heated, and it is a noncompressible fluid. As a result, the pressure in the system increases as water is heated. If the system does not include an air-type compression tank to absorb excess pressure, the relief valve will discharge to relieve it.

7.1.0 Testing Gas-Fired Water Heaters

After installing the water heater, do the following:

- Examine the heater to ensure that there are no leaks or drips.

Water Heater Stands

Gas-fired water heaters draw air for combustion from about 2 inches above the floor. Sometimes water heaters are installed in locations where flammable liquids are stored (for example, in a garage where cans of gasoline or paint thinner may be stored). These liquids give off vapors that are heavier than air. The vapors sink, puddle near the floor, and are drawn into the water heater. Result—an explosion.

Manufacturers offer water heater stands that raise the heater a minimum of 18 inches off the floor. This raises the appliance above the area where flammable vapors puddle. The stand looks like a small table. Consult your local applicable code for guidelines on the use of water heater stands in your area.

- Apply a code-approved soap solution to joints and fittings and look for bubbles that would indicate a possible gas leak. If you find a leak, examine the fittings for problems and replace them.
- Light the pilot, and turn the control valve up and down to ensure that the burner works.
- Turn the control knob to low and wait to see if the burner will cycle off.

WARNING!

Always check the water heater temperature setting. A customer may be seriously scalded by water that's too hot. At 125°F, it takes about 1½ to 2 minutes for hot water to cause scalding. At 155°F, it takes about 1 second.

7.2.0 Testing Electric Water Heaters

After installing an electric water heater, do the following:

- Turn on the water and check for leaks.
- Use a meter to ensure that the heater is set at the proper voltage.
- Set the thermostat to the desired setting.

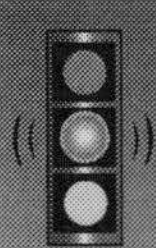

CAUTION

In spite of a manufacturer's quality control, fittings can have defects. Be sure to examine fittings for holes, cracks, or poor casting. Read and follow the manufacturer's installation instructions.

Review Questions

Sections 5.0.0–7.0.0

1. Water heater capacity is determined by recovery rate and _____.
   a. capacity of immersion element or burner
   b. discharge rate
   c. Btu output
   d. storage tank size

2. When installing or replacing an electric water heater, it is always safest to _____.
   a. install a heater stand
   b. check the voltage with a voltage tester
   c. assume that breakers and fuses are labeled correctly
   d. turn on the heating elements before putting water into the tank

3. A gas water heater flue may connect to an existing furnace chimney only if it is _____.
   a. above the furnace flue
   b. below the furnace flue
   c. installed with a tee connector directly into the furnace flue
   d. installed with a wye connector branching down to the furnace flue

4. The drip line for a gas water heater should be installed so that it _____.
   a. recycles back into the water tank
   b. connects directly into the floor drain
   c. maintains an air gap to prevent back siphoning
   d. includes a plug between the T/P relief valve and the outlet discharge

5. When installing water heaters in parallel, the pipes leading from both feed lines to the cold water inlet and the hot water outlet must be equal in length. If they are not, _____.
   a. most or all of the hot water will be drawn from one tank
   b. both water heaters will take longer to generate hot water
   c. water will be drawn off slowly from the heater while cold water entering the base of the tank keeps the gas control in the ON position
   d. it could cross-contaminate the water heater

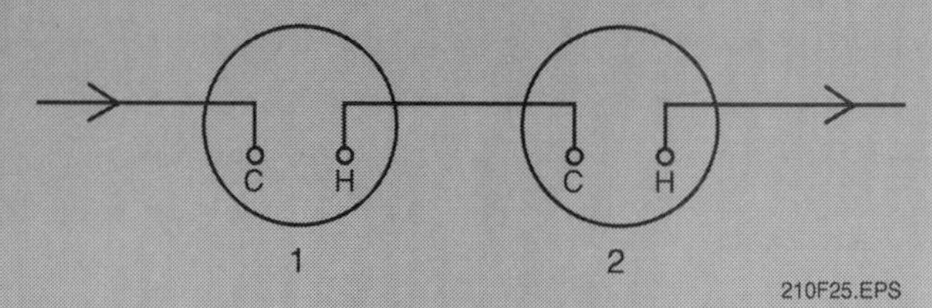

Figure 25 ◆ Review question series connection.

6. In *Figure 25*, a series connection, the water heater labeled 1 provides _____ while the water heater labeled 2 provides _____.
   a. first heat; second heat
   b. pre-heat; final heat
   c. initial heat; final heat
   d. post-heat; pre-heat

7. When installing a safety pan underneath a water heater, run a pipe with a diameter of _____ from the pan to the floor drain or waste opening.
   a. ¼ to ½ inch
   b. ½ to ¾ inch
   c. ¾ to 1 inch
   d. 1 to 1¼ inch

8. After replacing a gas water heater, you should test for gas leaks by _____.
   a. setting the thermostat too high
   b. lighting the pilot and burner
   c. setting the thermostat too low
   d. applying a soap solution to the joints and fittings

9. To ensure that the burner in a gas water heater works, _____.
   a. adjust the gas/air mixture in the main burner chamber
   b. set the thermostat to the desired setting
   c. light the pilot and turn the control valve up and down
   d. adjust the position of the baffles

10. After installing an electric water heater, a plumber should _____.
    a. apply a soap solution to the joints and fittings
    b. turn the control to low to see if it cycles off
    c. check that the heater is set at the proper voltage
    d. set the thermostat as high as possible

Summary

The water heaters most commonly installed in homes and commercial buildings are gas-fired, electric, or solar automatic storage tank heaters. These water heaters heat and store water at a thermostatically controlled temperature for delivery on demand. While there are some differences between gas-fired and electric water heaters, the basic components of both are the same: a heat source, a storage tank and its shell, temperature controls, and safety devices. Electric water heaters use two thermostat-controlled immersion elements to heat water. Solar water heaters collect the heat energy of sunlight and transfer it to water in one of several ways. Gas-fired heaters require a piping system that delivers fuel to the burner unit and a venting system that removes combustion products from the building. Indirect water heaters heat water with a boiler and a heat exchanger. Instantaneous water heaters heat water on demand, rather than storing it.

The installation of a water heater is a two-stage operation: rough-in plumbing followed by the installation of the heater. As part of the installation process, a safety-conscious plumber checks the water heater for damage and verifies that it will fit in the space planned for it. Ensure that gas heaters are properly vented. Follow the manufacturer's instructions and your local applicable code when hooking up the water lines and installing a T/P relief valve. When the installation is complete, test the water heater to make sure it works efficiently and safely. Instruct the building owner in the safe operation of the water heater, and provide copies of the manufacturer's operating and safety instructions. These steps will prevent expensive callbacks that will harm your reputation as a professional plumber, as well as the reputation of your company.

Notes

Trade Terms Introduced in This Module

Anti-siphon hole: An opening near the top end of the dip tube on a water heater that prevents hot water from siphoning out if an interruption occurs in the cold water supply.

Aquastat: A thermostat that regulates the temperature of hot water in a hot water boiler.

Baffle: A plate that slows or changes the direction of the flow of air, air-gas mixtures, or flue gases.

British thermal unit: The amount of heat needed to raise the temperature of one pound of water by 1°F. Abbreviated Btu.

Collector: In a solar water heater, a box that collects heat from the sun and transfers it to water as it flows through a pipe that runs through the box.

Condensing water heater: A gas water heater that features spark ignition, fans for induced air draft, and heat transfer surfaces that are larger than gravity-fed models.

Convection: The motion of water resulting from density differences caused by different temperatures throughout the water.

Dip tube: A device that prevents cold water from mixing with hot water; it delivers incoming cold water through the stored hot water to the bottom of the tank.

Dry fire: To operate a water heater with no water in the heater's storage tank.

Flammable Vapor Ignition Resistant (FVIR) water heater: A gas fired water heater that meets the ANSI standard for resisting the accidental combustion of vapors emitted by flammable liquids stored, used, or spilled near the water heater.

Flue: A heat-resistant enclosed passage in a chimney that carries away combustion products from a heat source to the outside.

Heat exchanger: A device that transfers heat from one liquid to another without the two liquids coming into contact with each other.

High-limit control: A safety device that protects against extreme water temperatures caused by a defective thermostatic control or by grounded water heater elements.

Immersion element: In an electric water heater, an electrical device inserted into the storage tank and used to heat the water.

Indirect water heater: A water heater in which a heat exchanger increases the water's temperature.

Instantaneous water heater: A water heater that heats water when it is used, rather than storing and heating water in a tank.

Jacket: The outer shell of a water heater, made of enamel baked on steel.

Main burner: The chamber in which the fuel gas mixes with air and is ignited. The combustion in the chamber heats the water in the tank to a preset temperature.

Pilot burner: A device that ignites the gas at the main burner when turned on by the thermostatic control. Also called a safety pilot.

Probe: A sensing element that is attached to the back of a thermostatic control immersed in a water heater storage tank.

Recovery rate: The rate at which cold water can be heated.

Relief valve opening: An opening on a hot water tank that provides access for installation of a temperature/pressure (T/P) relief valve.

Safety pilot: Another name for a pilot burner.

Scale: The crust on the inner surfaces of boilers, water heaters, and pipes, formed by deposits of silica and other contaminants in the water.

Solenoid valve: A valve opened by a plunger controlled by an electrically energized coil.

Storage tank: The enclosed tank that stores heated water; it allows heated flue gases to cover the bottom of the tank before they enter the vertical flue. Commonly called a center-flue tank.

Thermocouple: A small electric generator made of two different metals joined firmly together. It produces a small electric current that holds the safety shutoff gas valve open.

Thermostatic control: In a water heater, the control that adjusts the water temperature.

Troubleshooting Gas Water Heaters

Complaint	Possible Cause	Service Tip
No hot water	Pilot extinguished	Relight heater. Check and adjust pilot if necessary. Check thermocouple output and dropout reading. Clear pilot burner.
	Gas supply off	Restore gas supply; relight all appliances.
	Gas plug cock closed	Open cock fully and relight pilot.
	Control knob not in ON position	Turn knob to ON position and relight if needed.
	ECO has cut off pilot	Allow tank to cool. Relight heater. Dial down setting 10°F. Thermostat may be defective.
Water not hot enough	Control dial set too low	Dial control to higher setting. **AVOID SCALDING TEMPERATURES!**
	Control defective	Replace control if it is out of calibration. Test by turning to higher setting. **AVOID SCALDING TEMPERATURES!**
	Temperature differential too high	Replace the control if the differential exceeds 25°F.
Insufficient hot water	Control dial too low	Dial control to higher setting. **AVOID SCALDING TEMPERATURES!**
	Control defective	Replace control if it is out of calibration.
	Gas plug cock not fully open	Turn plug cock to open position.
	Knob on dial not in ON position	Turn control knob to ON position.
	Excessive demand	Check for change in hot water usage. Review methods for diversifying demands to accommodate heater operation. Always size heater properly to accommodate demand required.
	Heater fouled with scale	Remove scale, or replace heater.
	Temperature differential too high	Replace the control if the differential exceeds 25°F.
	Heater appears to have a slower recovery	Check gas input. If needed, adjust gas pressure or replace main burner orifice. Heater may have been installed in warmer months when water supply may have been warmer.
	Flue baffle not in flue	Install flue baffle.
	Dip tube missing or in hot side of heater	Install dip tube in cold side of heater.

210A01.EPS

Complaint	Possible Cause	Service Tip
Water too hot/heater too noisy	Control dial too high	Dial control down.
	Heater stratification (or stacking)	Check for leaky faucets in the building.
	Sizzle	Dripping condensation. A normal condition when heating a cold tank.
	Rumble**	Sediment buildup. Drain sediment or remove it using deliming solution.
	Ticking	Expansion or contraction of flue, tank, or piping. Normal condition.
	Combustion sounds—burner using too much primary air	Install and regulate air shutter. If occurring in winter, dial control down based on occupants' demands. If gas heater is so equipped, adjust air shutter to blue condition but not to the point that it is noisy or lifting off burner ports. Install a water mixing valve if necessary.
	Combustion sounds—wrong burner orifices	Ensure that orifices match the gas firing rate.
Water leaks	Joint leakage	Repair faulty installation.
	Condensation	Insulate pipes if necessary.
	Relief valve discharge	Relief valve may be wrong rating. Install water hammer arrestors so that relief valve does not open because of pressure surges in building system. Continuous leakage should be checked for signs of valve failure. Check for the installation of a backflow preventer in the piping to the heater; if one is installed, install a pressure relieving device or an expansion tank in line to the heater. Do not allow regular relief valve cycling to serve as a solution to this problem.
	Drain cock leaking	If valve will not close, turn valve wide open to allow for sediment discharge, which may be preventing proper closure. If valve will still not close, replace it.
	Dripping noises on burner	Dripping condensation. A normal condition when heating a cold tank.
Gas odors	Gas valve or piping connections	Check for leaks.
	Vent blockage	Check for backdraft, blockages, excessive soot, or improperly installed vent piping.
	Overfiring	Adjust input.
	Draft diverter not installed	Install draft diverter.

NOTES:

* Heater stratification occurs when water is drawn off slowly from the heater and the cold water entering the base of the tank keeps the gas control in the "on" position. The water at the top of the tank becomes extremely hot.

** Sediment collects at the base of the heater. As the flame heats the tank base, the water trapped below the sediment becomes superheated and flashes into steam, especially when water is drawn off.

210A02.EPS

Resources & Acknowldgments

Additional Resources

This module is intended to be a thorough resource for task training. The following reference works are suggested for further study. These are optional materials for continued education rather than for task training.

The Hot Water Handbook: An Advanced Primer on Domestic Hot Water. 1998. George Lanthier and Robert Suffredini. Arlington, MA: Firedragon Enterprises.

Planning and Installing Solar Thermal Systems: A Guide for Installers, Architects, and Engineers. 2004. German Solar Energy Society (DGS). Sterling, VA: Earthscan.

Residential Hot Water Systems: Repair and Maintenance. 1987. John E. Traister. Englewood Cliffs, N.J.: Prentice-Hall.

References

"CPSC Splits Vote on Gas Water Heaters," Consumer Product Safety Commission Press Release #95-043, December 15, 1994. Washington, DC: Consumer Product Safety Commission. http://www.cpsc.gov/cpscpub/prerel/prhtml95/95043.html, reviewed November 2004.

Dictionary of Architecture and Construction, Third Edition, 2000. Cyril M. Harris, ed. New York: McGraw-Hill.

"Water Heaters to Be Resistant," Robert P. Mader, *Contractor Magazine*, 2003. Cleveland, OH: Penton Media. http://www.contractormag.com/articles/newsarticle.cfm?newsid=272, reviewed November 2004.

Figure Credits

Gastite Division of Titeflex Corporation	Module divider
A.O. Smith Corporation	210F01
Bradford White Corporation	210F03–210F07, 210F20
Rheem Manufacturing Company	210F09, 210F11
Doucette Industries, Inc.	210F14
Tyco Valves & Controls	210F16 (photos)
Marotta Controls Inc.	210F16 (diagram)
Eemax, Inc.	210F18

CONTREN® LEARNING SERIES — USER UPDATE

The NCCER makes every effort to keep these textbooks up-to-date and free of technical errors. We appreciate your help in this process. If you have an idea for improving this textbook, or if you find an error, a typographical mistake, or an inaccuracy in NCCER's Contren® textbooks, please write us, using this form or a photocopy. Be sure to include the exact module number, page number, a detailed description, and the correction, if applicable. Your input will be brought to the attention of the Technical Review Committee. Thank you for your assistance.

Instructors – If you found that additional materials were necessary in order to teach this module effectively, please let us know so that we may include them in the Equipment/Materials list in the Annotated Instructor's Guide.

Write: Product Development and Revision
National Center for Construction Education and Research
P.O. Box 141104, Gainesville, FL 32614-1104

Fax: 352-334-0932

E-mail: curriculum@nccer.org

Craft ____________ Module Name ____________

Copyright Date ____________ Module Number ____________ Page Number(s) ____________

Description

____

____

____

(Optional) Correction

____

____

(Optional) Your Name and Address

____

____

Plumbing Level Two

02211-05

Fuel Gas Systems

02211-05
Fuel Gas Systems

Topics to be presented in this module include:

Overview

Plumbers design, size, install, and test safe and efficient fuel gas and fuel oil systems. Plumbers must understand the characteristics of the three types of fuel commonly used today: natural gas, liquefied petroleum gas (LP gas), and fuel oil. These fuels have several elements in common. They each create heat and energy, although the temperatures, amount of energy, and combustion byproducts vary depending on the fuel.

These fuels have important differences too. Factors specific to natural gas include meters, valves, regulators, unions, anodes, protective coatings, and purging. Materials and considerations specific to LP gas include storage containers, valves, regulators, vents, gauges, unions, installation plans, and purging. Materials and factors specific to fuel oil include filters, pumps, fuel gauges, valves, vents, unions, installation drawings and layout, storage tanks, and installation procedures.

Installations include plans and specifications of the elements of the fuel system. Plumbers determine the size of the pipe in the piping system by determining gas demand for each outlet. Plumbers test service line installations and the roughed-in piping for leaks, and they apply fire-stopping materials and devices. While there are general installation practices basic to safe appliance installation, plumbers refer to local code, the gas or oil supplier, and the manufacturers' instructions for detailed information. Safe fuel gas and fuel oil piping systems require proper installation techniques, quality materials, and strict adherence to code.

Focus Statement

The goal of the plumber is to protect the health, safety, and comfort of the nation job by job.

Code Note

Codes vary among jurisdictions. Because of the variations in code, consult the applicable code whenever regulations are in question. Referring to an incorrect set of codes can cause as much trouble as failing to reference codes altogether. Obtain, review, and familiarize yourself with your local adopted code.

Objectives

When you have completed this module, you will be able to do the following:

1. Identify the major components of the following fuel systems and describe the function of each component:
   - Natural gas
   - LP gas (liquefied petroleum gas)
   - Fuel oil
2. Identify the physical properties of each type of fuel.
3. Identify the safety precautions and potential hazards associated with each type of fuel and system.
4. Connect appliances to the fuel gas system properly.
5. Apply local codes to various fuel gas systems.
6. Design, size, purge, and test fuel gas systems.
7. Demonstrate familiarity with applicable fuel gas codes.

Trade Terms

Condensate	Mercaptan
Drip	Meter index
Flash point	Purge
Gravity feed	Regulator
Home run	Specific gravity
Insulating union	Suction line
Manometer	Viscosity

Required Trainee Materials

1. Appropriate personal protective equipment
2. Pencil and paper
3. Copy of local applicable code

Prerequisites

Before you begin this module, it is recommended that you successfully complete *Core Curriculum; Plumbing Level One; Plumbing Level Two,* Modules 02201-05 to 02210-05.

This course map shows all of the modules in the second level of the Plumbing curriculum. The suggested training order begins at the bottom and proceeds up. Skill levels increase as you advance on the course map. The local Training Program Sponsor may adjust the training order.

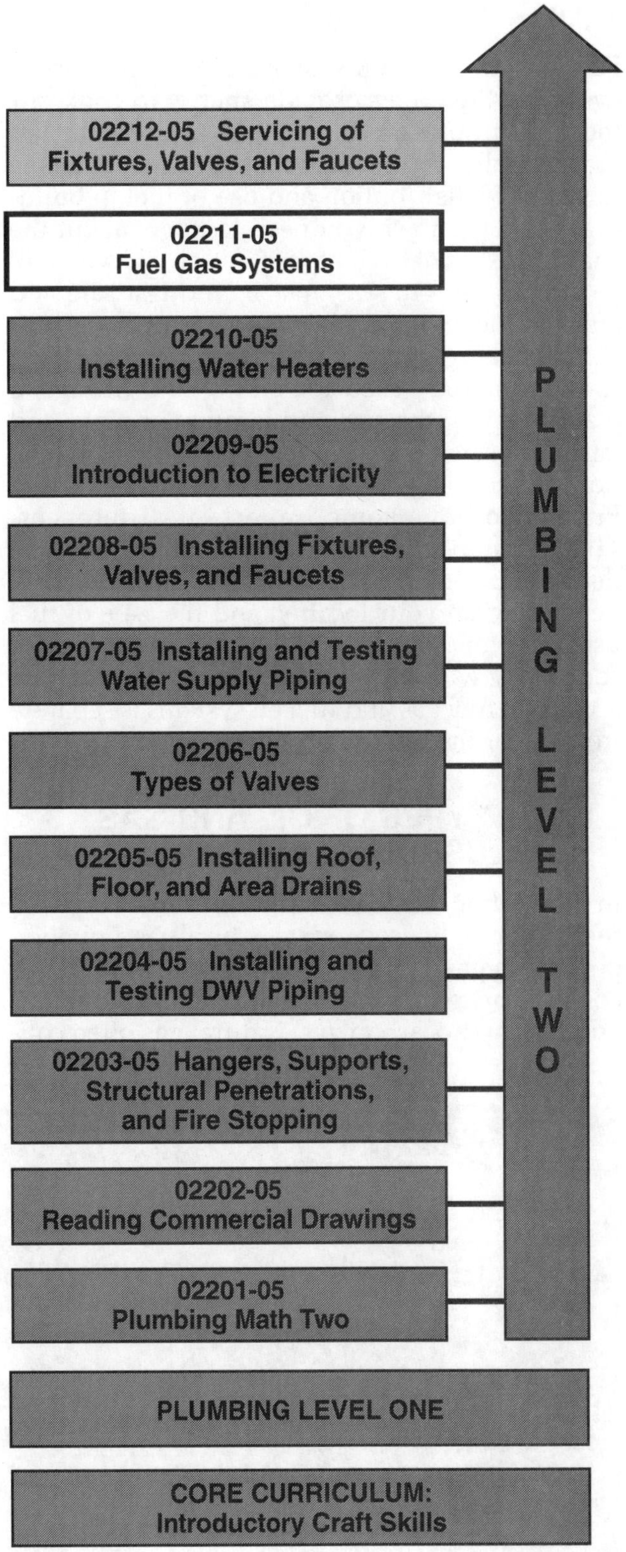

PLUMBING LEVEL TWO

02212-05 Servicing of Fixtures, Valves, and Faucets

02211-05 Fuel Gas Systems

02210-05 Installing Water Heaters

02209-05 Introduction to Electricity

02208-05 Installing Fixtures, Valves, and Faucets

02207-05 Installing and Testing Water Supply Piping

02206-05 Types of Valves

02205-05 Installing Roof, Floor, and Area Drains

02204-05 Installing and Testing DWV Piping

02203-05 Hangers, Supports, Structural Penetrations, and Fire Stopping

02202-05 Reading Commercial Drawings

02201-05 Plumbing Math Two

PLUMBING LEVEL ONE

CORE CURRICULUM: Introductory Craft Skills

211CMAP.EPS

1.0.0 ◆ INTRODUCTION

Fuel gas and fuel oil systems are a vital part of everyday life. They provide energy to cook our food, heat our homes, and power our vehicles. Traditionally, the plumber has been responsible for the safe distribution and use of fuel in buildings. As a plumber, you need to know about the characteristics of each type of fuel so that you will be able to design, size, install, and test safe and efficient fuel gas and fuel oil systems.

Fuels have several elements in common, such as the materials used for piping and fittings, a plan outlining the size and location of pipes and tanks, the safe venting of the products of combustion, and local codes that regulate each fuel's use. Fuels also have some important differences. These common elements and differences are detailed in this module. Applicable codes vary depending on your location and the type of fuel used, so you should read your local applicable code along with this module. Often, the code that applies to fuel gas and fuel oil systems is separate from either the building or plumbing code.

2.0.0 ◆ TYPES OF OIL AND GAS USED AS FUELS

In this section, you will learn about three types of fuels commonly used today: natural gas, liquefied petroleum gas (LP gas), and fuel oil. Natural gas consists primarily of methane, a colorless and odorless hydrocarbon gas. Natural gas often contains small amounts of other gases such as ethane, propane, butane, and pentane. LP gas, also known as LPG or bottled gas, is most often produced from vapors during petroleum refining. The most commonly used LP gases are propane or a mixture of propane and butane. Fuel oil is a hydrocarbon that is distilled from crude oil. All of these fuels create heat and energy when burned, though the temperatures, amount of energy, and combustion byproducts vary depending on the fuel. The types of fuel used in your area may depend on factors such as code standards, availability, and cost.

2.1.0 Natural Gas

Natural gas is classified as either dry or wet. Dry gas contains few or no liquid hydrocarbons. Wet gas contains liquid hydrocarbons, or **condensate.** Gasolines and other liquid products may be made from this natural gas condensate. Although most natural gas does not require much processing, some gas must be dehydrated. This process removes any water and separates condensate from the gas.

Mercaptans, or compounds containing sulfur in place of oxygen, are usually added to natural gas to give the gas its familiar odor. This is done to help detect gas leaks because natural gas is generally odorless.

Natural gas is lighter than air, which means that it tends to rise when released into the air. It has the lowest ratio of Btus per cubic foot of gas. This means that pipes carrying natural gas at its nor-

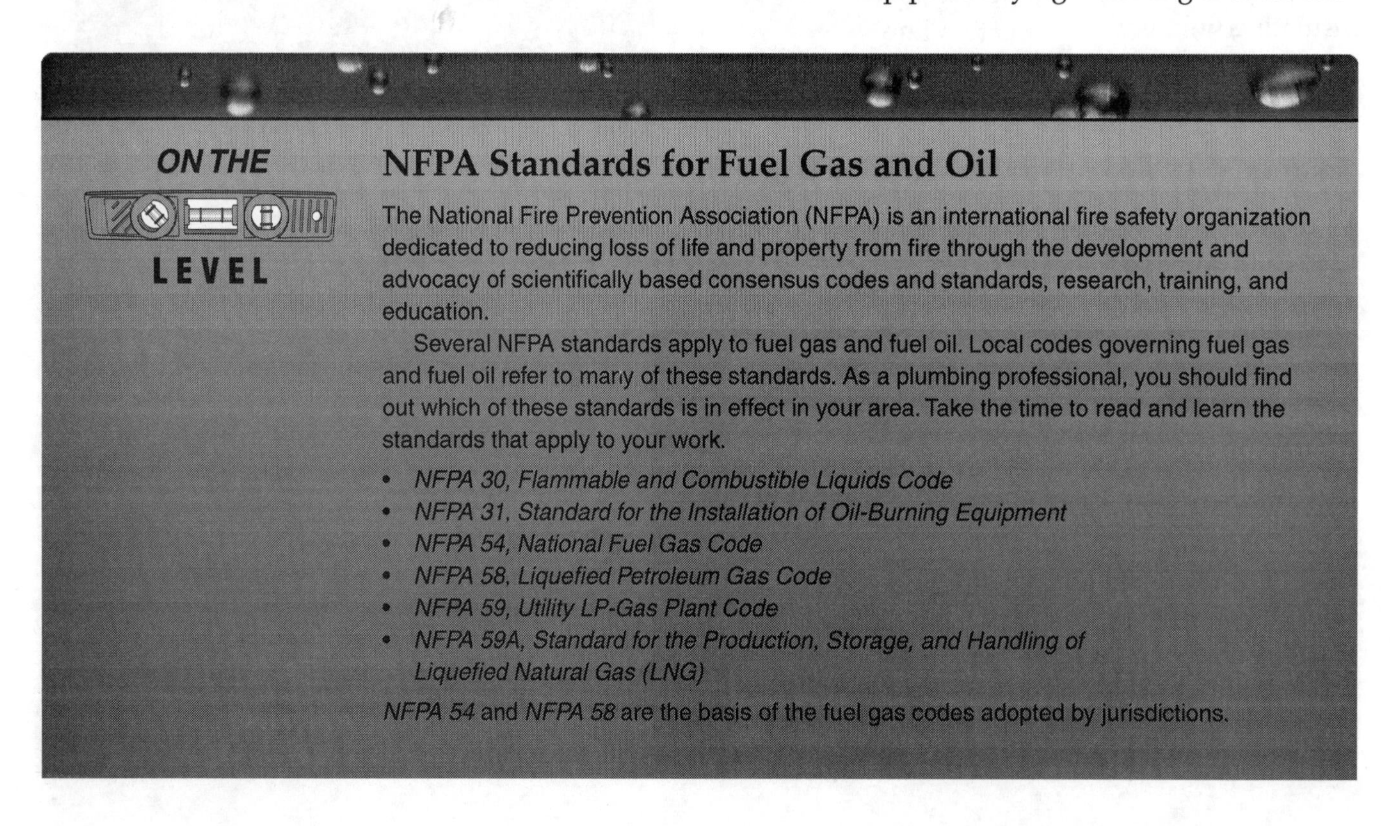

NFPA Standards for Fuel Gas and Oil

The National Fire Prevention Association (NFPA) is an international fire safety organization dedicated to reducing loss of life and property from fire through the development and advocacy of scientifically based consensus codes and standards, research, training, and education.

Several NFPA standards apply to fuel gas and fuel oil. Local codes governing fuel gas and fuel oil refer to many of these standards. As a plumbing professional, you should find out which of these standards is in effect in your area. Take the time to read and learn the standards that apply to your work.

- *NFPA 30, Flammable and Combustible Liquids Code*
- *NFPA 31, Standard for the Installation of Oil-Burning Equipment*
- *NFPA 54, National Fuel Gas Code*
- *NFPA 58, Liquefied Petroleum Gas Code*
- *NFPA 59, Utility LP-Gas Plant Code*
- *NFPA 59A, Standard for the Production, Storage, and Handling of Liquefied Natural Gas (LNG)*

NFPA 54 and *NFPA 58* are the basis of the fuel gas codes adopted by jurisdictions.

mal pressure for use would be larger than the pipes used for other fuel gases or fuel oil.

Because natural gas is lighter than air, the model plumbing codes restrict its use under a building. If a leak occurred, natural gas would pool under a building's floors or floor slabs. Numerous gas explosions over the years would not have happened if modern plumbing codes had been observed at the time the piping was installed.

Propane and butane are forms of liquefied natural gas that are easily transported and stored. They are often used for fuel in areas where other gaseous fuels are not readily available. Mercaptans are added to propane and butane just like they are to other natural gases.

Natural gas is distributed by pipes from the gas well to the gas appliances that use it. Distribution of natural gas is regulated by the U.S. Department of Energy, the NFPA, and the *National Fuel Gas Code*, as well as by state and local plumbing codes. These regulations describe the exact conditions that must be met and the equipment that must be used in distributing gas to consumers.

The difference among the various gas piping systems is the pressure of the gas, which ranges from thousands of pounds per square inch (psi) in the gas well to less than 0.5 psi at the appliances. **Regulators** control the pressure at each stage from the gas choke on the wellhead to the final regulator at the gas meter. Natural gas is sold by the cubic foot, and the gas meter measures cubic feet of gas used.

2.2.0 Liquefied Petroleum Gas (LP Gas)

LP gas is heavier than air and is colorless, odorless, and highly explosive. Bottled gas may be either a liquid or a vapor. Compressed under high pressure, the vapor may be changed into a liquid. This makes LP gas easy to store and transport. When the liquid is allowed to return to normal atmospheric pressure and temperature, it changes into a vapor. Bottled gas can be used as fuel only when it is in its vapor state. Because of its qualities, LP gas is a convenient source of fuel for remote areas.

Because it is heavier than air, LP gas tends to pocket in depressions when released into the air. For this reason, local codes often forbid the use of LP gas appliances in basements.

LP gas is sold by the gallon but is calculated by the pound, and it is usually distributed by special tanker trucks that meter the gas flow and refill bottles or tanks at individual sites.

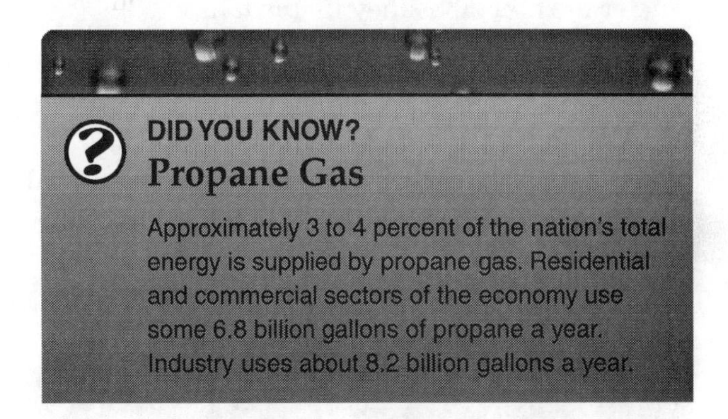

DID YOU KNOW?

Propane Gas

Approximately 3 to 4 percent of the nation's total energy is supplied by propane gas. Residential and commercial sectors of the economy use some 6.8 billion gallons of propane a year. Industry uses about 8.2 billion gallons a year.

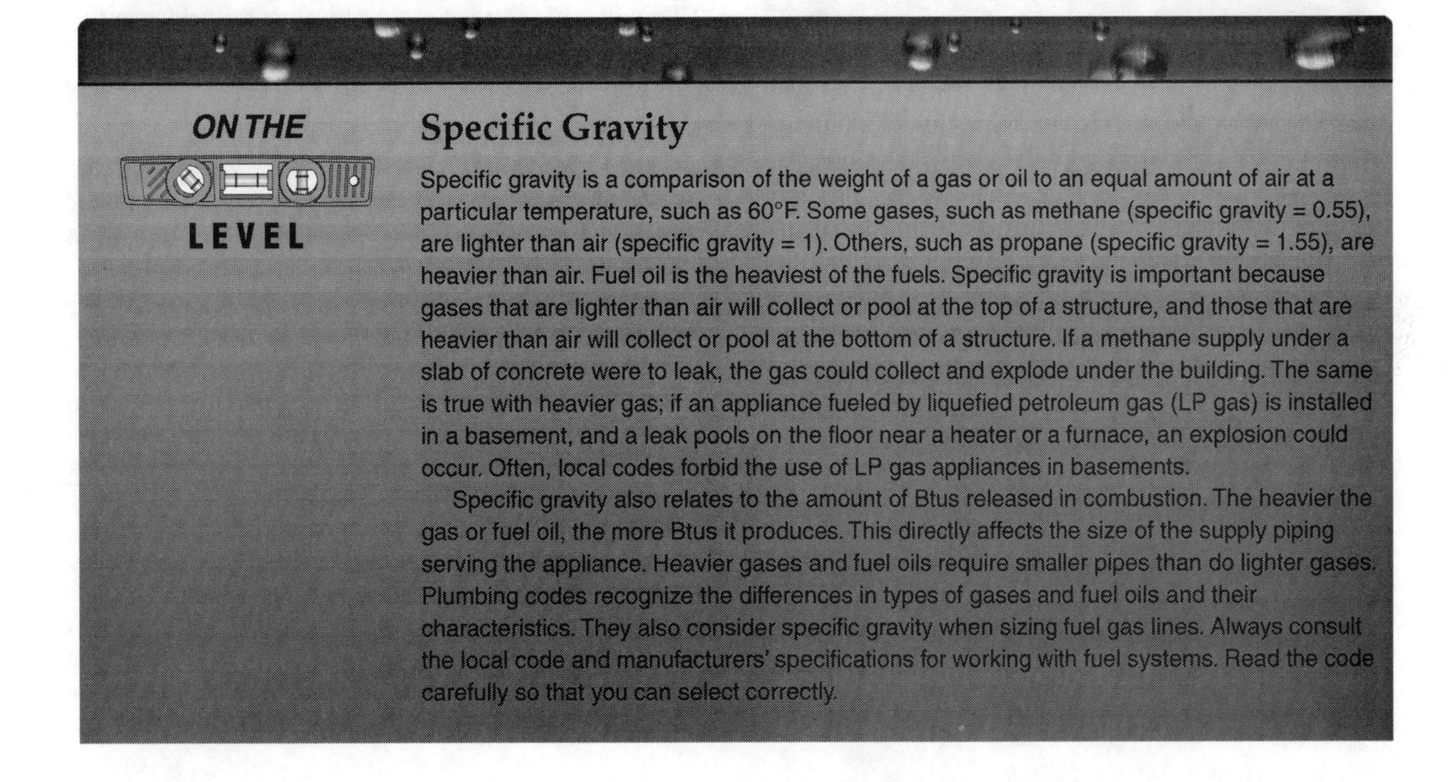

Specific Gravity

Specific gravity is a comparison of the weight of a gas or oil to an equal amount of air at a particular temperature, such as 60°F. Some gases, such as methane (specific gravity = 0.55), are lighter than air (specific gravity = 1). Others, such as propane (specific gravity = 1.55), are heavier than air. Fuel oil is the heaviest of the fuels. Specific gravity is important because gases that are lighter than air will collect or pool at the top of a structure, and those that are heavier than air will collect or pool at the bottom of a structure. If a methane supply under a slab of concrete were to leak, the gas could collect and explode under the building. The same is true with heavier gas; if an appliance fueled by liquefied petroleum gas (LP gas) is installed in a basement, and a leak pools on the floor near a heater or a furnace, an explosion could occur. Often, local codes forbid the use of LP gas appliances in basements.

Specific gravity also relates to the amount of Btus released in combustion. The heavier the gas or fuel oil, the more Btus it produces. This directly affects the size of the supply piping serving the appliance. Heavier gases and fuel oils require smaller pipes than do lighter gases. Plumbing codes recognize the differences in types of gases and fuel oils and their characteristics. They also consider specific gravity when sizing fuel gas lines. Always consult the local code and manufacturers' specifications for working with fuel systems. Read the code carefully so that you can select correctly.

2.3.0 Fuel Oil

Fuel oil is a vital source of energy for residential and industrial use. It is transported and stored in a liquid state. However, it must be changed to a vapor or gas in order to be burned.

Fuel oils are most often identified by their **flash point** and **viscosity.** The flash point is the temperature at which fuel oil will ignite. Viscosity is a measure of the fuel's flowing quality or resistance to flow. Oil with a high viscosity is very thick and resists flowing. Oil with a low viscosity is thin and flows easily.

Fuel oil burner systems are designed to handle different grades, or kinds, of oil. The most common grades are No. 1 to No. 5. The higher the number, the higher the flash point and viscosity. State and local codes regulate the materials that can be used in fuel oil installations. Fuel oil is sold by the gallon, and it is distributed by tankers that meter the fuel oil as they fill the tanks.

3.0.0 ◆ COMMON FACTORS IN FUEL SYSTEMS

This module covers three separate fuels (natural gas, LP gas, and fuel oil), each of which has many different properties, uses, and installation procedures. However, the three also have many factors in common, which are detailed in this section.

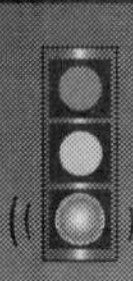

NOTE

Many jurisdictions require plumbers to have a separate gas-fitter license or certification to install and repair fuel gas systems. Refer to your local applicable code. Always ensure that you have the proper training and licenses to do your assigned work.

3.1.0 Materials

All three fuels are transported through pipe and piping systems. However, there are some differences among the piping systems for the different types of fuels. Always refer to your local applicable code for the guidelines that apply in your area.

3.1.1 Natural Gas Piping

Steel pipe (Schedule 40) is the most commonly used piping material in natural gas systems, although other piping materials, such as copper and plastic, are common. Steel pipe is durable, versatile, and strong. As you learned in *Plumbing Level One*, steel pipe is manufactured in both black iron and galvanized forms. Black iron is less expensive because it is not coated with zinc. The carbon contained in steel gives black iron pipe its color. Black iron is used in installations where corrosion will not affect its uncoated surfaces. If black iron pipe is used in underground installations, it must be coated, wrapped, and cathodically protected, which means it must be made nonreactive to electricity.

Galvanized steel pipe is dipped in molten zinc, which gives the pipe a shiny, silver color. The zinc protects the pipe's surfaces from corrosion and abrasive materials. Galvanized pipe may be installed above or below the ground.

Steel pipe joints may be screwed or welded. All screw fittings should be made of either malleable iron or steel. Some codes require welded joints for piping used on fuel gas systems with pressures higher than 5 psi. Contact the natural gas or LP gas supplier in your area for applicable requirements.

Copper pipe (type K or L) can be used in gas system installations if the gas is not corrosive. Copper pipe can be connected by flared, silver soldered, or brazed joints. Soft copper tubing with flared joints may be used outside a building, when provisions are made to prevent joint separation. Flared joints are not recommended for concealed locations. Threaded joints are used for thick-walled copper or brass piping (not tubing) only. Copper tubing (L or K) may be joined by brazing with a low-phosphorous brazing alloy with a melting point of 1,000°F.

Polyethylene (PE) plastic pipe and fittings are gaining dominance in natural gas distribution. This flexible pipe is most often used underground outside a building. It is usually connected by mechanical joints. Be sure to consult the local gas code for the rules regulating the use of plastic pipe in your area.

Corrugated stainless steel tubing (CSST), which is flexible tubing covered with a plastic jacket, is used for gas piping. CSST systems are becoming more popular because of the lower labor costs involved in using flexible tubing, which requires few joints. However, the tubing costs more than other materials, and no single standard of manufacturing has been agreed upon. Each CSST system is proprietary. This means that a manufacturer's pipe and fittings are suitable for use only with that manufacturer's system.

Typically, CSST systems are manifold systems; the gas is distributed from a large pipe with many openings to the appliances by CSST tubing with no other branches. Each run of pipe is a **home run** from the manifold to the appliance. The manifold design allows the manufacturer to make just a few sizes of piping; generally, they are ⅜-inch, ½-inch, ¾-inch, and 1-inch nominal size. Naturally, these sizes limit this material to smaller commercial jobs and residential uses where larger pipe sizes would not be required. Special installation requirements are noted in manufacturers' installation manuals.

All manufacturers require that installers be trained for certification in the use of their CSST systems. In-wall protection of CSST is specified in your local applicable code. Install in-wall protection according to your local applicable code.

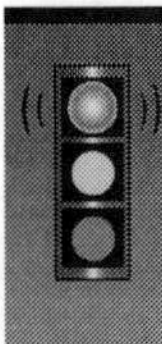

WARNING!

Never join copper used in gas piping by soldering. If a fire occurs, the heat could melt the joint and add fuel to the fire. Solder joints also sometimes deteriorate underground, which may cause a costly gas leak.

3.1.2 LP Gas Piping

LP gas systems use steel or copper piping, depending on the circumstances. For LP gas systems, most codes require piping or tubing from a tank to have a working pressure equal to or greater than the working pressure of the tank. This prevents overpressurizing and possibly overloading the lines, which may result in a break or a leak. All piping on the low-pressure side of the regulator should have a minimum working pressure of 125 psig (pounds per square inch gauge pressure). Steel, seamless copper, soft flexible copper (type K or L), or brass tubing may be used for piping LP gas. Steel and copper pipes are used most often.

3.1.3 Fuel Oil Piping

In fuel oil systems, soft copper tubing (type K or L) is the most commonly used piping material. Most areas allow copper tubing to be connected only with flare joints. The size of copper tubing most often recommended in fuel oil codes is ⅜-inch outside diameter (OD). The only type of steel pipe permitted in fuel oil piping systems is black iron.

3.2.0 Design and Sizing

Every installation should include plans and specifications of the elements of the fuel system. These are usually required to get a permit from the local jurisdiction. Even when a permit is not required, a sketch of the installation is a vital tool for the organized plumber. Each plan should include the following elements (see *Figure 1*):

- Pressure of the gas being carried
- Location and Btu requirements of the fixtures being served
- Lengths and sizes of piping from the meter or tank to the appliances being served
- Location of regulators or pumps
- Type and size of the exhaust venting system
- Specific types and brands of the tanks, valves, regulators, and equipment being installed

Two critical factors in planning the pipe and vent installation are the location and elevation of the tanks in relation to the equipment.

The factors that come into play when designing gas systems include the following:

- The kind of gas being delivered
- The **specific gravity** (ratio of the weight of a given volume of gas compared to the weight of the same volume of air under the same conditions) of the gas
- The Btu rating of the gas
- The developed length of the run from the meter or regulator to the farthest equipment (in the case of the example in *Figure 1*, the developed length is 80 feet)
- The Btu demand of the equipment to be served
- The type of piping being used (for example, CSST has its own sizing charts and pressures)

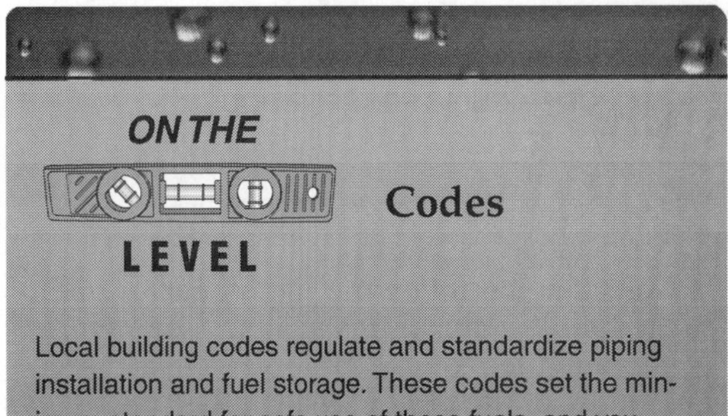

Codes

Local building codes regulate and standardize piping installation and fuel storage. These codes set the minimum standard for safe use of these fuels, and you should always consult them before you design a fuel system. Together, the National Fire Protection Association (NFPA) and the American Gas Association (AGA) produce the *National Fuel Gas Code*. This code is used by many government agencies along with the model plumbing codes including the Building Officials and Code Administrators (BOCA), the Southern Building Code Conference (SBCC), and the International Association of Plumbing and Mechanical Officials (IAPMO). These codes outline acceptable materials, fittings, valves, and installation procedures and provide charts showing minimum pipe sizes for the fuel gases. Ask your local administrative authority for the name of the code you should consult.

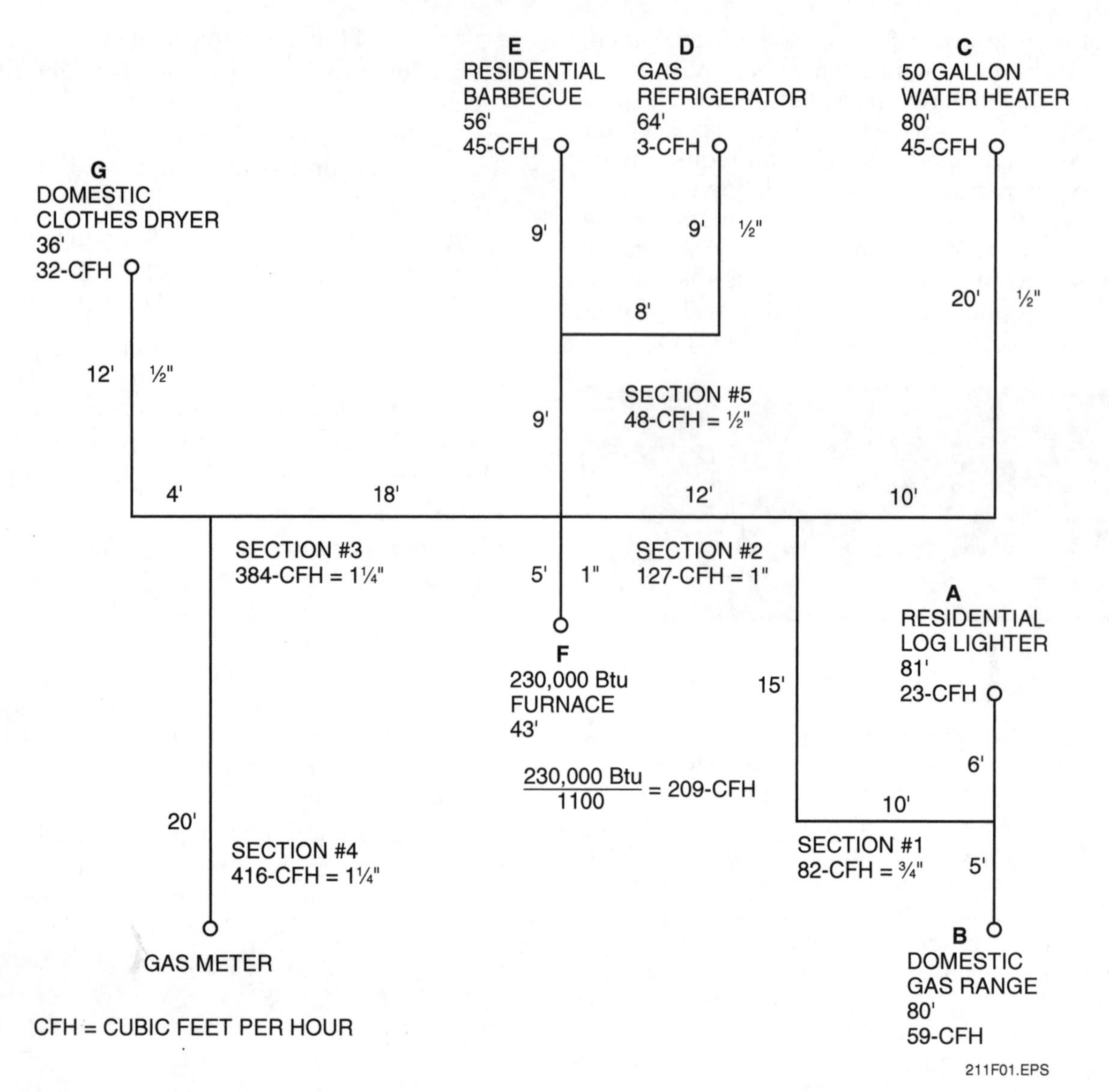

Figure 1 ◆ Typical natural gas piping plan.

Consider the following hypothetical example for a simple natural gas system (see *Figure 2*). Determine the required pipe size of each section and outlet of the piping system. Assume the system uses Schedule 40 metallic pipe and the designated pressure drop is ½ inch w.c. (water column). The gas has a specific gravity of 0.60 and a heating value of 1,000 Btu per cubic foot.

Begin by determining the maximum gas demand for each outlet. The demand can be expressed as the consumption in Btus per hour (Btu/hr) divided by the Btus of gas in Btu per cubic foot (Btu/cu ft). The following formula shows the demand expressed mathematically:

$$\frac{\text{Consumption (in Btu/hr)}}{\text{Btu of gas (in Btu/cu ft)}} = \text{Gas demand}$$

The consumption is determined from the appliance's rating plate input, or from sizing tables in your local fuel gas code (see *Table 1*).

Maximum gas demand for Outlet A:

$$\frac{\text{Consumption (in Btu/hr)}}{\text{Btu of gas (in Btu/cu ft)}} = \frac{35{,}000}{1{,}000} = 35 \text{ cu ft/hr}$$

Maximum gas demand for Outlet B:

$$\frac{\text{Consumption (in Btu/hr)}}{\text{Btu of gas (in Btu/cu ft)}} = \frac{75{,}000}{1{,}000} = 75 \text{ cu ft/hr}$$

Maximum gas demand for Outlet C:

$$\frac{\text{Consumption (in Btu/hr)}}{\text{Btu of gas (in Btu/cu ft)}} = \frac{35{,}000}{1{,}000} = 35 \text{ cu ft/hr}$$

Maximum gas demand for Outlet D:

$$\frac{\text{Consumption (in Btu/hr)}}{\text{Btu of gas (in Btu/cu ft)}} = \frac{100{,}000}{1{,}000} = 100 \text{ cu ft/hr}$$

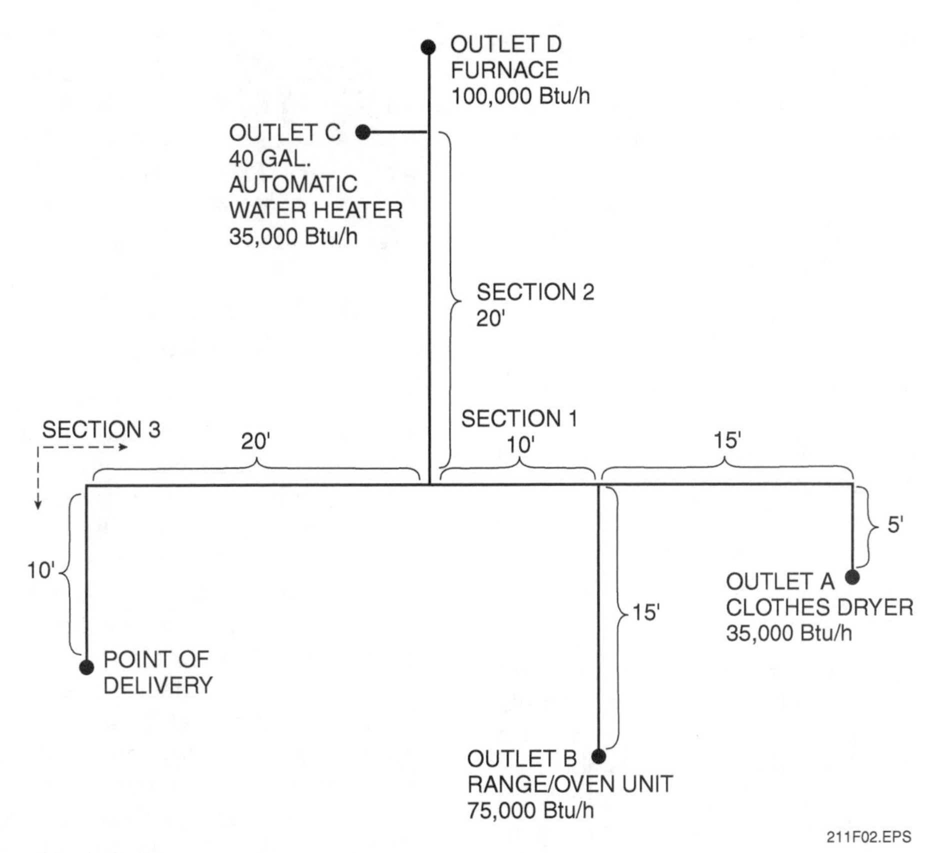

Figure 2 ◆ Piping plan for a hypothetical fuel gas installation using steel pipe.

Table 1 Size of Gas Piping, Schedule 40 Metallic Pipe

Gas	Natural
Inlet Pressure	0.50 psi or less
Pressure Drop	1/2 inch w.c.
Specific Gravity	0.60

	Pipe Size (in)										
Nominal	1/3	3/8	1/2	3/4	1	1-1/4	1-1/2	2	2-1/2	3	4
Actual ID	0.364	0.493	0.622	0.824	1.049	1.380	1.610	2.067	2.469	3.068	4.026
Length (ft)	**Max. Capacity in cu ft of Gas per hr**										
10	32	72	132	278	520	1,050	1,600	3,000	4,800	8,500	17,500
20	22	49	92	190	350	730	1,100	2,100	3,300	5,900	12,000
30	18	40	73	152	285	590	890	1,650	2,700	4,700	9,700
40	15	34	63	130	245	500	760	1,450	2,300	4,100	8,300
50	14	30	56	115	215	440	670	1,270	2,000	3,600	7,400
60	12	27	50	195	195	400	610	1,150	1,850	3,250	6,800
70	11	25	46	96	180	370	560	1,050	1,700	3,000	6,200
80	11	23	43	90	170	350	530	990	1,600	2,800	5,800
90	10	22	40	84	160	320	490	930	1,500	2,600	5,400
100	9	21	38	79	150	305	460	870	1,400	2,500	5,100
125	8	18	34	72	130	275	410	780	1,250	2,200	4,500
150	8	17	33	64	120	250	380	710	1,180	2,000	4,100
175	7	15	28	59	110	225	350	650	1,050	1,850	3,800
200	6	14	26	55	109	210	320	610	980	1,700	3,500

Next, determine the length of pipe from the point of delivery to the most remote outlet (Outlet A). Refer to *Figure 2*. The distance is 60 feet. Now refer to *Table 1*. Find the row indicating 60 feet. Read across the row until you come to the closest approximate output rating for each appliance. Read up along the column to determine the correct pipe size for each branch:

- Outlet A, supplying 35 cu ft/hr, requires a ⅜-inch pipe.
- Outlet B, supplying 75 cu ft/hr, requires a ½-inch pipe.
- Section 1, supplying Outlets A and B (110 cu ft/hr), requires a ½-inch pipe.
- Section 2, supplying Outlets C and D (135 cu ft/hr), requires a ¾-inch pipe.
- Section 3, supplying Outlets A, B, C, and D (245 cu ft/hr), requires a 1-inch pipe.

Note that gas demands for sections are calculated by adding the demands of the outlets they supply.

If the gas has a different specific gravity from that in the table provided, the local applicable fuel gas code will provide a table of multipliers by which you can multiply the output rating to obtain the pipe size (see *Table 2*). Later in your training, you will learn how to size more complex piping systems.

Table 2 Specific Gravity Multipliers

Specific Gravity	Multiplier
0.35	1.31
0.40	1.23
0.45	1.16
0.50	1.10
0.55	1.04
0.60	1.00
0.65	0.96
0.70	0.93
0.75	0.90
0.80	0.87
0.85	0.84
0.90	0.82
1.00	0.78
1.10	0.74
1.20	0.71
1.30	0.68
1.40	0.66
1.50	0.63
1.60	0.61
1.70	0.59
1.80	0.58
1.90	0.56
2.00	0.55
2.10	0.54

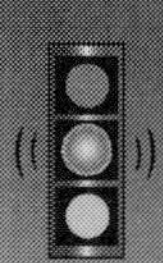

CAUTION

Do not size natural gas systems using sizing tables for LP gas systems, or vice versa. The pressures, pressure drops, and specific gravity are different for each system.

3.3.0 Manufacturer's Installation Instructions

The petroleum industry is constantly changing. Equipment and appliances are always being improved to increase safety and efficiency. That is why plumbers must always refer to the manufacturer's instructions supplied with fuel gas system components. The instructions outline the clearances required, special considerations for inlets and outlets, and many other factors that will ensure proper installation.

Often, the manufacturer's warranty is based on the assumption that these installation instructions have been followed. Failing to follow the instructions, therefore, may invalidate the product's warranty as well as put the customer at risk. Ensure that you collect the manufacturer's instructions as each appliance is delivered. Store the documents in a safe and easily accessible location, and refer to them throughout the installation process. Provide all of the manufacturer's instructions to the customer when the installation is complete.

3.4.0 Testing

Service line installations from the curb to the structure are tested before the trench is backfilled. When the rough-in of a piping system is completed, use a threaded cap or plug on all openings, and leave them closed until the appliance is connected. Never use wooden plugs, corks, or other improvised methods of closing pipe ends.

Next, test the roughed-in piping for leaks. It is important to do this test now, before walls and ceilings are finished. Otherwise, holes will have to be cut in them to find and fix leaks.

To check for leaks, fill the piping with the fuel gas, air, or an inert gas such as helium, neon, nitrogen, or xenon. These inert gases are very stable and have extremely low combustion rates. Never use oxygen to test or **purge** gas lines. To find or check for a leak, apply a soap-and-water solution to connections when the pipe is filled with air or inert gas. Bubbles indicate a leak.

Inside the building, a **manometer** is also used to test for leaks in the piping. The manometer is a U-shaped tube made of a transparent material. It is usually filled with water, but may be filled with mercury to measure higher pressures. The manometer contains an amount of water that lays at equal depths on both sides of the tube when not connected to a gas line (see *Figure 3*).

A scale in inches and fractions of an inch is marked on the manometer. When one end of the manometer is connected to the gas line (usually by a small rubber hose), the pressure forces the water out of its balanced, or even, position. The water line on the other side of the tube, which is not connected to the gas line, rises. Find the gas pressure by reading the scale. This reading is quite accurate. If the pressure drops gradually, the gas distribution system contains a leak.

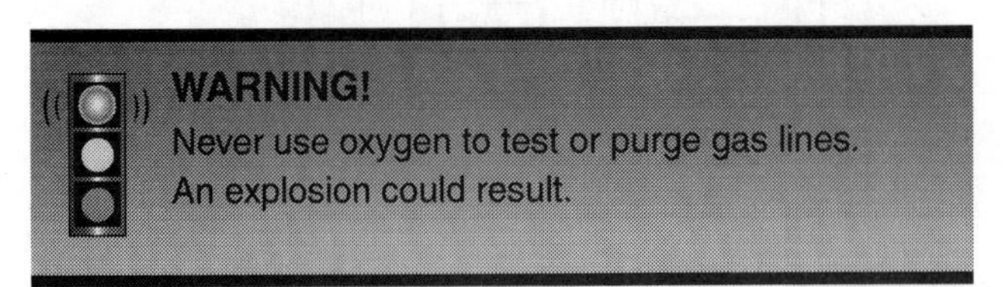

WARNING!
Never use oxygen to test or purge gas lines.
An explosion could result.

Local authorities often accept the use of sensitive gauges other than the manometer to pressure test piping systems. Always check with these authorities to learn their inspection and testing requirements.

3.5.0 Combustion Air

Every fuel-burning appliance must be supplied with combustion air, which provides three things:

- Oxygen to support combustion
- Dilution air for the venting system
- Ventilation air for cooling the equipment enclosure

Combustion air can sometimes be supplied from infiltration of air into the space where the equipment is located. Whether this can be done is often based on the cubic feet of space in the room and the extent to which it is airtight. Modern buildings are often so tightly sealed that infiltration is not possible, and in those cases, ducting of outside air is required. *Figure 4* illustrates some common ducting options. Note that all openings in attics must be extended above insulation. Consult your local code to ensure that the system receives adequate combustion air.

Although ducting is not often in the scope of the plumber's work, it would be irresponsible of a plumber to connect a device to the fuel system without determining that combustion air had been considered and was present in some form. A starved burner can result in the buildup of carbon monoxide, poor or nonexistent venting, and burner flameouts. These conditions can result in property damage or even death.

3.6.0 Venting

Heating appliances that burn fuel must be vented to the exterior (see *Figure 5*). The combustion products of a fuel-burning appliance are dangerous. They can cause injury or death if they are not disposed of properly. The vent must not be smaller than the vent collar of the appliance.

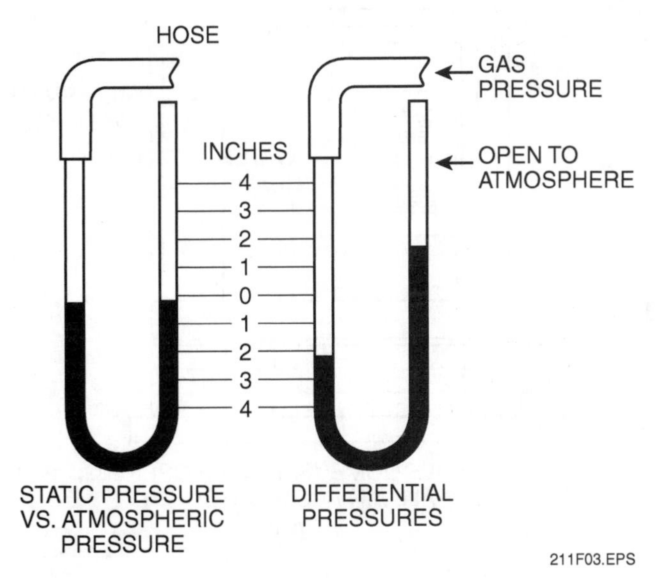

Figure 3 ◆ Water manometer.

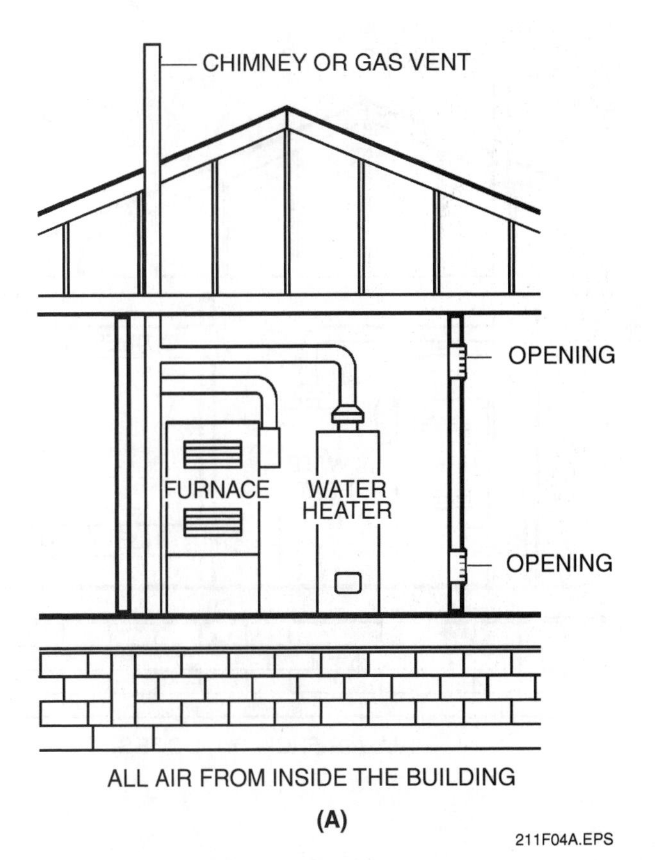

Figure 4 ◆ Air supply methods (1 of 2).

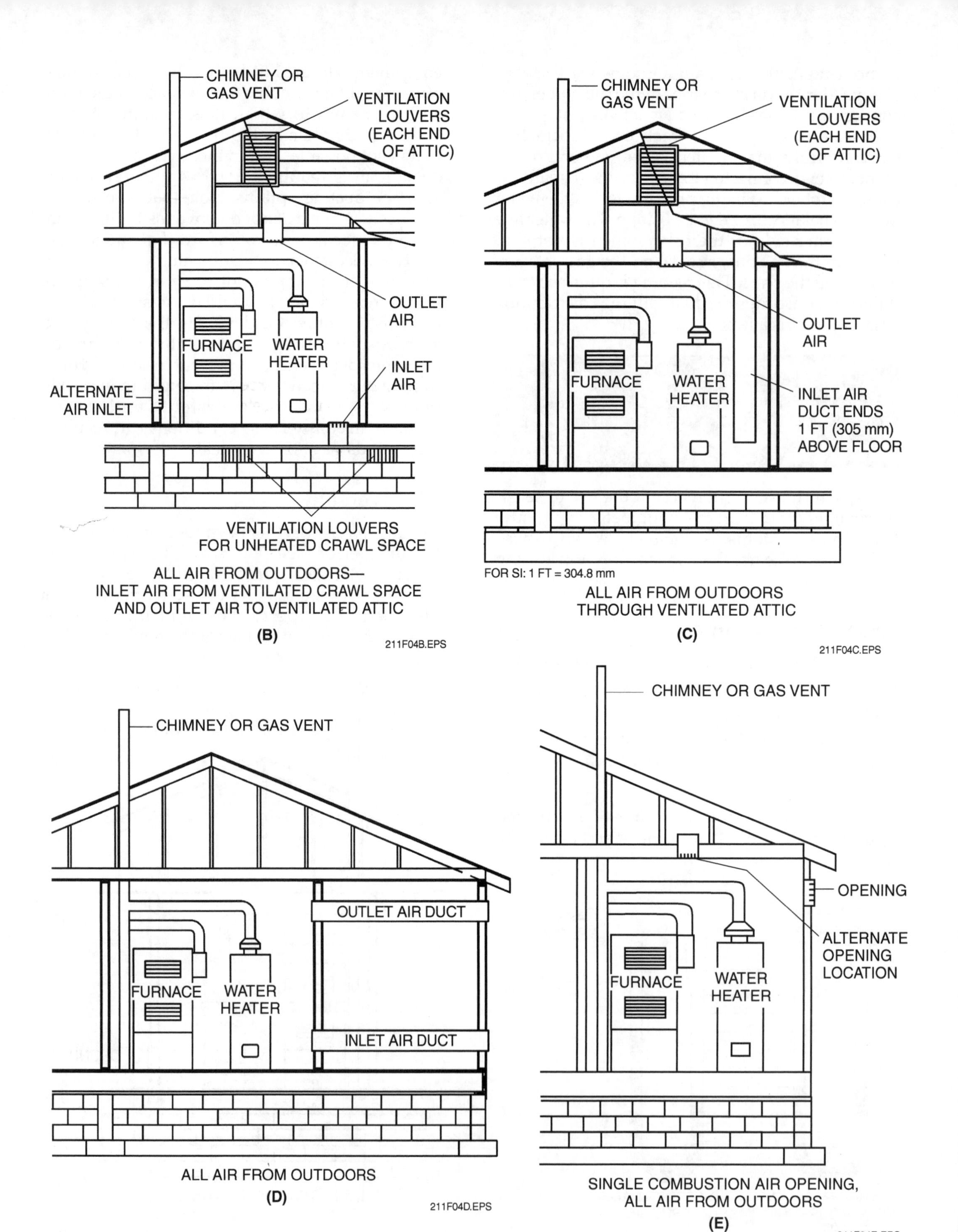

Figure 4 ◆ Air supply methods (2 of 2).

Type B gas vents are used for gas-fueled appliances (those using natural gas and LP gas), and type L vents are used for oil-fueled appliances. Vents are part of an engineered system, and the manufacturer's instructions are vital for proper installation.

Examples of Type B gas-fueled appliances include the following:

- Central furnaces (warm-air types)
- Low-pressure boilers (hot water and steam)
- Water heaters
- Duct furnaces
- Unit heaters
- Vented room heaters (with appropriate input compensation)
- Floor furnaces (with appropriate input compensation)
- Conversion burners (with draft hoods)

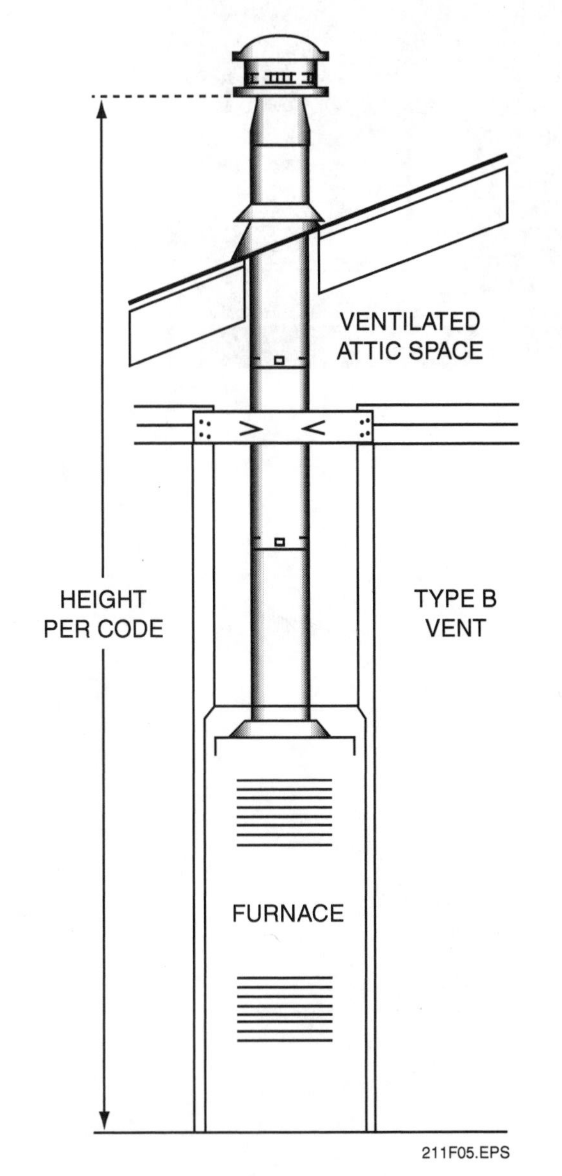

Figure 5 ◆ Typical vent for furnace.

Local codes give specific guidelines and restrictions for the installation of gas and oil venting, and they also must be consulted.

3.7.0 Appliances

The four basic categories of gas-fueled appliances are described as follows:

- Category I – An atmospheric noncondensing gas appliance that operates with a nonpositive vent pressure
- Category II – An atmospheric condensing gas appliance that operates with a nonpositive vent pressure
- Category III – A forced or induced noncondensing gas appliance that operates with a positive vent pressure
- Category IV – A forced or induced condensing gas appliance that operates with a positive vent pressure

With Category I and III appliances, water will not collect internally or within the vent during continuous operation. Water may collect internally or within the vent during continuous operation of Category II and Category IV appliances. Nonpositive pressure means the vent is able to operate by natural (gravity) draft. Positive pressure is introduced in a nonpositive system by adding a fan or burner that produces additional vent pressure to cause flow. Positive pressure is also found in systems where the internal static flue gas pressure is greater than the atmospheric pressure. In such systems, vent joints must be sealed to prevent leakage.

Some general installation practices are basic to safe appliance installation. Refer to the local code, the gas or oil supplier, and the manufacturer's instructions for more detailed information.

The authority having jurisdiction should approve all appliances, materials, equipment, and procedures used. In the case of LP gas, for instance, they should be approved by the local gas code, by a nationally recognized authority such as the *National Fuel Gas Code*, or both. In general, approval is based on tests performed by nationally recognized testing laboratories such as the one operated by the AGA or Underwriters Laboratories (UL). These laboratories certify that the appliance, material, or equipment meets the minimum requirements of a national standard.

Before installation, determine whether the appliance is designed to operate on the gas to which it is to be connected. Never attempt to convert natural gas and LP gas appliances to burn fuel oil.

All appliances have a manufacturer's label (see *Figure 6*) that provides critical information:

- Type of gas required
- Btus the appliance uses (to ensure the proper size of supply pipe is used)
- Manufacturer's name
- Testing laboratory (UL, AGA) that lists the appliance
- Minimum clearance to combustible materials

Be sure to read and follow the manufacturer's installation procedures. All manufacturers' installation, operation, and maintenance instructions should be left at the job location.

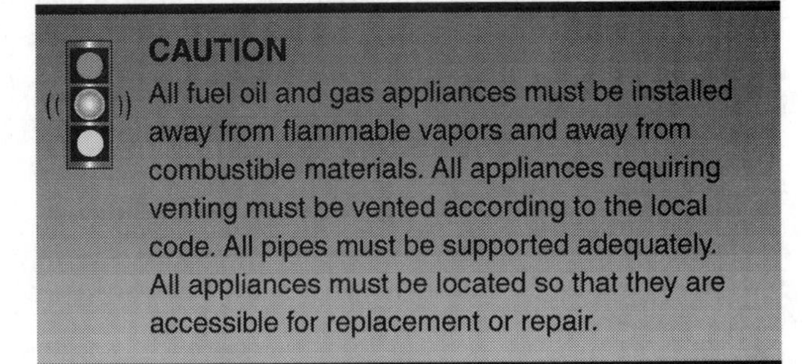
CAUTION

All fuel oil and gas appliances must be installed away from flammable vapors and away from combustible materials. All appliances requiring venting must be vented according to the local code. All pipes must be supported adequately. All appliances must be located so that they are accessible for replacement or repair.

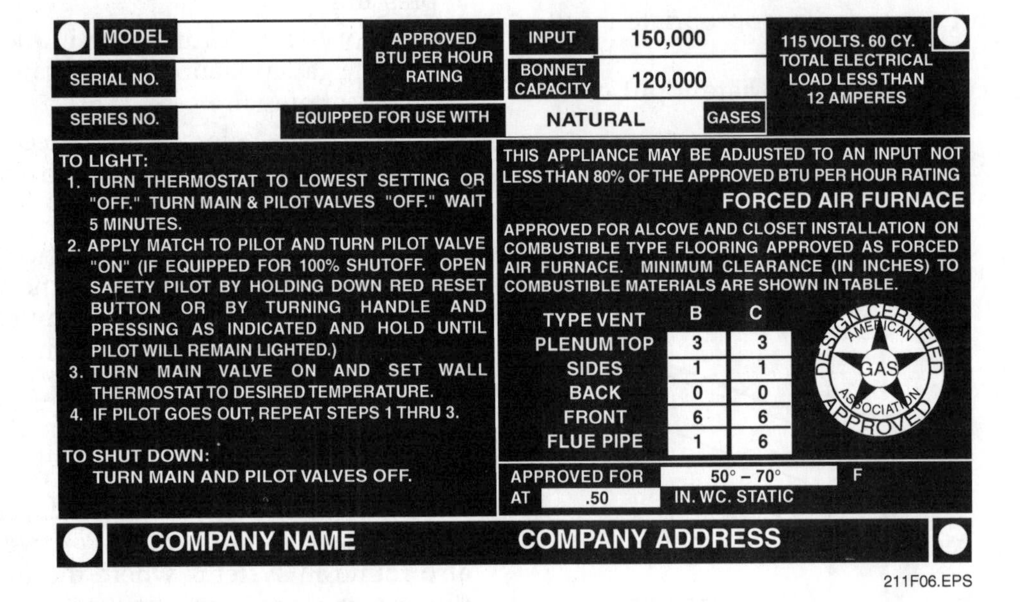

Figure 6 ◆ Typical appliance label.

Review Questions

Sections 1.0.0–3.0.0

1. _____ is lighter than air and has the lowest ratio of Btus per cubic foot of gas of the fuel gases.
   a. Liquefied petroleum
   b. Hydrogen
   c. Methane
   d. Oxygen

2. Because of the tendency of _____ to collect under floors or floor slabs in situations where a pipe might develop a leak, piping for this type of fuel is never installed underneath a building.
   a. natural gas
   b. propane
   c. fuel oil
   d. hydrogen

3. _____ control gas pressure all along the natural gas pipeline, reducing it from thousands of pounds per square inch at the source to just less than 0.5 psi at the appliance connection.
   a. Gas meters
   b. Gas chokes
   c. Regulators
   d. Pressure valves

4. \Appliances using liquefied petroleum gas can be installed safely in a basement.
   a. True
   b. False

5. Fuel oils are rated by both flash point and _____.
   a. resistance
   b. specific gravity
   c. viscosity
   d. compression

6. Fuel oil supply piping lines are most commonly made using _____.
   a. galvanized steel
   b. black iron
   c. polyethylene
   d. type K or L copper

7. Each of the following should be included in the plans and specifications for a fuel system *except* _____.
   a. pressure of the gas being carried
   b. location of regulators or pumps
   c. capacity of the meter in cubic feet per hour
   d. type and size of the exhaust venting system

8. Helium, neon, nitrogen, and xenon are examples of _____ gases used for testing lines for leaks.
   a. inert
   b. pressurized
   c. manometer
   d. sensitized

9. A starved appliance will draw oxygen out of the surrounding atmosphere and produce lethal quantities of _____ if not ducted properly to provide combustion air.
   a. methane
   b. carbon dioxide
   c. carbon monoxide
   d. propane

10. _____ requirements to remove combustion products are determined by code based on the type of fuel used.
    a. Infiltration
    b. Venting
    c. Ducting
    d. Combustion

4.0.0 ◆ FACTORS SPECIFIC TO NATURAL GAS, LP GAS, AND FUEL OIL SYSTEMS

In the previous section, you learned about factors that are common to LP gas and fuel oil systems. However, many materials, installation and testing procedures, and other factors are specific to the type of fuel used. These are detailed in this section.

4.1.0 Factors Specific to Natural Gas

Following are the materials, installation considerations, and other factors specific to natural gas. These factors include meters, valves, regulators, unions, anodes, protective coatings, and purging.

4.1.1 Meters

Gas meters are used to measure the amount of natural gas used by consumers. Meters are located where they can be read easily and where their connections are readily accessible for service. Location, space requirements, dimensions, and the type of installation should be acceptable to the local gas supplier. Before locating the meter, contact the local gas supplier for specifications so that the meter can be located and sized properly. Frequently, meters are supplied and installed by the gas company. Select the meter based on the load information provided by the gas utility. Refer to your local applicable code or the local gas utility for the practice in your area.

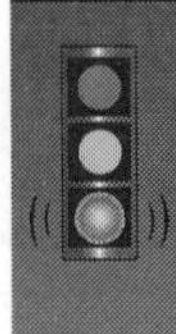

NOTE

Meters can be used to test for leaks. To do this, begin by shutting off the system and purging the gas using methods specified in your local applicable code. Then refill the system with gas, ensuring that all gas appliances are off. When the system is filled, watch the meter. If one or more meter gauges turn, indicating gas flow, you will know that the system has a leak somewhere. Use code-approved leak detection methods to find the leak.

Gas meters used in residential and light commercial systems are classified as either diaphragm meters (see *Figure 7*) or rotary meters (see *Figure 8*). In a diaphragm meter, a flexible sheet of material, called a diaphragm, moves in direct proportion to the amount of gas flowing through the meter. Diaphragm meters are also referred to as positive displacement meters, meaning that no gas is allowed to pass through the diaphragm. Diaphragm meters are available in aluminum-case, steel-case, and ductile iron-case designs.

Rotary meters work according to the principle of positive displacement. As gas flows through the meter, it passes through a set of vanes, gears, or pistons. Each turn of the vane, gear, or piston momentarily "chops" the flowing gas into precisely measured segments of known volume. By counting the number of turns of the vanes, gears, or pistons, the exact volume of gas flowing through the meter can thus be calculated.

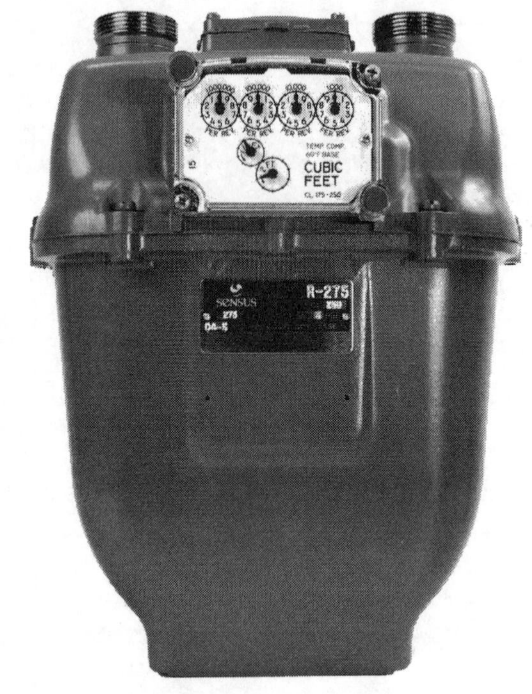

211F07.EPS

Figure 7 ◆ Diaphragm gas meter.

211F08.EPS

Figure 8 ◆ Rotary meter.

Both diaphragm and rotary meters use a display called a **meter index** to indicate the amount of gas that has passed through the meter (see *Figure 9*). The measurement is recorded in cubic feet. The meter is located to permit easy access so it can be read for customer billing. Different colors on the meter index indicate different pressures. For example, white indicates low pressure, yellow indicates 2 pounds, and red indicates 5 pounds. Pressure higher than 5 pounds would have a special index called a BPI or a VPI. The color scheme was developed by the AGA.

4.1.2 Valves

Your local code and the NFPA *National Fuel Gas Code* will specify the types of valves that can be installed in fuel gas systems to shut off the gas flow. Typically, a flat-head cock valve with a lock wing is located before every meter (see *Figure 10*). Other types of shutoff valves, such as the lever-handled gas cock valve (see *Figure 11*) and the ball valve (see *Figure 12*), are positioned in front of the union and are used to connect all gas appliances. Install valves at every point where safety, convenience of operation, and maintenance require them.

Many local codes allow the use of ball valves that meet related American Society of Mechanical Engineers (ASME), American National Standards Institute (ANSI), or AGA classifications. Refer to the manufacturer's specifications to determine the valve's rating. Ensure that the valve is properly rated for the application.

Plug valves are widely used in gas lines to provide both on/off and throttling control. Plug valves provide positive shutoff when closed. Plug valves operate in one of two ways, depending on the model of valve. In some designs, rotation of the valve handle clockwise raises the cylindrical or cone-shaped plug out of its seat, increasing the flow. Rotating the valve handle counter-clockwise lowers the plug back into its seat, restricting flow. In other models, the cone-shaped plug has a horizontal opening through which flow is permitted until the opening is turned away from the flow. These types of plug valves are nonrising stem, quarter-turn valves.

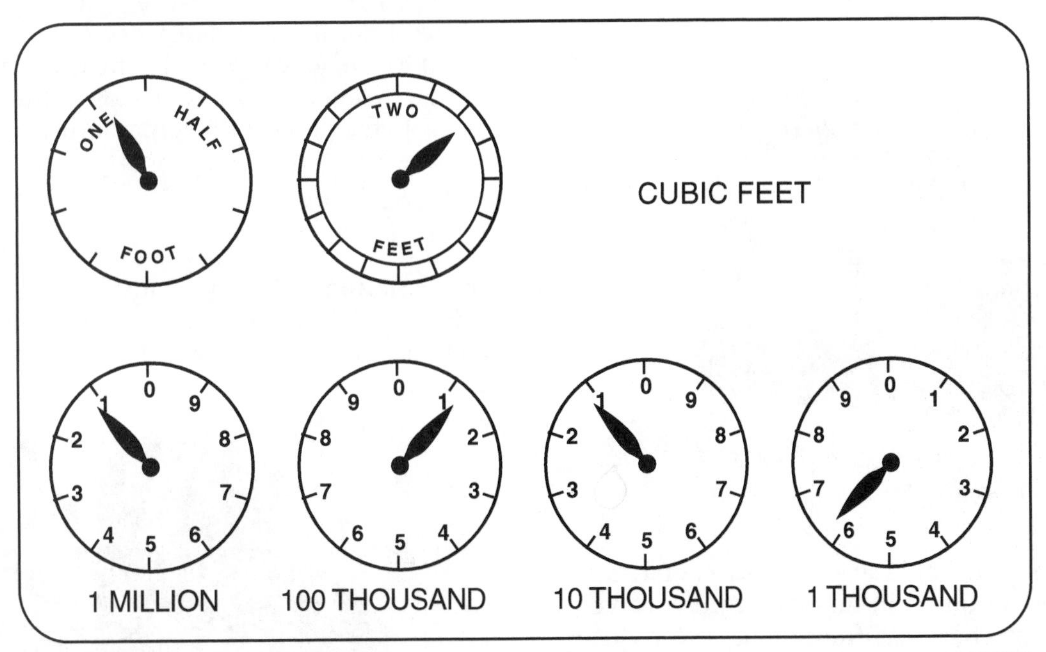

Figure 9 ◆ Close-up of a meter index showing a reading of 1,116,000 cubic feet.

DID YOU KNOW?

Tin-case gas meters were once very common. Gas companies are gradually replacing all tin-case meters with aluminum-case meters, though you may still encounter them on service calls to older buildings. Tin-case meters can be identified by their stabilizer bars, installed above the meter casing. The bar helps keep the inlet and outlet connections in place.

Figure 10 ◆ Flat-head gas cock valve.

Figure 11 ◆ Lever-handle gas cock valve.

Figure 12 ◆ Ball valve.

Some older plug valves use metal-to-metal seals without lubrication. Non-lubricated plug valves may feature mechanical lifting devices to reduce the torque required to turn them. Some non-lubricated plug valve models use elastomeric, or rubber-like, sleeves or coatings that reduce friction when the plug is turned. Note that non-lubricated plug valves often suffer from chafing or sticking. They should be carefully maintained according to the manufacturer's specifications.

Lubricated plug valves, as their name suggests, use a special type of grease as a lubricant between the plug and the valve seat. The grease reduces wear and eliminates sticking. The lubricant also reduces leakage around the plug.

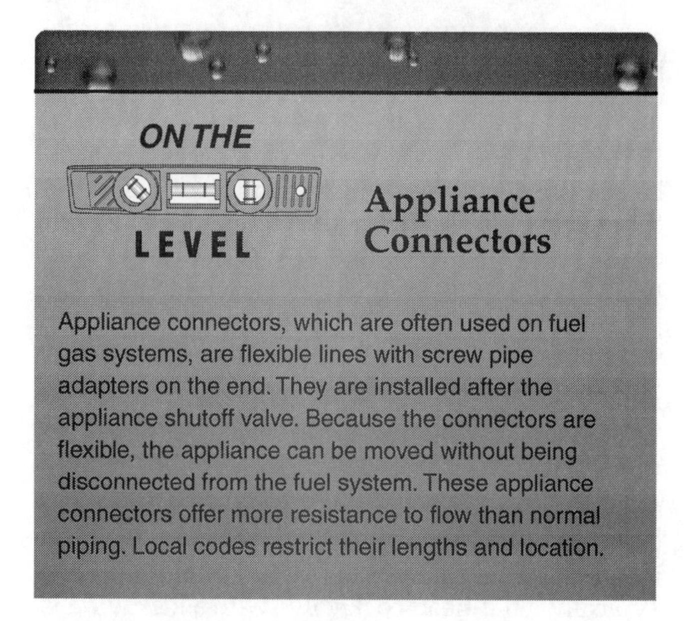

4.1.3 Regulators

The regulator is located in or near the gas meter (see *Figure 13*). It controls the amount of gas pressure from the meter into the building. It is self-regulating. If the gas pressure is low in the main service line, the regulator opens, allowing gas flow into the building. As the pressure increases in the service line, the regulator closes, cutting off gas flow into the building. Regulators are self-regulating once they have been properly adjusted and set.

4.1.4 Unions

An **insulating union** is always located on the gas main side of the meter. The union prevents the conduction of electricity through the customer's ser-

Figure 13 ◆ Regulator.

vice line to the gas main. Other unions are required to connect appliances to the customer's gas distribution system. These unions should be the ground joint (metal-to-metal) type. Unions used for appliance connections are always located between the shutoff valve and the appliance. Codes prohibit concealed unions. Some codes may require the gas company to install unions on natural gas lines.

4.1.5 Anodes

As you have already learned, anodes are attached to steel piping systems to protect them from corrosion (see *Figure 14*). Any steel or wrought iron pipe 2 inches or smaller in diameter and 100 feet or less in length must have a magnesium anode that weighs at least 5 pounds when installed. The anode should be buried a minimum distance of 2 feet from the pipe at or below trench depth. An anode should be located on the device to be protected. Plastic pipe systems use anodeless risers (see *Figure 15*).

Usually, the gas utility locates the meter at the building. As a result, the line from the street to the building belongs to the utility and is isolated from the building piping. In such cases, the anode is the property of the utility and is maintained by it.

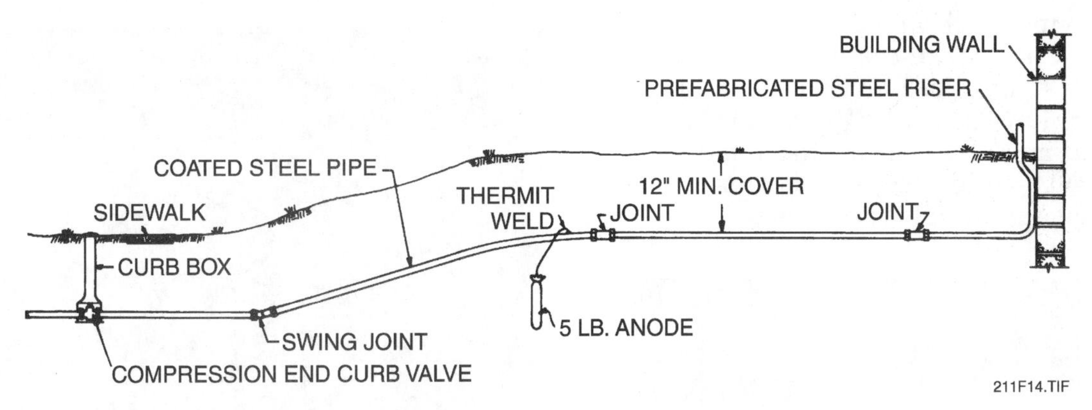

Figure 14 ◆ Anode as part of a system from gas main to meter.

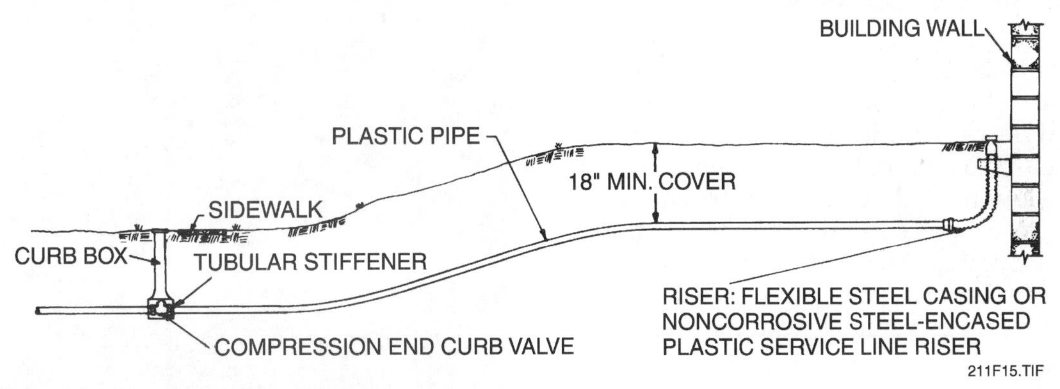

Figure 15 ◆ Anodeless riser for plastic pipe.

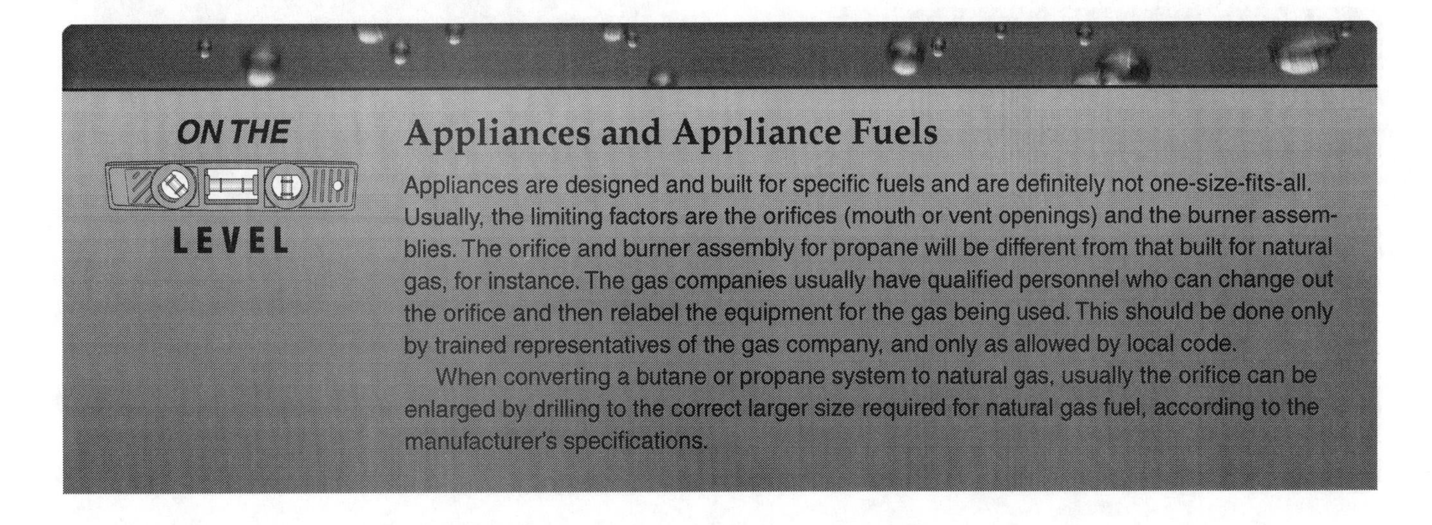

Appliances and Appliance Fuels

Appliances are designed and built for specific fuels and are definitely not one-size-fits-all. Usually, the limiting factors are the orifices (mouth or vent openings) and the burner assemblies. The orifice and burner assembly for propane will be different from that built for natural gas, for instance. The gas companies usually have qualified personnel who can change out the orifice and then relabel the equipment for the gas being used. This should be done only by trained representatives of the gas company, and only as allowed by local code.

When converting a butane or propane system to natural gas, usually the orifice can be enlarged by drilling to the correct larger size required for natural gas fuel, according to the manufacturer's specifications.

If the system uses plastic pipe, each steel or malleable iron coupling must be individually protected by a zinc or magnesium anode. The anode is connected to the gas line or coupling by welding. Clean the anode wire connectors and the metal pipe or connector before welding. Both the anode and its connecting wire are to be coated with a proper coating material to protect the service line. Pipes over 100 feet in length require additional anodes. Consult your local applicable code for the requirements that apply in your area.

4.1.6 Protective Coatings

All underground steel piping systems must be coated with a protective coating. This coating extends at least 2 inches on vertical sections of pipe above the finished grade. The coating system consists of a primer and a coating that retards corrosion. Consult your local gas supplier for specifications and an approved list of coatings.

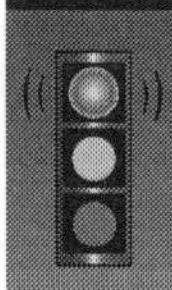

WARNING!

Always test gas supply piping for leaks before applying protective coatings. Otherwise, an explosion could result.

Also consult the project specifications and plans. The local gas company may use a particular brand that is not included in the specification and may be hard to obtain. However, the specification may allow a local supplier to provide its own brand. On LP gas systems, the owner may have meters located at the street. If so, the owner is responsible for ensuring that protective coating is applied on the pipes running from the street to the structure.

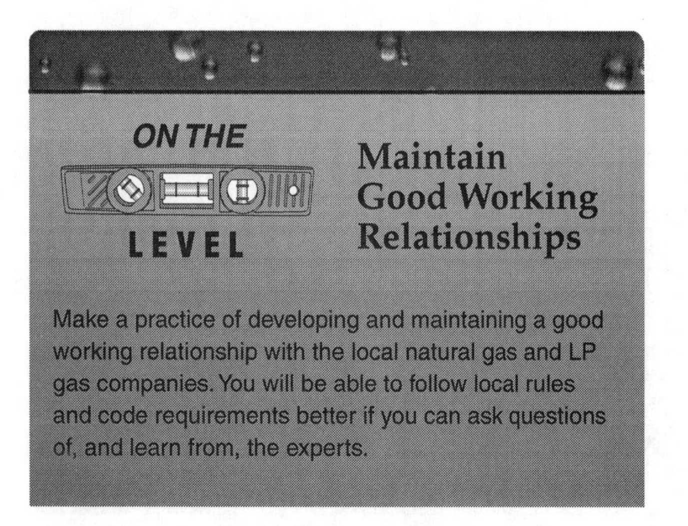

Make a practice of developing and maintaining a good working relationship with the local natural gas and LP gas companies. You will be able to follow local rules and code requirements better if you can ask questions of, and learn from, the experts.

4.1.7 Purging

After the piping has been tested, the entire gas distribution system must be purged with inert gas to remove air and other unwanted substances. Never purge piping into the combustion chamber of an appliance. The open end of the piping system being purged should never be discharged into confined spaces or areas where there is a source of ignition.

After purging the gas piping, purge all appliances and light their pilots. Refer to the manufacturer's specifications for the particular appliance and to your local code for the appropriate purging methods.

CAUTION

You must have a plan of action in place before purging the lines so that you can perform this procedure safely. Your plan should cover the following procedures:

- Controlling the purging rate
- Using proper ventilation
- Eliminating hazardous conditions

The mixture of gas and air released near the end of the purge is highly flammable. It must be dispersed before the appliance is ignited. Establish a set waiting time before lighting the appliance. Limit the amount of air being released so that the installer has plenty of time to react when just the fuel gas is coming out of the valve.

When there are concentrations of heavier-than-air fuel gases on the floor, or natural gas near the ceiling, fans may be needed to blow these concentrations outside. A hose temporarily connected to the end of the piping and run to the exterior may be an alternative to fans. After purging, check the pilots for trapped air pockets.

As on any work site, you must be aware of your working area and equipment, and eliminate or correct conditions that can injure you or your coworkers. Always check with your supervisor if you have any questions about this or any other procedure. After purging and when any raw pilot is lighted, a plan should be in place to check the pilots for a specified period of time to ensure that no air pockets have moved into the pilot and extinguished the flame. Ensure that no heat sources are present when purging gas.

4.2.0 Factors Specific to LP Gas

The materials, installation considerations, and other factors specific to LP gas include storage containers, valves, regulators, vents, gauges, unions, installation plans, and purging. Each is covered in detail below.

4.2.1 Storage Containers

LP gas storage containers are sized according to the amount of gas required (see *Figure 16*). Cylinders provide small amounts of gas for domestic use. Cylinders, which can be lifted, are checked by weight in pounds. The capacities of large tanks that are too heavy to be carried are expressed in gallons. All LP gas storage containers must be UL-listed.

Most local codes require that LP gas storage containers be finished in a white or silver heat-reflecting surface. The finish reduces the amount of heat that the container would otherwise absorb, thus helping to prevent a pressure buildup within the container.

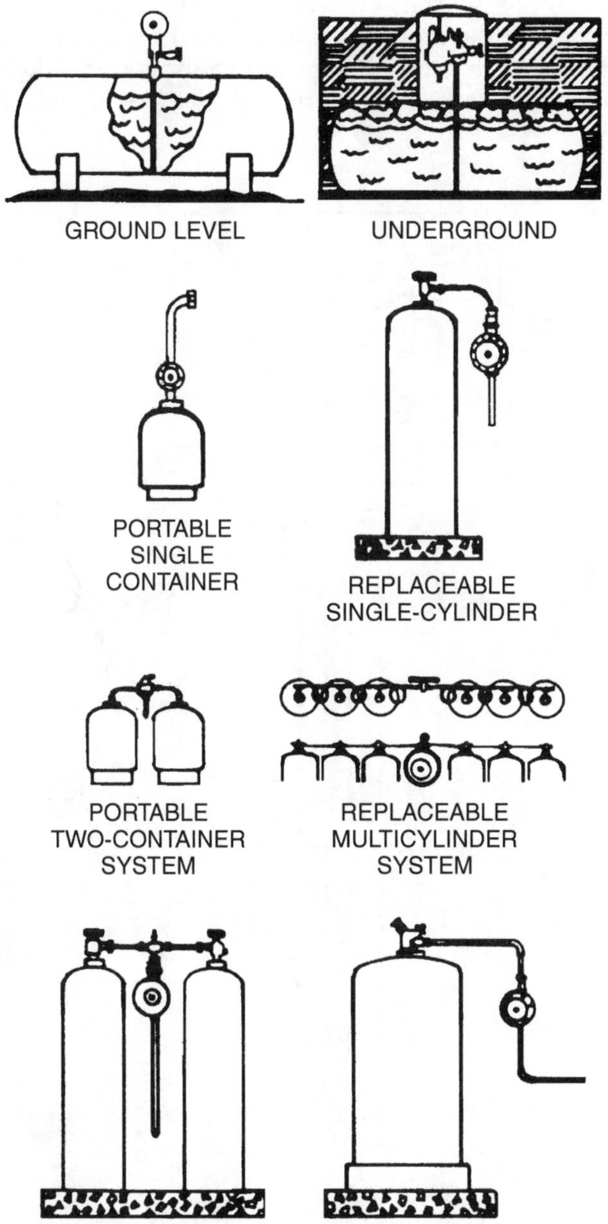

Figure 16 ◆ Different types of LP gas storage containers.

4.2.2 Valves

A typical LP gas system uses a variety of valves. Because LP gas absorbs heat, it expands and increases its pressure on the inside of the container. Install relief valves on all tanks or gas containers to prevent them from exploding in the event of a pressure increase (see *Figure 17*). Relief valves are sized according to the needed discharge capacity of the container.

Excess flow valves are designed to prevent the escape of gas when pipes or hose lines break (see *Figure 18*). The valve actually shuts off the flow of gas if the flow becomes excessive. This prevents an entire tank from emptying into the atmosphere. Excess flow valves should have a rated closing flow (maximum allowed) that is 50 percent greater than the normal anticipated flow. The size of the pipe on either side of the flow valves should never exceed the size of the pipe of the excess flow valves. These valves should be tested at the time of installation and at least once a year after that.

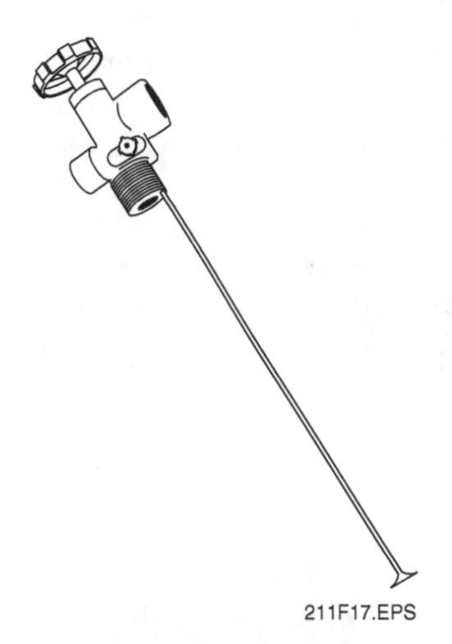

Figure 17 ◆ Relief valve.

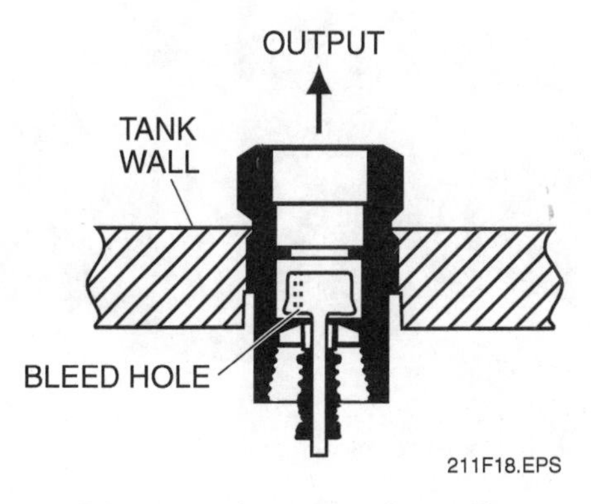

Figure 18 ◆ Schematic of an excess flow valve.

Special valves called overfill protection devices (commonly called OPD valves) are required on bottled-gas containers to protect against overfill. These valves must be located on the service outlet (building side) of the container. They must be accessible at all times.

4.2.3 Regulators

Pressure regulators must be located on the outlet side (appliance side) of the LP gas container (see *Figure 19*). Regulators control the amount of gas pressure from the storage container to the appliances. Regulators are self-controlling. If the demand for gas is increased (if all the appliances are being used, for instance), the regulator allows greater gas flow into a building. As the need for gas decreases, the regulator restricts the gas flow. Like meters and lines, regulators have maximum capacities. Ensure that the regulator you select is sized for the system.

Two regulators installed in the same service line increase the pressure control. The use of two regulators is known as two-stage regulation. The second regulator must be installed where the service line enters the building. Two-stage regulation is not used in high-pressure systems.

The first regulator reduces the pressure in the service line to 5–20 psig. It is referred to as the high-pressure regulator. The second regulator further reduces the flow of gas to a water column pressure of about 11 inches. This is usually measured with a manometer. The two-stage system gives more accurate control of gas pressure because of the reduced line pressure at the second regulator.

Two-stage regulation systems offer several advantages over systems that use only one regulator. They provide uniform pressure to all appliances. Because regulator freeze-up is more common in single-stage systems, two-stage systems typically require fewer service calls. Finally, because the second regulator reduces the gas pressure, service size minimums are specified in the local applicable code.

All regulators must be vented. Venting allows an increase in gas pressure caused by a malfunction in a regulator to be vented into the atmosphere without causing the service line to rupture. If a gas regulator is located inside the building, it should be vented to the outside.

4.2.4 Gauges

Gauges are used on large LP gas storage tanks to measure the amount of liquid in the container. A float gauge uses an indicator calibrated (marked) by percentage of total tank capacity (see *Figure 20*).

Figure 19 ◆ Pressure regulator.

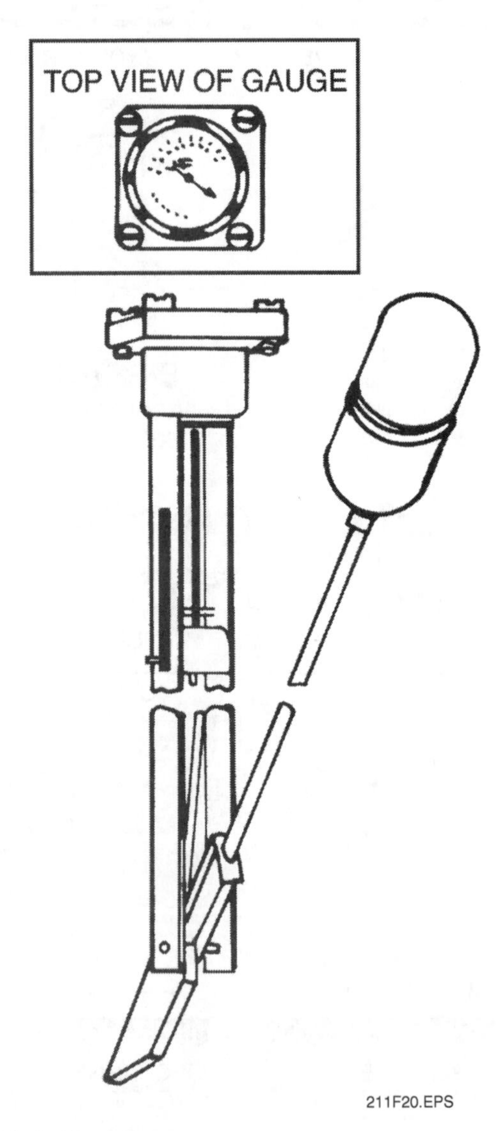

Figure 20 ◆ Float gauge.

4.2.5 Unions

Unions are required to connect all appliances to the gas supply line. Unions make it possible to disconnect or remove an appliance. These unions should be of the ground joint (metal-to-metal) type. Concealed unions are prohibited. Unions must be located between the shutoff valve and the appliance.

4.2.6 Installation Plans

Residential plumbing drawings often do not include LP gas system instructions. Even if the gas supply lines are indicated, the local LP gas supplier must be consulted before the installation begins. The LP gas supplier will usually provide the following installation information:

- The size of the system required
- The location of the storage container
- The type of storage container needed
- What kind of regulator is needed and where it should be placed
- The customer service line requirements
- What inspections and tests are required and when they should be requested
- The appliance location and the type of venting system

After obtaining this information, mark it on a set of plans and use the annotated plans as a reference during the gas system installation. On construction drawings, gas supply lines can be identified by a solid line marked with the letter G.

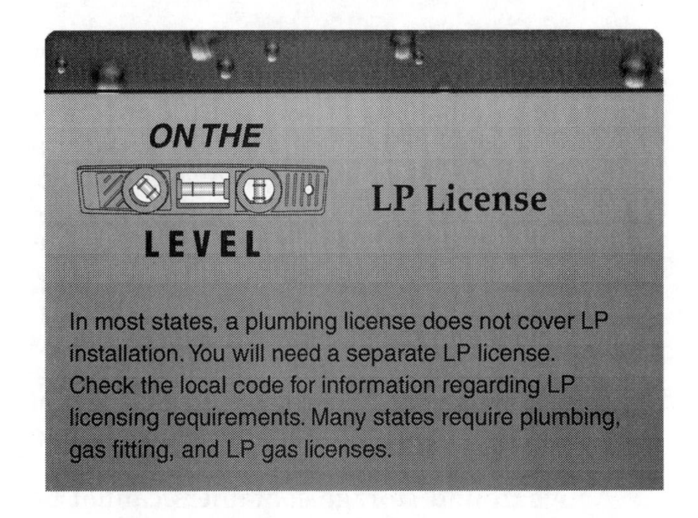

ON THE
LEVEL
LP License
In most states, a plumbing license does not cover LP installation. You will need a separate LP license. Check the local code for information regarding LP licensing requirements. Many states require plumbing, gas fitting, and LP gas licenses.

If the system is not low-pressure, indicate the pressure on the plans.

4.2.7 Aboveground Installations

The plumber's responsibilities begin with sizing and locating the storage tank or cylinder. Consult your local LP gas supplier for specifications in your area. Storage tanks can be located above or below the ground (see *Figure 21*). Mount and secure horizontal tanks on saddles (supports) that allow for expansion and contraction. Ensure that the saddles are large enough and positioned to distribute the container's weight equally. Follow the manufacturer's corrosion-prevention guidelines to ensure that the portion of the container that rests on the saddles or a foundation is protected from rusting. If required, install an anode on the tank for this purpose.

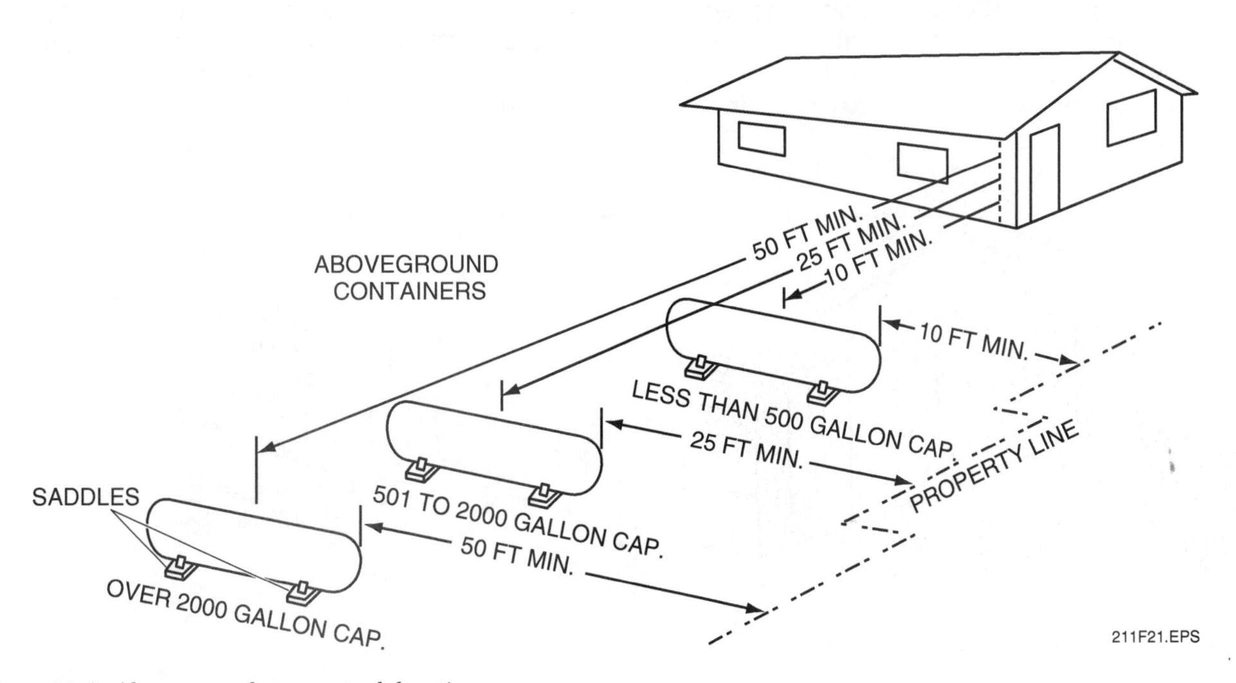

Figure 21 ◆ Aboveground storage tank locations.

Consult your local code and gas supplier for specific regulations governing aboveground storage tank installations in your area. Containers must be finished with a heat-reflecting surface, in either white or silver. Aboveground containers with a capacity of 1,200 gallons or greater must be electrically grounded. Place aboveground containers well away from traffic areas. Usually, cylinders can be located according to the customer's preference and considering accessibility to service trucks. A typical cylinder location is illustrated in *Figure 22*. Remember to review local and state regulations and *NFPA 58* before locating the cylinders.

Aboveground storage containers cannot be used for all kinds of LP gas. For example, at 30°F and at atmospheric pressure, butane remains a liquid. So aboveground containers are not suitable for butane, which, like propane, only burns in vapor form. Butane was never used for household heating in the North and was sold for that purpose only in certain areas in the South. Because it does not make economic sense to distribute butane only to certain Southern markets and propane to everyone else, the LP industry no longer distributes butane for home heating use. For retail household use and for most industrial uses, propane is the standard.

4.2.8 Underground Installations

Underground containers must be placed so that the top of the container is at least 2 feet below ground. Set underground containers on a firm earthen foundation. If containers are anchored with straps or cables, they can be surrounded by packed or tamped sand or soft earth.

To prevent corrosion, prime and coat underground containers with a protective coating. In areas where soft soils and flooding are common, the tank must be secured to prevent floating. Consult your local LP gas supplier for more specific requirements in your area, and locate storage tanks on the customer's property according to local specifications. Place regulators, service piping, and vents on LP gas systems according to local and state requirements. Exterior piping must be as direct as possible and must be installed at least 2 feet below the ground.

WARNING!

Use only UL-listed underground gas storage tanks. UL-listed tanks have a coating that prevents corrosion. Underground tanks that are not UL-listed may fail, causing a fire or explosion.

Grade is very important to keep in mind when installing LP gas distribution systems. Remember that LP gas vapor will return to a liquid at a given pressure and temperature. Grade lines allow LP gas to flow back into the underground piping and change back into a vapor. Some codes, including the *National Fuel Gas Code* and the *International Fuel Gas Code*, that allow **drips** to be installed on underground lines allow storage of liquefied petroleum until it changes back to a vapor. Liquefied LP gas flowing into a burner is very dangerous. Consult your local applicable code for approved methods for preventing the flow of liquefied LP gas into burners.

Drips must be constructed on horizontal and vertical runs of pipe or tubing. The drip pipe should be the same size as, or larger than, the service line. A large drip pipe provides the needed capacity without using a long piece of pipe.

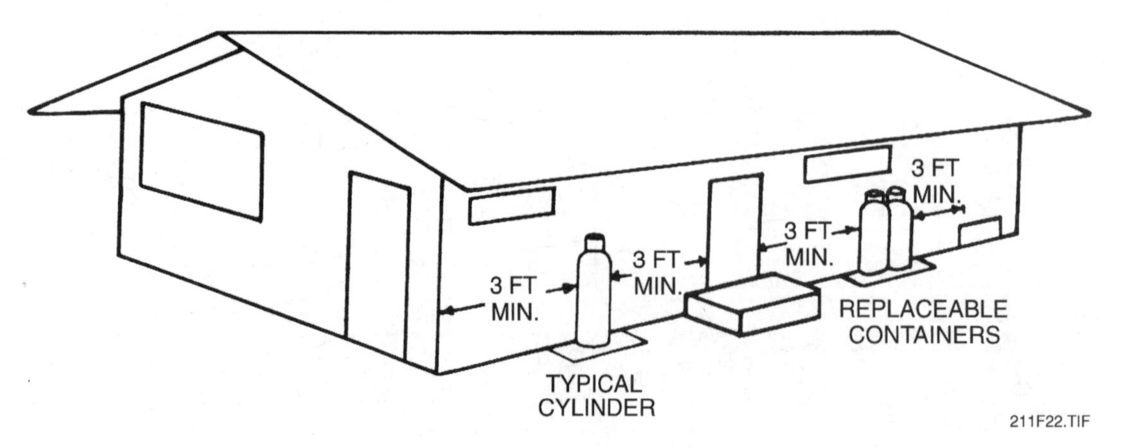

Figure 22 ◆ Cylinder locations.

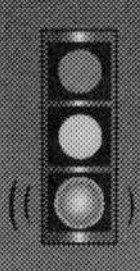

NOTE

Underground and aboveground tanks require overfill protection to prevent liquefied gas from entering the appliance. Plumbers must be specially licensed to install overfill valves. Refer to the product specifications. Install the recommended overfill protection according to the manufacturer's instructions.

4.2.9 Installations Inside a Building

LP gas regulations for installation inside a building are very specific. Never run interior LP gas piping in or through air ducts, clothes chutes, chimneys or flues, gas vents, ventilating ducts, or masonry blocks. Always ensure that piping and appliances, and piping under a building, are installed with sufficient ventilation. When installing piping aboveground, bury a drip underground to ensure vaporization of any liquid condensate.

Grade all LP gas piping at least ¼ inch per 10 feet, pitched toward the storage tanks. Where LP gas piping is installed without the proper slope or grade, provide drips at the lowest point in the line.

Install branch lines from the top or sides of lines. Whenever possible, use 45-degree ells on vertical gas piping. A 90-degree ell always offers the possibility of creating a trap, allowing LP gas to condense back into liquid form.

4.2.10 Purging

After the piping has been tested, all LP gas distribution systems, like natural gas distribution systems, must be purged. The same precautions that apply to purging a natural gas system apply to LP gas. Many codes allow the purging to be done using fuel gas. Introduce the gas at one end of the system in a moderately rapid and continuous flow, while allowing air to vent out the other end of the system.

When purging any fuel gas system, do not vent the system into a confined space or an area with ignition hazards. Refer to your local applicable code for proper safety precautions, and be sure to follow them carefully. Ensure that the ventilation is adequate, that the purging rate is controlled by an operator, and that all hazardous conditions have been eliminated. Ensure that the vent is not left unattended during the purging process. When the system has been completely purged, seal the vent.

Some piping must be purged with inert gas to remove air. The total length of pipe to be purged with inert gas will vary depending on the diameter of the pipe. Your local applicable code will provide information on how to purge using inert gas. Be sure to follow the code's requirements closely.

After properly purging the piping, purge all appliances and light their pilots. Refer to the appliance manufacturer's specifications and your local applicable code for the appropriate purging methods.

LP gas storage tanks must also be purged. Purging is accomplished by bleeding the air out of a vent on top of the tank during filling. Most local LP gas suppliers are responsible for purging the tanks when they refill them.

4.3.0 Factors Specific to Fuel Oil

The sections below review the materials, installation considerations, and other factors that are specific to fuel oil.

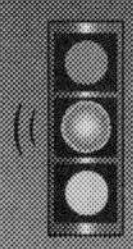

CAUTION

If threaded steel pipe is used for fuel oil supply lines, use a premium grade thread sealant specifically labeled for fuel oil and petroleum products. Welding is generally preferred over threaded pipe for steel fuel oil supply piping.

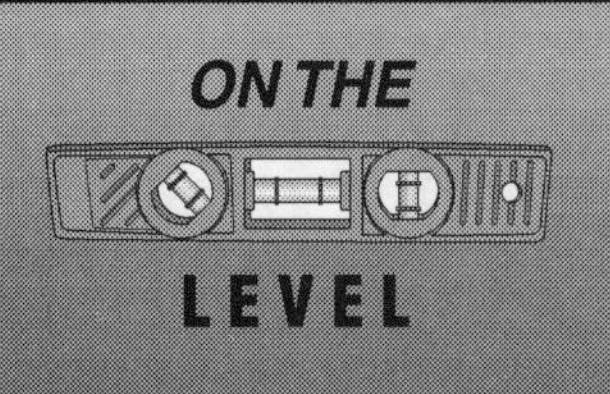

Nonresidential Fuel Oil Tanks

Above- and belowground nonresidential fuel oil tanks are generally required to be double-walled or to have some other approved form of leak protection. Aboveground tanks may also require a containment basin. All nonresidential tanks must be equipped with leakage-monitoring devices. Consult the current EPA regulations concerning the installation of fuel oil tanks. The pressure rating of the tank will depend on the height of the fill and vent line.

All underground fuel oil piping must be contained and fitted with a leak-monitoring system. Consult the tank suppliers and current EPA regulations for installation guidelines.

4.3.1 Filters

Filters are placed in the fuel oil line (also called the **suction line**) between the storage tank and the burner to collect water and moisture that might be mixed with the fuel oil. A fuel filter must be located as close as possible to the burner.

4.3.2 Pumps

Pumps move the fuel oil through the piping system. They can push fuel, pull fuel, or both. A single-stage pump is used in **gravity feed** or low-lift systems. This kind of pump usually allows gravity to get the fuel to the pump, and then the pump

Where Does Fuel Oil Come From?

The raw materials used to produce fuel oil come from oil and are called hydrocarbons, also known as petroleum. Petroleum requires considerable processing and refining before the raw materials can be converted from their original form into finished products such as gasoline, motor oil, and fuel oil.

The hydrocarbons used to produce fuel oil are found beneath the earth's surface. Rocks trap the petroleum deposits and prevent the oil from dissipating or migrating into surrounding areas. These trapped deposits are called pools.

Petroleum consists of hydrogen and carbon atoms linked together to form chains of molecules. The basic hydrocarbon unit is the methane molecule, which consists of one atom of carbon (C) and four atoms of hydrogen (H):

```
     H
     |
H —  C — H
     |
     H
```

This molecule may be combined with itself to form different chains of molecules, such as propane:

```
     H     H     H
     |     |     |
H —  C  —  C  —  C — H
     |     |     |
     H     H     H
```

In general, short hydrocarbon chains are in the form of gases, whereas liquid petroleum is composed of longer, heavier chains:

```
   H    H    H    H    H    H    H    H    H    H
   |    |    |    |    |    |    |    |    |    |
H— C —  C —  C —  C —  C —  C —  C —  C —  C —  C — H
   |    |    |    |    |    |    |    |    |    |
   H    H    H    H    H    H    H    H    H    H
```

Crude oil is the natural form of oil as it is found in the earth. It is a chemically complex substance. Two of the fundamental classifications of crude oils, or crudes, are paraffin base (having a high wax content) and asphalt base (having a high asphalt content).

Crudes are also categorized according to their sulfur content. Sulfur has a distinct smell, so crudes are referred to as sweet (containing very little sulfur) or sour (containing a high amount of sulfur). The sweet crudes have less impact on air quality and pollution problems than the sour crudes do.

Crude oil varies in color, texture, and density. Heavy crude oil may be so thick that it is solid at room temperature. This type of crude oil requires extensive refining before any useful product can be made from it. Light crude oil, as its name suggests, is lighter in weight and easier to refine. Some crude oils need no processing at all.

Most crude oil comes from the earth mixed with foreign materials, such as water or particulates. These must be removed before the oil can be sold. Processed crude oil is converted into gasoline, fuel oil, lubricants, synthetics, and other petroleum products.

pushes the oil into the burner. A single-stage pump should always be used on single-pipe or gravity feed systems (see *Figure 23*).

The two-stage pump pulls the fuel oil to the pump and then pushes it into the burner and returns the excess to the storage tanks located below the pump (see *Figure 24*). A two-stage pump should always be used in fuel piping systems that contain two or three pipelines.

Pumps must be sized for proper capacity and for suction lift from the bottom of the tank. Suction lines must be as short as possible. Consult the pump manufacturer's specifications for installation requirements. Fuel oil transfer sets may be required for larger installation. Piping size is determined by burner firing rate requirements, not by rule-of-thumb methods. Tank test pressure must include the fill-pipe height.

4.3.3 Fuel Gauges

A fuel gauge is used to measure the capacity of the fuel oil or LP gas storage tank. A variety of fuel gauges are manufactured; use a type approved in your area.

4.3.4 Valves

Valves are used to shut off the oil flow. Gate valves are usually located before (or in front of) the pump, on the return line, after the appliance, and between the suction line and the storage tank. Every appliance should have a shutoff valve on the fuel line before the union connection. Valves should be placed everywhere that safety, convenience of operation, and maintenance demand.

4.3.5 Vents

Open or automatically operated vents are placed on all fuel oil storage tanks. They prevent excessive pressure from building up inside the tank. Most vents are constructed of galvanized iron pipe 1¼ inches in diameter. A vent cap is usually placed on a vent to prevent water from entering the tank.

SINGLE-PIPE SYSTEMS

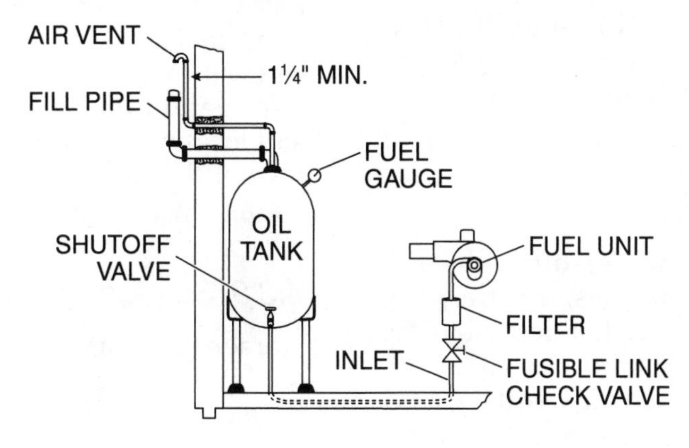

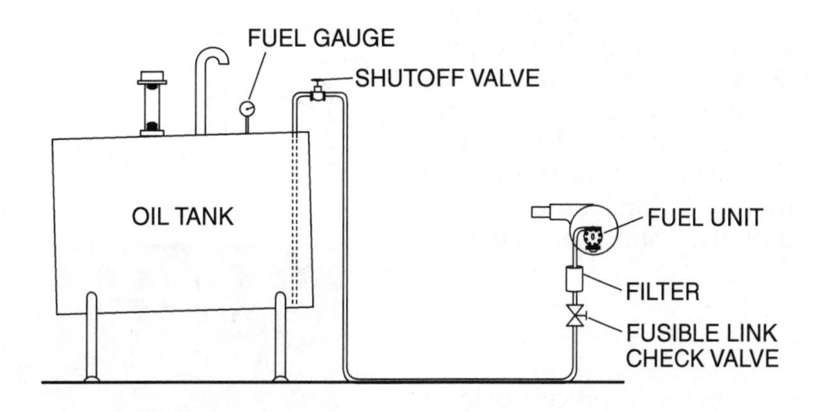

INSTALL IN ACCORDANCE WITH LOCAL AND UNDERWRITERS REGULATIONS

211F23.EPS

Figure 23 ◆ Gravity feed system.

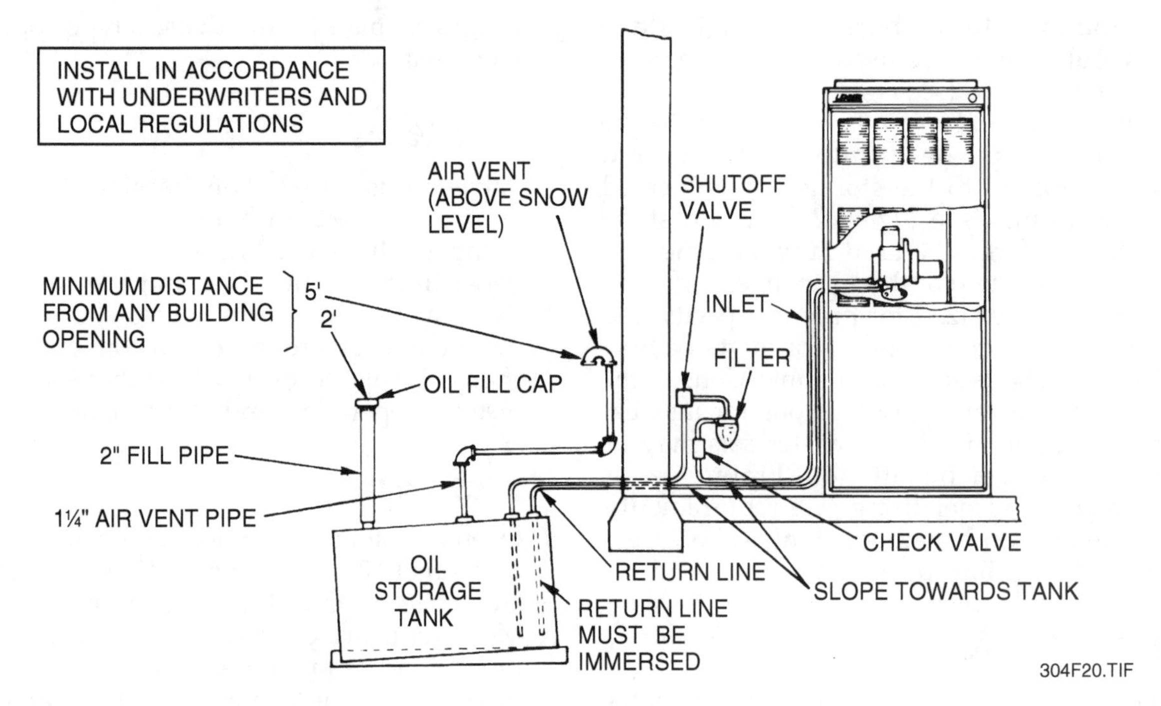

Figure 24 ◆ Two-stage pump system.

4.3.6 Unions

Unions are required between the shutoff valve and the appliance. Concealed unions are prohibited. Unions make it possible to disconnect and remove an appliance.

4.3.7 Installation Drawings

Residential plumbing drawings often do not include fuel oil distribution lines, but they do show the location of appliances. Before designing the fuel oil system, be aware of the local code regulations, the supplier's requirements, and the customer's preferences.

Before starting the installation, determine the fuel oil storage tank location by consulting the supplier and the property owner. Keep in mind the requirements of your local applicable code. Determine the locations of appliances and venting systems, and the type of fuel oil system required for each particular installation. After obtaining the necessary information for the installation, mark the information on a set of plans. Refer to these plans as needed during the installation of the fuel oil system.

4.3.8 Layout

As you have learned, layout is the process of locating the exact position of pipes, fittings, appliances, and other fuel system materials. The layout is determined by information in the construction drawings and codes. The fuel oil supplier often has specific requirements. Examples of layout work include the location of holes to be drilled, sleeves to be set, and supports. Other layout information includes size of pipe, type of pipe, and locations of branches. Remember that fittings restrict the flow of fuel, so design the system to use as few fittings as possible.

4.3.9 Storage Tanks

Storage tanks are used to store the fuel oil (see *Figure 25*). These tanks are made of steel and are supported on concrete, masonry, or steel. Be sure to test the tank to the fill pipe. Test the tank to the fill pipe opening or the vent opening.

Your responsibilities as a plumber begin with sizing and locating the fuel oil storage tank. Consult your local fuel oil supplier for the specifications in your area. Fuel oil storage tanks can be located above or below the ground or inside the building. They can also be enclosed or unenclosed.

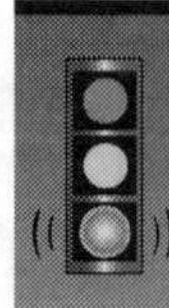

NOTE
Ensure that storage tanks are sized properly. An undersized storage tank can cause major problems. Consult your tank supplier and your local applicable code to ensure that the tank you select is the proper size.

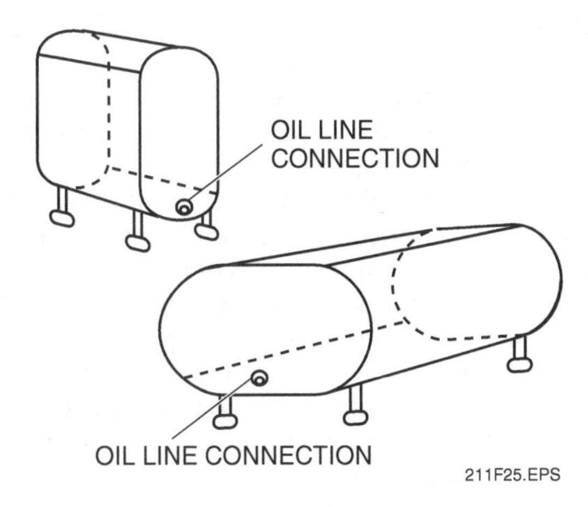

Figure 25 ◆ Typical storage tanks.

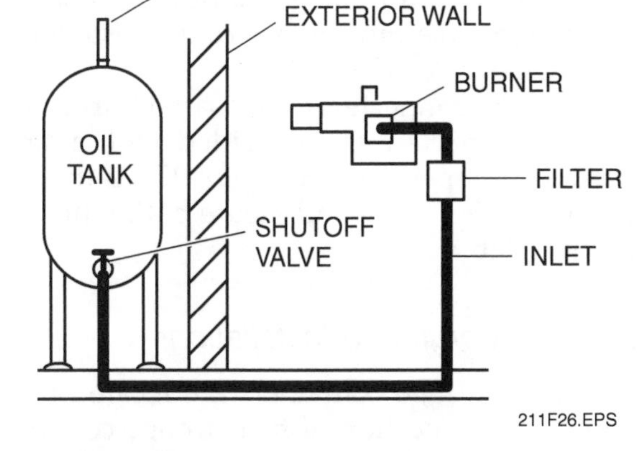

Figure 26 ◆ Typical aboveground system.

4.3.10 Aboveground Installations

Refer to the specifications that apply in your area regarding aboveground tank installations (see *Figure 26*). Begin by sizing the system to determine the tank size. In any event, the tank cannot be larger than what is permitted by code. Locate the storage tank in relation to the appliances and the applicable venting specifications.

Locate tanks away from property lines or public traffic areas (refer to *Table 3*). Do not use more than two tanks of locally approved size or more than two tanks that together equal the locally approved size.

Place a shutoff valve between the tank and the suction line. Size the vent pipe on a tank according to *Table 4*. Size and locate the fill opening to

Table 3 Location of Fuel Oil Storage Tanks

Capacity of tank (in gallons)	Minimum distance (in feet) from property line that is or can be built upon, including the opposite side of a public way	Minimum distance (in feet) from nearest side of any public way or from nearest important building on the same property
275 or less	5	5
276 to 750	10	5
751 to 12,000	15	5
12,001 to 30,000	20	5
30,001 to 50,000	30	10
50,001 to 100,000	50	15
100,001 to 500,000	80	25
500,001 to 1,000,000	100	35
1,000,001 to 2,000,000	135	45
2,000,001 to 3,000,000	165	55
3,000,000 and up	175	60

Table 4 Size of Vent Pipe on Fuel Oil Storage Tanks

Capacity of tank (in gallons)	Approximate Imperial gallon	Diameter of vent, iron pipe size (in inches)
500 or less	500 or less	1¼
501 to 3,000	501 to 2,500	1½
3,001 to 10,000	2,501 to 8,300	2
10,001 to 20,000	8,301 to 16,600	2½
20,001 to 35,000	16,601 to 29,000	3

Note: Where tanks are filled by the use of a pump through tight connections, a vent pipe no smaller than the discharge of the pump shall be used.

permit easy filling with minimal spillage. Consult your local code and fuel gas supplier for specific information governing aboveground installations in your area.

Two tanks connected to the same burner may be cross-connected and provided with a single vent and fill opening (see *Figure 27*). If this connection is used, place and secure both tanks on a common slab.

4.3.11 Underground Installations

Many fuel storage containers are located below ground level. Location of fuel storage containers depends on the owner's preference. Tanks installed underground must be covered with at least 2 feet of dirt. Ensure that underground containers are set on a firm earthen foundation. They must then be surrounded with clean sand, dirt, or gravel and must be well tamped in place.

To prevent corrosion, prime and coat tanks with an approved protective coating. Place tanks into the ground gently to prevent damage. Underground tanks must have an open or automatically operated vent. When oil supply tanks are lower than the burner, ensure that the supply line is sloped toward the tank. All tanks must have an oil-level gauge. Extend suction lines and fuel return lines from the top of the tank to no higher than 4 inches from the bottom.

Consult your local fuel oil supplier for specific code requirements in your area. The suction line from the tank to the building must be as direct as possible. Remember that fittings provide resistance to the flow of oil. Therefore, if possible, fittings should be eliminated from the suction line. Use valves, vents, fittings, and pipes that are approved by your local code. Oil lines entering a building through a masonry wall must be run through a sleeve made from larger pipe.

Pipes for oil distribution should not be smaller than ⅜-inch iron pipe or ⅜-inch-OD copper tubing. The only exception to this is when the top of the tank is below the level of the fuel pump. In such cases, ¼-inch- or 5⁄16-inch-OD tubing may be used. Never use a single-line system on a buried tank. If possible, provide a loop of tubing in the suction line before the pump. This permits easier servicing and cuts down on vibration.

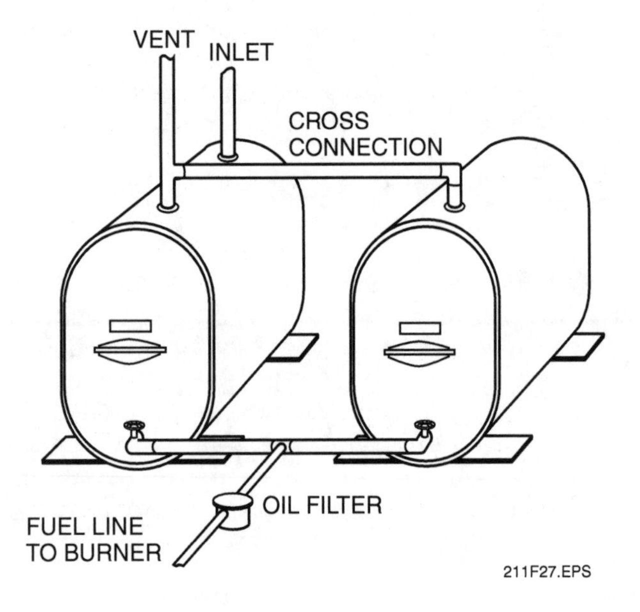

Figure 27 ◆ Cross-connected tanks.

4.3.12 Installations Inside a Building

Fuel oil piping inside a building must not be run in or through an air duct, masonry block, or brick. If piping must enter a building through concrete, the pipe must be run through a sleeve, which should be encased in concrete. Fuel oil lines, especially those made of copper, must never be exposed to traffic areas. Avoid running fuel lines above the burner level because air has a tendency to collect at the highest point and stop the flow of fuel.

Fuel filters must be placed as close to the burner as possible. A ball check valve must also be placed near the burner and in an accessible position (see *Figure 28*). Codes in some areas suggest a valve on the return line to prevent siphoning from the storage tank into the burner. Care must be taken to open this valve before the burner is turned on to prevent the pump seals from rupturing.

Fuel oil regulations within a building are specific in most areas. Ensure that the lines running to and from the burner do not prevent repairs to, or tests of, the burner. Verify that the fuel pump is the correct pump for the system. Design the pumping system according to the manufacturer's specifications. Protect vent and fill openings from the elements, such as rain and snow.

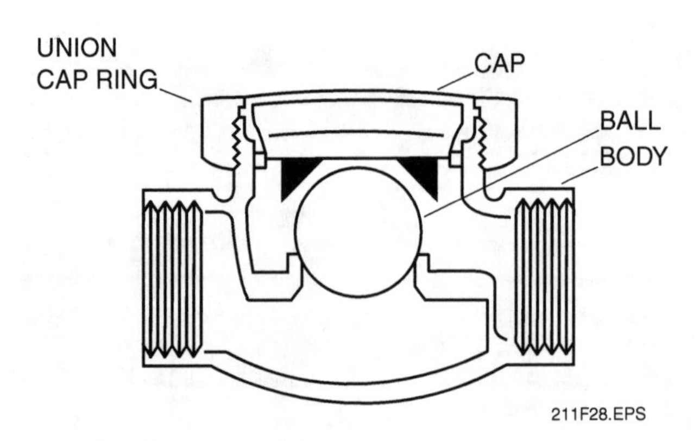

Figure 28 ◆ Ball check valve.

4.3.13 *Enclosed Storage Tank Installations Inside a Building*

Fuel oil tanks may be installed within a building. These installations are specified as either enclosed or unenclosed. The local codes will specify the maximum size of tank that can be unenclosed, so any tank larger than that must be enclosed (see *Figure 29*).

Regardless of the enclosure, the size of a supply tank located above the lowest part (cellar or basement) of the building is limited by code. Refer to your local applicable code for specific requirements in your area. The floor, walls, and ceiling of the enclosure must have a fire resistance rating of not less than 3 hours. This means that it would take 3 hours before a fire could penetrate the walls.

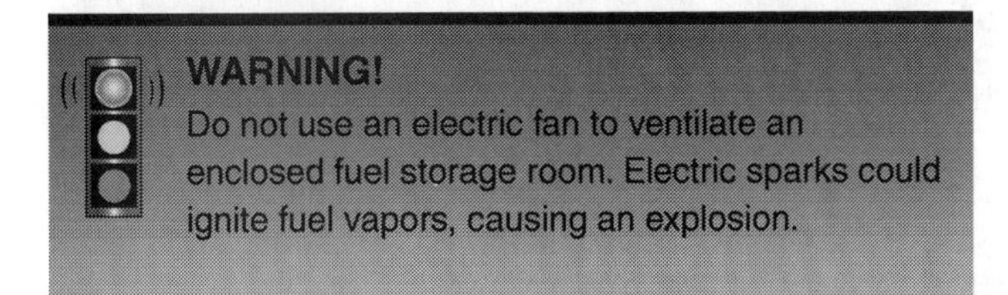

WARNING!

Do not use an electric fan to ventilate an enclosed fuel storage room. Electric sparks could ignite fuel vapors, causing an explosion.

The tank must be supported at least 6 inches above the floor. Refer to your local applicable code for support and height requirements in your area. Provisions must be made to properly ventilate the enclosed room. Ensure that enclosed tanks are properly vented to the outside atmosphere using a separate vent line. Provide enclosed tanks with capacity-gauging devices to allow measurement of fuel inside.

4.3.14 *Unenclosed Storage Tank Installations Inside a Building*

Some tanks may be installed inside a building without being enclosed (see *Figure 30*). Your local applicable code will indicate the maximum size of tank that may be installed unenclosed, as well as other standards applying to the installation of unenclosed tanks inside a building.

The supply tank must be of a size and shape that can be installed or removed from the building as a unit. The sizes of supply tanks placed above the lowest part of the building must conform to the standards in your local applicable code.

Unenclosed tanks cannot be closer than 2 feet to any source of heat. They must be supported and secured. Install a shutoff valve between the tank and the supply line from the tank. Equip the tank with a vent and a fill gauge. Plug any unused openings on the tank to prevent fuel oil vapors from escaping.

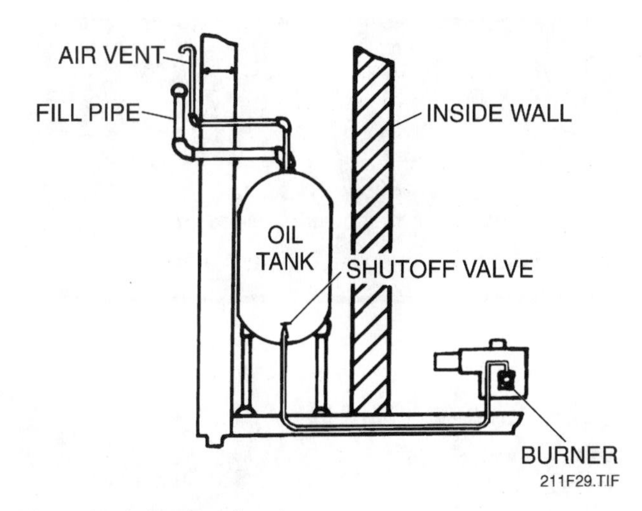

Figure 29 ◆ Enclosed tank.

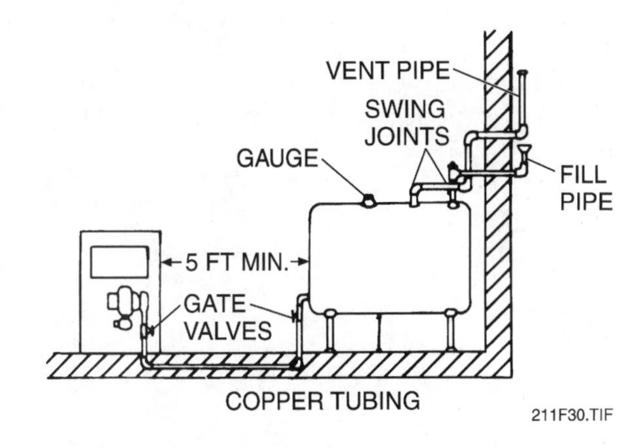

Figure 30 ◆ Unenclosed tank.

4.4.0 Fire Stopping

As you have already learned, fire codes require the installation of fire-stopping material in penetrations through fire-rated structural members, as well as through fire-rated walls, floors, and ceilings. Fire-stopping systems are designed to seal structural penetrations completely in the event of a fire. This applies to fuel gas and fuel oil lines as well.

In addition to preventing the spread of flame, fire-stopping materials must be smoke-, gas-, and watertight. Refer to the local applicable code for the fire-stopping requirements that apply in your area. When applying fire-stopping materials and devices, be sure to follow the manufacturer's instructions carefully. Use only approved fire-stopping materials and sealants, such as those recognized by UL or the American Society for Testing and Materials (ASTM). Select fire-stopping materials that have the same fire rating as the structural member.

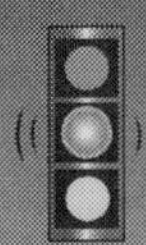

CAUTION

Each building type has its own fire code that must be followed during construction. The fire-stopping requirements vary among these different codes. Ensure that the fire stopping you install is in accordance with the applicable code for the type of building.

Ensure that the fire-stopping materials will not degrade over time from exposure to moisture or climate. Consult the local applicable fire code for approved methods and materials before cutting, drilling, or notching structural members. Your fire code may require a qualified or experienced professional to install fire-stopping materials. Always follow the manufacturer's instructions when applying fire-stopping materials.

Review Questions

Section 4.0.0

Match the type of fuel with its correct valve installations.

1. _____ Natural gas
2. _____ LP gas
3. _____ Fuel oil

a. Install relief valves on all tanks or containers to prevent explosions caused by a pressure increase.

b. Valves are not required in residential systems that operate at less than 6 psi, though codes may require a check valve before the meter.

c. Typically, a flat-head cock valve with a lock wing is located before every meter.

d. A gate valve is usually located on the return line, after the appliance, and between the suction line and the storage tank.

4. In a natural gas piping system, a(n) _____ controls the amount of gas pressure from the meter into the building.
   a. flat-head cock valve
   b. union
   c. anode
   d. regulator

5. An insulating _____ is always located on the gas main side of the meter to prevent the conduction of electricity through the customer's service line to the gas main.
   a. valve
   b. purge
   c. union
   d. regulator

6. Protective coatings on underground steel piping systems must extend at least 3 inches above the finished grade on vertical sections of pipe.
   a. True
   b. False

7. Pressure regulators must be located on the _____ side of an LP gas container.
   a. inlet
   b. high-pressure
   c. outlet
   d. low-pressure

Review Questions

8. Unions must be located _____.
   a. in a concealed position
   b. at the inlet of the gas supply
   c. upstream of the supply shutoff valve
   d. between the supply shutoff valve and the appliance

9. The indicator on an LP gas storage tank float gauge is calibrated by _____.
   a. meter capacity in cubic feet per hour
   b. percentage of total tank capacity
   c. tank output valve rating
   d. gas flow rate

10. An aboveground LP gas tank with a capacity of between 501 and 2,000 gallons must be located at least _____ feet from the building and at least _____ feet from the property line.
    a. 10; 10
    b. 25; 25
    c. 40; 40
    d. 55; 55

11. Ensure that lines have proper _____ to permit LP gas to flow back into underground piping and change back into a vapor.
    a. layout
    b. piping
    c. grade
    d. venting

12. Depending on the system, a single-stage or two-stage _____ is required to move fuel oil through the piping system.
    a. filter
    b. pump
    c. valve
    d. union

13. A _____ valve must be placed between the tank and the suction line.
    a. fill-opening
    b. one-stage
    c. spillage
    d. shutoff

14. Underground storage containers must have an open or automatically operated _____.
    a. valve
    b. vent
    c. regulator
    d. gauge

15. Support fuel oil _____ on concrete, masonry, or steel.
    a. distribution lines
    b. canisters
    c. tanks
    d. saddles

16. Oil tanks installed underground must be covered with at least _____ feet of dirt.
    a. 5
    b. 4
    c. 3
    d. 2

17. For fuel oil storage tanks installed inside a building, the surrounding area must have a fire resistance rating of not less than _____ hours.
    a. 2
    b. 3
    c. 4
    d. 5

Summary

Fuel gas systems and fuel oil systems play important roles in modern society. They provide energy for our homes and supply power for our vehicles. Each of the three fuels—natural gas, LP gas, and fuel oil—have many things in common as well as specific differences. Plumbers must be aware of these similarities and differences so that they can properly plan, design, install, and test the systems and appliances that use these fuels.

Plumbers are responsible for ensuring that fuel gas and fuel oil installations are safe and efficient. Safe fuel gas and fuel oil piping systems are a result of proper installation techniques, quality materials, and strict adherence to code. Anything less than a proper piping installation can place the customer, as well as the plumber, in a potentially dangerous situation. Plumbers must always follow the standards of their profession and the requirements of their local applicable code when designing and installing fuel gas and fuel oil systems.

Notes

Trade Terms Introduced in This Module

Condensate: Water or other fluid that has been condensed from steam or from a vapor.

Drip: An extension of piping created by connecting a T, nipple, and cap that is installed on underground lines to allow storage of liquefied petroleum until it changes back into a vapor.

Flash point: The temperature at which vapors given off by a fuel may be ignited.

Gravity feed: A system that depends on gravity to bring fuel to the pump. Gravity describes the tendency of objects or substances to move downward.

Home run: A run of pipe from a distribution point such as a manifold to the appliance with no branches. It serves only one appliance.

Insulating union: A nonconductive union placed on the gas main side of the meter to prevent electricity from being conducted through the customer's service line back to the gas main.

Manometer: A U-shaped tube containing fluid and marked with a graduated measuring scale, used for measuring slight changes in a low-pressure system such as a heating, venting, and air conditioning system. One end of the tube is connected to the low-pressure source. Slight changes in pressure cause the fluid in the tube to move up or down. Also called a draft gauge.

Mercaptan: A chemical compound that has sulfur rather than oxygen, typically added as an odorant to natural gas.

Meter index: An indicator used in diaphragm and rotary meters to indicate the amount of gas that has passed through the meter, in cubic feet.

Purge: The removal of all unwanted substances from a gas piping line.

Regulator: A device designed to control pressure.

Specific gravity: A measure of the weight of a given volume of gas expressed as a ratio to the weight of the same volume of air under the same conditions.

Suction line: A line with negative pressure that will draw a liquid or gas.

Viscosity: A measure of a liquid's resistance to flow. The higher the viscosity, the slower a liquid flows.

Resources & Acknowledgments

Additional Resources

2003 International Fuel Gas Code. Falls Church, VA: International Code Council and American Gas Association.

Fuel Gas Systems. 1983. Donald L. Wise, ed. Boca Raton, FL: CRC Press.

National Fuel Gas Code Handbook, Third Edition, 1996. Theodore C. Lemoff, ed. Quincy, MA: National Fire Protection Association.

References

2003 International Fuel Gas Code. Falls Church, VA: International Code Council and American Gas Association.

Dictionary of Architecture and Construction, Third Edition, 2000. Cyril M. Harris, ed. New York: McGraw-Hill.

Figure Credits

Source	Figures
Topaz Publications	Module divider
International Code Council *2003 International Plumbing Code. Copyright 2003. Falls Church, Virginia: International Code Council, Inc. Reproduced with permission. All rights reserved.*	211F02, 211F04, Table 2
Sensus Metering Systems	211F07
Total Control Systems	211F08
American Valve, Inc.	211F10, 211F11
Watts Regulator Company	211F12
Marshall Gas Controls	211F13, 211F19
National Fire Protection Association *Reprinted with permission from NFPA 58-2004,* LP Gas Code, *Copyright © 2004, National Fire Protection Association, Quincy, MA. This reprinted material is not the complete and official position of the NFPA on the referenced subject, which is represented only by the standard in its entirety.*	211F21, 211F22
National Fire Protection Association *Reprinted with permission from NFPA 54-2002,* National Fuel Gas Code, *Copyright © 2002, National Fire Protection Association, Quincy, MA. This reprinted material is not the complete and official position of the NFPA on the referenced subject, which is represented only by the standard in its entirety.*	Table 1

CONTREN® LEARNING SERIES — USER UPDATE

The NCCER makes every effort to keep these textbooks up-to-date and free of technical errors. We appreciate your help in this process. If you have an idea for improving this textbook, or if you find an error, a typographical mistake, or an inaccuracy in NCCER's Contren® textbooks, please write us, using this form or a photocopy. Be sure to include the exact module number, page number, a detailed description, and the correction, if applicable. Your input will be brought to the attention of the Technical Review Committee. Thank you for your assistance.

Instructors – If you found that additional materials were necessary in order to teach this module effectively, please let us know so that we may include them in the Equipment/Materials list in the Annotated Instructor's Guide.

Write: Product Development and Revision
National Center for Construction Education and Research
P.O. Box 141104, Gainesville, FL 32614-1104

Fax: 352-334-0932

E-mail: curriculum@nccer.org

Craft ______________ Module Name ______________

Copyright Date ______________ Module Number ______________ Page Number(s) ______________

Description

____

____

____

____

(Optional) Correction

____

____

____

(Optional) Your Name and Address

____

____

____

02212-05

Servicing of Fixtures, Valves, and Faucets

02212-05

Servicing of Fixtures, Valves, and Faucets

Topics to be presented in this module include:

Overview

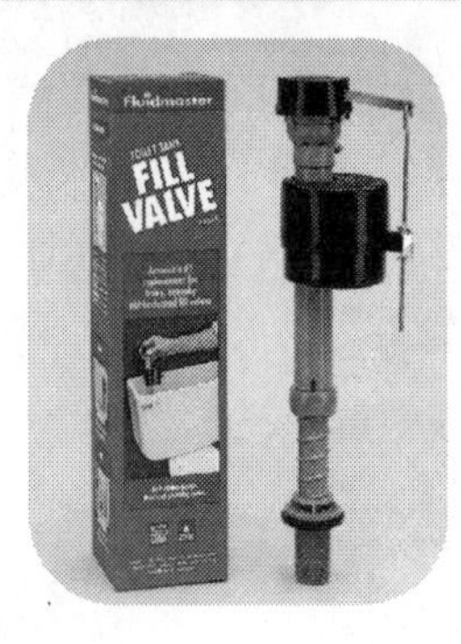

Plumbers use a variety of fixtures and faucets for plumbing installations and must understand how to service and repair them. Fixtures are receptacles connected to drainage systems that receive and discharge liquid and other wastes; valves and faucets control the flow of fluids and gases. When responding to service calls, plumbers follow general safety guidelines, including wearing the appropriate personal protective equipment, turning off necessary electrical circuits and valves, moving water, covering or moving obstacles, placing ladders carefully, and cleaning up the work area.

Valves come in a variety of sizes and styles. Plumbers must understand how the basic categories of valves operate so that they can troubleshoot common problems with each. Valves can be divided into basic categories, including globe and gate valves, flushometers, float-controlled valves, tank flush valves, balancing valves, temperature and pressure valves, backflow preventers, cartridge faucets, rotating ball faucets, and ceramic disc faucets.

Plumbers develop the knowledge to determine when to repair and when to replace the fixture, valve, or faucet. Typically, if a leak in a fixture is the result of a problem with the valve or faucet, plumbers will repair or replace the valve or faucet. If the problem is with the fixture itself, however, plumbers will replace the fixture rather than repair it. By recognizing common problems and identifying possible causes, plumbers can accurately determine appropriate solutions.

Focus Statement

The goal of the plumber is to protect the health, safety, and comfort of the nation job by job.

Code Note

Codes vary among jurisdictions. Because of the variations in code, consult the applicable code whenever regulations are in question. Referring to an incorrect set of codes can cause as much trouble as failing to reference codes altogether. Obtain, review, and familiarize yourself with your local adopted code.

Objectives

When you have completed this module, you will be able to do the following:

1. Identify common repair and maintenance requirements for fixtures, valves, and faucets.
2. Identify the proper procedures for repairing and maintaining fixtures, valves, and faucets.

Trade Terms

Packing extractor
Removable seat wrench
Reseating tool
Screw extractor
Tap
Twist packing

Required Trainee Materials

1. Appropriate personal protective equipment
2. Pencil and paper
3. Copy of local applicable code

Prerequisites

Before you begin this module, it is recommended that you successfully complete *Core Curriculum; Plumbing Level One; Plumbing Level Two,* Modules 02201-05 through 022011-05.

This course map shows all of the modules in the second level of the *Plumbing* curriculum. The suggested training order begins at the bottom and proceeds up. Skill levels increase as you advance on the course map. The local Training Program Sponsor may adjust the training order.

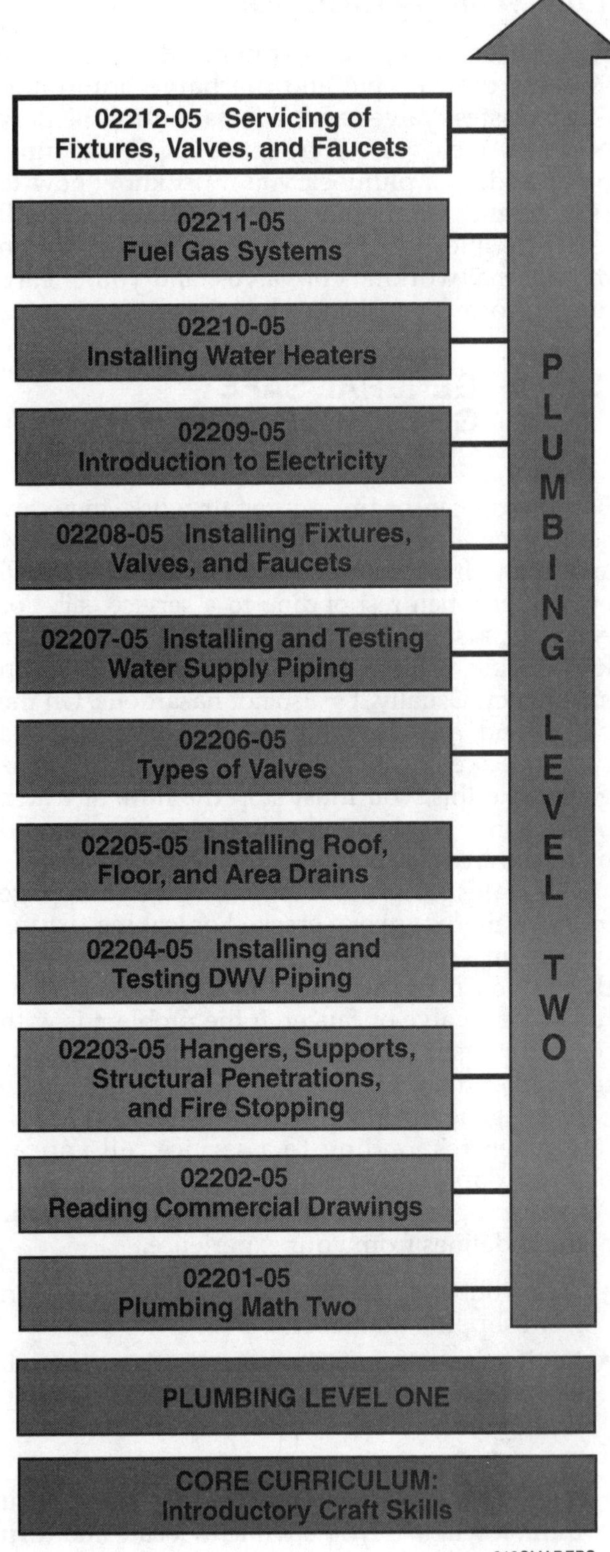

1.0.0 ◆ INTRODUCTION

Fixtures are receptacles connected to drainage systems that receive and discharge liquid and other wastes. Valves and faucets control the flow of fluids and gases. They serve important functions, and, as a plumber, you must know how to service and repair them. In this module you'll learn troubleshooting procedures that focus on the internal workings of valves, and you'll learn how to correct problems when they occur.

2.0.0 ◆ GENERAL SAFETY GUIDELINES FOR SERVICE CALLS

Before you can begin any repair work, there are some general safety guidelines you should know and follow. It is impossible to predict what you will encounter when responding to a service call. For example, a small leak from a faucet spout into a kitchen sink, while wasteful and annoying, is not an emergency. Usually it's also not hazardous. On the other hand, a valve leaking a lot of water above a suspended ceiling is usually hazardous. In an emergency like this, you must stop the flow of water, immediately minimize the potential damage, and make the area safe before you make the repairs.

Generally, repairs to fixtures involve a stoppage in the drain fixture or a cracked or leaking fixture. If a leak in a fixture is the result of a problem with the valve or faucet, you will usually repair or replace the valve or faucet. If the problem is with the fixture itself, you will generally replace the fixture rather than repair it.

Some general safety guidelines for you to follow when responding to a service call appear below. Adapt these guidelines to the job at hand. As you progress in your career, you'll add your own guidelines from your experience.

- Wear rubber-soled shoes or boots for protection from slipping and electric shock.
- Notify the owner and occupants that you will be shutting off electrical and water service within the building.
- Turn off electrical circuits.
- Shut off a valve upstream from the leak. If you think it will take you a while to locate and turn off this valve, direct the leak into a suitably sized container to minimize damage until you can turn the water off.
- Remove excess water.
- Move furniture, equipment, or any other obstacles away from the work area.
- Cover furniture, floors, and equipment to protect them from any damage that could occur as a result of your work on the valve.
- Place ladders carefully, bracing them if necessary. Wear appropriate personal protective equipment for working on ladders.
- When finished, turn the water back on to test your repair. Notify the owner and occupants that service is back on.
- Do your work neatly, and clean up when you are done.

3.0.0 ◆ SERVICING FIXTURES, VALVES, AND FAUCETS

Once you take the necessary safety precautions, you are ready to inspect and repair the valve. While valves and faucets come in many styles, they can be divided into the following categories for repair work purposes:

- Globe valves
- Gate valves
- Flushometers
- Float-controlled valves (ball cocks)
- Tank flush valves
- Balancing valves
- Temperature and pressure (T&P) valves
- Backflow preventers
- Cartridge faucets
- Rotating ball faucets
- Ceramic disc faucets

A number of valves, especially those designed for commercial installations, are electronically controlled. The operation of these types of valves is controlled by a light beam. You will learn more about pop-up valves and electronic controls later in this section.

3.1.0 Globe Valves

Internally, globe valves, angle valves, and compression faucets are all designed with the same basic parts (see *Figure 1*). Therefore, the types of problems and their solutions are similar. These problems and their likely causes are presented in *Table 1*.

When repairing valves, always use the correct size wrenches, and make sure their jaws are clean and smooth to prevent scratching the finished surfaces. Do not use Channellock® pliers or pipe wrenches to repair valves.

If you have problem 1 or 3 in *Table 1*, remove the valve bonnet, which covers and guides the stem, with a correctly sized wrench. Once you've removed the bonnet and stem assembly, inspect the seat disc (washer) and the seat. You can solve most of these cases by replacing the disc and resurfacing or replacing the seat.

Table 1 Troubleshooting Globe and Angle Valves and Compression Faucets

Problems	Possible Causes
1. Drip or stream of water flows when valve is closed	Worn or damaged seat Worn or damaged seat disc
2. Leak around stem or from under knob	Loose packing nut Defective packing Worn stem
3. Rattle when valve is open and water is flowing	Loose seat disc Worn threads on stem
4. Difficult or impossible to turn handwheel or knob	Packing nut too tight Damaged threads on stem

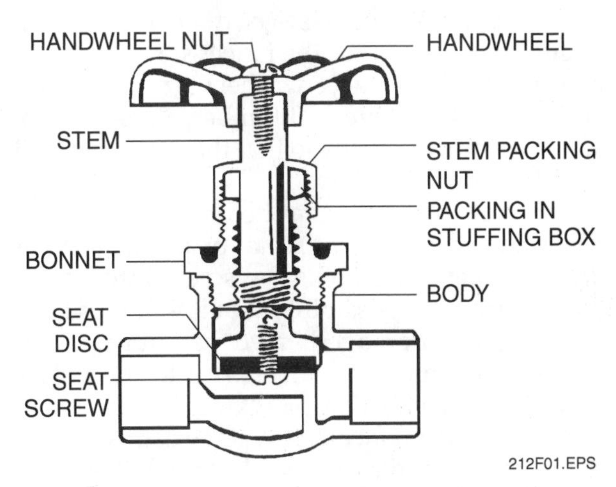

Figure 1 ◆ Basic parts of globe and angle valves and compression faucets.

A screw holds the seat disc (washer) in place. If you can't remove the screw or if the screw breaks off, replace the stem or use a **screw extractor** (see *Figure 2*) to remove the broken screw.

When using a screw extractor, do not attempt to extract a broken screw by attaching the extractor bit to a drill. Follow these steps to remove a broken screw using a screw extractor:

Step 1 Drill out a hole in the screw using a drill bit. The correct size for the drill bit is stamped on the extractor.

Step 2 Apply penetrating oil and wait for it to permeate the hole.

Step 3 With a hammer or mallet, drive the extractor firmly into the hole.

Step 4 Use a tee tap wrench to turn the extractor counterclockwise. Apply only the amount of torque that would normally be applied to the screw size you are trying to extract or you may break the extractor.

Once you have removed the old screw use a **tap** to recut or clean out the threads.

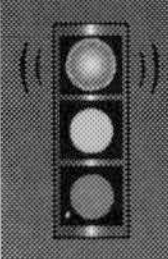

WARNING!

Never attempt to extract a broken screw by attaching the extractor bit to a drill. Improper use of a screw extractor can cause injury as well as damage the equipment. Always follow the manufacturer's instructions, and wear appropriate personal protective equipment.

Electrical shocks are not limited to electricians! Always protect yourself against accidental contact with energized sources when working on plumbing installations and service calls. Water conducts electricity.

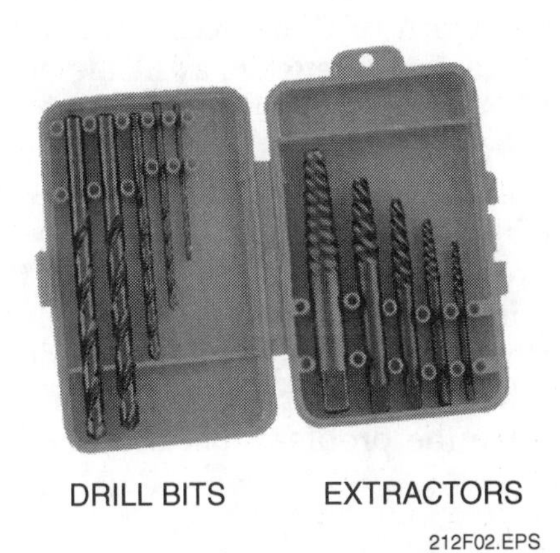

Figure 2 ◆ Screw extractors with corresponding drill bits.

DID YOU KNOW?

American Standard

American Standard is the world's largest producer of bathroom and kitchen fixtures and fittings. It is also a leading producer of air conditioning and heating systems.

American Standard resulted when the American Radiator Company merged with the Standard Sanitary Manufacturing Company in 1929. At first, the company was called the American Radiator & Standard Sanitary Corporation. In 1967, it changed its name to American Standard.

Standard Sanitary was formed in 1899. This company pioneered improvements such as the one-piece toilet, built-in tubs, combination faucets (which mix hot and cold water), and brass fittings that don't tarnish or corrode.

To replace the disc, select the correct disc and secure it to the end of the faucet stem. There are a variety of disc (faucet washer) sizes and shapes available (see *Figure 3*). Plumbers generally carry an assortment of discs so that they are prepared to handle most repair jobs.

Replace the screw that holds the disc in place. Use screws with a soft metal or plastic plug near the tip. This plug locks the screw into position, eliminating the possibility that it will loosen.

To repair a damaged valve seat, fit a cutter into a **reseating tool** and insert it into the valve or faucet body (see *Figure 4*). When you turn the handle, the cutter refaces the valve seat. After a few turns of the handle, inspect the seat. If it is smooth, stop. If it is not smooth, give the handle another turn or two and inspect the seat again.

Some compression valves are equipped with replaceable seats. Simply replace the seat when it becomes worn or damaged. Three of the many **removable seat wrenches** available are shown in *Figure 5*. A few of the many replaceable seats available appear in *Figure 6*. Service and repair plumbers should make a habit of carrying an assortment of seats with them on the job.

Before you reinstall the stem, check the amount of play in the threads that join the stem to the bonnet. If these threads are badly worn, the stem may rattle when the valve is opened. Replacing the stem may solve the problem. Because many different types of stems are available, you'll need to know the name of the valve manufacturer to get the right replacement. You may have to take the old stem to the supplier for a direct comparison to the replacement part. A few of the many types of stems available are shown in *Figure 7*. If the stem threads in the valve body are badly worn, you should probably replace the entire valve.

Refer to *Table 1*. To solve problems 2 and 4, remove the packing nut, which holds the packing around the stem in place, and inspect the packing. Leaks around the stem generally result from wear of the packing or the stem shaft. If the stem shaft is excessively worn, replace the stem. If the problem is the packing, replacement of the packing and lubrication should solve the problem. Use a **packing extractor** to remove the packing (see *Figure 8*). Before you select the replacement packing, examine the size and shape of the area under the packing nut. Various sizes and shapes of preformed packing are available (see *Figure 9*). If the correct size and shape of preformed packing is not available, you can use **twist packing** (see *Figure 10*). Install the twist packing by wrapping it around the stem to fill the packing area. When you tighten the packing nut, the packing forms into the required shape. Twist packing is not as durable as preformed packing, which is made specifically to fit the valve.

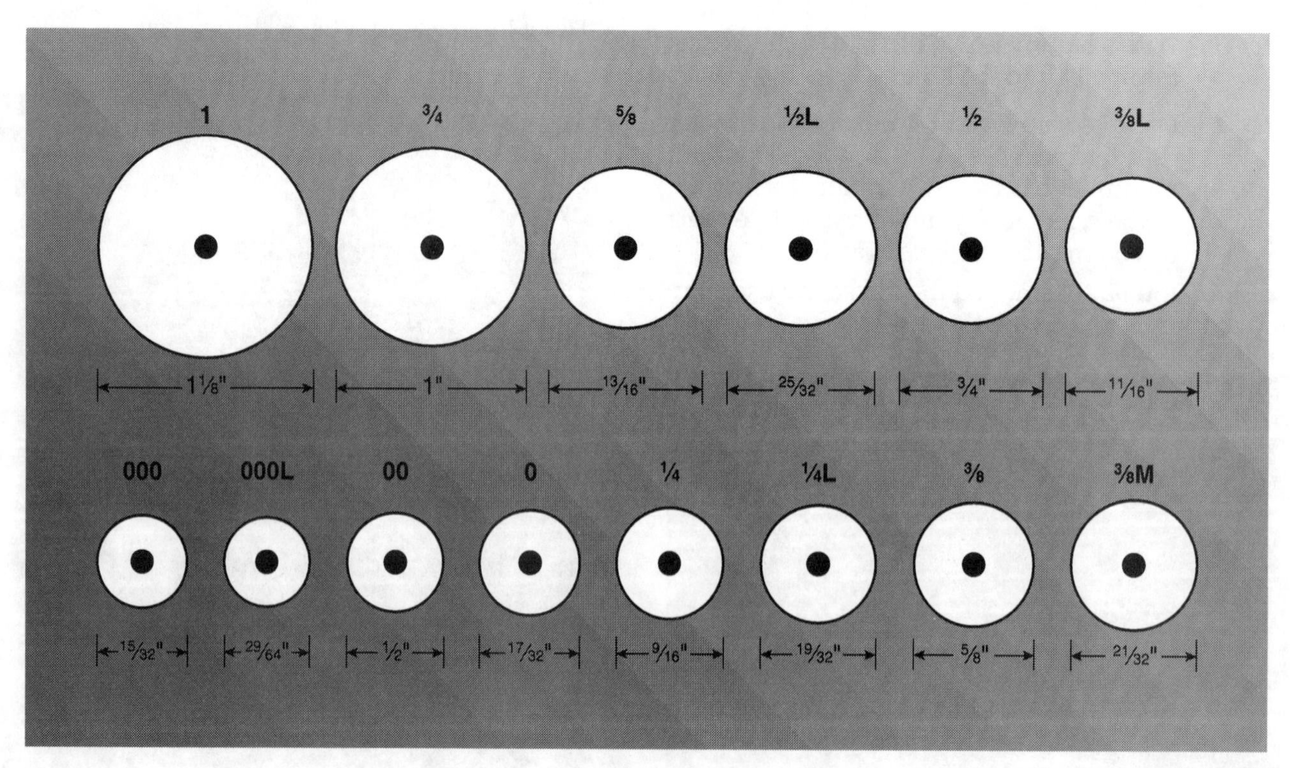

Figure 3 ◆ Seat disc sizes.

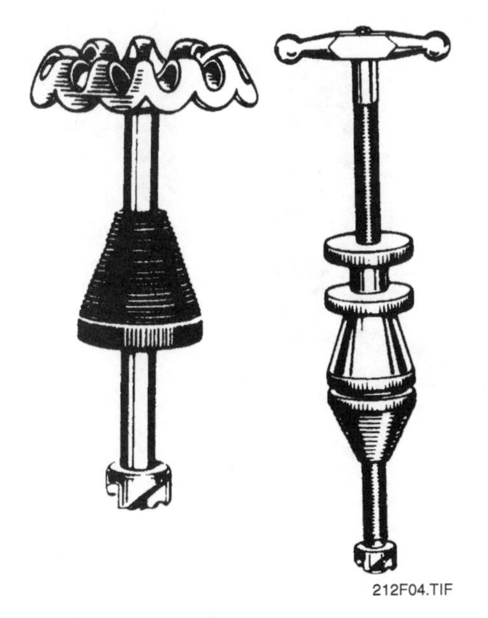

Figure 4 ◆ Reseating tool.

ON THE LEVEL

Cleaning Up is Part of the Job

Whether you're installing or repairing valves, you must always clean up your work site. A clean work site is safer for you and for any trades that come in after you. Cleaning up becomes especially important if you are called in to repair a major leak. No one wants to pay to repair a valve or a faucet and then have to replace furnishings or carpet because you were careless.

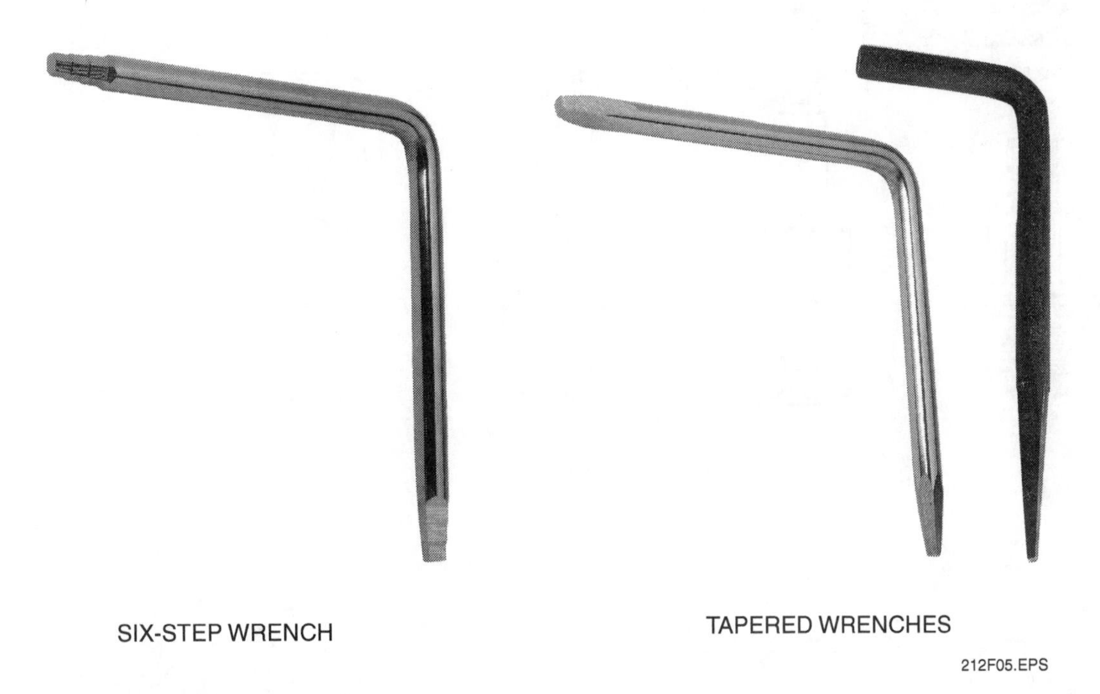

Figure 5 ◆ Removable seat wrenches.

Figure 6 ◆ Replaceable seats.

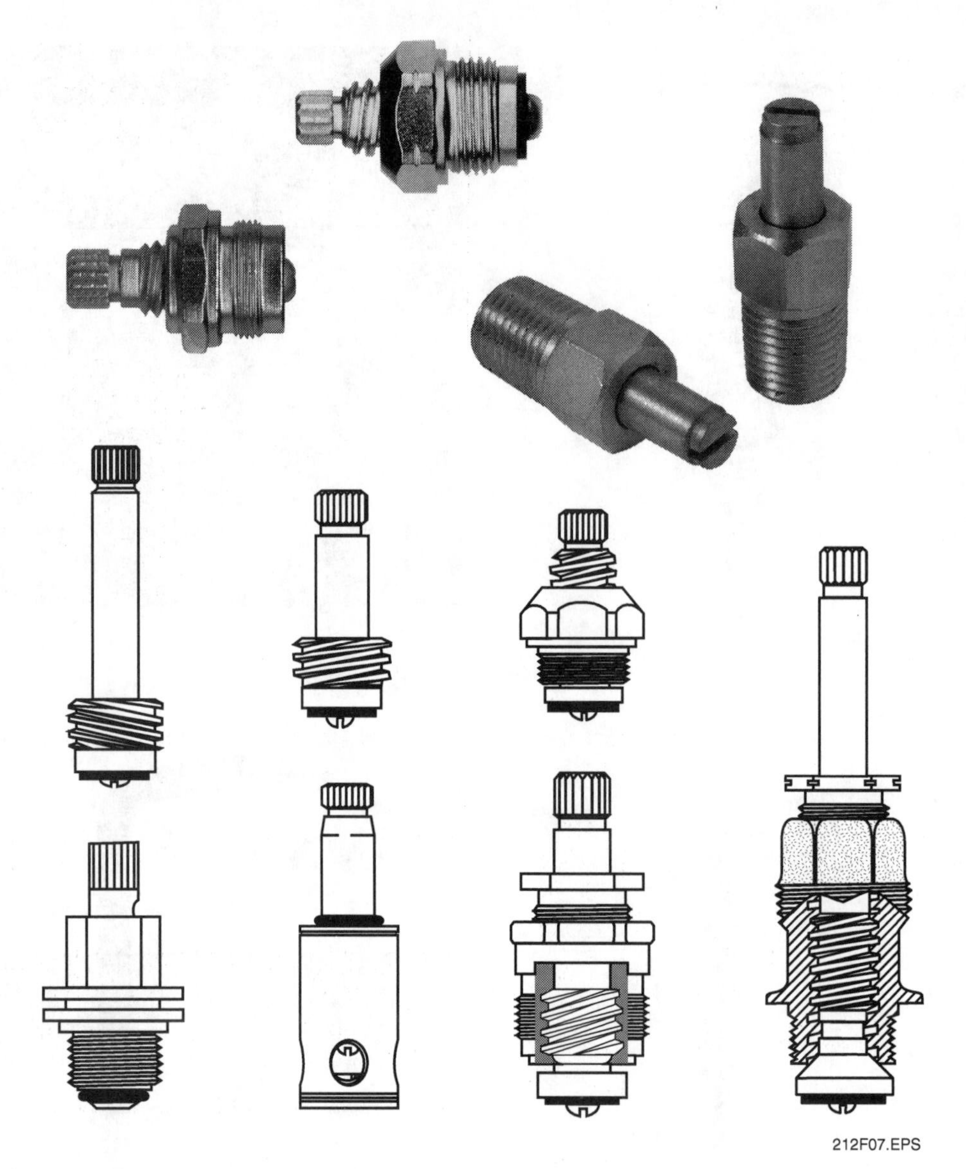

212F07.EPS

Figure 7 ◆ Stems.

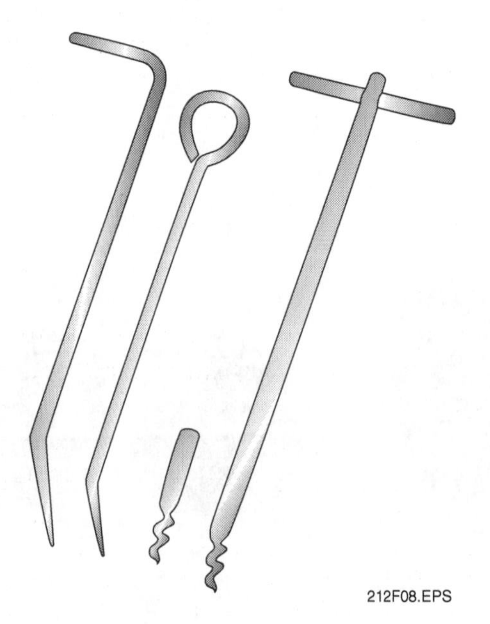

212F08.EPS

Figure 8 ◆ Packing extractors.

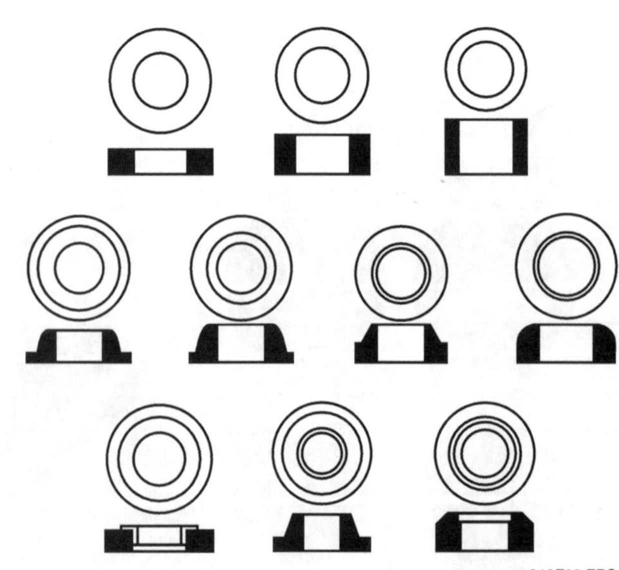

212F09.EPS

Figure 9 ◆ Preformed packing.

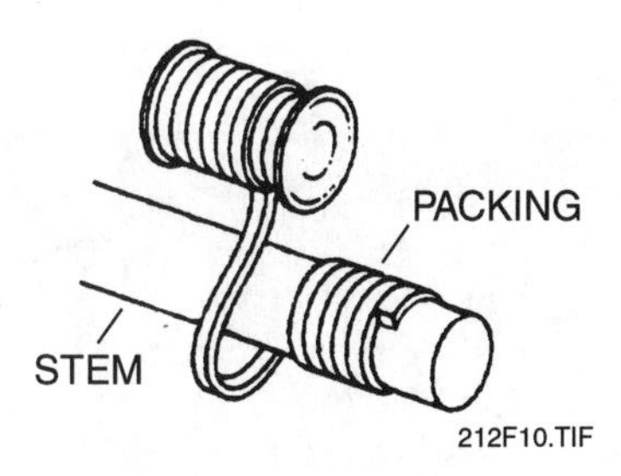

Figure 10 ◆ Twist packing.

If the stem is difficult to turn, the reason may be one of the following:

- The packing nut is too tight. Loosening the nut and lubricating the stem where it passes through the packing should solve the problem. A special lubricant that resists high temperatures is available. In addition to being heat resistant, it is also waterproof.
- The threads on the stem are damaged. If the threads in the base of the valve are not damaged, replacing the stem should solve this problem. If these threads are damaged, replace the valve.

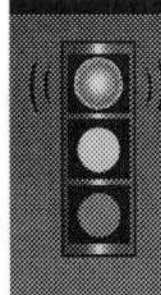

WARNING!

Slips, trips, and falls constitute the majority of general industry accidents, according to the Occupational Safety and Health Administration (OSHA). They cause 15 percent of all accidental deaths, and they are second only to motor vehicles as a cause of fatalities. Slips, trips, and falls happen quickly. Take whatever time is necessary to make your work area safe.

3.2.0 Gate Valves

Some of the problems common to globe valves are also common to gate valves (see *Figure 11*). A leak around the stem is the result of either packing wear or wear on the stem. Use the same procedure described for globe valves to solve this problem.

The wedge-shaped disc in the gate valve is designed to be either fully opened or fully closed. Gate valves are not intended to throttle the volume of liquid flowing through the pipe. When gate valves are opened only partway, the gate has a tendency to vibrate or chatter, which causes the edge to erode. Repair of the valve is not likely to solve this problem. More than likely you will have to replace the valve.If a gate valve fails to stop the flow of water, one of three possible problems may be the cause:

- The gate is worn.
- The valve seat is worn.
- Some foreign material is preventing the gate from seating.

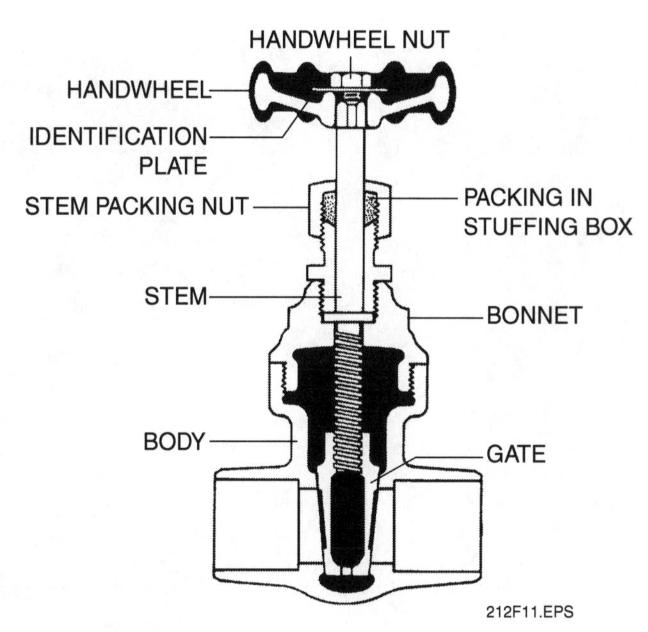

Figure 11 ◆ Gate valve.

To determine which problem exists, remove the bonnet and inspect the gate and seat. You may be able to solve the problem by scraping out mineral deposits and carefully cleaning the mating surfaces. You may have to replace the gate, and if the seat area is worn, you'll have to replace the valve.

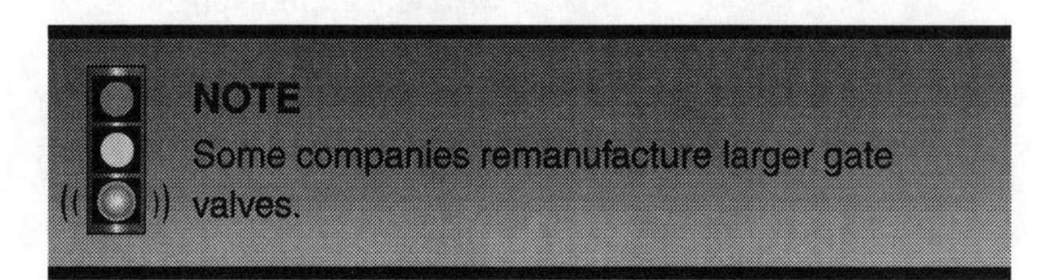

NOTE

Some companies remanufacture larger gate valves.

3.3.0 Flushometers

A flushometer connects directly to the water supply and requires no storage tank, which allows for repeated, rapid flushing. The troubleshooting guide in *Table 2* applies to flushometers (see *Figure 12*). Five common problems are associated with flushometers:

- Leakage around the handle
- Failure of the vacuum breaker
- Malfunction of the control stop, which regulates the flow of water
- Failure of the flushometer to completely close
- Leakage in the diaphragm that separates the upper and lower chambers

Kits that contain the components you'll need to repair each of these defects are available. Even though all flushometers contain the same basic components, specific parts will vary from manufacturer to manufacturer. Always follow the manufacturer's specifications and instructions when installing replacement parts.

Table 2 Troubleshooting Guide for Standard and Electronic Eye Flushometers

Problem	Cause	Solution
1. Nonfunctioning valve	Control stop or main valve closed	Open control stop or main valve
2. Not enough water	Control stop not open enough Urinal valve parts installed in closet parts Inadequate volume or pressure	Adjust control stop to siphon fixture Replace with proper valve Increase pressure at supply
3. Valve closes off	Ruptured or damaged diaphragm	Replace parts immediately
4. Short flushing	Diaphragm assembly and guide not hand-tight	Tighten
5. Long flushing	Relief valve not seating	Disassemble parts and clean
6. Water splashing	Too much water is coming out of faucet	Throttle down control stop
7. Noisy flush	Control stop needs adjustment Valve may not contain quiet feature The water closet may be the problem	Adjust control stop Install parts from kit Place cardboard under toilet seat to separate bowl noise from valve noise—if noisy, replace water closet
8. Leaking at handle	Worn packing Handle gasket may be missing Dried-out seal	Replace assembly Replace Replace
9. Valve fails to close off	Damaged or dirty valve Insufficient pressure	Clean and replace, then flush lines Ensure that control stop is fully open

Electronic Eye Flushometers

Note: These are general guidelines. It is best to follow the specific manufacturer's guidelines when troubleshooting electronic eye flushometers.

Problem	Cause	Solution
1. Valve won't flush	Sensor not sensing presence of user; range is too short	Increase the range
2. Valve won't flush or flushes only when someone walks past	Sensor is focusing on a reflective wall or mirror; range is too long	Decrease the range
3. Unit flashes when user steps into range	Low batteries	Replace batteries
4. Valve won't shut off	Dirt or debris clogging diaphragm orifice Seal damaged or worn Overtightened solenoid valve Malfunction in electronic module	Clean Replace Loosen or replace Replace
5. Water flow to fixture too low	Control stop improperly adjusted Not enough flow or pressure	Adjust Increase flow or pressure
6. Water flow to fixture too high	Control stop improperly adjusted Improper diaphragm installed in valve	Adjust Replace with proper diaphragm

Repair or Replace?

Often replacing a defective valve makes more sense than trying to repair it. It can be difficult to find the right part components needed for repair. You can save yourself and your customer a lot of time and money by simply replacing defective valves. Every situation is different, so be sure to think about cost effectiveness and customer service when deciding whether to repair or replace a valve or faucet. Determine how long service will be turned off for the repair. Notify the owner and occupants of this information, and also when the service has been both turned off and restored.

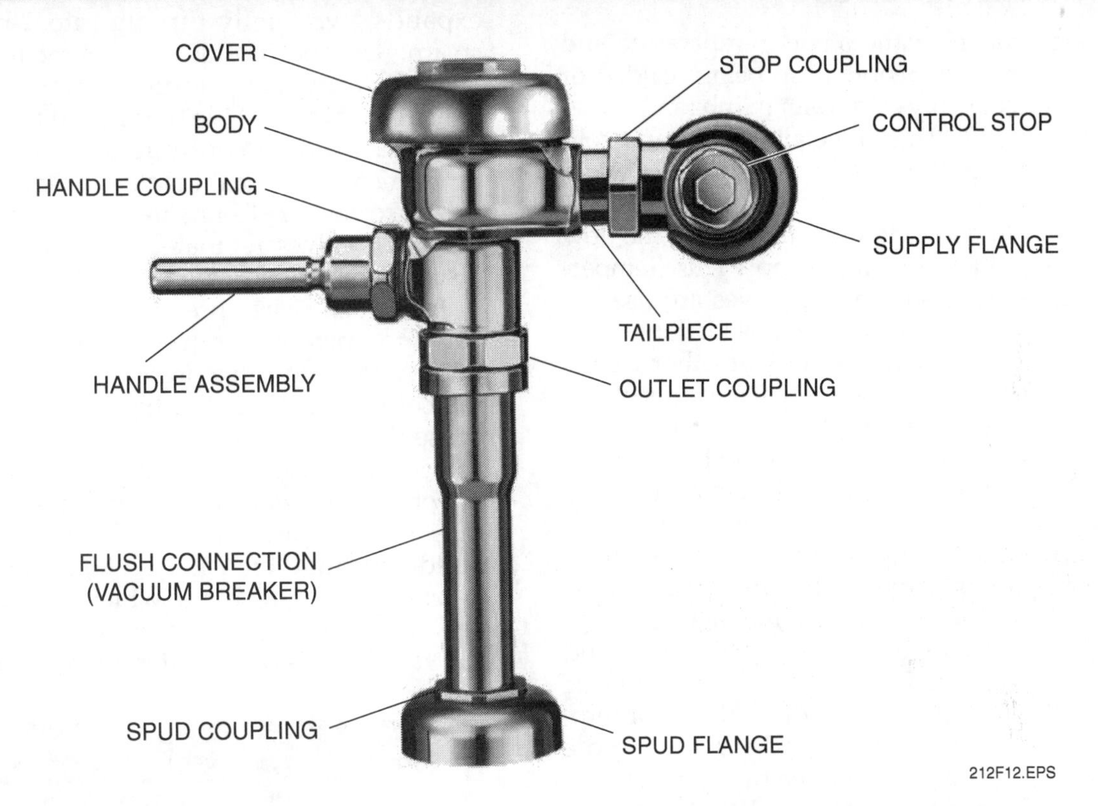

Figure 12 ◆ Diaphragm type flushometer.

3.4.0 Float-Controlled Valves (Ball Cocks)

The valve that controls the water level in a water closet tank is a float-controlled valve, also called a ball cock (see *Figure 13*). Manufacturers provide repair kits, and you must refer to their specifications and instructions when making repairs. If the stem, valve body, or float mechanism is damaged, you may have to replace the entire float valve. Replacement parts are available for leaking floats or broken float rods.

Figure 13 ◆ Float-controlled valve.

3.5.0 Tank Flush Valves

Tank flush valves are available in many styles (see *Figure 14*). If the valve and the lever that operates the valve are badly corroded, replace both parts. However, most problems occur with the component parts: the tank ball, the flapper tank ball, the connecting wires or chains, and the guide. Inspect all of these parts to determine whether to repair or replace the entire assembly. If the valve seat is corroded, you can use a reseating tool to restore it.

3.6.0 Balancing Valves

A balancing valve balances both temperature and pressure. It protects people from being scalded or getting a blast of really cold water in tubs and showers (see *Figure 15*). This valve allows hot and cold water to mix together so that a comfortable water temperature is delivered to the faucet or shower head. It is designed to immediately block the flow of hot water if the cold water supply fails. Temperature- and pressure-balancing valves are used in both residential and commercial installations.

Leaks can occur when solder or other debris gets between the washers and seat surfaces of this valve. To repair a leak, replace all washers. You should also inspect the top surface of both the hot and cold seat and replace them if damaged.

If the hot and cold water are not properly balanced or the water temperature changes without the handle being moved, the problem is most likely debris in the pressure-balancing spindle. To correct this problem, open the valve halfway and remove the handle. Tap the spool with a plastic hammer. If this does not solve the problem, remove the spool assembly and tap the handle end against a solid object to free the piston. Soaking the assembly in vinegar will help remove small particles of lime and scale buildup.

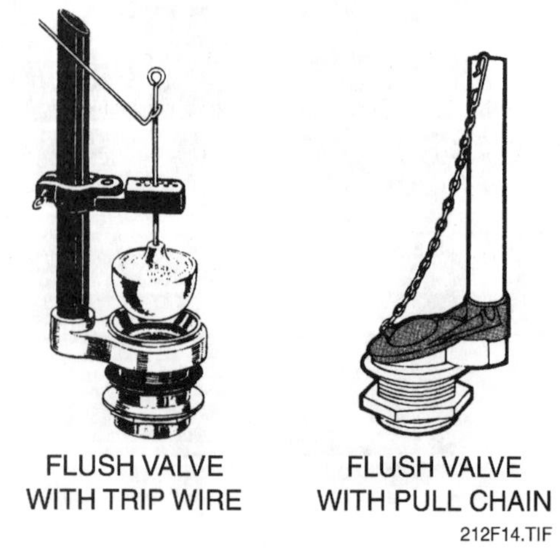

Figure 14 ◆ Tank flush valves.

3.7.0 Temperature and Pressure (T&P) Valves

The T&P valve is designed to open and vent heated water (the temperature part of the valve) and air (the pressure part of the valve) into the atmosphere (see *Figure 16*). This venting action brings the water temperature and pressure back to safe levels. Therefore, the T&P valve is the principle safety feature of water heaters.

As water is heated, its volume expands. A simple example of this principle is a tea kettle filled with boiling water. The heated water rapidly expands, eventually turning into a blast of hot steam. The thermostat on a water heater maintains the water temperature to prevent a buildup of heat. However, if the thermostat fails, the water temperature will continue to increase and the rapid increase of heat inside the heater can result in an explosion. To ensure safety, codes require that T&P valves regulated for residential use be installed on water heaters.

Too much water pressure can cause the T&P valve to operate with too much force. To correct this problem, install a pressure reducing valve on the cold water side. Too high a temperature can cause the T&P valve to operate too frequently. To correct this problem, check the thermostat. You may have to lower the setting or replace the thermostat. Drips are typically the result of mineral buildup around the valve seat. Lift the stem lever to let the water flush out, which should also flush out debris. Rotate the lever to reseat the stem. If the valve seat is damaged, replace the valve.

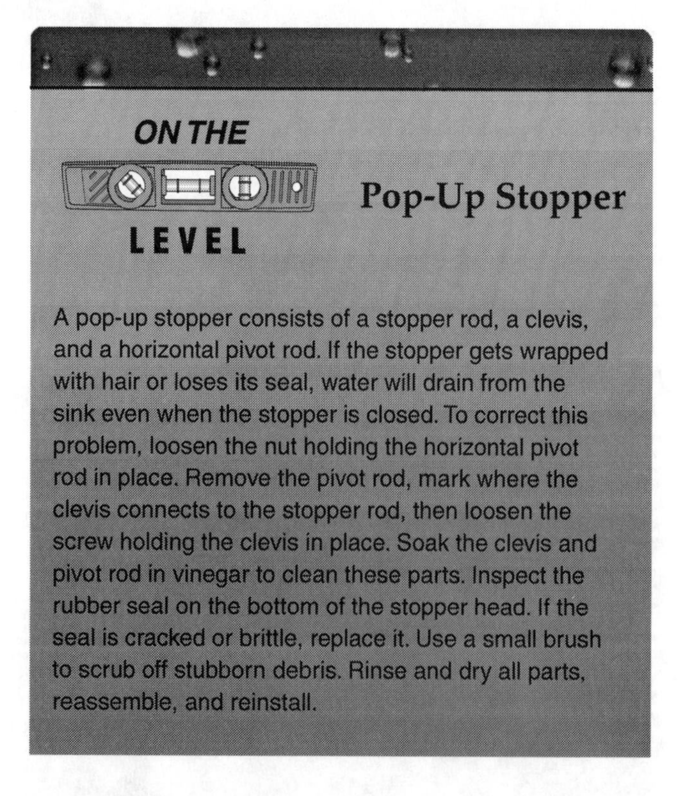

Pop-Up Stopper

A pop-up stopper consists of a stopper rod, a clevis, and a horizontal pivot rod. If the stopper gets wrapped with hair or loses its seal, water will drain from the sink even when the stopper is closed. To correct this problem, loosen the nut holding the horizontal pivot rod in place. Remove the pivot rod, mark where the clevis connects to the stopper rod, then loosen the screw holding the clevis in place. Soak the clevis and pivot rod in vinegar to clean these parts. Inspect the rubber seal on the bottom of the stopper head. If the seal is cracked or brittle, replace it. Use a small brush to scrub off stubborn debris. Rinse and dry all parts, reassemble, and reinstall.

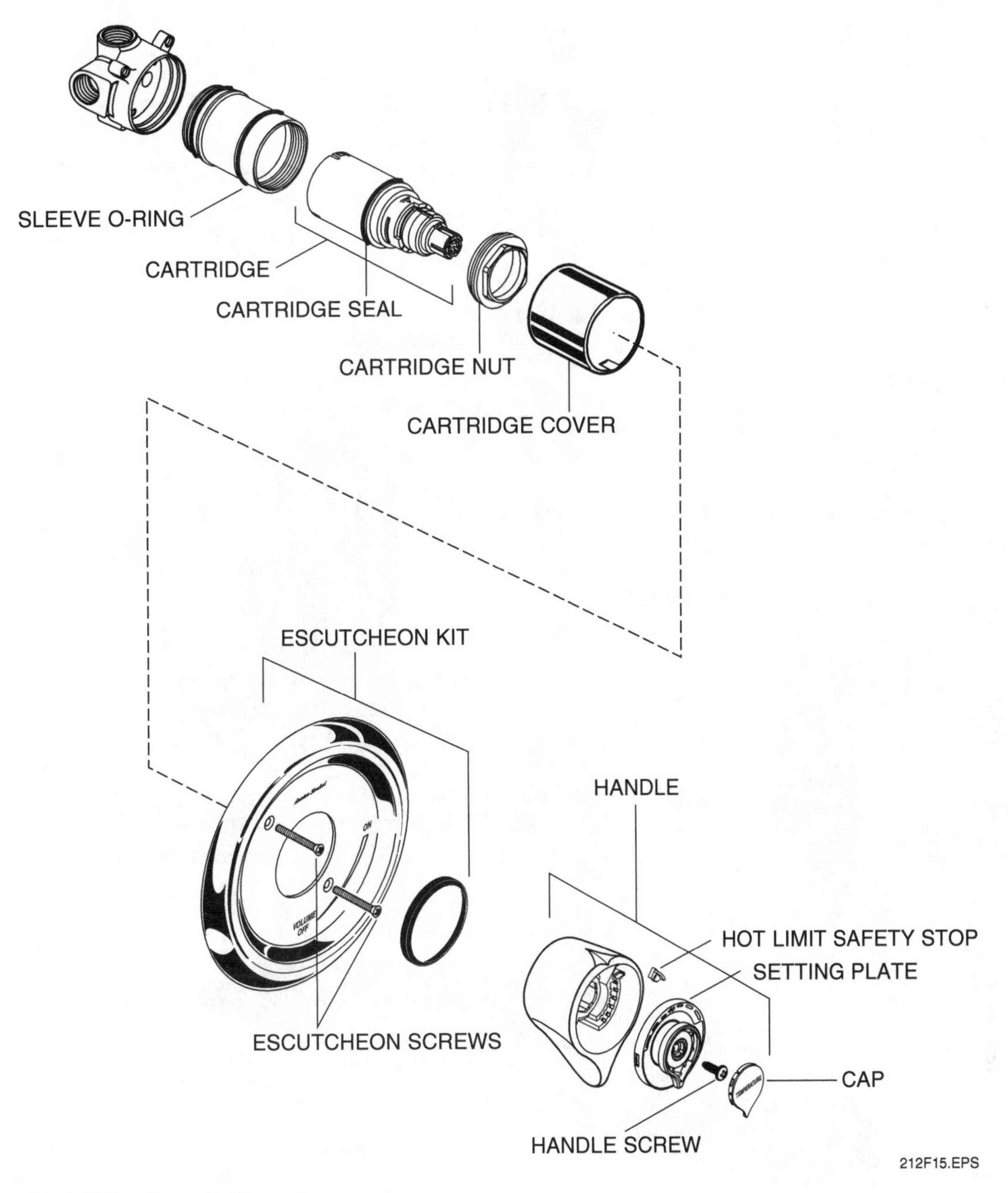

Figure 15 ◆ Balancing valve for a shower.

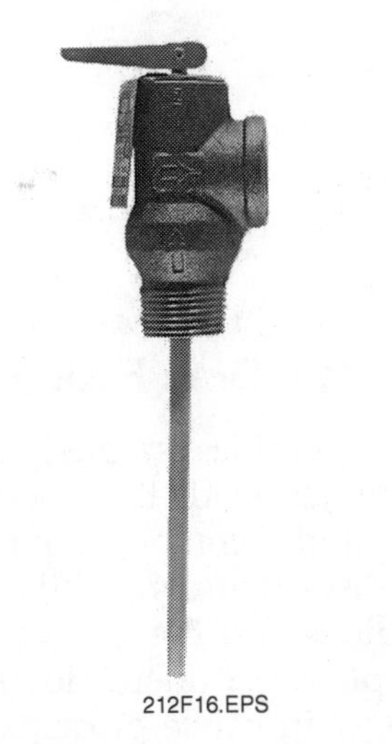

Figure 16 ◆ Temperature and pressure valve.

3.8.0 Cartridge Faucets

You can identify a cartridge faucet by the metal or plastic cartridge, or sealed unit, inside the valve body (see *Figure 17*). The cartridge is located between the spout assembly and the handle. Many single-handle faucets and showers are cartridge designs. Replacing a cartridge is a fairly easy repair that will fix most faucet or shower drips.

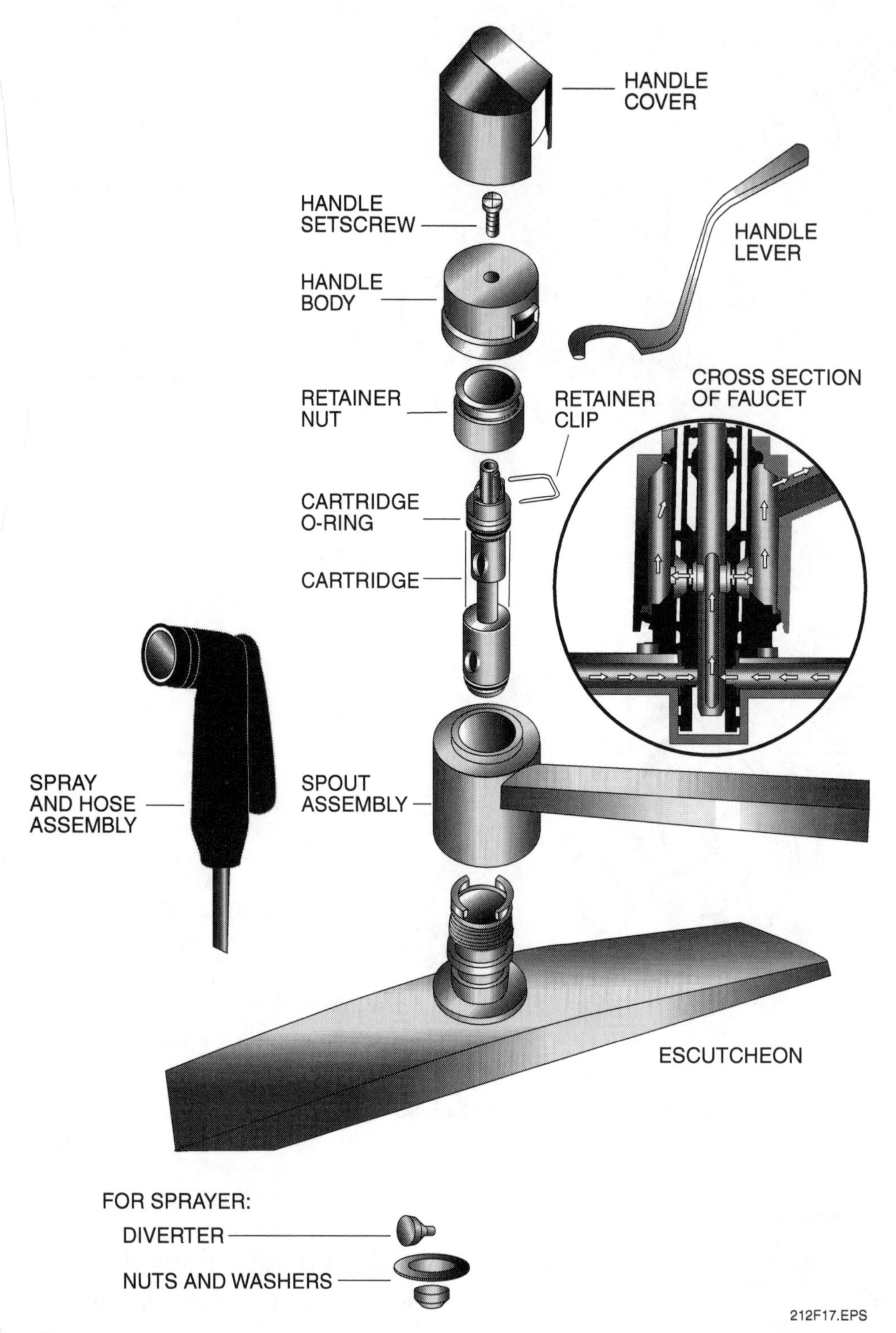

Figure 17 ◆ Cartridge faucet.

3.9.0 Rotating Ball Faucets

A rotating ball faucet has a single handle. You can identify it by the metal or plastic ball inside the faucet body (see *Figure 18*). Drips usually occur as a result of wear on the valve seals. Replacement valve seals and springs are readily available. Leaks from the base of the faucet are usually caused by worn spout O-rings. You should replace these parts.

3.10.0 Ceramic Disc Faucets

A ceramic disc faucet is a single-handle faucet that has a wide cylinder inside the faucet body (see *Figure 19*). The cylinder contains a pair of closely fitting ceramic discs that control the flow of water. The top disc slides over the lower disk. These discs rarely need replacing. Mineral deposits on the inlet ports are the main cause of drips. Cleaning the inlets and replacing the seals eliminates most drips.

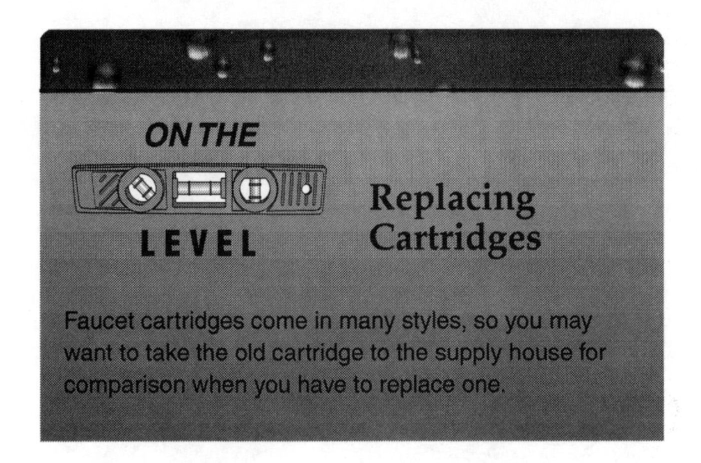

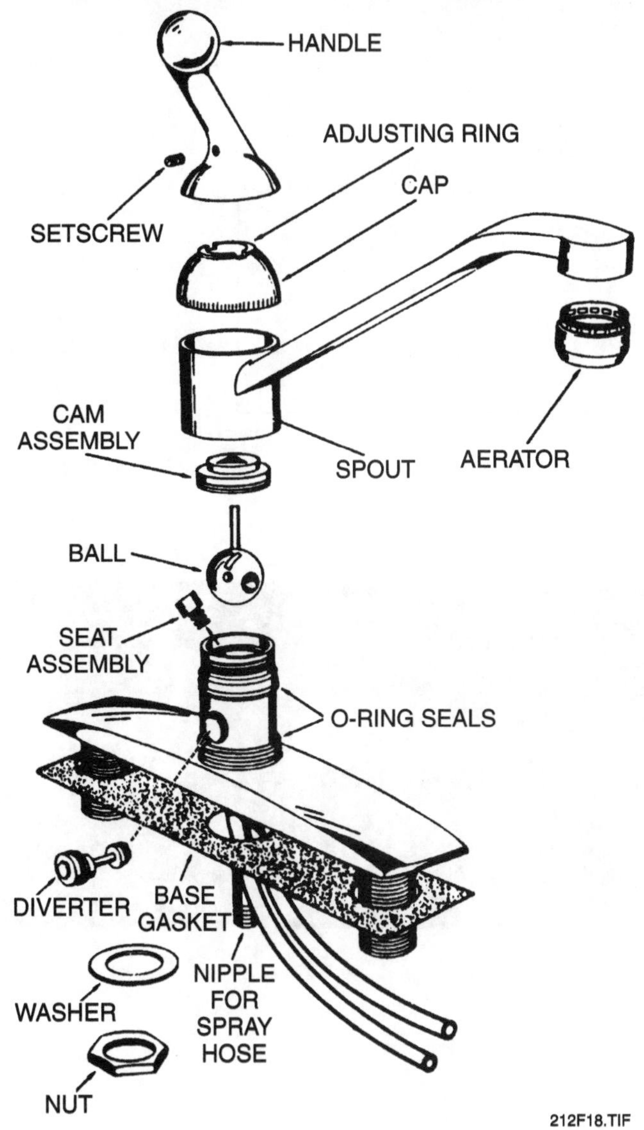

Figure 18 ◆ Rotating ball faucet.

3.11.0 Faucet Water Filters

Water filters come in undercounter mountings, countertop mountings, and fixture mountings. Some faucets combine the functions of water delivery and filtration in one unit. Although styles vary, all filters work on the same basic principle. Water intended for drinking or washing passes through a cartridge, which usually contains carbon, to remove unwanted tastes, odors, and chemicals such as chlorine. Always follow the manufacturer's installation and maintenance instructions. In general, most problems result from filters not being changed regularly. In addition, customers may complain of a lower water flow. Filters do restrict the flow of water from the faucet to allow contact time with the filter. It's a trade-off for cleaner, better tasting water.

3.12.0 Electronic Controls

Electronic, or hands-free, controls are available for faucets, toilets, and urinals. These controls emit a constant beam of infrared light, which operates the fixture. Water flows into sinks and lavatories when the user blocks the beam. Toilets and urinals flush when the user no longer blocks the beam. The relatively high cost of these controls generally limits their use to commercial and institutional installations, where they may be installed new or retrofitted onto existing plumbing. The controls save water and energy and are more hygienic than hand-operated controls.

Some electronic controls use battery power to operate flush valves and faucets, and most problems result when the battery power is exhausted or when a battery is improperly installed. If the problem is not with the battery, the problem may be a clogged filter or component that has been installed backwards. Clean and reinstall clogged filters, and ensure that all components are installed properly.

The installation and maintenance instructions for electronic controls vary from one manufacturer to another. Always follow the manufacturer's installation and maintenance instructions. (For general troubleshooting guidelines for electronic flushometers, refer to *Table 2*.)

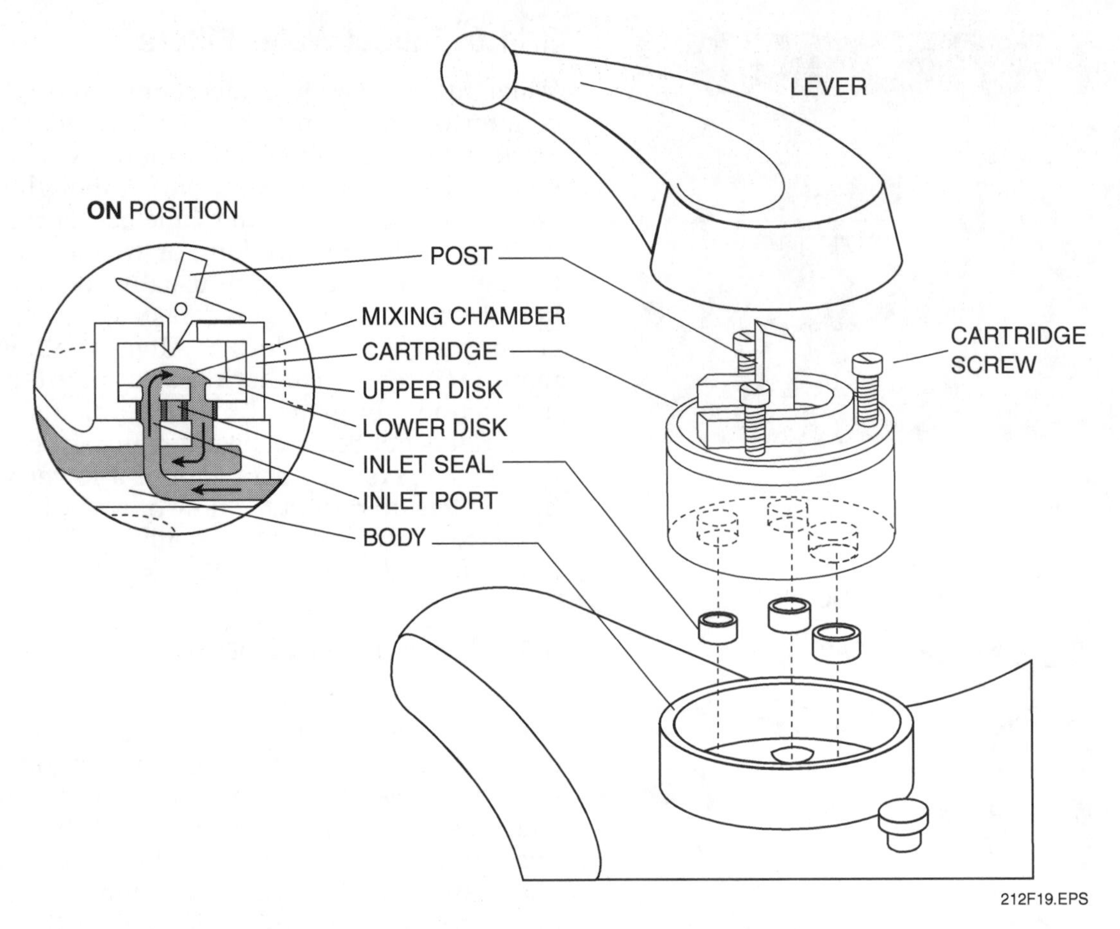

Figure 19 ◆ Ceramic disc faucet.

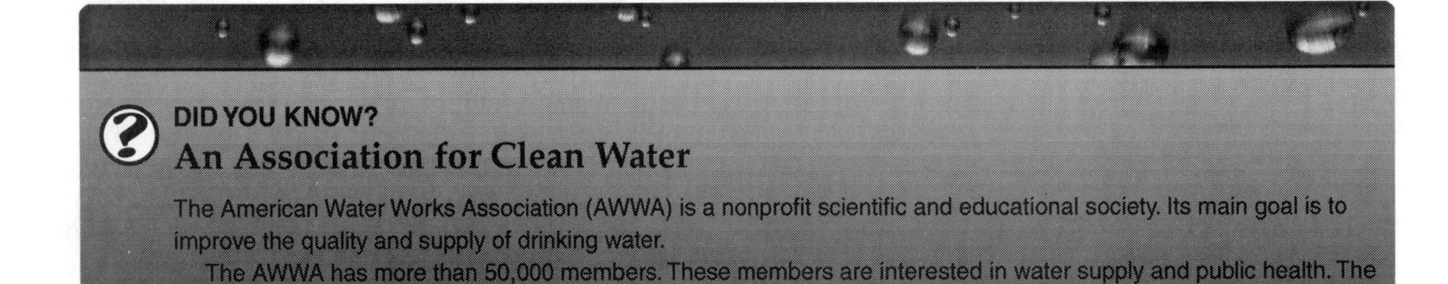

DID YOU KNOW?

An Association for Clean Water

The American Water Works Association (AWWA) is a nonprofit scientific and educational society. Its main goal is to improve the quality and supply of drinking water.

The AWWA has more than 50,000 members. These members are interested in water supply and public health. The membership also includes more than 4,000 utilities that supply water to about 180 million people in North America.

Summary

Customers may not spend much time thinking about fixtures, valves, and faucets, but they rely on them to operate the plumbing efficiently. When any of these parts stop working you'll be called in to repair or service these parts. You will have to minimize damage from major leaks. You must also work safely to prevent injuries to yourself and to co-workers. Finally, you must develop the judgment you'll need to decide when to make repairs and when to replace a fixture, valve, or faucet entirely.

Valves come in a wide variety of sizes and styles, but are divided into several categories for repair work purposes. In many commercial and institutional installations, some valves are operated via electronic controls. Because valves and faucets operate under pressure and have moving parts, they can wear out or break down. By understanding how valves work, you can repair them and get more use from them, although you'll sometimes have to replace the valve. In some cases, you'll use special tools. In other cases, you'll use a kit supplied by the manufacturer to replace broken parts.

As always, you must think safety first, not only for yourself and your co-workers, but for the customers who will use the fixtures you repair. This module also covered some basic safety rules to follow when making repairs.

Notes

Review Questions

1. Before attempting to repair a leaking valve or faucet, it is recommended that you _____.
   a. place packing around the leak
   b. wrap the leaking fixture with Teflon tape
   c. shut off the valve upstream from the leak
   d. remove any traps near the leaking fixture

2. _____ is a common cause of leaking around the stem or under the knob of a globe valve.
   a. A worn or damaged seat disc (washer)
   b. A worn or damaged seat
   c. Worn threads on the stem
   d. Defective packing

3. _____ is necessary to repair a globe or angle valve that rattles when the valve is opened and water is flowing.
   a. Rethreading the stem
   b. Replacing the disc and resurfacing the seat
   c. Repacking and tightening the stem
   d. Replacing the entire valve

4. _____ may be required to repair a corroded valve seat in a globe valve.
   a. A new tank ball
   b. A reseating tool
   c. Repositioning of the handle assembly
   d. Readjustment of connecting wires or chains

5. If the correct size and shape of preformed packing is *not* available, you can use _____ packing.
   a. rope
   b. corded
   c. twine
   d. twist

6. If a stem is difficult to turn because of damaged threads in the base of the valve, you should _____.
   a. replace the valve
   b. repack the stem
   c. install a new packing nut
   d. resurface the seat

7. If a flushometer is flushing too long, the most likely cause is _____.
   a. the relief valve is not seating
   b. the diaphragm is ruptured
   c. the control stop is malfunctioning
   d. inadequate water pressure

8. If a flushometer is leaking at the handle, one possible cause is _____.
   a. a ruptured or damaged diaphragm
   b. high water pressure or volume
   c. worn packing
   d. an improperly seating relief valve

9. In a balancing valve, if the water temperature changes *without* the handle being moved, the problem is most likely _____.
   a. a loose handle
   b. debris in the pressure-balancing spindle
   c. a broken thermostat
   d. too much cold water flowing to the fixture

10. In a T&P valve, drips are usually the result of _____.
    a. an imbalance between the temperature and pressure sides
    b. cracks in the drip tube
    c. a seal installed upside down
    d. buildup of minerals around the valve seat

11. Many single-handle faucets and showers are cartridge designs.
    a. True
    b. False

12. Leaks from the base of a rotating ball faucet are usually caused by _____.
    a. loose packing
    b. corrosion of the valve seat
    c. worn spout O-rings
    d. mineral deposits on the inlet ports

Review Questions

13. Drips in ceramic disc faucets can be repaired by _____.
    a. replacing the seals
    b. replacing the discs
    c. repacking the faucet
    d. installing a globe faucet

14. Water intended for drinking or washing passes through a cartridge, which usually contains _____ to remove unwanted tastes.
    a. capsicum
    b. zinc
    c. talc
    d. carbon

15. Electronic controls on waste fixtures operate by emitting a(n) _____ beam of _____ light.
    a. intermittent; infrared
    b. constant; infrared
    c. intermittent; ultraviolet
    d. constant; ultraviolet

Trade Terms Introduced in This Module

Packing extractor: A tool used to remove packing from a valve stem.

Removable seat wrench: A tool used by plumbers to replace valve and faucet seats.

Reseating tool: A tool used by plumbers to reface a valve seat.

Screw extractor: A tool designed to remove broken screws.

Tap: A tool used for cutting internal threads.

Twist packing: A string-like material used to pack valve stems when preformed packing material is not available. It is not as durable as preformed packing.

Resources & Acknowldgments

Additional Resources

This module is intended to be a thorough resource for task training. The following reference works are suggested for further study. These are optional materials for continued education rather than for task training.

Installing & Repairing Plumbing Fixtures. 1994. Peter Hemp. Newton, CT: Taunton Press.

Plumbing Fixtures and Appliances. 2001. Washington, D.C.: International Pipe Trades Joint Training Committee.

Plumbing Fixtures and Fittings. 2003. Boca Raton, FL: Catalina Research.

Piping and Valves. 2001. Frank R. Spellman and Joanne Drinan. Lancaster, PA: Technomic.

References

Dictionary of Architecture and Construction, Third Edition, 2000. Cyril M. Harris, ed. New York: McGraw-Hill.

2003 National Standard Plumbing Code. Falls Church, VA: Plumbing-Heating-Cooling Contractors—North America.

Figure Credits

Courtesy of Delta Faucet Company (Michael Graves Collection)	Module divider
Coyne & Delaney Company	212F01
Robert Bosch Tool Corporation	212F02
BrassCraft Manufacturing Company	212F05–212F07
Sloan Valve Company	212F12
Fluidmaster, Inc.	212F13
American Standard	212F15
Watts Regulator Company	212F16
Kohler Co.	212F18

CONTREN® LEARNING SERIES — USER UPDATE

The NCCER makes every effort to keep these textbooks up-to-date and free of technical errors. We appreciate your help in this process. If you have an idea for improving this textbook, or if you find an error, a typographical mistake, or an inaccuracy in NCCER's Contren® textbooks, please write us, using this form or a photocopy. Be sure to include the exact module number, page number, a detailed description, and the correction, if applicable. Your input will be brought to the attention of the Technical Review Committee. Thank you for your assistance.

Instructors – If you found that additional materials were necessary in order to teach this module effectively, please let us know so that we may include them in the Equipment/Materials list in the Annotated Instructor's Guide.

Write: Product Development and Revision
National Center for Construction Education and Research
P.O. Box 141104, Gainesville, FL 32614-1104

Fax: 352-334-0932

E-mail: curriculum@nccer.org

Craft ______ Module Name ______

Copyright Date ______ Module Number ______ Page Number(s) ______

Description

______

______

______

______

(Optional) Correction

______

______

______

(Optional) Your Name and Address

______

______

______

Glossary of Trade Terms

Accessibility requirements: The requirements outlined in some building codes and by ANSI (the American National Standards Institute) that affect physically challenged people and their access to public buildings and facilities.

Addendum: A notification issued prior to a bid for changes in a construction drawing.

Air test: A test to detect leaks and to determine a pipe's ability to withstand air pressure.

Alternate: An illustration on a plan showing changes resulting from the use of different materials, methods, or layouts.

Alternating current (AC): An electrical current that changes direction in cycles.

Ammeter: A test instrument used to measure current flow.

Ampere (amp): The unit of measurement for current flow. The magnitude is determined by the number of electrons passing a point at a given time.

Analog meter: A meter that uses a needle to indicate a value on a scale.

Angle valve: A valve with flow characteristics similar to a globe valve, with a valve body that allows it to be used in a 90-degree elbow configuration.

Anti-siphon hole: An opening near the top end of the dip tube on a water heater that prevents hot water from siphoning out if an interruption occurs in the cold water supply.

Approved submittal data: A drawing that includes rough-in dimensions, manufacturer's specifications, and other information for a fixture that has been approved for use in an installation.

Aquastat: A thermostat that regulates the temperature of hot water in a hot water boiler.

Architectural drawing: A construction drawing that uses plan views, elevations, sections, and details to show the overall shape, size, and appearance of a building.

As-built drawing: A drawing of record for a project, showing the final configuration that incorporates all of the approved changes from addenda, change orders, and RFIs.

ASME Boiler and Pressure Vessel Code: Publication that establishes criteria that govern the construction and maintenance of pressure vessels and related devices, such as steam safety valves.

Baffle: A plate that slows or changes the direction of the flow of air, air-gas mixtures, or flue gases.

Ball-check valve: A valve that allows one-way fluid flow in water supply or drainage lines by using an internal ball and seat.

Batter board: One of a pair of horizontal boards that are nailed at right angles to each other to three posts set beyond the corners of a building elevation. Strings fastened to these boards indicate the exact corner of a building.

Bench mark: In surveying, a marked reference point on a permanent, fixed object such as a metal disk set in concrete, the elevation of which is known, and from which the elevation of other points may be obtained.

Blade: The longer of two sides of a framing square.

Blocking: Pieces of wood used to secure, join, or reinforce structural members or to fill spaces between them.

Blowdown: The difference between the opening and closing pressures in a steam safety valve.

Bolted bonnet: A valve cover that guides and encloses the stem. It is used for larger valves or for high-pressure applications.

Bonnet: A valve cover that guides and encloses the stem.

Box-out: A short length of pipe or a wooden box attached to a roof deck before concrete is poured to preserve roof drain openings.

British thermal unit: The amount of heat needed to raise the temperature of one pound of water by 1°F. Abbreviated Btu.

Building cleanout: A cleanout located in a building drain immediately prior to its exit from a building, usually required by code in buildings that have basements.

Burn an inch: Practice of deducting an inch from the reading (measurement) to account for the room required to grip the folding rule with your fingers.

Butterfly valve: An on/off or throttling valve that uses a disc and rotates 90 degrees from fully closed (perpendicular to flow) to fully opened (parallel to flow) to control fluid flow.

Callback: Term that refers to being called back to the worksite to repair or replace defective materials or to correct faulty workmanship.

Can wash drain: A can that is placed upside down over a drain to allow water to squirt upward, washing the inside of the container.

Carrier fittings: The support apparatus for wall-hung bathroom fixtures.

Change order: A document issued to indicate modifications to a plan once construction has begun.

Chase: A hollow, enclosed area between two back-to-back commercial restrooms used to house plumbing connections.

Circuit: A loop of electrical current that contains a voltage source, a load, and conductors.

Circuit diagram: A schematic drawing that indicates the path that electricity flows in a circuit.

Civil drawing: A construction drawing that provides the overall shape of the building site, the location of the building, and other planned structures.

Clamping ring: A clamp that secures subsurface flashing or a waterproof membrane to the drain body. Also called a flashing clamp.

Clamp-on ammeter: A current meter in which jaws placed around a conductor measure the amount of current flow.

Clarification: Documents issued in response to an RFI.

Closet bolt: A bolt with a large-diameter, low circular head that is cupped on the underside so that it is sealed against the surface when the bolt is tightened. It is used to fasten a water closet bowl to the floor.

Collector: In a solar water heater, a box that collects heat from the sun and transfers it to water as it flows through a pipe that runs through the box.

Compressor: A machine for compressing air or other gases.

Condensate: Water or other fluid that has been condensed from steam or from a vapor.

Condensing water heater: A gas water heater that features spark ignition, fans for induced air draft, and heat transfer surfaces that are larger than gravity-fed models.

Conductor: A material that readily conducts electricity; also, the wire that connects components in an electrical circuit.

Constant: A number that never changes when used in an equation. For example, when finding the travel of a 45-degree right triangle, always use the constant 1.414.

Contactor: A control device consisting of a coil and one or more sets of contacts used as a switching device in high-voltage circuits.

Continuity: A continuous current path. Absence of continuity indicates an open circuit.

Contour line: An element of a civil drawing that indicates areas of identical elevation, used to determine grade.

Convection: The motion of water resulting from density differences caused by different temperatures throughout the water.

Convention: A standard labeling and marking system used to provide information in plans and schedules.

Coordination drawing: A drawing that is used to identify the proposed locations of all trade work, in order to identify potential conflicts.

Cross pattern: A method for tightening bolts or screws in an even sequence, tightening each bolt or screw a little at a time to avoid putting too much pressure on any one bolt or screw.

Cross threading: A condition that occurs when the initial female thread on a pipe or fitting does not properly mesh with the initial male thread. This causes the connection to jam and can strip the threads.

Curb box: A vertical sleeve that provides access to a buried curb stop.

Curb stop: In a water service pipe, a control valve for a building's water supply, usually placed between the sidewalk and the curb.

Cure: A natural process in which excess moisture in concrete evaporates and the concrete hardens.

Current: The rate at which electrons flow in a circuit, measured in amperes.

Cut sheet: Another term for approved submittal data.

Cycle: In AC electricity, a single repetition of alternating current flow.

Daily log: A description of all work accomplished during the day on a project, maintained by the site supervisor.

Deck clamp: A ring that fastens to a drain body with several bolts.

Demand: The total water requirement for a water supply system, including the pipes, fittings, outlets, and fixtures used in the system.

Digital meter: A meter that provides a direct numerical reading of the value measured.

Dip tube: A device that prevents cold water from mixing with hot water; it delivers incoming cold water through the stored hot water to the bottom of the tank.

Direct current (DC): An electrical current that flows in one direction.

Dome: A ventilated drain cover that prevents debris from entering the piping system and helps reduce the chance of pipe clogs. Also called a grate or strainer.

Drain body: The basic element of all drains, the drain body funnels water into the piping system.

Drain receiver: A type of flange that distributes the roof drain's weight over a large area. It also supports a drain in an oversized or off-center roof opening.

Drip: An extension of piping created by connecting a T, nipple, and cap that is installed on underground lines to allow storage of liquefied petroleum until it changes back into a vapor.

Dry fire: A term that refers to turning on an electric water heater with no water in the storage tank, causing the elements to burn out.

Electromagnet: A coil of wire wrapped around a soft iron core. When a current flows through the coil, magnetic flux lines are created.

Electromotive force (EMF): Another term for voltage.

Elevation: A drawing that shows a vertical projection of a building.

Escutcheon: A flange on a pipe used to cover a hole in a floor or wall through which the pipe passes.

Expansion joint: A joint or gap between adjacent parts of a building that permits the parts to move as a result of temperature changes or other conditions without damage to the structure.

Extension: A part that helps set a fixture, such as a drain, at the proper grade.

External thread: A valve end connection that has the threads on the outside.

Finish work: The completion phase of any construction project. Generally the most visible part of the work, it includes paint, stain, trim, and fixtures.

Fire stopping: A system of materials and fittings used to prevent the passage of flame, smoke, and/or toxic byproducts of fire through structural penetrations.

Flammable Vapor Ignition Resistant (FVIR) water heater: A gas fired water heater that meets the ANSI standard for resisting the accidental combustion of vapors emitted by flammable liquids stored, used, or spilled near the water heater.

Flanged end: A valve end connection that uses bolts and gaskets to connect the valve to a companion pipe flange.

Flash point: The temperature at which vapors given off by a fuel may be ignited.

Flashing clamp: Another term for clamping ring.

Float-controlled valve (ball cock): A valve installed in water closet flush tanks to maintain a constant water level in the tank.

Floor sink: A type of floor drain that is coated with a porcelain or easy-to-clean finish. Most often used in areas requiring a high level of sanitation.

Flue: A heat-resistant enclosed passage in a chimney that carries away combustion products from a heat source to the outside.

Flush valve: A device located at the bottom of a tank for flushing water closets and similar fixtures.

Flushing floor drain: A floor drain that is equipped with an integral water supply connection, enabling flushing of the drain receptor and trap.

Framing square: Specialized square with tables and formulas printed on the blade for making quick calculations, such as finding travel and determining area and volume.

Frequency: In AC electricity, a measure of the number of cycles per second.

Frost line: The depth of frost penetration into the soil. The depth varies in different parts of the country. Water supply and drainage pipes should be set below this line to prevent freezing.

Full-load amps (FLA): The normal running current of a compressor motor.

Grate: Another term for dome.

Gravity feed: A system that depends on gravity to bring fuel to the pump. Gravity describes the tendency of objects or substances to move downward.

Ground fault: A type of short circuit caused by a live conductor touching another conducting substance.

Heat exchanger: A device that transfers heat from one liquid to another without the two liquids coming into contact with each other.

High-limit control: A safety device that protects against extreme water temperatures caused by a defective thermostatic control or by grounded water heater elements.

Home run: A run of pipe from a distribution point such as a manifold to the appliance with no branches. It serves only one appliance.

Huddling chamber: A chamber within a safety valve that harnesses the expansion forces of air or steam to quickly open (pop) or close the valve.

Hydrostatic test: A test to determine a pipe's ability to withstand internal hydrostatic (water) pressure and to detect possible leaks in the system.

Hydrostatic test pump: A pump used to test hydrostatic pressure.

I-beam: A rolled or extruded structural metal beam having a cross-section that looks like the capital letter I.

Immersion element: In an electric water heater, an electrical device inserted into the storage tank and used to heat the water.

Indirect water heater: A water heater in which a heat exchanger increases the water's temperature.

Induction: The generation of an electrical current by a conductor placed in a magnetic field. The motion of either the conductor or the magnetic field causes the current flow.

In-line ammeter: A current reading meter that is connected in series with the circuit under test.

Instantaneous water heater: A water heater that heats water when it is used, rather than storing and heating water in a tank.

Insulating union: A nonconductive union placed on the gas main side of the meter to prevent electricity from being conducted through the customer's service line back to the gas main.

Insulator: A device that inhibits the flow of electrical current. An insulator is the opposite of a conductor.

Internal thread: A valve end connection that has the threads inside.

Intumescent: Having the ability to expand when exposed to flame.

Invert elevation: In plumbing, the lowest point or the lowest inside surface of a channel, conduit, drain, pipe, or sewer pipe. It is the level at which fluid flows.

Jacket: The outer shell of a water heater, made of enamel baked on steel.

Ladder diagram: A simplified schematic diagram in which the load lines are arranged like the rungs of a ladder between vertical lines representing the voltage source.

Landscape drawing: A construction drawing that indicates how decorative elements will be arranged around the building.

Lever-and-weight swing-check valve: A type of swing-check valve that uses an external counterweight connected to an internal swing-check disc, frequently used on lines with pulsating flow.

Lift rod assembly: A mechanism consisting of the clevis and clevis screw that is used to adjust the operation of a pop-up plug.

Lift-check valve: A type of valve that prevents flow reversal in a piping system. It is used for gas, water, steam, or air, and for lines where frequent fluctuations in flow occur.

Line-duty device: A protective device that is placed directly in line with a voltage source.

Load: A device that converts electrical energy into another form of energy, such as heat, mechanical motion, or light.

Load factor: A specified percentage of total flow from connected fixtures that is likely to occur at any point along the DWV system, usually provided in the local applicable code.

Locked-rotor amps (LRA): The initial current surge in a compressor motor.

Magnetic flux lines: Lines of force in magnetized iron.

Magnetic starter: An electrically operated switching device used to start large motors.

Main burner: The chamber in which the fuel gas mixes with air and is ignited. The combustion in the chamber heats the water in the tank to a preset temperature.

Manometer: A U-shaped tube containing fluid and marked with a graduated measuring scale, used for measuring slight changes in a low-pressure system such as a heating, venting, and air conditioning system. One end of the tube is connected to the low-pressure source. Slight changes in pressure cause the fluid in the tube to move up or down. Also called a draft gauge.

Mechanical drawing: A construction drawing that indicates the location and size of the HVAC system, as well as other features such as elevators, conveyors, and escalators.

Mercaptan: A chemical compound that has sulfur rather than oxygen, typically added as an odorant to natural gas.

Meter index: An indicator used in diaphragm and rotary meters to indicate the amount of gas that has passed through the meter, in cubic feet.

Multimeter: A test instrument capable of reading voltage, current, and resistance.

Nonrising stem: A type of stem in which neither the handwheel nor the stem rises when the valve is opened. It is suitable for areas where space is limited.

North arrow: An element on a construction drawing that indicates compass north.

Offset: A combination of elbows or bends that brings one section of pipe out of line but into a line parallel with the other section.

Ohm: The unit of measurement for electrical resistance.

Outside screw and yoke stem: A type of valve stem suitable for use with corrosive fluids because it does not come in contact with the fluid line. Often abbreviated OS&Y.

Packing: A material that seals the valve stem and prevents fluids from leaking up through it.

Packing extractor: A tool used to remove packing from a valve stem.

Parallel circuit: A circuit in which each load is connected directly to the voltage source.

Percentage of grade: The slope of a line of pipe expressed as the fall in feet, divided by the run in feet, multiplied by 100.

Perineal bath: A therapeutic bathtub used to treat diseases and injuries of the groin.

Pilot burner: A device that ignites the gas at the main burner when turned on by the thermostatic control. Also called a safety pilot.

Pilot-duty device: A protective device in which a wire coil opens an electric motor's contacts when the current rises above a pre-set level.

Pipe dope: A compound applied to pipe joints to allow them to be tightened to the point that they are leakproof.

Plan view: A drawing that shows the horizontal projection of a building.

Plug valve: A valve that provides both on/off and throttling control by raising and lowering a cylindrical or cone-shaped plug into the flow. Widely used in sanitary plumbing systems.

Pole: In a magnet, the ends of the magnetic flux lines, identified as either north or south. In a battery, the direction of current flow toward a positive (+) or away from a negative (–) charge.

Pop-up drain: A mechanism that allows a user to open or close a lavatory drain by pulling on a lift knob connected to a pivot rod.

Power: The amount of energy, measured in watts, consumed by an electrical load. Power equals voltage multiplied by current.

Pre-construction plan: A thorough outline of an entire installation process for a plumbing system.

Prefabricated: Assembled prior to delivery to the project site.

Pressure regulator valve: A valve that reduces water pressure in a building.

Pressurestat: A pressure-sensitive switch used to protect compressors.

Probe: A sensing element that is attached to the back of a thermostatic control immersed in a water heater storage tank.

Purge: The removal of all unwanted substances from a gas piping line.

Pythagorean theorem: In any right triangle, the longest leg squared is equal to the sum of the other two legs squared. As applied to plumbing, the theorem is $\text{Travel}^2 = \text{Run}^2 + \text{Offset}^2$.

Ratio: A relationship between numbers.

Rebar: A ribbed steel bar that provides a good bond when used as a reinforcing bar in concrete.

Recovery rate: The rate at which cold water can be heated.

Rectifier: A device that converts AC voltage to DC voltage.

Regulator: A device designed to control pressure.

Relay: A magnetically operated device consisting of a coil and one or more sets of contacts.

Relief valve opening: An opening on a hot water tank that provides access for installation of a temperature/pressure (T/P) relief valve.

Removable seat wrench: A tool used by plumbers to replace valve and faucet seats.

Request for information (RFI): A document issued to the project architect or engineer to identify an error on, or ask a question about, a construction drawing.

Reseating tool: A tool used by plumbers to reface a valve seat.

Resistance: The opposition to current flow, such as in a load.

Right triangle: Any triangle with a 90-degree angle.

Rise: The vertical measurement as opposed to the horizontal measurement of a run of pipe.

Rising stem: A type of valve in which both the handwheel and the stem rise.

Roll: In a rolling offset, the displacement of the offset that is perpendicular to both the rise and the run.

Rolling offset: An offset in which the two parallel sections of pipe on either end of the offset are not in the same vertical or horizontal plane.

Run: The distance between where a run of pipe starts to change direction and where it returns to its original direction.

Safety pilot: Another name for a pilot burner.

Sanitary increaser: An enlargement of a stack vent used in cold climates to prevent water vapor from condensing and freezing inside the vent opening.

Scale: The crust on the inner surfaces of boilers, water heaters, and pipes, formed by deposits of silica and other contaminants in the water.

Schedule: A detailed list of components included on a construction drawing.

Screw extractor: A tool designed to remove broken screws.

Screwed bonnet: A two-piece unit that is used in low-pressure applications and where frequent disassembly of the valve is not required.

Section: A hypothetical view that appears to be a vertical slice through a building.

Sediment bucket: A removable device inside a drain body that traps small solids that pass through the grate, or dome, to keep the solids out of the piping. Also called a sediment trap.

Series circuit: A circuit that provides a single path for current flow.

Series-parallel circuit: A circuit that consists of both a series and a parallel design.

Setpoint: A pre-selected temperature in a thermostatic switch that causes the switch to close when exceeded.

Sewer tap: The inlet of a private sewage disposal system.

Sheathing: The covering (usually over wood boards, plywood, or wallboard) that is placed over the exterior framing or rafters of a building. It provides a base for the application of exterior cladding.

Sheetrock®: A proprietary name for gypsum board. Also called wallboard.

Shoring: The act of using timbers set diagonally to temporarily hold up a wall. In excavations, the use of such timbers to stabilize the ditch sides is employed to prevent cave-ins.

Short circuit: A high current flow caused by a conductor bypassing the load.

Site plan: Another term for civil drawing.

Sizing: The process of calculating the proper sizes for the drains, stacks, sewer lines, and vents in a DWV system.

Slab-on-grade: A term describing a building in which the base slab is placed directly on grade without a basement.

Sleeve: A piece of piping through which water supply or drain, waste, and vent piping is inserted when that piping penetrates a building's structural elements such as concrete footings or floors. The sleeve protects the piping from being damaged by concrete.

Slope: The ground elevation or level planned for or existing at the outside walls of a building or elsewhere on the building site.

Slow-blow fuse: A fuse with a built-in time delay.

Slurries: Liquids that contain large amounts of suspended solids.

Solder end: A type of end connection for valves. The smooth ends are soldered to copper and brass pipes that have no internal or external threads.

Solenoid: An electromagnetic coil used to control a mechanical device.

Solenoid valve: A valve opened by a plunger controlled by an electrically energized coil.

Soleplate: A horizontal timber that serves as a base for the studs in a stud partition.

Solid-wedge gate valve: A type of gate valve that is used for steam, hot water, and other services where shock is a factor.

Specific gravity: A measure of the weight of a given volume of gas expressed as a ratio to the weight of the same volume of air under the same conditions.

Split-wedge gate valve: A type of gate valve that is used on lines that require a more positive closure—for example, low-pressure and cold water lines. It is also used on lines containing volatile fluids.

Storage tank: The enclosed tank that stores heated water; it allows heated flue gases to cover the bottom of the tank before they enter the vertical flue. Commonly called a center-flue tank.

Street fitting: A fitting with male threads on one end and female threads on the other.

Structural drawing: A construction drawing that indicates the structure of a building's framework.

Structural penetration: A modification to a structural member in the form of notching, drilling, or boxing.

Suction line: A line with negative pressure that will draw a liquid or gas.

Supply stop valve: A valve used to disconnect the hot or cold water supply to individual fixtures such as water closets, lavatories, and sinks. Also called the supply stop.

Swing-check valve: A type of check valve that features a low-flow resistance, making it suited for lines containing liquids or gases with low to moderate pressure.

Tap: A tool used for cutting internal threads.

Thermocouple: A small electric generator made of two different metals joined firmly together. It produces a small electric current that holds the safety shutoff gas valve open.

Thermostatic control: In a water heater, the control that adjusts the water temperature.

Throw: A single operation of a switch, from one position to another.

Time-critical material: Material that must arrive at a job site at a specified time and that must be installed before any other work can proceed.

Title block: An element of a construction drawing that includes information such as scale, revision date, drawing number, and the architect's or engineer's name.

Tongue: The shorter of two sides of a framing square.

Top plate: The top horizontal member of a frame building to which the rafters are fastened.

Transformer: An electromagnet that is used to raise or lower voltage.

Trap primer: A device or system of piping to maintain a water seal in a trap.

Trap primer valve: A valve that allows a small amount of water into a trap when the valve is operated.

Travel: The longest leg of a right triangle. It is also the center-to-center distance between fittings.

Trim: The parts of a valve that receive the most wear and tear: the stem, disc, seat ring, disc holder, wedge, and bushings.

True offset: The longest of the legs of a triangle formed by a rolling offset.

Tub waste: The drain fixture and fittings that take wastewater from the tub.

Turbine: A set of rotary blades spun by the high-speed flow of steam, water, or gas.

Twist packing: A string-like material used to pack valve stems when preformed packing material is not available. It is not as durable as preformed packing.

Union bonnet: A type of valve construction in which a union nut connects the bonnet to the valve body; best suited for applications requiring frequent disassembly of the valve.

Vernier: An auxiliary scale that slides against, and is used in reading, a primary scale. The scale makes it possible to read a primary scale much closer than one division of that scale.

Viscosity: A measure of a liquid's resistance to flow. The higher the viscosity, the slower a liquid flows.

Volt: The unit of measurement for voltage.

Voltage: A measure of the electrical potential for current flow. Also known as electromotive force (EMF).

Voltage drop: Voltage measured across any load.

Water supply fixture unit (WSFU): The measure of a fixture's load, varying according to the amount of water, the water temperature, and the fixture type.

Watts: The unit of measure for power consumed by a load.

Wax seal: A ring made of heavy-duty wax or rubber that fits between the bottom of the water closet bowl and the floor.

Wire drawing: A condition caused when gate valves open only part way, causing the disc, or gate, to vibrate, which erodes the disc edge.

Index

Index